Nicotine Psychopharmacology

Nicotine Psychopharmacology

Molecular, Cellular, and Behavioural Aspects

Edited by

S. Wonnacott

Senior Research Fellow, Department of Biochemistry, University of Bath

M. A. H. Russell

Reader in Addiction and Honorary Consultant Psychiatrist, Health Behaviour Unit, Institute of Psychiatry and Maudsley Hospital

and

I. P. Stolerman

Reader in Behavioural Pharmacology, Institute of Psychiatry

Oxford New York Tokyo

OXFORD UNIVERSITY PRESS

1990

Oxford University Press, Walton Street, Oxford OX2 6DP
Oxford New York Toronto
Delhi Bombay Calcutta Madras Karachi
Petaling Jaya Singapore Hong Kong Tokyo
Nairobi Dar es Salaam Cape Town
Melbourne Auckland
and associated companies in
Berlin Ibadan

Oxford is a trade mark of Oxford University Press

Published in the United States
by Oxford University Press, New York

British Library Cataloguing in Publication Data
Nicotine psychopharmacology.
1. Nicotine. Pharmacological aspects
I. Wonnacott, S.
615'.78
ISBN 0–19–261614–5

Library of Congress Cataloging in Publication Data
Nicotine psychopharmacology: molecular, cellular, and behavioural aspects/edited by S. Wonnacott, M.A.H. Russell, and I.P. Stolerman.
1. Nicotine–Physiological effect. 2. Psychopharmacology.
I. Wonnacott, S. (Susan) II. Russell, M.A.H. (Michael Anthony Hamilton)
III. Stolerman, Ian P.
[DNLM: 1. Nicotine. 2. Psychopharmacology. QV 137 N6617]
RA1242.N5N54 1990 615'.78—dc20 89-22839 CIP
ISBN 0–19–261614–5

Typeset by Colset Private Ltd, Singapore
Printed in Great Britain by
Courier International Ltd
Tiptree, Essex

Preface

Nicotine is one of the most widely used psychoactive drugs. Evidence of tobacco smoking has been traced back as far as the Mayan civilization in Mexico (*c.* AD 600) and, since European navigators first explored the Americas, tobacco usage has spread throughout the world. Nicotine has been the subject of considerable research by clinicians and basic scientists interested in drug dependence. Nicotine has also been exploited as a research tool for investigating the functioning of the nervous system, and this use has continued since the early studies of classical physiologists at the turn of the century. Recent technological advances in electrophysiology and molecular biology, for example, have led to new insights into the mechanisms of the action of nicotine in the central nervous system.

The aim of this volume is to bring together a very wide range of material from different areas of research relevant to understanding the psychopharmacology of nicotine. The areas covered range from studies of psychological and behaviour effects of nicotine in humans and animals, to its actions at the cellular and molecular levels. Many internationally recognized experts have contributed to the book, and we are most grateful for the excellence of their contributions. As well as providing a detailed resource for neuroscientists working on nicotinic mechanisms, each review has been written with the intention of making its content accessible to specialists in other areas of nicotine research, and to neurobiologists in general. Jargon has been avoided wherever possible, and descriptions and explanations are provided for specialized techniques, such as psychological tests for memory and performance, electrophysiological recording techniques and molecular biology experiments. The provision of extensive cross-referencing between chapters is intended to facilitate using the book to provide an integrated view of the current information and theories of nicotine dependence and its actions on brain function and in neurological disease.

Each chapter is a comprehensive review of the current state of knowledge about a particular aspect of nicotine psychopharmacology. The first section of the book surveys evidence that effects of nicotine, including its addictive properties, are the basis for human use of tobacco. It begins with the definition of drug dependence and proceeds to review the clinical and experimental evidence that tobacco smoking is a manifestation of nicotine dependence. Psychological effects of nicotine that may provide a basis for its human use provide the main theme of the final chapter of this section. The second section of the book reviews recent research findings from neurobiological approaches to understanding how nicotine acts in the brain. The distribution of nicotine in the body, its delivery to the brain, and its breakdown and elimination are reviewed comprehensively. The cellular and molecular mechanisms of the actions of nicotine in the brain are described from the perspectives of neurophysiology, biochemistry, and molecular biology. These chapters focus on the brain areas influenced by nicotine, the neurotransmitter and hormone responses to nicotine, and the receptor molecules through which nicotine acts. The main behavioural effects of nicotine are then surveyed and their relationships to nicotinic receptors and dependence are discussed. The final section of the book turns to clinical implications of the actions of nicotine in the central nervous system. Recent advances in the knowledge of nicotine and its receptors in the brain are discussed with respect to Parkinson's disease, Alzheimer's disease and tobacco smoking. These articles reflect the increasing interest in nicotine as both an investigative and therapeutic agent in the study of certain neurological diseases involving the cholinergic system. The final article reviews the role of nicotine in tobacco use and attempts to link findings from behavioural studies with the advances in knowledge at the cellular level. From this a better understanding of the nature of nicotine dependence in human smokers emerges, which paves the way for new pharmacological approaches to assist smoking cessation in the future.

Finally, we would like to acknowledge the help of everyone who has assisted in the preparation of this volume, including colleagues who have given their time in providing advice, secretaries, and the publishers.

Bath and *London*
1989

S. W.
M. A. H. R.
I. P. S.

Contents

Analytical table of contents

Contributors

Luigi F. Agnati
The Department of Human Physiology, University of Modena, Modena, Italy.

Kurt Andersson
Department of Histology and Neurobiology, Karolinska Institute, Stockholm, Sweden.

Neal L. Benowitz
Division of Clinical Pharmacology and Experimental Therapeutics, University of California, San Francisco, San Francisco General Hospital Medical Center, San Francisco, California 94110, USA.

Paul B. S. Clarke
Department of Pharmacology and Therapeutics, McGill University, Montreal, Quebec, Canada, H3G 1Y6.

Peter Eneroth
Unit for Applied Biochemistry, Research Center, Huddinge Hospital, Huddinge, Sweden.

Kjell Fuxe
Department of Histology and Neurobiology, Karolinska Institute, Stockholm, Sweden.

Steven R. Goldberg
National Institute on Drug Abuse, Addiction Research Center, Behavioral Pharmacology Laboratory, Baltimore, Maryland 21224, USA.

Anders Härfstrand
Department of Histology and Neurobiology, Karolinska Institute, Stockholm, Sweden.

Jack E. Henningfield
National Institute on Drug Abuse, Addiction Research Center, Biology of Dependence and Abuse Potential Assessment Laboratory, Baltimore, Maryland 21224, USA.

Peyton Jacob
Division of Clinical Pharmacology and Experimental Therapeutics, University of California, San Francisco, San Francisco General Hospital Medical Center, San Francisco, California 94110, USA.

Jerome H. Jaffe
Addiction Research Center, National Institute on Drug Abuse, Rockville, Maryland, USA.

Kenneth J. Kellar
Department of Pharmacology, Georgetown University, Schools of Medicine and Dentistry, Washington, DC 20007, USA.

Miguel Perez de la Mora
Instituto de Fysiologia Celular, Universidad Nacional Autonoma de México, Mexico.

Herve Porchet
Division of Clinical Pharmacology and Experimental Therapeutics, University of California, San Francisco, San Francisco General Hospital Medical Center, San Francisco, California 94110, USA.

**Charles Reavill*
Department of Psychiatry, Institute of Psychiatry, De Crespigny Park, London SE5 8AF, UK.

Michael A. H. Russell
ICRF Health Behaviour Unit, Institute of Psychiatry, 101 Denmark Hill, London SE5 8AF, UK.

Ian P. Stolerman
Department of Psychiatry, Institute of Psychiatry, De Crespigny Park, London SE5 8AF, UK.

†Michael D. B. Swedberg
National Institute on Drug Abuse, Addiction Research Center, Behavioral Pharmacology Laboratory, Baltimore, Maryland 21224, USA.

David M. Warburton
Department of Psychology, University of Reading, Whiteknights, Reading, RG6 2AL, UK.

Susan Wonnacott
Department of Biochemistry, University of Bath, Bath BA2 7AY, UK.

*Smith Kline & French Research Ltd., Welwyn, Hertfordshire, AL6 9AR, U.K.

†Ferrosan, CNS Division 2860, Soeborg, Denmark

Abbreviations

ACh	acetylcholine
ACTH	adrenocorticotrophic hormone
AMPT	alpha-methyl-para-tyrosine
APA	American Psychiatric Association
CCK	cholecystokinin
CNS	central nervous system
CNV	contingent negative variation
COHb	carboxyhaemoglobin
CRF	corticotrophin releasing factor
dlGN	dorsal lateral geniculate nucleus
DMPP	dimethylphenylpiperazinium
DOPA	3, 4-dihydroxyphenylalanine
DOPAC	dihydroxyphenylacetic acid
DSM	Diagnostic and Statistical Manual
EEG	electroencephalogram
ERPs	event related potentials
FSH	follicle stimulating hormone
GABA	γ amino butyric acid
ICD	International Classification of Diseases
IPN	interpeduncular nucleus
LC	locus coeruleus
MGN	medial geniculate nucleus
mHb	medial habenula
5HT	5-hydroxytryptamine
LH	luteinizing hormone
LHRH	luteinizing hormone releasing hormone
LPZ	lateral pallisade zone
MBTA	(4-N-maleimide) benzyltrimethyl-ammonium iodide
MPP^+	1-methyl-4-phenylpyridine
MPTP	1-methyl-4-phenyl-1, 2, 3, 6-tetrahydropyridine
MPZ	medial pallisade zone
SNC	substantia nigra zona compacta
SPA	(1-1, 2 diphenyl-1-dimethylaminoethan-HCL)
THA	tetrahydroaminoacridine
TPMP	triphenyl-methylphosphonium
TRH	thyrotropin releasing hormone
TSH	thyroid stimulating hormone
VTA	ventral tegmental area
WHO	World Health Organization

1 Tobacco smoking and nicotine dependence

Jerome H. Jaffe

1.1 Introduction

Over the past 20 years there has been a major reassessment of the way that the medical and behavioural sciences view the habitual use of tobacco. For more than 400 years there was little question that the use of tobacco could be a habit that is exceedingly persistent and resistant to change, but until recently there was strong reluctance to consider the tobacco habit a form of drug dependence, analogous in most respects to the chronic use of morphine, cocaine, or alcohol. The situation has changed. Prestigious national and international organizations now recognize heavy tobacco use to be a form of dependence (or addiction), a dependence that is the major factor keeping hundreds of millions of people smoking a substance that causes disease and death (Royal College of Physicians 1977; Surgeon General 1979, 1985, 1986; APA 1980, 1987; WHO 1985).

In May of 1988, the Surgeon General of the United States released a report entitled *Nicotine Addiction*. The 600-page report, which incorporated the contributions of more than 50 scientists, and cited more than 2500 published scientific papers, drew the following three major conclusions:

1. Cigarettes and other forms of tobacco are addictive.
2. Nicotine is the drug in tobacco that causes addiction.
3. The pharmacological and behavioural processes that determine tobacco addiction are similar to those that determine addiction to drugs such as heroin and cocaine.

This chapter (which was completed before the release of the 1988 US Surgeon General's report) summarizes the rationale for the earlier separation of tobacco use from other varieties of drug dependence. The chapter also summarizes some of the findings from the large body of recent research which have forced the scientific community to include this most common form of drug dependence with the others.

1.2 Nicotine as a drug of dependence

1.2.1 *The exclusion of tobacco use as drug dependence*

In the light of current views about tobacco dependence, it is reasonable to ask how it was possible to consider the behaviour as anything other than a form of drug dependence. Two major beliefs about addiction or drug dependence seem to have played a role. Firstly, marked drug tolerance and dramatic, obvious, and severe withdrawal phenomena were considered inherent aspects of the 'true addictions'. Secondly, addictions (morbid cravings for drugs) were associated primarily, if not exclusively, with drugs which at some stage in their use induce states of intoxication, thereby affecting the users' judgement, personality, and will-power. It is not difficult to see in the writings of respected scholars and researchers of the not too distant past how these beliefs could lead to a sharp distinction between tobacco use and other patterns of drug taking that were readily categorized as addictions.

Sir Humphrey Rolleston (Rolleston 1926), who headed the committee which set the tone for the British response to opiate dependence for many years to come, wrote:

> The Ministry of Health's Departmental Committee on Morphine and Heroin Addiction (1926) defined an addict as a 'person who, not requiring the continued use of a drug for the relief of the symptoms of organic disease, has acquired, as a result of repeated administration, an overpowering desire for its continuance, and in whom withdrawal of the drug leads to definite symptoms of mental or physical distress or disorder.' That smoking produces a craving for more when an attempt is made to give it up . . . is undoubted, but it can seldom be accurately described as overpowering, and the effects of its withdrawal, though there may be definite restlessness and instability, cannot be compared with the physical distress caused by withdrawal in morphine addicts. To regard tobacco as a drug of addiction may be all very well in a humorous sense, but it is hardly accurate

Louis Lewin (1931), sometimes called the Father of Psychopharmacology, was unimpressed by either the degree to which nicotine tolerance develops or the severity of the tobacco withdrawal syndrome.

> It is, moreover, common knowledge that the use of tobacco for smoking and chewing does not necessitate a progressive increase of the dose as in the case in other toxic substances and that the symptoms due to withdrawal of tobacco, if they occur at all, are easily overcome. These latter consist of an extreme feeling of discomfort and eventually bad humour and dejection. It is very exceptionally that graver symptoms occur.

Lewin gave significant weight to the differences between the effects of nicotine and other psychoactive drugs on mental functions.

> Smoking does not call forth an exultation of internal well-being as does the use of wine, but it adjusts the working condition of the mind and the disposition of many mentally active persons to a kind of serenity or 'quietism' during which the activity of thought is in no way disturbed, and from a physical point of view a certain calmness of movement occurs.

Paradoxically, perhaps, from a present-day perspective, Lewin recognized the power over behaviour which could be exerted by nicotine.

There is no hope for the inveterate cigarette-smoker. Not even the prospect of premature death serves to bring about a lessening of the passion, even if the organism has given warning in this respect. I have frequently made this observation in many cases where I have been consulted. *Volenti non fit injuria!* Such foolish folk, who would rather die of nicotinism than curb their passion for smoking, are also subject to predestination. Their ashes supply them with the final object of their desires.

The central importance of physical dependence and of psychotoxicity also appeared to have formed the basis for the efforts in 1952 and again in 1957 of the WHO Expert Committees on Drugs Liable to Produce Addiction to propose formal criteria for differentiating a drug 'habit' from a drug 'addiction'. According to those criteria, both conditions are a result of repeated consumption of a drug. Drug addiction was seen as 'a state of periodic or chronic *intoxication* produced by the repeated consumption of a drug', and the urge to use the drug was seen as 'an overpowering desire or need (compulsion)' in contrast to only a 'desire' in the case of a habit; furthermore, the compulsion to use implied a willingness to violate socially accepted norms ('obtain it by any means'). In an addiction, tolerance and physical dependence were present, while in a habit there was no physical dependence and no abstinence syndrome. In an addiction, the behaviour was harmful to the individual and society; in a habit the harm was only to the individual (WHO 1957).

From our present-day perspective, it is apparent that the WHO Expert Committees tried to make use of three often non-correlated dimensions in making the distinction between addictions and habits:

(1) the degree to which the drug comes to exert control over behaviour (or the individual loses flexibility with respect to drug use);

(2) the severity of the manifestations of physical dependence;

(3) the degree to which the use of the drug produces psychotoxicity and causes adverse effects to the individual and/or society.

The emphasis placed by these WHO Expert Committees on the concept of drug 'intoxication' rather than on a more morally-neutral term such as 'reinforcing effects' of the drug should not be seen as either arbitrary or moralistic. The words used to describe self-administered drugs or the people who use them are not always comparable in different languages. The English word 'addiction' (derived from a Latin legal term for consigning a person as slave over to a master) implies only that the individual is controlled (enslaved) by the drug-using behaviour. In other languages, however, the very words used for addiction carry additional implications. Thus, in Hungarian the word for addict means literally one who uses stupefying drugs; in Chinese the character for addiction conveys the idea of hidden or secret sickness. In the Spanish 'estupefaciente' and the French 'drogue', the notion of intoxication is incorporated into the very words used for drugs which are self-administered for non-therapeutic purposes.

1.2.2 *A Surgeon General's report*

In the 1964 report of the Advisory Committee of the US Surgeon General on Smoking and Health, these WHO Expert Committee criteria were used as the basis for viewing tobacco use as a 'habit' rather than an addiction. The committee was also influenced by another consideration that may have been only implied by the then operative 1957 WHO Expert Committee criteria. The prevailing view among psychiatrists was that 'addiction to potent drugs is based upon serious personality defects from underlying psychologic or psychiatric disorders which may become manifest in other ways if the drugs are removed'. It was obvious to the committee that this is not the case with the overwhelming majority of tobacco smokers. However, the primary reasons for the decision to call heavy, even compulsive smoking a 'habit' but not an 'addiction' were:

(1) the absence of clear evidence of nicotine-induced physical dependence in animal models;
(2) the belief that symptoms observed when smokers stop tobacco use were 'secondary to the deprivation of a desired object or habitual experience' (and not a specific nicotine withdrawal syndrome);
(3) the variable duration of these cessation symptoms;
(4) the non-fatal nature of sudden cessation of tobacco use;
(5) the belief that the obvious tolerance of smokers for nicotine was of low grade and of a distinct variety since excessive doses could still elicit toxic effects;
(6) the variety of interventions which seemed to be equally efficacious (or non-efficacious) in helping motivated smokers give up the behaviour. The absence of 'psychotoxic' effects was not emphasized.

It is worth noting that this historic document expressed little doubt that the primary reason so many people smoked tobacco was to obtain nicotine. However, the door was left open to the possibility that other irritants in smoke, which contribute to the sensory experience, could play some role in the behaviour.

By 1964 (too late for use in the US Surgeon General's report), the problems inherent in attempting to use several uncorrelated dimensions to separate 'habits' from 'addictions' led the WHO Expert Committee to recommend abandoning both terms in favour of the concept of *drug dependence* and to emphasize that what constituted dependence was to be spelled out for each drug or drug category. Drug dependence was defined as 'a state, psychic and sometimes also physical, resulting from the interaction between a living organism and a drug, characterized by behavioural and other responses that always include a compulsion to take the drug on a continuous or periodic basis in order to experience its psychic effects, and sometimes to avoid the discomfort of its absence. Tolerance may or may not be present. A person may be dependent on more than one drug' (Eddy *et al.* 1965).

Despite this broader definition, in several subsequent reports WHO Expert Committees developed eight categories of dependence-producing drugs, but made no reference to tobacco or nicotine. It was not until the report of 1974 (WHO 1974) that the committee stated that:

> . . . though not listed above, it [tobacco] clearly is a dependence-producing substance with a capacity to cause physical harm to the user, and its use is so widespread as to constitute a public health problem. However, unlike the types of dependence-producing drug just noted, it produces relatively little stimulation or depression of the central nervous system, or disturbances in perception, mood, thinking, behaviour or motor function. Any such psychotoxic effects produced by tobacco, even when it is used in large amounts, are slight compared with those of the types of dependence-producing drugs listed above. . . . Attention has been restricted to the use of dependence-producing drugs capable of exerting major psychotoxic effects

Thus, it can be seen that the notions about the importance of severe withdrawal symptoms and drug-induced psychotoxic effects served to separate tobacco dependence from other drug-using patterns which, compared with the tobacco habit, often exerted less persistent and less pervasive control over the behaviour of the user. In the WHO-sponsored ninth version of International Classification of Diseases (ICD-9) (WHO 1978), tobacco dependence is listed separately from other drugs because of the absence of 'psychotoxic effects'. But even the mention of tobacco dependence is significant. Something happened between the 1960's and 1974. Just what happened may never be entirely clear, but several findings, decisions, and events seem worth noting. The failure of smokers to respond to health pronouncements with a wholesale abandonment of smoking underscored the tenacity of the behaviour. Researchers in the field of the addictions took note. Pharmacologists began emphasizing the importance of nicotine as the addictive agent in tobacco (Armitage *et al.* 1968; Jarvik 1970), and around the same time an influential review focused attention on behavioural factors in the pharmacology of addictive processes (Schuster and Thompson 1969). The Addiction Research Unit at the Institute of Psychiatry in London created a unit to study tobacco dependence. In 1971, Michael Russell, a member of that unit, published a short, but powerful summary of the argument for nicotine as a dependence-producing drug. Russell's paper in turn influenced Brecher (1972) in the United States, who wrote a widely read book (*Licit and Illicit Drugs*) expressing a similar view, under the sponsorship of the respected Consumers Union. Simultaneously, in the United States, in the early 1970's there was increased boldness in describing tobacco use along with other forms of drug dependence in standard texts of pharmacology (see Jaffe 1975), and the committee to revise the American Psychiatric Association's (APA) Diagnostic and

Statistical Manual (DSM) was at least debating the need to include tobacco dependence among the diagnostic possibilities. In 1977, the Royal College of Physicians described tobacco smoking as a form of drug dependence comparable to other forms of addiction. An early proposal to create special criteria for tobacco dependence because of the large number of dependent smokers in the population was abandoned (see Jaffe and Jarvik 1978), but tobacco dependence was included in DSM-III in 1980 (APA 1980). Emboldened by these developments in the academic and professional world, the National Institute on Drug Abuse made a decision in the late 1970's to increase funding for research on tobacco use and dependence. All of these events and decisions helped to fuel the great increase in research on tobacco dependence, only some of which is highlighted in this introductory chapter.

1.2.3 *Recent changes in the concept of drug dependence*

The concept of tobacco use as a form of drug dependence cannot be discussed without some consideration of the way in which ideas about drug dependence have continued to evolve. For a variety of reasons, the criteria for diagnosing drug dependence incorporated into DSM-III in 1980 continued to give inordinate weight to evidence of withdrawal symptoms (physical dependence), thus continuing the confusing overlap between biological adaptations to the administration of a drug and drug dependence as a bio-behavioural syndrome in which physical dependence, when present, is but one element. Meanwhile, a WHO working group on nomenclature of drug and alcohol-related problems was developing a terminology and a diagnostic framework that could better incorporate newer research findings and evolving concepts. The most important of these concepts were that:

(1) dependence is a clustering of phenomena — cognitive, behavioural, and physiological;

(2) learning is a central element in the dependence process;

(3) multiple criteria are necessary for its assessment and evidence of physical dependence (neuroadaptation) is just one and not necessarily the most important;

(4) the dependence syndrome is not absolute but exists in different degrees.

These ideas, as well as some suggested 'dimensions' for deciding to what degree dependence exists, were incorporated into a WHO memorandum (Edwards *et al.* 1981). Thus, in a movement away from the effort to develop criteria for types of dependence based on pharmacological categories, this WHO group proposed a general model of dependence which was applicable across a broad range of pharmacological agents. For the first time in 15 years, WHO definitions and terminology coincided with that in at least one standard pharmacology textbook (Jaffe 1975, 1980). After considerable debate, most of the ideas of the WHO memorandum, including the notion of a syndrome that varies in intensity, were incorporated into a revised version of the APA's Diagnostic and Statistical Manual (DSM-III-R) in 1987. The DSM-III-R criteria for drug dependence (Table 1.1) were designed to a considerable extent to assess the phenomena or dimensions of the drug-dependence syndrome as described in the 1981 WHO memorandum (Edwards *et al.* 1981). Evidence of physical dependence and tolerance, once the central feature of dependence in previous conceptualizations, now became one of several characteristic features. The material to be summarized here will touch upon the degree to which tobacco use meets this set of criteria. Because of its historical significance, and because of older perceptions of the central importance of withdrawal symptoms, evidence for tobacco physical dependence will be considered first.

Table 1.1. Diagnostic criteria for psychoactive substance dependence of the American Psychiatric Association (1987)

A. At least three of the following:
(1) substance often taken in larger amounts or over a longer period than the person intended
(2) persistent desire or one or more unsuccessful efforts to cut down or control substance use
(3) a great deal of time spent in activities necessary to get the substance (for example, theft), taking the substance (for example, chain smoking), or recovering from its effects
(4) frequent intoxication or withdrawal symptoms when expected to fulfil major role obligations at work, school or home (for example, does not go to work because hung over, goes to school or work 'high', intoxicated while taking care of his or her children), or when substance use is physically hazardous (for example, drives when intoxicated)
(5) important social, occupational or recreational activities given up or reduced because of substance use
(6) continued substance use despite knowledge of having a persistent or recurrent social, psychological or physical problem that is caused or exacerbated by the use of the substance (for example, keeps using heroin despite family arguments about it, cocaine-induced depression, or having an ulcer made worse by drinking)
(7) marked tolerance: need for markedly increased amounts of the substance (i.e. at least a 50 per cent increase) in order to achieve intoxication or desired effect, or markedly diminishing effect with continued use of the same amount

Note: The following items may not apply to cannabis, hallucinogens, or phencyclidine (PCP):
(8) characteristic withdrawal symptoms (see specific withdrawal syndromes under Psychoactive Substance-induced Organic Mental Disorders)
(9) substance often taken to relieve or avoid withdrawal symptoms

B. Some symptoms of the disturbance have persisted for at least one month, or have occurred repeatedly over a longer period of time

1.3 Criteria for drug dependence

1.3.1 *Withdrawal symptoms*

1.3.1.1 *The evidence for a tobacco (nicotine) withdrawal syndrome* As early as 1942, Johnston reported that injections of nicotine relieved withdrawal symptoms experienced when a cigarette smoker stopped smoking. In 1945, it was reported that some smokers given low nicotine cigarettes experienced symptoms, now recognized as tobacco-withdrawal symptoms (Finnegan *et al.* 1945). However, carefully controlled experiments on tobacco withdrawal comparable to those used to establish the nature and specificity of opiate, barbiturate, and alcohol withdrawal syndromes were not carried out until many years later.

Over the past decade there have been numerous studies on the characteristics and time course of the tobacco withdrawal syndrome and its relationship to reduction in nicotine intake. These studies were aided in substantial measure by better methods to measure body levels of nicotine, assess nicotine intake, observe and record withdrawal phenomena, and manipulate levels of nicotine in the body. Many experiments have been carried out using nicotine polacrilex (nicotine gum); several have employed either intravenous nicotine, nicotine nasal solutions, nicotine patches which permit nicotine to be absorbed through the skin, or in at least one, a cigarette-like device which delivers nicotine vapour. Nevertheless, the demonstration of which signs and symptoms are specific to nicotine deprivation remains partially incomplete because of limitations on the forms and doses of nicotine that can be given safely to human volunteers, and because not all researchers use the same instruments to observe and record the signs and symptoms of withdrawal. Only a few experiments have involved repetitive intravenous nicotine doses, or nicotine inhaled in a form that is rapidly absorbed into the circulation. Either of these methods would mimic better the sudden sharp increases and brain levels of nicotine which occur when cigarette smoke is inhaled than do buccal, nasal, or transdermal routes.

Despite these limitations there is remarkable consistency in the signs and symptoms reported from various parts of the world when cigarette smokers stop smoking. These include craving for tobacco (nicotine), irritability, impatience, frustration or anger, anxiety, difficulty concentrating, restlessness, decreased heart rate, increased appetite, weight gain, depression, disturbed sleep, constipation, difficulty in socializing, decreased levels of adrenaline and noradrenaline in urine, altered electroencephalographic patterns while awake (increased slow waves) and while asleep (increased REM sleep), and decrements in performance on a variety of cognitive tasks (Myrsten *et al.* 1977; Elgerot 1978; Guilford 1966; Ryan 1973; Larson and Silvette 1975; Shiffman and Jarvik 1976; Wesnes and Warburton 1984; Jarvis *et al.* 1982; West *et al.* 1984; Hughes *et al.* 1984; Hughes and Hatsukami 1986; Fagerstrom 1988;

Henningfield and Jasinski 1988). Some of these symptoms are also seen after cessation of smokeless tobacco (for example, snuff) (Hatsukami *et al.* 1987) although overall syndrome severity may not be as great; also, some of the symptoms may be due to the ending of direct effects of nicotine. Only some of these symptoms are listed as criteria in DSM-III-R (APA 1987) because that document gave greater weight to symptoms observed in recent controlled prospective studies in the United States.

The consistency of the pattern of tobacco withdrawal symptoms in different countries argues for a biologically based and largely unlearned syndrome. However, the most persuasive evidence that components of the syndrome are due to withdrawal of nicotine is that in double-blind studies, nicotine is far superior to placebo in suppressing or preventing the appearance of many of the signs and symptoms (Hughes *et al.* 1984; Schneider *et al.* 1984; West *et al.* 1984; Fagerstrom 1988; Stitzer and Gross 1988; Henningfield and Jasinski 1988).

In placebo-gum controlled studies, certain signs and symptoms appear to be more consistently suppressed than others by currently available dosage forms of nicotine gum. Irritability and impatience were reduced in at least five independent studies; anxiety and difficulty concentrating were reduced in at least two studies. Other symptoms where nicotine versus placebo-gum differences in severity were significant in at least one study include depression, hunger, somatic complaints, and sociability. At least three studies of nicotine gum report a reduction in urge to use or craving for tobacco (Hjalmarson 1984; Schneider 1986; Stitzer and Gross 1988) but a number of other studies have not found such a reduction. Reduction in craving was seen with transdermal nicotine (Rose *et al.* 1985*a*). However, as pointed out by West (1988) much depends on how the questions are asked. Smokers may report severe craving endorsing the highest category on the craving scale, but if asked, they may also report that craving is helped by the use of the gum or other nicotine delivery forms.

Equally persuasive support for the view that these various symptoms are related to nicotine withdrawal is the observation that some former smokers who stop smoking cigarettes with the aid of nicotine gum develop physical dependence on the gum. When placebo gum is substituted for nicotine gum in long-term users, nicotine withdrawal symptoms emerge (West and Russell 1985*a*; Hughes *et al.* 1986). In one double-blind cross-over study of 8 smokers, who were using about 10 pieces of gum per day but had not smoked cigarettes for at least one month, placebo substitution resulted in irritability, anxiety, restlessness, impatience, difficulty concentrating and hunger, as well as craving for tobacco. Observer-rated withdrawal discomfort increased for 7 out of 8 subjects. In the two subjects who had the highest ratings of withdrawal discomfort during placebo substitution, one relapsed to cigarettes, the other to known nicotine gum. Most subjects could discriminate placebo from nicotine gum on the basis of withdrawal discomfort, but not on the basis of taste or side-effects. The magnitude of withdrawal discomfort was comparable to that seen after smoking cessation (Hughes *et al.* 1986). In gum users who had stopped smoking for a year or more, nicotine withdrawal symptoms, without craving for cigarettes, were observed (West and Russell 1985*a*). This pattern of suppression and re-emergence of symptoms can leave little doubt that components of the tobacco withdrawal syndrome are due to nicotine withdrawal and not merely due to the emotional experience associated with the loss of 'a desired object'.

1.3.1.2 *Aspects of tobacco withdrawal in need of further research* Several aspects of the tobacco withdrawal syndrome are not yet fully explained. Firstly, why is there such an inconsistent relationship between apparent tobacco intake and the severity of the syndrome? Secondly, why is there such variability between individuals in the manifestations of withdrawal? Thirdly, why has it sometimes been so difficult to reduce craving for cigarette smoke (as opposed to other symptoms) with nicotine itself? Fourthly, what explains the great variability in duration of withdrawal symptomatology? Fifthly, why is the severity of tobacco withdrawal a weak predictor of successful long-term abstinence?

While most studies have shown that difficulty in quitting and the severity of tobacco withdrawal are directly related to the number of cigarettes smoked (Hughes *et al.* 1981; Surgeon General 1985), others have found no relationship (Hughes and Hatsukami 1986; West and Russell 1985*b*). A partial explanation for this inconsistent dose–withdrawal severity relationship so typical of other forms of drug dependence may lie in the tremendous discrepancy between nicotine intake as estimated from the number of cigarettes smoked and intake estimated from more direct biological measures. As a device for delivering nicotine the cigarette permits an exquisite degree of control and the possibility of varying intake of smoke constituents over a wide range. Smokers can vary not only frequency of each cigarette, but also puff frequency, puff volume, depth of inhalation, duration of inhalation, degree of dilution of smoke, and how close to the tip each cigarette is smoked (see Fig. 1.1). As a result, there can be wide variation in nicotine intake among smokers smoking a comparable number of cigarettes of the same brand.

Russell and coworkers (Russell 1986) have found almost no correlation between plasma levels of nicotine in the afternoon and the nicotine content of the brand smoked. While the findings are not yet consistent, several groups have found correlations between biological measures of nicotine intake, such as levels of nicotine or its inactive metabolite cotinine, and either difficulty in quitting or response to treatment with nicotine gum (Hall *et al.* 1985; Zeidenberg *et al.* 1977). West and Russell (1985*b*) found that pre-abstinence plasma nicotine intake significantly predicted craving, hunger, restlessness, inability to concentrate and overall withdrawal severity. Expired carbon monoxide level, which represents a more indirect estimate of nicotine intake,

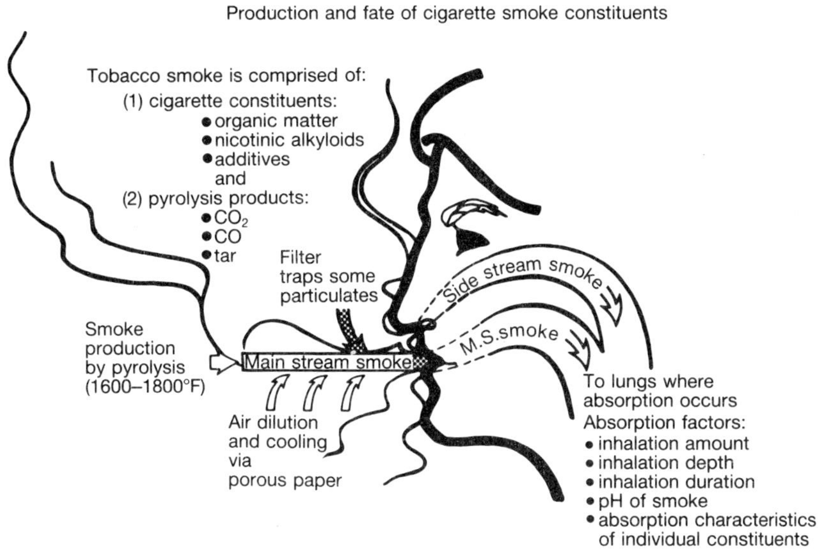

Fig. 1.1. This figure shows the complexity of the process by which nicotine is extracted from a cigarette. The amount of nicotine ultimately absorbed is as much a function of the behaviour of the smoker as of the characteristics of the cigarette (Henningfield 1984). M. S. = main stream.

predicted only craving and restlessness. Usual daily cigarette consumption did not significantly predict any withdrawal effects. Pomerleau *et al.* (1983) also reported a correlation between cotinine levels and abstinence-induced craving.

On the other hand Hatsukami *et al.* (1987) reported that nicotine plasma concentrations were correlated with craving for cigarettes but not with other symptoms of withdrawal such as caloric intake, confusion, or sleep disturbance. Measures of total smoke exposure were not correlated with craving or other withdrawal phenomena (puff volume was not measured). Given the difficulty in accurately measuring body levels of nicotine, it is probably unreasonable to expect to see the relationships between dose and withdrawal severity observed when known doses of an opioid or alcohol are given under laboratory settings, but obviously more work in this area is needed for an understanding of the determinants of tobacco withdrawal severity. No doubt variability in actual nicotine intake provides part of the answer to the question of variability in symptom severity among smokers. But there are obviously other factors. Some differences in severity may be related to reported motives for smoking (Williams 1979; West and Russell 1985*b*). Pomerleau and Pomerleau (1984) have suggested that different smokers may adjust nicotine levels so as to affect different neurotransmitter systems (including acetylcholine, catecholamine, and opioid systems) depending on the stressors to which they are responding. This might help to explain the interaction between the effects smokers seek from nicotine and what they experience during withdrawal. Environment clearly affects the intensity of withdrawal. For example, one group of researchers reports that hospitalized subjects appear to experience fewer withdrawal symptoms than outpatient subjects (Hughes and Hatsukami 1986).

There is obviously considerable variability. Shiffman (1979) reported that almost one fourth of smokers experience no withdrawal symptoms. Schachter (1982) conducted informal retrospective interviews with smokers who successfully quit without formal help and reported that 29 per cent of heavy smokers and 88 per cent of light smokers claimed to have had 'no problems'. However, it is not clear that the situation is radically different from other forms of dependence (for example, opioid or alcohol dependence). Even where the drug dosage has been carefully controlled and observations are made under laboratory conditions, there is biological variability between individuals which cannot as yet be accounted for. Difficulties in stopping smoking are discussed further in section 1.3.5.

The degree of tolerance and physical dependence on nicotine prior to abstinence can be estimated from answers to special questionnaires (Fagerstrom 1988; Tønnesen 1988). Those who appear to be more physically dependent by these criteria may benefit more from use of nicotine gum.

The persistence of craving for cigarette smoke when buccal nicotine is substituted for a smoker's usual brand of cigarette can be explained, at least in part, by considerations of dose. Nicotine gum typically produces nicotine plasma levels which are only about 1/3 those seen after inhalation of cigarette smoke (Russell 1986, 1988). However, even when much higher doses are used, producing higher blood levels of nicotine, nicotine gum does not significantly reduce desire to smoke in subjects who are not trying to quit. Such doses do reduce smoking behaviour (Nemeth-Coslett *et al.* 1987).

The failure of higher plasma nicotine levels to reduce craving may be due in part to the rate at which blood and brain levels rise. When tobacco smoke is inhaled, a small 'bolus' of nicotine is rapidly absorbed from the alveoli and may pass quickly into the brain. Russell and Feyerabend (1978) proposed that, for many smokers, dependence on tobacco is in large measure a result of the reinforcing effects of these boluses of nicotine that may be experienced thousands of times in the course of a year. For many cigarette smokers (if not all smokers) what is important may be this peak nicotine effect, and what is craved may be the perception of high peak levels of nicotine in the brain. It is not clear why such peaks should be so reinforcing (see below), but it is conceivable that only when these peak tissue levels are followed by brief decreases in tissue level do certain neurotransmitter effects consistently occur. Russell (1988) has pointed out that the nicotine receptors involved in the effects produced by rapid increases in nicotine brain level may be distinct from those

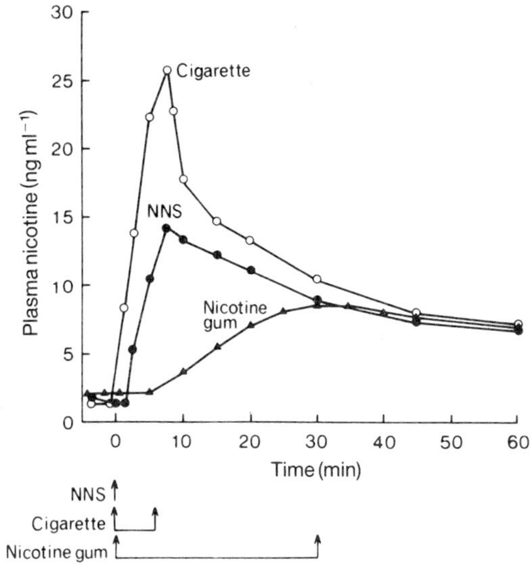

Fig. 1.2. Blood nicotine concentrations before, during, and after a single 2 mg dose of nasal nicotine solution (NNS), compared with concentrations from a cigarette and from a piece of 2 mg gum (Russell 1988).

which produce some of the other nicotine effects (see Russell, chapter 11, this volume); and that tolerance appears to develop at different rates to nicotine's various effects. As shown in Fig. 1.2, neither nicotine nasal spray nor nicotine gum can mimic the rise and fall of nicotine plasma levels produced by smoking, although if used in sufficient amounts both forms can probably prevent the troughs in nicotine levels which are associated with some aspects of the increased urge to smoke and with nicotine withdrawal symptoms. In an uncontrolled study, use of a nicotine nasal spray seemed to reduce craving for nicotine (Jarvis *et al.* 1987); reductions in craving are also reported with transdermal nicotine (Rose *et al.* 1985*a*). Yet it remains to be explained why smokers in a research setting do not exhibit reliably the expected reductions in puffing behaviour when given a series of rapid doses of nicotine intravenously (Kumar *et al.* 1977), or nicotine infusions producing high plasma levels of nicotine (West 1988).

Some researchers believe that the failure of intravenous nicotine to reduce craving sharply may be due to doses of nicotine too low to produce the high brain levels produced by inhaled boluses, but it is conceivable that findings on peripheral sensory stimuli associated with drug use and smoking can account better for this apparent discrepancy between results and our initial expectations.

Rose and coworkers (Rose *et al.* 1984, 1985*c*; Rose 1988) have recently reported on the importance of sensory cues such as taste, and effects of nicotine on the respiratory tissues, in the reinforcing effects of tobacco smoke in man. These researchers showed that when the familiar smells and tracheal sensations produced by each puff of tobacco smoke are blocked with a local anaesthetic there is less reduction in craving produced by inhalation of smoke than in the control condition where subjects did report the typical reduction in craving. Thirty minutes later there was no difference in craving between the conditions. Further, experienced smokers whose taste and olfactory sensations were blocked with local anaesthetic reported substantial enjoyment of puffs of a citric acid aerosol which produces tracheal sensations mimicking cigarette smoke, rating these puffs more similar to their own brands than a very low tar/nicotine cigarette. In still another experiment, a shallow puff of cigarette smoke which gave strong tracheal sensations was preferred to a very dilute but deep puff which appeared to result in nicotine absorption. Taken together these short-term studies do not necessarily show that nicotine is not a critical element, but they do emphasize the way in which sensory stimuli associated with nicotine smoke can come to have important reinforcing properties. Alternatively, as Rose (1988) has suggested, taste and oropharyngeal stimuli may be reinforcers independent of any association with the effects of nicotine. Taste may be an inherent 'hard-wired' reinforcer. People given adequate parenteral nutrition may still feel unsatisfied, craving the taste and feel of food in the mouth.

Animal experiments support the view that the sensory and olfactory stimuli associated with tobacco-using behaviour function as conditioned stimuli due to their previous association with nicotine. In animals trained to self-administer drugs such as cocaine (or nicotine), stimuli associated with drug administration (for example, a light which comes to signal drug administration) can become powerful reinforcers of drug-seeking behaviour, and for a time can maintain responding behaviour at high rates even when the drug is no longer administered. When such a signal is made part of a behavioural chain so that the animal works to get a series of signals, and only infrequently the primary drug reinforcer, this 'secondary' reinforcer (i.e. the light) can markedly influence the chain of drug-seeking behaviour (Goldberg 1973; Goldberg *et al.* 1975). When the secondary reinforcer is withdrawn, rates of responding decrease even though the drug continues to be given at the same frequency. These effects of secondary stimuli (lights) have been replicated in monkeys trained to self-administer nicotine intravenously (Goldberg *et al.* 1981; see Swedberg *et al.*, chapter 2, this volume). It is quite probable that the cues evoked by the inhalation of smoke and the peripheral and central actions of nicotine, including subjective effects, become linked through learning. The removal of secondary reinforcers probably accounts for the finding that when subjects are asked to smoke through a transparent screen in which the cigarette is mounted the number of puffs falls to about 60 per cent of ad lib smoking baseline values (Gritz 1980). It is also likely that internal cues associated with negative emotional states, including nicotine withdrawal, can become linked to the behaviour associated with their alleviation (nicotine intake). Thus the urge to use tobacco (nicotine) might be expected to be triggered by a wide range of internal and external cues.

The persistence of craving for cigarettes in the presence of reasonable plasma levels of nicotine and the remarkable reinforcing properties of sensory stimuli associated with smoking have prompted some workers (West 1988) to postulate that tobacco dependence, at least in these respects, is distinct from other forms of drug dependence. If it is, the differences are only of degree. With both opioids and alcohol, craving for preferred patterns of drug intake persists in the presence of high plasma levels of pure drug. Craving for intravenous heroin often persists even when body levels of methadone produce average opioid receptor occupation exceeding that which occurs under typical self-administration situations (Martin *et al.* 1973; Childress *et al.* 1986). It is an axiom among those trained in the use of methadone to treat intravenous heroin users, that it is necessary to help patients deal with persistent craving for intravenous heroin even after their doses of methadone are high enough to produce both clear-cut opioid effects and substantial tolerance-induced blockade of the effects of heroin. Among heroin users in a laboratory setting craving for the drug was higher when subjects were allowed to self-administer the drug, and knew it was available, than when they definitely knew it was unavailable, or when they knew that they would be unable to experience its effects because they had been given an opioid antagonist (Meyer and Mirin 1979). Some of this waxing and waning in the urge to use a drug (the term craving is typically used when the experienced urge is intense) is clearly a result of learned (conditioned) associations between stimuli associated with drug use and withdrawal. Craving can be evoked in laboratory and natural settings by presentation of such stimuli. Craving for opioids is not simply a concommitant of evoked conditioned withdrawal, but represents a distinct type of learned (conditioned) response, as does conditioned euphoria (Childress *et al.* 1986). Further, as mentioned above, in human volunteers the cognitive appreciation of drug availability can serve to increase craving, as can emotional states such as anger, depression, or anxiety (McLellan *et al.* 1986).

These conditioned or learned links that evoked conditioned opioid withdrawal or conditioned opioid craving can be remarkably persistent and can be further influenced by emotional states at the time of stimulus presentation. But it is not clear just how long the conditioned 'highs' can persist in the absence of the primary reinforcer. For example, heroin users given the opportunity to inject saline often find the effects pleasant at first, but when the act is repeated it becomes increasingly dysphoric (McLellan *et al.* 1986). While it may at first seem paradoxical, taking away a drug of dependence does not always increase craving, and giving it does not inevitably suppress craving. Indeed, in non-dependent cocaine users, craving for cocaine is accentuated within minutes after an intravenous bolus dose of cocaine (Jaffe *et al.* 1989).

What has been found in opioid and tobacco dependence is true also for alcohol dependence. In the case of those dependent on alcohol, the likelihood of experiencing craving for alcohol is sometimes influenced more by beliefs (expectations) about whether alcohol has been given, and by the severity of their dependence on alcohol, than by the pharmacological effects of the substance administered (Hodgson *et al.* 1979; Kaplan *et al.* 1983; Laberg 1986). Thus there are findings that smokers may, for a time, smoke lettuce leaf or ultra-low nicotine tobacco rather than not smoke, or that those who believe that the gum they are chewing contains nicotine experience less craving than those who believe the gum has no nicotine, or that tobacco-associated sensory cues exert a powerful influence on a smoker's sense of craving, satisfaction, or perception of withdrawal severity; such findings do not necessarily show that tobacco (or nicotine) dependence is unique among the drug-dependence disorders. Indeed, these findings may underscore the similarities between tobacco dependence and other forms of drug dependence.

1.3.1.3 *The variable duration of withdrawal* The reasons for the variable duration of tobacco withdrawal are simply not clear at present. Controlled studies usually report that after peaking within the first 48 hours most symptoms begin to subside after about 2–3 weeks. Some smokers report feeling fully recovered after a few weeks, while others complain about difficulties that resemble withdrawal symptoms (for example, irritability, difficulty concentrating) persisting for many months (Ryan 1973; Hughes and Hatsukami 1986; West *et al.* 1987; Stitzer and Gross 1988). Increased hunger, appetite, and associated weight gain typically persist for weeks or months before leveling out, with former smokers weighing 2–3 kg more on average than those who continue to smoke. Conceivably, some of the reported post-cessation difficulties are due to conditioned withdrawal symptoms. Similar variability has been reported for the benzodiazepine, alcohol, and opioid withdrawal syndromes. While these syndromes typically resolve within several weeks, in the case of opioids and alcohol, there is laboratory evidence of a protracted withdrawal syndrome lasting up to six months (see Jaffe 1975).

1.3.1.4 *Is withdrawal related to relapse?* While at first it might be self evident that those who experience the most severe drug withdrawal would have the most difficulty in giving up the drug, this is not always the case. Indeed, in the case of alcohol and opioid dependence 70–90 per cent of those who seek inpatient treatment usually manage to get through the first week or two when withdrawal is usually most severe. This is also true of treatments for most forms of drug dependence, including tobacco. As illustrated in Fig. 1.3 (from Hunt *et al.* 1971), relapse to drug use after initial cessation takes place at a high rate over the first 8–12 weeks after which the rate of relapse slows, reaching an asymptote at about one year.

Many factors affect the motivation to tolerate both acute and subacute symptoms. The intensity of acute tobacco withdrawal does not predict reliably who will achieve abstinence for a period of a few days in a smoking cessation

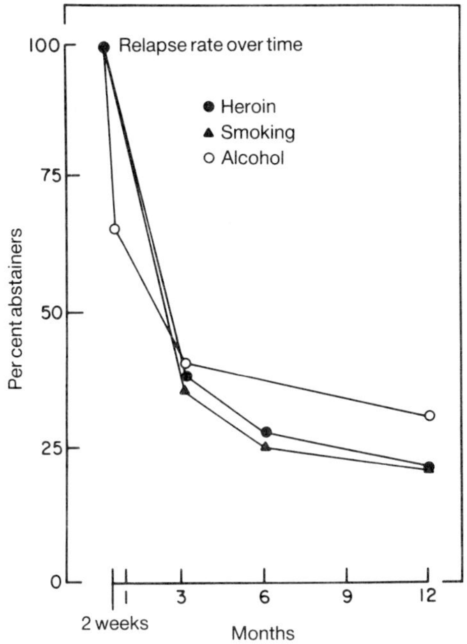

Fig. 1.3. Relapse rate over time for heroin, smoking, and alcohol (Hunt *et al.* 1971).

programme, but there does appear to be a relationship between some components of withdrawal and the capacity to become and remain abstinent for a period of several weeks. There have now been a number of controlled studies showing that nicotine gum increases the probability of remaining abstinent for more than a few days (Fagerstrom 1984; Hall *et al.* 1985; Hjalmarson 1984; Russell 1986; Jarvis *et al.* 1982; Sachs 1986; Schneider *et al.* 1984; Schneider 1986; Tónnesen 1988; Stitzer and Gross 1988). The major impact of the gum on relapse rates is seen during the first several weeks, after which relapse rates tend to be parallel for both those given the gum and those given placebo. The most reasonable inference from the data is that the nicotine gum produces its effects by suppressing elements of nicotine withdrawal syndrome (Lam *et al.* 1987). A logical extension of this conclusion is that nicotine withdrawal plays a role in relapse.

1.3.2 *Do people use tobacco to get nicotine?*

To show that there is nicotine physical dependence, and that people smoke to get nicotine to avoid withdrawal, does not explain why people smoke (or use smokeless tobacco) in the first place, why they often return to tobacco use sometimes months after successfully coping with withdrawal, or why there are some intermittent tobacco users who are clearly not physically dependent. Four distinctly different lines of evidence point to the central role of nicotine as the prime reason people use tobacco.

The first may be designated historical/commercial. Smoking materials that do not contain nicotine have never been successful in the marketplace (non-tobacco cigarettes delivering drugs of abuse are obviously exceptions). Even tobacco cigarettes delivering very low amounts of nicotine have failed to capture more than a tiny fraction of the market. Almost all of those who do smoke such cigarettes were formerly users of higher-nicotine brands who became concerned about health hazards of smoking. In one study where smokers were recruited into a program that offered progressive money incentives and health messages designed to motivate them to switch to very low tar/nicotine cigarettes, only 6 of 121 participants adopted brands delivering 0.2 mg or less of nicotine. The effect of the incentives was to reduce the cost of such cigarettes to about 15 cents per pack (Jaffe *et al.* 1982).

The second line of evidence is the way that cigarette smokers respond to manipulations of the nicotine delivered by their cigarette, their capacity to experience the effects of nicotine, their blood levels of nicotine, or of their rates of excretion of nicotine. The techniques for such manipulations and for measuring their effects are manifold. Smokers can be given cigarettes delivering higher or lower amounts of nicotine or they can be given diluted smoke; they can be pretreated with oral, intranasal, transdermal, intravenous, or intrapulmonary nicotine; they can be given nicotine antagonists, or they can be given substances such as ammonium chloride to change the acidity of their urine, thus altering the rate of nicotine excretion. The responses to such manipulations have included mood, craving for cigarettes, performance on tests, patterns of smoke inhalation, and various physiological and biological measures. The tendency of smokers to adjust or regulate their smoking behaviour to keep their body nicotine levels within broad limits is frequently called 'titration'. Studies of nicotine regulation have been reviewed in detail by Gritz (1980) and more recently by Russell (1986, 1987, 1988) and West (1988) and is the subject of chapter 11, this volume.

Some of the findings that have been repeatedly replicated are the following. When smokers are given cigarettes which deliver less nicotine (as measured by a smoking machine) they change the way they smoke the new cigarettes so as to extract more nicotine. They increase puff volume or duration of inhalation, and/or smoke more of the cigarettes. Smoking more cigarettes does not appear to be the primary mechanism for adjusting to the decrease. Compensation is far from perfect, but over a rather wide range of nominal nicotine deliveries there is no substantial decline in smokers' plasma nicotine until they are asked to smoke ultra-low nicotine (<0.2 mg) cigarettes, at which time nicotine plasma levels fall (but by far less than would have been predicted by the decrease in nominal delivery), and smokers seem considerably less satisfied with the cigarettes (Ebert *et al.* 1983; West *et al.* 1984; Russell 1986, 1988; Benowitz *et al.* 1983, 1986*a*). When asked to smoke fewer cigarettes, smokers also extract more nicotine from each one (Benowitz *et al.* 1986*b*).

Smokers also smoke more when the acidity of their urine is lowered, causing them to excrete somewhat more nicotine in the urine (Benowitz 1986). Smokers who are persuaded with various incentives to switch to very low-nicotine cigarettes tend to change back when the incentives are removed, even if they have smoked the lower-nicotine cigarettes for three months or more (Kanzler *et al.* 1983).

When pretreated with drugs which block or interfere with the effects of nicotine in the central nervous system (for example, mecamylamine) smokers behave as if the nicotine in the cigarettes had been reduced and smoke more (Stolerman *et al.* 1973; Pomerleau *et al.* 1983; Nemeth-Coslett *et al.* 1986). This finding is analogous to the initial increase of opioid self-administration when an opioid antagonist is given. The effect on smoking is not observed when a peripherally-active nicotine blocker is given, demonstrating the critical role of nicotine's effects in the central nervous system.

Smokers will also alter the way they smoke when given unusually high-nicotine delivery cigarettes so as to avoid sharp rises in nicotine plasma levels, supporting the older observations that tolerance to aversive toxic effects, while considerable, can be readily overcome by dose increases (Russell 1986). When smokers are given nicotine in other forms prior to an opportunity to smoke tobacco, the effects on smoking are less predictable and complex. In a laboratory setting nicotine gum produced a dose-related decrease in the number of cigarettes smoked per session and puffs of smoke per session, and increased the interval between cigarettes. The increase in expired carbon monoxide (a marker for depth of inhalation) was decreased, but the gum had little effect on desire to smoke (Nemeth-Coslett *et al.* 1987). Smokers given nicotine through a skin patch decrease the nicotine content of the first few puffs of smoke inhaled from a smoke mixing machine 90 minutes later (Rose *et al.* 1985*a*). Nicotine infusions can also decrease subsequent ad lib smoking (Lucchesi *et al.* 1967; Russell 1988). In one experiment (Kumar *et al.* 1977) boluses of nicotine given intravenously did not affect tobacco puffing behaviour, but they did so in another study (Henningfield and Jasinski 1988).

A third way of looking at nicotine as a reinforcer is to give subjects the opportunity to self-administer nicotine intravenously, and to compare both their subjective responses and patterns of administration to patterns of self-administration observed with drugs such as cocaine and opioids. Henningfield and coworkers (1983) gave subjects with and without histories of drug abuse an opportunity to press a lever to self-administer nicotine or saline intravenously. The patterns of lever pressing were similar to those seen with more common reinforcers such as food or money, and the patterns shared some characteristics with more thoroughly studied drugs of abuse. At the end of three-hour sessions the subjects reported 'liking' the effects of nicotine, although they also reported dysphoria and nausea during the course of the session. They continued to self-administer nicotine despite these effects. Subjects also self-administered saline, but the saline effects produced neither subjective effects nor orderly patterns of self-administration. Two subjects, who did not have histories of illicit drug use, showed gradual increases in the number of 1.5 mg nicotine injections self-administered over the course of seven test sessions.

A fourth line of evidence that nicotine is the pharmacological reinforcer in tobacco smoking comes from what subjects say about the effects of nicotine given by various routes. Nicotine, in doses absorbed by tobacco users, produces actions in the central nervous system, presumably at one or more specific receptors (see Stolerman, chapter 8, this volume). However, descriptions of the subjective effects produced by inhaled tobacco, or nicotine by any route, are far less dramatic than those provided for other dependence-producing drugs. When nicotine is absorbed rapidly, as after inhalation or intravenous administration, it produces vague and transient feelings of dizziness or 'light-headedness' (nicotine buzz) which smokers find slightly pleasant. Tolerance develops rapidly to this effect, so that regular smokers experience it only after a few hours of abstinence, or only with the first cigarette of the day (West and Russell 1987; Russell 1988). Similar light-headed or dizzy feelings are produced by nicotine nasal solutions, but not by nicotine gum. The absence of any marked subjective mood alteration with each dose of nicotine, after the first of the day, suggests that the continued use of nicotine must be maintained by more subtle subjective effects, or because it has some functional utility for the users. These useful effects may be quite distinct from suppression of nicotine withdrawal, and may include increased alertness or decreased distractibility, anti-anxiety and anti-stress effects, improved cognitive performance, suppression of weight gain, and decreased irritability, among many possibilities (Balfour 1984; Warburton and Wesnes 1979; Wesnes and Warburton 1984; Pomerleau and Pomerleau 1984).

Most smokers report that smoking is pleasurable (Ikard *et al.* 1969). In his pioneering uncontrolled study Johnston (1942) reported that he and his subjects not only found injections of nicotine pleasant, but that after a while he came to prefer them to cigarettes. In a single-blind study, Jones *et al.* (1978) gave intravenous nicotine to smokers and non-smokers, one bolus dose per hour over four hours. Seven of the eight smokers 'reported the nicotine injections to be rapidly followed by a pleasurable, stimulant-like sensation that many of them termed a "rush". Four of the non-smokers reported similar effects. Four of the smokers so enjoyed this feeling that they requested substantially higher doses . . .'. When Henningfield and coworkers administered nicotine intravenously to subjects with and without histories of illicit drug use, subjects with histories of drug abuse exhibited dose-related increases in scores on scales of a questionnaire that had been shown to be sensitive to euphorigenic effects of opioids and amphetamines. Subjects' 'drug-liking' scores were also dose-related. Scores on scales sensitive to the effects of other categories of drug (for example, LSD, barbiturates) were unchanged. Morever, when subjects were asked to identify the substance given from a list of psychoactive compounds, they frequently identified higher nicotine doses as either cocaine or amphetamine (Henningfield *et al.* 1985; Henningfield and Jasinski 1988). Although the effects of nicotine were qualitatively similar to those of cocaine, the magnitude of the euphorigenic effects was much less. In contrast, nicotine gum does not produce dose-related increases in 'liking'; in one study 'liking' for gum was inversely related to nicotine dose (Nemeth-Coslett *et al.* 1987). Thus, despite the observation that about one-third of those who quit smoking with the help of gum persist in using it for six months or longer, nicotine delivered by gum, in contrast to more rapidly absorbed forms of nicotine, is a very weak primary reinforcer, and at high doses may have aversive effects.

While the evidence is reasonably solid that cigarette smokers regulate nicotine intake to avoid substantial decreases in plasma levels and prevent withdrawal effects, the evidence that smoking occurs solely or primarily as a response to changes in the body's nicotine level (Schachter 1978) simply does not account for all of the data. As noted previously, nicotine (like other psychoactive drugs) can produce a number of effects that users could find helpful in responding to everyday environmental demands and stresses. Nicotine appears to facilitate task performance by improving reaction time and information processing, and by reducing the effects of distracting stimuli (see Warburton, chapter 3, this volume). Some of these effects can be shown in non-smokers (Wesnes and Warburton 1984). Nicotine may reduce aggression, tension, and anxiety, and can alter pain threshold in smokers, but it is not clear to what extent it has these actions in non-smokers. Nicotine suppresses weight gain. In animals, nicotine may improve memory (see Stolerman, chapter 8, this volume). Some of these actions may be related to nicotine-induced alterations in neurotransmitter and neuropeptide release in the CNS. It has been proposed that the smoker learns to adjust nicotine levels to produce the nicotine CNS effects most useful at a given time (Pomerleau and Pomerleau 1984). As a result of repeated nicotine-induced reductions in stress, or improvement in performance, these affective states or cognitive demands may become internal cues for smoking. Thus some cigarette smoking is not related at all to decreases in nicotine levels or avoidance of withdrawal, but may occur in response to a wide variety of environmental demands and emotional cues (Warbuton and Wesnes 1979; Balfour 1984; Pomerleau and Pomerleau 1984).

The overall reinforcing effects of a drug involve both positive reinforcement and negative reinforcement (suppression of withdrawal and alleviation of distress). One way to compare the reinforcing effects of tobacco with other commonly used psychoactive drugs is to examine the proportions of daily users of a drug to the number of people who have ever tried the drug. Such data are available from Johnston *et al.* (1987) who conducted surveys of drug use among a panel of High School seniors in the United States. These data show that 92 per cent of seniors have had some experience with alcohol and 5 per cent report themselves to be daily users. The daily use/ever-used proportion is 5.4 per cent. Similarly, 15.2 per cent of seniors have tried cocaine, but only 0.3 per cent report daily use; the proportion is 2 per cent. For opioids other than heroin the proportion is 1 per cent, for marijuana the proportion is 6.6 per cent. For cigarettes the proportion is 25 per cent, several times larger than that for alcohol which is also sold legally. The significance of daily use rates of drugs that differ in social acceptability and legality is exceedingly difficult to weigh and giving them too much consideration is inappropriate. All of these findings taken as a whole amply support the central role of nicotine in tobacco use, especially cigarette smoking. However, they also underscore two other significant points. Firstly, cigarette smoking is more than nicotine dependence and once it is firmly established even nicotine itself cannot supply all of the effects which the smoker associates with and seeks from cigarette smoke. Secondly, while nicotine is psychoactive, it is not a drug which reliably produces the kinds of euphoria and mood alteration typical of other drugs of abuse. Nor do tobacco users seek such mood alterations from either tobacco products or nicotine preparations used in treating tobacco dependence. In this sense, the 'absence of psychotoxicity' is not an irrational basis for considering tobacco (nicotine) dependence as somewhat distinct from other forms of drug dependence.

1.3.3 *Is the average cigarette smoker dependent on tobacco (nicotine)?*

Obviously, much depends on how dependence is defined. Let us accept for purposes of debate the following core ideas or propositions about drug dependence put forth by the WHO working group in 1980: that

> drug dependence is a syndrome manifested by a behavioural pattern in which the use of a given psychoactive drug or class of drugs, is given a much higher priority than other behaviours that once had higher value . . . [that] the intensity of the syndrome is measured by the behaviours that are elicited in relation to using the drug and by the other behaviours that are secondary to drug use . . . [that] no sharp cut-off point can be identified for distinguishing drug dependence from non-dependent but recurrent drug use . . . [and that] at a time when neither the drug-using behaviour nor the neuroadaptive [physical dependence] syndrome is manifest (drug dependence in partial remission) the continued presence of the dependence syndrome may be manifest in the degree to which drug-related cues may elicit a desire for the drug and the effort that must be expended in order to resist the urge to continue drug use. (Edwards *et al.* 1981.)

In DSM-III-R the same central concept of drug dependence is worded slightly differently,

> '. . . the essential feature of this disorder is a cluster of cognitive, behavioural, and physiological symptoms that indicate that the person has impaired control of psychoactive substance use and continues use of the substance despite adverse consequences.' (APA 1987.)

Let us further assume that the criteria for drug dependence in DSM-III-R (Table 1.1) were intended to probe the dimensions of the core concepts just summarized. To determine whether a tobacco user is dependent we can then ask: to what degree does the typical smoker meet these DSM-III-R criteria?. The preceeding discussion has summarized some of the evidence that nicotine is a psychoactive drug, that tolerance and a variety of physical dependence (neuroadaptation) to nicotine develops, and that most tobacco smokers studied smoke to get nicotine, in order to obtain its effects on the central nervous system as well as to avoid withdrawal. In DSM-III-R, if smokers have four or more typical withdrawal symptoms when they stop using nicotine they meet the criteria for the presence of the nicotine withdrawal syndrome. But what is the evidence that tobacco users meet the additional criteria for drug dependence?

1.3.4 *Do smokers take tobacco in 'larger amounts or over a longer period' than intended?*

This criterion was formulated primarily to assess dependence on alcohol. The typical smoker is generally aware of his/her usual daily level of intake, often making provision by buying a number of packs at a time. In this sense smokers do not commonly use more than they intended to use. Yet it is also true that beginning smokers do not plan to become dependent; few even plan to be daily smokers. They believe they will be able to stop when they want to do so. Few deliberately set out to be smoking forty cigarettes per day by the time they are thirty years-old. Once the teenage years pass, there are not many occasional smokers. In the United Kingdom, of those who smoke, about 80 per cent do so on a daily basis. At least 50 per cent of male smokers and 40 per cent of female smokers smoke at least 20 cigarettes per day (Ashton and Stepney 1982). Data from a National Household Survey in the US show that 90 per cent of men and 70 per cent of females smoke at least 20 cigarettes per day, and 19 per cent of men and 11 per cent of women smoke 40 or more cigarettes per day. Over the period from 1978–80, among both men and women, the percentages of current smokers smoking 20 or more cigarettes has increased, as has the percentage smoking 40 or more cigarettes per day. These shifts could reflect the tendency of those who smoke the fewest cigarettes to stop entirely, or a tendency of all smokers to increase the amount smoked. The shift to lower tar/nicotine brands could have contributed to an increase in cigarettes smoked. Henningfield (1986) reports that the amount of tobacco consumed daily (either smokeless or in the form of cigarettes) increases over a period of years, suggesting considerable tolerance or need to use more to prevent withdrawal symptoms, or both. Typically, tobacco smokers do not go on binges and wake up a day or two later regretting all the extra cigarettes. Judging from the three out of four smokers who say they would like to stop, it seems fair to infer that over a period of weeks most smokers would say they smoked more than they intended to. However, it is not clear how many smokers meet this specific criterion as set forth in DSM-III-R.

1.3.5 *Do cigarette smokers have 'persistent desire or one or more efforts to cut down or control substance use'?*

Repeated surveys have found that, at some point in time, most smokers report they would like to stop smoking but have been unable to do so. In surveys in the United Kingdom, 59 per cent of smokers have said they would like to stop and 70 per cent have tried (Ashton and Stepney 1982). In the United States, surveys carried out from 1978–1980 found that 59 per cent of smokers had made at least one serious unsuccessful attempt to stop at some time in their lives (Surgeon General 1985). Other surveys have found that 42 per cent of smokers have tried to stop at least three times (Hughes *et al.* 1987). The probability that a daily cigarette smoker will be successful in stopping does not appear to be high. Russell (1986), using data from the United Kingdom and Ireland, has stated that 80–85 per cent of teenagers who smoke more than a few cigarettes go on to become regular smokers as adults, and that only one in three women smokers and one in four male smokers succeeds in stopping permanently before age 60.

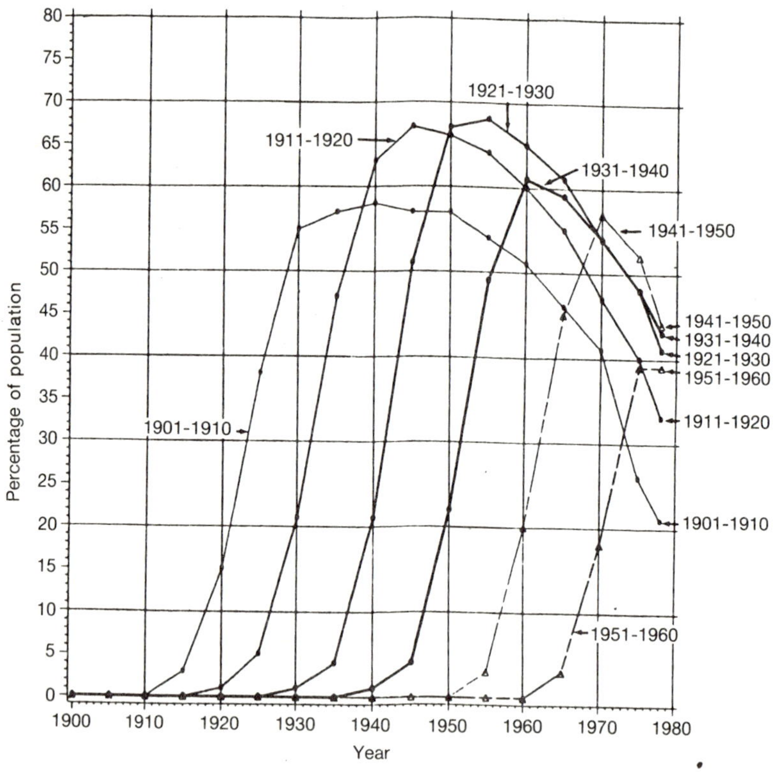

Fig. 1.4. Changes in the prevalence of cigarette smoking among successive groups of U. S. men, 1900–78 (Surgeon General 1985).

Although 30 million Americans describe themselves as ex-smokers (at the time of any given survey), 50 million are current smokers. Household Survey data from the United States indicate that for any given age the prevalence of cigarette smoking declines by about 1 per cent per year (Fig. 1.4). For men born between 1911–20, about 65 per cent were cigarette smokers in 1945; by 1975 the figure was about 35 per cent (Surgeon General 1985). Since only smokers still alive could be interviewed and since it is not clear how many non-smokers at the time of the survey were permanently non-smokers, Russell's pessimistic estimate may not be far from describing the natural history of cigarette smoking in the United States.

The difficulties of successful cessation can also be gauged from studies of smokers who have tried to stop (and from the problems of motivating smokers to make the attempt). Typically, smokers who try to stop with or without formal help can abstain for at least a few days. Many cessation programs have high initial success (abstinence) rates often as high as 80 per cent, but the return to smoking (relapse) is rapid, especially in the first three months (see Fig. 1.3). Therefore, considering difficulty in quitting, the material here will focus on abstinence rates at 6 months and 1 year.

In a study of smokers in Greenfield, Iowa, where an entire community attempted to give up smoking, 37 per cent of smokers who were interviewed claim to have made the attempt. Seven months after 'cold turkey day' only 14 per cent of men and 3.9 per cent of women who had been smokers claimed to be abstinent (Ryan 1973). Since these claims were not validated, there is a possibility that the actual percentages of those who successfully quit were even lower.

Even when smoking cessation programs are available on work-site, the percentage of smokers who elect to participate varies widely, and so does the percentage of participants who are successful over the long term. When white collar smokers at two London work-sites were invited to participate in a smoking cessation program, 58 per cent elected to attend. Of those who did, 12 per cent were abstinent at one year, as compared to 1 per cent abstinence among those who elected to attend, were invited, but did not attend (Sutton and Hallett 1986). These figures illustrate both the difficulty in long-term cessation among those who try to stop, and the persistence of the behaviour without some intervention. Work-site cessation programs in the United States are often less successful in recruiting smokers. In one hospital only 4 per cent of employees who smoked participated; they showed the typical 67 per cent initial cessation rates at the end of an eight week program (Kanzler *et al.* 1976). It is estimated that average participation rates for work-site programs are around 10 per cent (Orleans and Shipley 1982; Surgeon General 1985), but special incentives supplied by an employer can sometimes raise these rates substantially. Blue collar workers may be even less likely to participate in work-site smoking cessation programs. At 28 military bases in Canada, average participation was fewer than 10 smokers per base. Impact of work-site programs have been reviewed in a report of the US Surgeon General (1985).

What happens when smokers are given unequivocal advice to quit by their physicians? Much depends on what the physician says or does. Russell and coworkers (1979, 1983) conducted two large studies which involved a number of primary care physicians, with patients randomly assigned to different conditions. In the first study patients were given either two minutes of firm advice to stop smoking and a leaflet, or in the control condition simply asked about the extent of their smoking. Brief advice yielded a one-year abstinence rate of 5.1 per cent compared to 0.3 per cent for the control condition. In the second, larger study with 1938 cigarette smoking patients, there were 3 groups:

(1) non-intervention controls;

(2) firm advice to quit plus a booklet;

(3) firm advice, booklet, and offer of nicotine gum.

The proportions in each group abstinent at four months and still abstinent at one year were 3.9 per cent, 4.1 per cent and 8.8 per cent respectively. About half of those offered did not try the gum. In the small group (8 per cent) who used more than 1 box of the gum (105 pieces) the success rate was 24 per cent. The 24 per cent success rate among this group can be viewed as an index of the tenacity of the syndrome among those motivated to give up smoking, or as a sign of hope that intervention efforts are improving. In a smaller sample of patients seen by general practitioners, Fagerstrom (1984) also found that one year abstinence rates are substantially increased by both a general practitioner's follow-up phone call, and by the prescription of nicotine gum.

In the United States, a large multi-site study recruited more than 12 000 men who had one or more cardiovascular risk factors into an intervention program making use of individual and group formats, medical examinations, and structured maintenance and extended intervention efforts for those smokers who did not stop smoking after initial efforts. Of the original 4103 smokers, 47 per cent reported they had stopped smoking after four months and 46 per cent were reported to have quit after four years. Among those who smoked between 20 and 39 cigarettes, the one-year abstinence rate was 43 per cent; for those smoking 40 or more, the rate was 31 per cent. While these rates did not show the further attrition often seen in less extended programs, after several more years of persistent efforts by the program the majority of the heavier smokers were still smoking. According to the report, some 57–69 per cent of those who smoked 20 or more cigarettes per day continued to smoke (Hughes *et al.* 1981). However, Jarvis *et al.* (1984) re-examined these data and stated that the thiocyanate cut-off levels used to verify non-smoking were artificially elevated by including levels of pipe and cigar smokers in the calculation of average levels for non-smokers. Jarvis' re-analysis suggested that the one year abstinence rates for the program were closer to 20 per cent.

The many studies on the response of smokers to formal intervention programs are not covered in detail, because those who participate are not a representative sample of cigarette smokers either in terms of their motivations to stop or their degree of dependence. The high relapse rates among those who are initially successful in stopping after participation in formal treatment efforts (Fig. 1.3) illustrates both a basic characteristic of the tobacco dependence syndrome and its similarity in this respect to other varieties of drug dependence. It is fair to conclude that many tobacco users will meet the criterion of unsuccessful efforts to cut down or control drug use, although only a few studies have established the role of nicotine in this phenomenon. The inability or unwillingness of tobacco users to stop using despite clear cut adverse effects is considered further as a separate criterion.

1.3.6 *Are tobacco smokers frequently unable to fulfil major role obligations because of nicotine intoxication or withdrawal symptoms?*

If there are such smokers they are exceedingly rare. I know several smokers at the executive level who, while attempting to quit, were unable to carry out their responsibilities because of an inability to concentrate enough to organize their thoughts and produce written material against a deadline. However, when compared with the impairment of role obligations produced by alcohol (or other forms of drug) intoxication and withdrawal, the role impairment attributable to tobacco dependence *per se* is minimal. In this respect tobacco dependence is unique among the major classes of dependence-producing drugs.

1.3.7 *Do smokers spend a great deal of time in activities necessary to get the drug, using it, or recovering from its effects?*

Tobacco is still relatively cheap. It is rarely necessary to rob or steal to get the money for it. Further, because tobacco (nicotine) does not produce intoxication, until recently many smokers were permitted to smoke while they worked, ate, travelled, or played. Under these circumstances there was little attention paid to the 10 or more minutes out of each hour spent smoking a cigarette. With the proliferation of no-smoking work-sites and smoking-restricted environments, the smoker's need to go frequently to an arca sct aside for smoking may bring into sharper focus the expenditure of timc associatcd with tobacco dependence. In general, except in the case of chain smokers, this criterion is not commonly used in making the diagnosis of tobacco dependence.

1.3.8 *Do smokers give up or reduce important social, occupational, or recreational activities because of the need to use tobacco?*

As long as society remains permissive and tolerant of tobacco use and smoking, those smokers who can tolerate a period of abstinence of a few hours should continue to find little interference with daily life. At most, they may have to miss part of a long movie to get a cigarette. Those who find difficulty in not smoking for a few hours may need to cut back on airline travel on those short flights where there is a complete ban on smoking.

1.3.9 *Do cigarette smokers continue to use tobacco despite knowledge of having a persistent or recurrent social, psychological, or physical problem caused or exacerbated by use of the substance?*

Most smokers are aware that tobacco smoking has been linked to a wide range of life-threatening medical disorders, but they keep smoking nevertheless. Some seem able to deny to themselves the relevance of the medical findings; others make efforts to quit, some do so successfully. The persistence of smoking among patients with life-threatening illnesses may be one of the better ways to judge to what degree tobacco use can control behaviour (or to what degree the users lose their capacity to modify their behaviour). A small sampling of findings on quitting among smokers with smoking-related medical problems, or at high risk for such problems, illustrates this point.

In a multiclinic study in Great Britain, smokers with smoking-related illnesses attending a thoracic specialty clinic were advised to stop smoking, and were randomly assigned to groups given different degrees of advice and/or nicotine gum. After correcting for false reporting of abstinence, at the end of one year about 9–10 per cent of smokers were abstinent. Men with heart disease appeared to be somewhat more successful (21 per cent abstinence) than those with pulmonary disease (Raw and Friend 1983).

Even more illustrative of the persistence of smoking in the face of smoking-related illnesses are findings on smokers who have developed major heart disease or cancer. Between 50 and 77 per cent of patients who have survived a myocardial infarction or coronary by-pass surgery are smokers again (or still) one year later (Burling *et al.* 1984; Salonen *et al.* 1985). Similarly, of patients who are still smoking at the time of diagnosis or surgery for lung cancer, 50–60 per cent continue or resume smoking (Johnston-Early *et al.* 1980).

While these abstinence rates are far higher than spontaneous quit rates and substantially higher than treatment-aided quit rates, the fact that more than 50 per cent of smokers with life-threatening smoking-related disorders continue to smoke underscores the degree to which dependence on this drug can produce a shift in values. Perhaps the most dramatic testimony to the tenacity of tobacco dependence is its persistence among smokers who have undergone serious and unavoidable mutilating surgery for such diseases as peripheral arterial disease, and cancer of the mouth or larynx. For example, 40 per cent of patients who were smokers at the time of laryngectomy tried to resume smoking after surgery; about half of these were consuming tobacco at the time of interview. Fourteen patients (11 per cent) of those smoking at time of surgery continued to smoke cigarettes; others were using pipes or cigars (Himbury and West 1985).

1.3.10 *How many of these criteria should be needed to establish dependence?*

According to the current diagnostic framework put forth in DSM-III-R, only three of the criteria for dependence (Table 1.1) need to be met to conclude that dependence is present to some degree. It is likely, therefore, that a high percentage of current daily smokers are tobacco dependent to some degree. DSM-III-R now recognizes a continuum of dependence severity. Patients exhibiting many symptoms in excess of those required to make the diagnosis, and where the symptoms markedly interfere with occupational or social activities or relationships with others, are considered severely dependent; those who have few symptoms in excess of those required for diagnosis are considered mildly dependent. Those with intermediate numbers of symptoms are moderately dependent. For nicotine dependence the requirement for impairment in occupational or social functioning is dropped, because there is no significant intoxication syndrome, and because of the ready availability of nicotine containing substances. However, even this special consideration does not address the question of equal weighting of criteria. Is a smoker who experiences withdrawal upon forced cessation, has tried once or twice to quit, smokes to avoid withdrawal, and continues to smoke following coronary artery by-pass or lung cancer surgery, only moderately dependent? If the same smoker had never tried seriously to quit or expressed a desire to cut down, would that smoker be only mildly dependent? Whether each of these criteria should be given equal weight, and how useful it is to set the cut-off point for dependence at the current level, is a subject of ongoing discussion and investigation.

1.4 Summary

Nicotine is a psychoactive (but not a psychotoxic) agent that can act as a behavioural reinforcer, increasing the likelihood that people who use it will continue to use it. Most of those who use tobacco do so primarily to get the effects of nicotine and to prevent the emergence of the nicotine withdrawal syndrome, but for those who have used tobacco for a long time other factors also become important parts of the experience. This is also true, however, for other drugs of abuse. The severity of the tobacco (nicotine) withdrawal syndrome is better correlated with biological indices of nicotine intake than with cruder measures, such as number of cigarettes smoked per day. The nicotine withdrawal syndrome is an obstacle to successful tobacco cessation. The use of available substitute forms of nicotine, such as nicotine gum, can suppress components of the withdrawal syndrome and increase the probability of achieving abstinence from tobacco over the short term and at longer term follow-up. Nicotine produces effects quite apart from suppression of nicotine withdrawal in smokers and possibly non-smokers as well. Some of these (calming, appetite suppression, better performance on some cognitive tasks, decreased distractibility) probably have utility for most smokers. Continued tobacco use is probably maintained in part by these subtle effects of nicotine, including its capacity to help users cope with stress of various kinds. As is the case with other dependence producing drugs, the effects of the drug (nicotine), as well as the elements of withdrawal syndrome, become linked through learning to environmental cues, and probably to internal cues associated with negative emotional states. These cues (including negative emotions), by evoking craving for the drug (tobacco) or feelings of withdrawal, may then play a role in sustaining drug-using (smoking) behaviour, and in relapse after abstinence.

Some of the sought-after nicotine effects may require rapid rates of increase in tissue nicotine levels; such rapid increases are difficult to achieve except by intravenous or intrapulmonary routes. Tolerance develops quickly to the initial subjective effects of intravenous doses, as well as to the readily perceived effects which are experienced when tobacco is inhaled after a few hours of abstinence. Although some smokers report pleasant effects and mood elevations on scales that are sensitive to euphorigenic drugs when pure nicotine is given intravenously, nicotine is different from other drugs associated with dependence in that users do not seek a state of intoxication. When given in the form of gum it does not produce does-related increases in euphoric effects. These pharmacological differences between nicotine and other agents have implications for public health policies for regulating pure nicotine preparations. Apart from these differences, for many smokers the complex and highly overlearned behaviour that is tobacco dependence is quite analogous to other forms of drug dependence, in that the use of the drug comes to have a higher value than other aspects of life. When the criteria recently developed to assess the presence of drug dependence are applied to tobacco users, a substantial percentage will meet criteria for dependence to some degree; like other forms of dependence, tobacco dependence can be exceedingly resistant to change despite real threats to health and to life.

References

APA (American Psychiatric Association) (1980). *Diagnostic and statistical manual of mental disorders*, Third edn. American Psychiatric Association, Washington, DC.

APA (American Psychiatric Association) (1987). *Diagnostic and statistical manual of mental disorders*, Third edn, Revised. American Psychiatric Association, Washington, DC.

Armitage, A.K., Hall, G.H., and Morrison, C.F. (1968). Pharmacological basis for the tobacco smoking habit. *Nature* **217**, 331–4.

Ashton, H. and Stepney, R. (1982). *Smoking: psychology and pharmacology.* Tavistock, London.

Balfour, D.J.K. (1984). The pharmacology of nicotine dependence: a working hypothesis. In *Nicotine and the tobacco smoking habit* (D.J.K. Balfour, ed.), pp. 101–2. Pergamon Press, Oxford.

Benowitz, N.L. (1986). The human pharmacology of nicotine. In *Research advances in alcohol and drug problems,* Vol. 9 (H. Kappell, ed.), pp. 1–52. Plenum Press, New York.

Benowitz, N.L., Hall, S.M., Herning, R.I., Jacob III, P., Jones, R.T., and Osman, A.L. (1983). Smokers of low-yield cigarettes do not consume less nicotine. *New England Journal of Medicine* **390**, 39–142.

Benowitz, N.L., Jacob III, P., Yu, L., Talcott, R., Hall, S., and Jones, R.T. (1986*a*). Reduced tar, nicotine, and carbon monoxide exposure while smoking ultralow but not low-yield cigarettes. *Journal of the American Medical Association* **256**, 241–6.

Benowitz, N.L., Jacob, III, P., Kozlowski, L.T., and Yu, L. (1986*b*). Influence of smoking fewer cigarettes on exposure to tar, nicotine and carbon monoxide. *New England Journal of Medicine* **315**(21), 1310–3.

Brecher, E.M. (1972). *Licit and illicit drugs.* Little, Brown, Boston.

Burling, T.A., Singleton, E.G., Bigelow, G.E., Baile, W.F., and Gottlieb, S.H. (1984). Smoking following myocardial infarction: a critical review of the literature. *Health Psychology* **3**(1), 83–96.

Childress, A.R., McLellan, A.T., and O'Brien, C.P. (1986). Conditioned responses on a methadone population. *Journal of Substance Abuse Treatment* **3**, 173–9.

Ebert, R.V., McNabb, McK. E., McCusker, K.T., and Snow, S.L. (1983). Amount of nicotine and carbon monoxide inhaled by smokers of low-tar, low-nicotine cigarettes. *Journal of the American Medical Association* **250**, 2840–2.

Eddy, N.B., Halbach, H., Isbell, H., and Seevers, M. (1965). Drug dependence: its significance and characteristics. *Bulletin of the World Health Organization* **32**, 721–33.

Edwards, G., Arif, A., and Hodgson, R. (1981). Nomenclature and classification of drug-and alcohol-related problems: a WHO Memorandum. *Bulletin of the World Health Organization* **59**, 225–42.

Elgerot, A. (1978). Psychological and physiological changes during tobacco-abstinence in habitual smokers. *Journal of Clinical Psychology* **34**, 759–64.

Fagerstrom, K.-O. (1984). Effects of nicotine chewing gum and follow-up appointments in physician-based smoking cessation. *Preventative Medicine* **5**, 517–27.

Fagerstrom, K.-O. (1988). Efficacy of nicotine chewing gum: a review. In *Nicotine replacement: a critical evaluation* (O.F. Pomerleau and C.S. Pomerleau, ed.), pp. 109–28, Alan R. Liss, Inc., New York.

Finnegan, J.K., Larson, P.S., and Haag, H.B. (1945). The role of nicotine in the cigarette habit. *Science* **102**, 94–6.

Goldberg, S.R. (1973). Comparable behavior maintained under fixed-ratio and second-order schedules of food presentation, cocaine injection or *d*-amphetamine injection in the squirrel monkey. *Journal of Pharmacological Experimental Therapeutics* **186**, 18–30.

Goldberg, S.R., Kelleher, R.T., and Morse, W.H. (1975). Second-order schedules of drug injection. *Federation Proceedings* **34**, 1771–6.

Goldberg, S.R., Spealman, R.D., and Goldberg, D.M. (1981). High rates of responding maintained by intravenous nicotine injections under second-order schedules. *Federation Proceedings* **40**, 297.

Gritz, E.R. (1980). Smoking behavior and tobacco use. In *Advances in substance abuse* (N.K. Mello, ed.), pp. 91–158. JAI Press, Inc., Greenwich, Connecticut.

Guilford, J.S. (1966). *Factors related to successful abstinence from smoking*. American Institutes for Research, Pittsburgh, Pennsylvania.

Hall, S.M., Tunstall, C., and Rugg, D. (1985). Nicotine gum and behavioral treatment in smoking cessation. *Journal of Consulting and Clinical Psychology* **53**, 256–8.

Hatsukami, D.K., Gust, S.W., and Keenan, R.M. (1987). Physiologic and subjective changes from smokeless tobacco withdrawal. *Clinical Pharmacology and Therapeutics* **41**, 103–7.

Hatsukami, D.K., Hughes, J.R., and Pickens, R.W. (1985). Blood nicotine, smoke exposure and tobacco withdrawal symptoms. *Addictive Behavior* **10**, 413–7.

Henningfield, J.E. (1984). Behavioral pharmacology of cigarette smoking. In *Advances in behavioral pharmacology* Vol. 4 (T. Thompson, P.B. Dews, and J.E. Barrett, ed.), pp. 131–210. Academic Press, New York.

Henningfield, J.E. (1986). How tobacco produces drug dependence. In *The pharmacologic treatment of tobacco dependence: proceedings of the world congress* (J. Ockene, ed.), pp. 19–31. Institute for the Study of Smoking Behavior and Policy, Harvard University, Cambridge, Massachusetts.

Henningfield, J.E. and Jasinski, D.R. (1988). Pharmacologic basis for nicotine replacement. In *Nicotine replacement: a critical evaluation* (O.F. Pomerleau and C.S. Pomerleau, ed.), pp. 35–61. Alan R. Liss, Inc., New York.

Henningfield, J.E., Miyasato, K., and Jasinski, D.R. (1985). Abuse liability and pharmacodynamic characteristics of intravenous and inhaled nicotine. *Journal of Pharmacology and Experimental Therapeutics* **234**, 1–11.

Henningfield, J.E., Miyasato, K., and Jasinski, D.R. (1983). Cigarette smokers self-administer intravenous nicotine. *Pharmacology, Biochemistry Behavior* **19**, 887–90.

Himbury, S. and West, R. (1985). Smoking habits after laryngectomy. *British Medical Journal* **29**, 514–5.

Hjalmarson, A.I.M. (1984). Effect of nicotine chewing gum in smoking cessation: a randomized, placebo-controlled, double-blind study. *Journal of the American Medical Association* **252**, 2835–8.

Hodgson, R.J., Rankin, H.J., and Stockwell, T.R. (1979). Alcohol dependence and the priming effect. *Behavioral Research and Therapy* **17**, 379–87.

Hughes, G.H., Hymowitz, N., Ockene, J.K., Simon, N., and Vogt, T.M. (1981). The multiple risk factor intervention trial (MRFIT). *Preventative Medicine* **10**, 476–500.

Hughes, J.R., Gust, S.W., and Pechacek, T.F. (1987). Prevalence of tobacco dependence and withdrawal. *American Journal of Psychiatry* **144**(2), 205–8.

Hughes, J.R., Hatsukami, D.K., Pickens, R.W., Krahn, D., Malins, S., and Luknic, A. (1984). Effect of nicotine on tobacco withdrawal syndrome. *Psychopharmacology* **83**, 82–7.

Hughes, J.R. and Hatsukami, D. (1986). Signs and symptoms of tobacco withdrawal. *Archives of General Psychiatry* **43**, 289–94.

Hughes, J.R., Hatsukami, D.K., and Skoog, K.P. (1986). Physical dependence on nicotine in gum. *Journal of the American Medical Association* **225**(23), 3277–9.

Hunt, W.A., Barnett, L.W., and Branch, L.G. (1971). Relapse rates in addiction programs. *Journal of Clinical Psychology* **27**, 455–6.

Ikard, F.F., Green, D.E., and Horn, D. (1969). A scale to differentiate between types of smoking as related to the management of affect. *International Journal of Addiction* **4**, 649–59.

Jaffe, J.H. (1975). Drug addiction and drug abuse. In *The pharmacological basis of therapeutics* (L.S. Goodman and A. Gilman, ed.), pp. 285–311. Macmillan, New York.

Jaffe, J.H. (1980). Drug addiction and drug abuse. In *The pharmacological basis of therapeutics* (A. Gilman, L.S. Goodman, and A.G. Gilman, ed.), pp. 284–324. Macmillan, New York.

Jaffe, J.H. and Jarvik, M.E. (1978). Tobacco use and tobacco use disorder. In *Psychopharmacology: a general of progress* (M.A. Lipton, A. DiMascio, and K.F. Killam, ed.), pp. 1665–75. Raven Press, New York.

Jaffe, J.H., Kanzler, M., Friedman, L., and Kaplan, T. (1982). Money and health messages as incentives for smoking low tar/nicotine cigarettes: changes in consumption and exhaled carbon monoxide. *British Journal of Addiction* **77**, 21–34.

Jaffe, J.H., Cascella, N., Kumor, K., and Sherer, M. (1989). Cocaine-induced cocaine craving. *Psychopharmacology* **97**, 59–64.

Jarvik, M.E. (1970). The role of nicotine in the smoking habit. In *Learning mechanisms in smoking* (W.A. Hunt, ed.), pp. 155–90. Aldine Publishing Co., Chicago, IL.

Jarvis, M.J., Hajek, P., Russell, M.A.H., West, R.J., and Feyerabend, C. (1987). Nasal nicotine solution as an aid to cigarette withdrawal. *British Journal of Addiction* **82**, 983–8.

Jarvis, M.J., Raw, M., Russell, M.A.H., and Feyerabend, C. (1982). Randomized controlled trial of nicotine chewing gum. *British Journal of Addiction* **285**, 537–40.

Jarvis, M., West, R.J., Tunstall-Pedoe, H., and Vesey, C. (1984). An evaluation of the intervention against smoking in the Multiple Risk Factor Intervention Trial. *Preventative Medicine* **13**, 501–9.

Johnston, L. (1942). Tobacco smoking and nicotine. *Lancet* **2**, 742.

Johnston, L.D., O'Malley, P.M., and Bachman, J.G. (1987). National trends in drug use and related factors among American high school students and young adults, 1975–1986. *Department of Health and Human Services* Publication Number (ADM) 87–1535, Rockville, MD.

Johnston-Early, A., Cohen, M.H., Minna, J.D., Paxton, L.M., Fossieck, B.E., Ihde, D.C., Bunn, P.A., Matthews, M.J., and Maruch, R. (1980). Smoking abstinence and small cell lung cancer survival: an association. *Journal of the American Medical Association* **244**, 2175–9.

Jones, R.T., Farrell III, T.R., and Herning, R.I. (1978). Tobacco smoking and nicotine tolerance. In *Self administration of abused substances: methods for study* (N.A. Krasnegor, ed.), pp. 202–8. NIDA Research Monograph Number 20, DHEW Publication Number (ADM) 78–727, ADAMHA, Rockville, MD.

Kanzler, M., Jaffe, J.H., and Nee, J. (1983). Low nicotine cigarettes; cigarette consumption and breath carbon monoxide after one year. *Clinical Pharmacology and Therapeutics*, **34**, 408–15.

Kanzler, M., Zeidenberg, P., and Jaffe, J.H. (1976). Response of Medical personnel to an on-site smoking cessation program. *Journal of Clinical Psychopharmacology* **32**, 670–4.

Kaplan, R.F., Meyer, R.E., and Stroebel, C.F. (1983). Alcohol dependence and responsivity to an ethanol stimulus as predictors of alcohol consumption. *British Journal of Addiction* **78**, 259–67.

Kumar, R., Cooke, E.C., Lader, M.H., and Russell, M.A.H. (1977). Is nicotine important to tobacco smoking? *Clinical Pharmacology and Therapeutics* **21**, 520–9.

Laberg, J.C. (1986). Alcohol and expectancy: subjective, psychophysiological and behaviorl responses to alcohol stimuli in severely, moderately and non-dependent drinkers. *British Journal of Addiction* **81**, 797–808.

Lam, W., Sacks, H.S., Sze, P.C., and Chalmers, T.C. (1987). Meta-analysis of randomized controlled trials of nicotine chewing gum. *Lancet*, July, 27–9.

Larson, P.S. and Silvette, H. (1975). *Tobacco: experimental and clinical studies (Supplement 3)*, Williams and Wilkins, Baltimore, MD.

Lewin, L. (1931). Phantastica: narcotic and stimulating drugs, their use and abuse. Kegan, Paul, and Trench, Ltd, London.

Lucchesi, B.R., Schuster, C.R., and Emley, G.S. (1967). The role of nicotine as a determinant of cigarette smoking frequency in man with observations of certain cardiovascular effects associated with the tobacco alkaloid. *Clinical Pharmacology and Therapeutics* **8**, 789–96.

Martin, W.R., Jasinski, D.R., Haertzen, C.A., Kay, D.C., Jones, B.E., Mansky, P.A., and Carpenter, R.W. (1973). Methadone – a reevaluation. *Archives of General Psychiatry* **28**, 286–95.

McLellan, A.T., Childress, A.R., Ehrman, R., O'Brien, O.P., and Pashko, S. (1986). Extinguishing conditioned responses during opiate dependence treatment: turning laboratory findings into clinical procedures. *Journal of Substance Abuse and Treatment* **3**, 33–40.

Meyer, RE. and Mirin, S.M. (1979). *The heroin stimulus: implications for a theory of addiction*. Plenum Publishing Corporation, New York.

Myrsten, A.L., Elgerot, A., and Edgren, B. (1977). Effects of abstinence from tobacco smoking on physiological and psychological arousal levels in habitual smokers. *Psychosomatic Medicine* **39**, 25–38.

National Institute on Drug Abuse (1987). *National household survey on drug abuse: population estimates 1985*, USDHHS, PHS, ADAMHA, Publication Number 87–1539, Rockville, MD.

Nemeth-Coslett, R., Henningfield, J.E., O'Keefe, M.K., and Griffiths, R.R. (1986). Effects of mecamylamine on human cigarette smoking and subjective ratings. *Psychopharmacology* **88**, 420–5.

Nemeth-Coslett, R., Henningfield, J.E., O'Keefe, M.K., and Griffiths, R.R. (1987). Nicotine gum: dose-related effects on cigarette smoking and subjective ratings. *Psychopharmacology* **92**, 424–30.

Orleans, C.S. and Shipley, R.H. (1982). Worksite smoking cessation initiatives: reviews and recommendations. *Addictive Behavior* **7**, 1–16.

Pomerleau, O.F., Fertig, J., Seyler, L.E., and Jaffe, J.H. (1983). Neuroendocrine reactivity to nicotine in smokers. *Psychopharmacology* **83**, 61–7.

Pomerleau, C.S., Pomerleau, O.F., and Majchrzak, M.J. (1983). Mecamylamine pretreatment increases subsequent nicotine self-administration as indicated by changes in plasma nicotine level. *Psychopharmacology* **83**, 61–7.

Pomerleau, O.F. and Pomerleau, C.S. (1984). Neuroregulators and the reinforcement of smoking: towards a biobehavioral explanation. Neuroscience and Biobehavior Review **8**, 503–13.

Raw, M. and Friend, J. (1983). The role of chest physicians as smoking cessation counsellors. In *Proceedings of the fifth world conference on smoking and health, Winnipeg, Canada, 1983* (W.F. Forbes, R.C. Frecker, and D. Nostbakken, ed.), pp. 427–31. Canadian Council on Smoking and Health, Ottawa.

Rolleston, H. (1926). Medical aspects of tobacco. *Lancet* **1**, 961–5.

Rose, J.E. (1988). The role of upper airway stimulation in smoking. In *Nicotine replacement: a critical evaluation* (O.F. Pomerleau and C.S. Pomerleau, ed.), pp. 95–106. Alan R. Liss, Inc., New York.

Rose, J.E., Herskovic, J.E., Trilling, Y., and Jarvik, M.E. (1985*a*). Transdermal nicotine reduces cigarette craving and nicotine preference. *Clinical Pharmacology and Therapeutics* **38**, 450–4.

Rose, J.E., Jarvik, M.E., and Ananda, S. (1985*b*). Nicotine preference increases after cigarette deprivation. *Pharmacology, Biochemistry Behavior* **23**, 289–93.

Rose, J.E., Tashkin, D.P., Ertle, A., Zinser, M.C., and Lafer, R. (1985*c*). Sensory blockade of smoking satisfaction. *Pharmacology, Biochemistry and Behavior* **23**, 289–93.

Rose, J.E., Zinser, M.C., Tashkin, D.P., Newcombe, R., and Ertle, A. (1984). Subjective response to cigarette smoking following airway anaesthetization. *Addictive Behavior* **9**, 211–5.

Royal College of Physicians (1977). *Smoking or health. The third report.* Pittman Medical Publishing Co. Ltd., Tunbridge Wells, Kent.

Russell, M.A.H. (1971). Cigarette smoking: natural history of a dependence disorder. *British Journal of Medical Psychology* **44**, 1–16.

Russell, M.A.H. (1986). The future of pharmacological treatment for smoking. In *Pharmacologic treatment of tobacco dependence* (J.K. Ockene, ed.), pp. 296–302. Harvard University Press, Institute for the Study of Smoking Behavior and Policy, Cambridge, Massachusetts.

Russell, M.A.H. (1987). Nicotine intake and its regulation by smokers. In *Tobacco smoking and nicotine: a neurobiological approach* (W.R. Martin, G.R. Van Loon, E.T. Iwamoto, and K.D. Davis ed.), pp. 25–50. Plenum Press, New York.

Russell, M.A.H. (1988). Nicotine replacement: the role of blood levels, their rate of change, and nicotine tolerance. In *Nicotine replacement: a critical evaluation* (O.F. Pomerleau and C.S. Pomerleau, ed.), pp. 63–94. Alan R. Liss, New York.

Russell, M.A.H. and Feyerabend, C. (1978). Cigarette smoking: A dependence on high-nicotine boli. *Drug Metabolism Reviews* **8**, 29–57.

Russell, M.A.H., Merriman, R., Stapleton, J., and Taylor, W. (1983). Effect of nicotine chewing gum as an adjunct to general practitioners' advice against smoking. *British Medical Journal* **287**, 1782–5.

Russell, M.A.H., Wilson, C., Taylor, C., and Baker, C.D. (1979). Effect of general practitioners' advice against smoking. *British Medical Journal* **2**, 231–5.

Ryan, F.J. (1973). Cold turkey in Greenfield, Iowa: a follow-up study. In *Smoking behavior: motives and incentives* (W.L. Dunn, Jr., ed.), pp. 231–41. Winston, Washington, DC.

Sachs, D.P.L. (1986). Cost benefit analysis of tobacco dependency treatment. In *The pharmacologic treatment of tobacco dependence: proceedings of the world conference congress* (J. Okene, ed.), pp. 270–80. Institute for the Study of Smoking Behavior and Policy, Harvard University, Cambridge, Massachusetts.

Salonen, J.T., Hamynen, H., and Heinonen, O.P. (1985). Impact of a health education program and other factors on stopping smoking after heart attack. *Scandinavian Journal of Social Medicine* **13**, 103–8.

Schachter, S. (1978). Pharmacological and psychological determinants of smoking. *Annals Internal Medicine* **88**, 104–14.

Schachter, S. (1982). Recidivism and self-cure of smoking and obesity. *American Psychologist* **37**, 436–44.

Schneider, N.G. (1986). Use of 2 mg and 4 mg gum in an individual treatment trial. In *The pharmacologic treatment of tobacco dependence: proceedings of the world congress* (J. Ockene, ed.), pp. 233–48. Institute for the Study of Smoking Behavior and Policy, Harvard University, Cambridge, Massachusetts.

Schneider, N.G., Jarvik, M.E., and Forsythe, A.B. (1984). Nicotine vs. placebo gum in the alleviation of withdrawal during smoking cessation. *Addictive Behavior* **9**, 149–56.

Schuster, C.R. and Thompson, T. (1969). Self-administration of and behavioral dependence on drugs. *Annual Reviews of Pharmacology* **9**, 483–502.

Shiffman, S.M. (1979). The tobacco withdrawal syndrome. In *Cigarette smoking as a dependence process* (N.A. Krasnegor, ed.), pp. 158–184. NIDA Research Monograph Number 23, DHEW Publication Number (ADM) 79–800, ADAMHA, Rockville, M.D.

Shiffman, S.M. and Jarvik, M.E. (1976). Smoking withdrawal symptoms in two weeks of abstinence. *Psychopharmacology* **50**, 35–9.

Stitzer, M.L. and Gross, J. (1988). Smoking relapse: the role of pharmacological and behavioral factors. In: *Nicotine replacement: a critical evaluation* (O.F. Pomerleau and C.S. Pomerleau, ed.) pp. 163–186. Alan R. Liss, Inc., New York.

Stolerman, I.P., Goldfarb, T., Fink, R., and Jarvik, M.E. (1973). Influencing cigarette smoking with nicotine antagonists. *Psychopharmacologia* **28**, 247–59.

Surgeon General (US Department of Health, Education and Welfare) (1964). *Smoking and health: report of the advisory committee to the Surgeon General*, pp. 349–58. PHS Publication Number 1103, Washington, DC.

Surgeon General (US Department of Health, Education and Welfare) (1979). *Smoking and health: a report of the Surgeon General.* DHEW Publication Number (PHS) 79–50066, Rockville, MD.

Surgeon General (US Department of Health and Human Services) (1985). *The health consequences of smoking. Cancer and chronic lung disease in the workplace*: a report of the Surgeon General. DHHS Publication Number (PHS) 85–50207, Public Health Service, Office on Smoking and Health, Rockville, MD.

Surgeon General (US Department of Health and Human Services) (1986). *Health consequences of involuntary smoking: a report of the Surgeon General.* DHHS Publication Number (CDC) 87–8398. Public Health Service, Office on Smoking and Health, Rockville, MD.

Surgeon General (US Department of Health and Human Services) (1988). *The health consequences of smoking: nicotine addiction. A report of the Surgeon General.* DHHS Publication Number (CDC) 88–8406. Public Health Service, Office on Smoking and Health, Rockville, MD.

Sutton, S. and Hallett, R. (1986). Nicotine gum in the workplace: preliminary report of two randomized trials. In *The pharmacologic treatment of tobacco dependence: proceedings of the world congress* (J. Ockene, ed.), pp. 249–53. Institute for the Study of Smoking Behavior and Policy, Harvard University, Cambridge, Massachusetts.

Tønnesen, P., Fryd, V., Hansen, M., Helsted, J., Gunnersen, A.B., Forchammer, H., and Stockner, M. (1988). Effect of Nicotine Chewing Gum in Combination with Group Counselling on the Cessation of Smoking. *New England Journal of Medicine* **318**(1), 15–18.

Warburton, D.M. and Wesnes, K. (1979). The role of electrocortical arousal in the smoking habit. In *Electrophysiological effects of nicotine* (A. Remond and C. Izard, ed.), pp. 183–99. Elsevier, Amsterdam.

Wesnes, K. and Warburton, D.M. (1984). Smoking, nicotine and human performance. In *Nicotine and the tobacco smoking habit* (D.J.K. Balfour, ed.), pp. 133–52. Pergamon Press Ltd., Oxford.

West, R.J. (1988). Nicotine: a dependence-producing substance. In *Nicotine replacement: a critical evaluation*, (O.F. Pomerleau and C.S. Pomerleau, ed.), pp. 237–260 Alan R. Liss, Inc., New York.

West, R., Hajek, P., and Belcher, M. (1987). Time course of cigarette withdrawal symptoms during four weeks of treatment with nicotine chewing gum. *Addictive Behavior* **12**, 199–203.

West, R.J., Jarvis, M.J., Russell, M.A.H., Carruthers, M.E., and Feyerabend, C. (1984). Effect of nicotine replacement on the cigarette withdrawal syndrome. *British Journal of Addiction* **79**, 215–9.

West, R. and Russell, M.A.H. (1985*b*). Pre-abstinence smoke intake and smoking motivation as predictors of severity of cigarette withdrawal symptoms. *Psychopharmacology* **87**, 334–6.

West, R. and Russell, M.A.H. (1985*a*). Effects of withdrawal from long-term nicotine gum use. *Psychiatric Medicine* **15**, 891–3.

West, R. and Russell, M.A.H. (1987). Cardiovascular and subjective effects of smoking before and after 24 hour abstinence from cigarettes. *Psychopharmacology* **92**, 118–21.

WHO (World Health Organization) (1952). *Expert Committee on Drugs Liable to Produce Addiction*. Third Report, World Health Organization Technical Report Series Number 57.

WHO (World Health Organization) (1957). *Seventh report of the expert committee on addiction-producing drugs*. WHO Technical Report Service Number 116.

WHO (World Health Organization) (1974). *Twentieth report of the expert committee on drug dependence*. WHO Technical Report Service Number 551.

WHO (World Health Organization) (1978). *Mental disorders: glossary to their classification in accordance with the ninth revision of the international classification of diseases, Geneva*.

WHO (World Health Organization) (1985). *WHO Programme on tobacco or health. Report by the director general*. Seventy-seventh Session, Executive Board, WHO, Geneva.

Williams, D.G. (1979). Different cigarette smoker classification factors and subjective state in acute abstinence. *Psychopharmacology* **64**, 231–5.

Zeidenberg, P., Jaffe, J.H., Kanzler, M., Levitt, M.D., Langone, J.J., and Van Vunakis, H. (1977). Nicotine: cotinine levels in blood during cessation of smoking. *Comprehensive Psychiatry* **18**, 93–101.

2 Nicotine dependency: animal studies

Michael D.B. Swedberg, Jack E. Henningfield and Steven R. Goldberg

2.1 Introduction

A central requirement in definitions of drug dependence is that a dependence-producing drug will engender behaviour aimed at securing additional quantities of the drug for consumption (Edwards *et al.* 1982; Meisch and Thompson 1974; Spealman and Goldberg 1978). Other concomitants may include the development of tolerance to effects of the drug, and withdrawal syndromes that occur when drug administration is discontinued. Strategies developed in behavioural pharmacology have led to the development of experimental animal models for human drug taking behaviour; such models have been used to generate a substantial body of data on the major pharmacological groups of dependence-producing drugs. An important distinction to be made is that physiological dependence (characterized by onset of a withdrawal syndrome when drug administration is discontinued) may be present, but is not a necessary component in the maintenance of self-administration behaviour. For example, even drugs which are capable of producing severe physiological dependence when high enough doses are taken (such as the opiates, alcohol, and barbiturates) may maintain self-administration behaviour when drug doses and frequency of access are employed which prevent the development of demonstrable physiological dependence (Goldberg 1976). This chapter will review and discuss recent studies of nicotine self-administration with emphasis on methodological and interpretive issues, studies which have led to variable interpretations of data regarding the efficacy of nicotine in maintaining self-administration behaviour in animals. The potential of nicotine to produce tolerance with chronic exposure, and withdrawal symptoms upon termination of chronic exposure, will also be reviewed.

2.2 Nicotine self-administration

Using operant methodologies originally developed by Skinner (1938), and employing schedules of reinforcement as described by Ferster and Skinner (1957), investigators have been able to demonstrate that laboratory animals can be trained to emit a response that results in the self-administration of a variety of drugs commonly abused by humans, including cocaine, *d*-amphetamine, and morphine (for example, Griffiths *et al.* 1980; Spealman and Goldberg 1978). These procedures have served as valuable tools for preclinical assessment of abuse liability of new compounds (Spealman and Goldberg 1978). Self-administration procedures were also used by a number of investigators in efforts to provide animal models of human nicotine taking behaviour. The two types of procedures most frequently employed involve self-administration of tobacco smoke through inhalation and self-administration of nicotine solutions by the intravenous route.

2.2.1 *Self-administration of tobacco smoke*

Toxicological studies have been conducted in which rats and dogs have been subjected to tobacco smoke as though in passive-smoking situations (Auerbach *et al.* 1977; Hammond *et al.* 1971; Nettesheim *et al.* 1977), or under conditions in which severe behavioural contingencies ensured tobacco smoke inhalation in baboons (Rogers *et al.* 1980, 1981). However, such studies were not designed to provide evidence as to whether nicotine or even tobacco smoke itself were functioning as reinforcing events in maintaining smoke inhalation.

Tobacco smoke has been shown to maintain smoking behaviour in rhesus monkeys, however. In one of the earliest studies, Jarvik (1967) trained rhesus monkeys to suck water through a tube, after which cigarette smoke was intermittently substituted for the water; later on water delivery from the tube was discontinued. Other monkeys were selected for their tendency to suck on their thumbs or toes. After initial training, the animals were able to inhale from two tubes protruding into their chambers, one tube delivering tobacco smoke, and the other air. At this point puffing behaviour was maintained by cigarettes alone. A tobacco burning machine produced smoke, the temperature of which was adjusted to maintain the highest rate of inhalation. The monkeys preferred tobacco smoke to air. However, the tobacco burning machine required lit cigarettes to be placed in it by the experimenter. Jarvik (1967) therefore devised an apparatus that could heat the tobacco so that it became volatile, and the vapour could be accessed at predetermined times by the monkey sucking on the tube. When tobacco vapour and hot air were the alternatives, the monkeys preferred the tobacco vapour. When cigar vapour and low-nicotine cigarette vapour were presented, these products maintained the same amount of puffing behaviour as did the standard cigarette vapour, whereas pipe-tobacco vapour was preferred the most. When tobacco smoke was compared to tobacco vapour, the real smoke was preferred. A problem with this study, and smoking studies in general, is that the concentration of nicotine required in the body for self-administration to occur was not determined, and that the effects of nicotine on rate of puffing *per se* was also unclear. For example, the effects of low-nicotine and standard cigarette vapour were similar, and the ability of nicotinic antagonists to reverse puffing was not addressed.

In an attempt to modify smoking behaviour by nicotinic antagonists, rhesus monkeys were trained to suck on a water delivery tube in order to drink (Glick *et al.* 1970). A cigarette smoke delivery tube was then substituted for the water tube, and water could be obtained subsequently from a water solenoid spout by sucking on the cigarette smoke tube. When puffing behaviour had been established the monkeys were presented with two tubes, one delivering tobacco smoke and the other air. From this point on, the monkeys earned all their water during daily 4-hour-long puffing sessions. Access to the water solenoid spout was contingent upon a sucking response, on either the tube delivering cigarette smoke or the tube delivering air. The number of puffs required to earn water was gradually increased from 1 to 30. Even though they could get access to water by puffing on the air tube, all four monkeys showed preference for the cigarette smoke tube. Preference for the cigarette smoke tube was reversed when a high enough dose of the centrally-acting nicotine receptor blocker mecamylamine was given intramuscularly prior to the smoking session. However, reversal of preference for smoke over air was also seen when high enough doses of the peripheral nicotine antagonist hexamethonium was administered, precluding a definitve assessment of the role of central nicotine receptors. Pretreatment with oral nicotine was found to decrease smoking in rhesus monkeys (Glick *et al.* 1971). Even though the studies by Glick *et al.* (1970, 1971) are suggestive of a role for nicotine in smoking behaviour, the fact that puffing behaviour was under control of water reinforcement leaves open the possibility that the reinforcing effects of the water presentations themselves may have confounded assessment of the reinforcing effects of nicotine.

In a more recent study, Ando and Yanagita (1981) induced and maintained smoking behaviour when a sweetened liquid was presented contingent upon puffing on cigarettes. Cigarette smoke was generated by an automatic 'smoking machine', and was accessible to the monkey by sucking on a tube. When the sweetened liquid was removed, 2 of 14 monkeys continued to smoke ('voluntary smoking'); in these monkeys substitution of low-nicotine or nicotine-free cigarettes decreased smoking (see Fig. 2.1). Two of the monkeys that did not show 'voluntary smoking' resumed smoking when the sweetened liquid was reintroduced. When nicotine-free cigarettes were presented to the 'voluntary smoking' monkeys, smoking behaviour decreased. The absence of a nicotine dose–response curve, and the limited success in inducing 'voluntary smoking', precluded a definitive assessment of the role of nicotine in maintenance of the smoking behaviour.

To summarize, these studies illustrate the use of operant contingencies to initiate and maintain tobacco smoking behaviour in non-human primates. Tobacco smoke may maintain puffing behaviour in monkeys, although

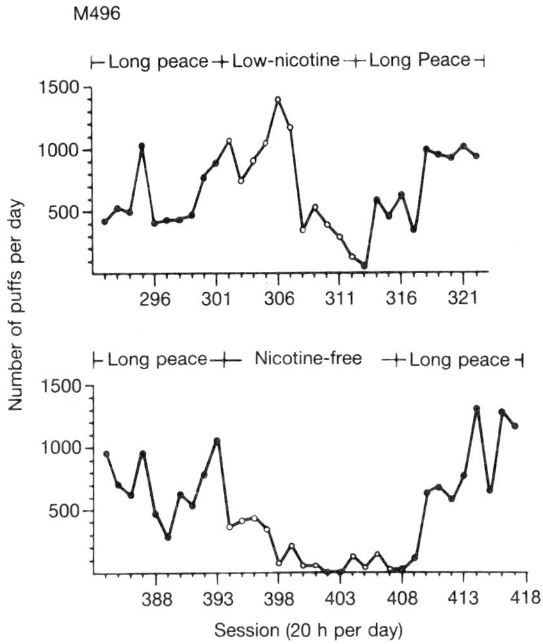

Fig. 2.1. Effects of substituting low-nicotine cigarettes and nicotine-free cigarettes for the standard nicotine cigarette ('Long peace') in a 'voluntary smoking' rhesus monkey. The abscissa shows number of puffs taken per day during daily 20 h-long sessions. The ordinate shows number of daily sessions. (From Ando and Yanagita 1981, with permission).

compared to intravenous self-administration of psychomotor stimulants, opiates, and sedatives, cigarette smoking is maintained only weakly and erratically. Although tobacco smoke appears to sustain sucking behaviour under some conditions, it is usually difficult to assess its reinforcing effects independent of other reinforcing events used to induce the behaviour. The degree to which the smoking behaviour results in inhalation of smoke, and absorption of behaviourally active amounts of nicotine, also remains unclear. Also, the general absence of attempts to reverse puffing behaviour by use of nicotinic antagonists precludes a definitive assessment of the importance of nicotine itself in maintaining smoking behaviour. In the one study in which nicotine antagonists were used (Glick *et al.* 1970), the finding that the centrally as well as the peripherally acting antagonists reversed puffing behaviour makes it impossible to establish a specific role for nicotine in the central nervous system in the maintenance of smoking behaviour.

2.2.2 *Intravenous nicotine self-administration*

Many of the factors which confound interpretation of results of the previously described smoking studies have been reduced by utilizing operant procedures in which specified doses of nicotine are made available for intravenous self-administration. Typically, an animal surgically prepared with an intravenous catheter (commonly in the jugular vein) is presented with a response lever which upon depression results in the delivery of a drug infusion by activation of an automated syringe pump. A drug is said to have reinforcing effects when the drug presented as a consequence of lever pressing increases the probability of the animal pressing the lever again (for example, Schuster and Thompson 1969). Demonstration that a drug will maintain self-administration behaviour over a range of conditions (for example, different schedules of drug access, different species) is a strong indication of the likelihood that the drug may be abused if made available to humans (Brady and Lukas 1984). Criteria for determination of whether or not a drug may have reinforcing effects include:

(1) that the drug maintains rates of responding similar to those of known drugs of abuse and of other reinforcing events, such as food presentation;

(2) that rates of responding vary systematically as a function of dose;

(3) that under appropriate conditions rates of responding are higher for the drug than for the drug vehicle alone;

(4) that specific antagonists reduce rates of responding towards those seen after vehicle alone.

Most drugs upon which humans may become dependent have also been demonstrated to maintain self-administration behaviour in animals (Griffiths *et al.* 1979, 1980; Schuster and Johanson 1974; Spealman and Goldberg 1978). Under certain conditions this is also true for nicotine.

Studies of the intravenous self-administration of nicotine have employed a number of different schedules of reinforcement. The main differences relate to the amounts of behaviour required, and to the timing of the delivery of the nicotine infusion in relation to the behavioural requirements. The schedules of reinforcement are:

(1) fixed-ratio schedules of reinforcement where a fixed number of responses is required before drug is delivered; for example, a fixed-ratio schedule in which a drug infusion is delivered immediately after one response has been emitted is called a fixed-ratio 1 schedule, and it is also referred to as a continuous reinforcement schedule;

(2) progressive-ratio schedules where the response requirement in a fixed-ratio schedule is continually increased until the animal stops responding;

(3) interval schedules of reinforcement where a drug infusion is produced by the first response to occur after a specified interval of time has elapsed since the previous infusion;

(4) concurrent schedules where the simultaneous availability of food and drug can under certain conditions generate what is referred to as schedule-induced self-administration behaviour;

(5) second-order schedules where responding under a schedule such as a fixed-ratio schedule results in presentation of a brief stimulus such as a light, and under a second schedule the brief stimulus is intermittently paired with intravenous infusion of drug.

When training animals to self-administer nicotine intravenously, two general approaches are typically used. One is to initially train the animal to self-administer a known drug of abuse, such as cocaine, and when self-administration has been established other drugs can then be substituted for the cocaine. The period of substitution can range from one single session to several months. Using this procedure a baseline of responding is usually maintained with the original training drug between substitutions with other drugs. Some studies initially train the animal to self-administer, for example, cocaine, and once behaviour has stabilized the drug to be studied is introduced without further exposure to the training drug. A second approach is to train the animal from the beginning with the drug to be studied.

Figure 2.2, from Goldberg and Spealman (1982), illustrates the development of self-administration behaviour in three squirrel monkeys with no previous drug history, under a time-based schedule of drug infusion in which the first lever press to occur after a 5 min period had elapsed produced an intravenous nicotine infusion (0.03 mg kg^{-1} per infusion); responding during the 5 min interval was recorded but had no programmed consequences (a fixed-interval schedule; see below). It can be seen in this figure that rates of responding around 0.1 response s^{-1} were maintained by nicotine infusions in these monkeys. These data demonstrate that it is not necessary to train animals to self-administer a known drug of abuse such as cocaine in order to establish nicotine self-administration behaviour.

2.2.2.1 *Fixed-ratio reinforcement schedules* The most common fixed-ratio schedule employed in nicotine self-administration studies has been the continuous reinforcement schedule, requiring one lever press per infusion. Early nicotine self-administration studies found that some rhesus monkeys responding under continuous reinforcement schedules would initiate intravenous self-administration of nicotine, whereas others required prolonged exposure to nicotine before self-administration was initiated (Deneau and Inoki 1967). The low levels of behaviour maintained and the limited amount of data reported by Deneau and Inoki (1967) make their results hard to evaluate against current criteria for self-administration behaviour (see above). Yanagita *et al.* (1974) reported that rhesus monkeys on a schedule of continuous reinforcement initiated intravenous nicotine self-administration when nicotine was available from the start of the experiment, but not when they were first

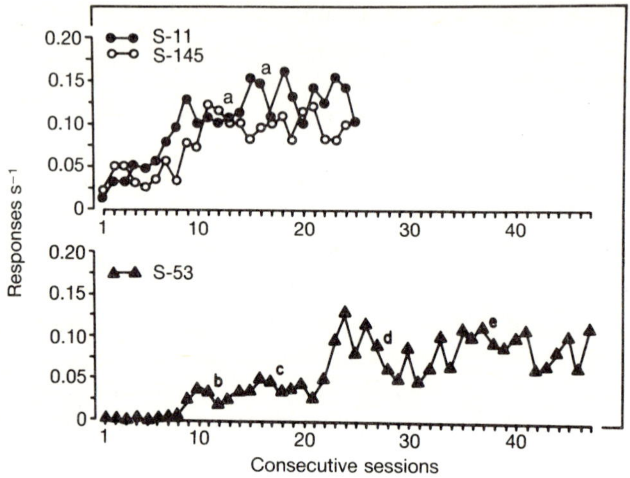

Fig. 2.2. Acquisition of responding maintained by intravenous infusions of nicotine (0.03 mg kg^{-1} per infusion) under a 5-min fixed-interval schedule. The ordinate shows consecutive sessions. The abscissa shows response rates. Monkeys S-11 and S-145 had previously responded under a fixed-interval schedule of food presentations, but responding had been extinguished before exposure to nicotine. If within 2 min after the 5 min fixed-interval a response had not occurred, an automatic infusion was presented (monkeys S-11 and S-145, upper curves). Automatic infusions were discontinued at [a]. Monkey S-53 was experimentally naive at the start of the study. For Monkey S-53 the initial fixed-interval was 30 s and an automatic infusion would be presented if no response had occurred within 4.5 min after the fixed-interval. At [b] and [c] the fixed-interval was increased to 1 and 3 min, respectively. At [d] the fixed-interval was increased to 5 min and any automatic infusions would occur 2 min after the fixed-interval. At [e] automatic infusions were discontinued. (From Goldberg and Spealman 1982, with permission.)

trained to self-administer the psychomotor stimulant SPA (1-1, 2 diphenyl-1-dimethylaminoethan-HCl); the low levels of behaviour maintained, the absence of dose–effect curves, and the lack of evaluation of the effects of a centrally acting nicotine antagonist such as mecamylamine with the monkeys that did self-administer nicotine, precludes evaluation of the reinforcing efficacy of nicotine in this study.

Using rats required to make one lever press to produce an infusion of nicotine (a fixed-ratio 1 schedule of reinforcement), Hanson *et al.* (1979) found that rats pretreated with one of several intravenous doses of nicotine, (0.004–0.06 mg kg^{-1} per infusion) every 30 min for two days, would subsequently self-administer the same dose of nicotine. The number of infusions self-administered first increased and then decreased with increasing doses of nicotine. Nicotine (0.06 mg kg^{-1}) pretreatment every 30 min for two days significantly increased subsequent self-administration of nicotine

(0.06 mg kg^{-1}) as compared to saline pretreatment. When saline infusions replaced nicotine (0.06 mg kg^{-1}) infusions, responding decreased. The centrally acting nicotine antagonist mecamylamine did not reliably block nicotine self-administration. The lack of reliable antagonism with mecamylamine reported by Hanson *et al.* (1979) indicates that factors other than nicotine may have contributed to the self-administration behaviour observed. Also, the use of different rats for each dose studied precludes an assessment of any differences in the reinforcing efficacy of different doses of nicotine within the same animal, further complicating a conclusion regarding the role of nicotine in this study.

It could be argued that a stimulatory effect of a drug, by way of a generally increased locomotor activity, may increase the number of infusions taken by an animal. If this was the case, then an increase in the number of infusions self-administered would actually be a reflection of an increased level of overall activity rather than, or in addition to, potentially reinforcing effects of the drug studied. To investigate this possibility regarding nicotine, female Wistar rats, with no history of either food or water derprivation, were studied under a schedule in which one lever press was required to produce each intravenous infusion of 0.03 mg kg^{-1} nicotine (a continuous reinforcement schedule; Cox *et al.* 1984). It was found that rats would significantly increase responding not only on the lever delivering nicotine, but also on an inactive control lever when nicotine was available. This finding indicated that a general increase in locomotor activity may also have contributed to increased response rates on the nicotine lever. When the nicotine dose per infusion was reduced from 0.03 to 0.003 mg kg^{-1} a significant differential increase in responding on the nicotine active lever was observed (Cox *et al.* 1984). Again, the absence of dose–effect curves and pretreatment with mecamylamine make these data hard to evaluate.

When comparing the ability of nicotine with that of cocaine, to maintain responding in rhesus monkeys when 10 lever presses were required per infusion (a fixed-ratio 10 schedule of reinforcement), Slifer and Balster (1985) found that the distribution of nicotine infusions within sessions was negatively accelerated, and resembled the results obtained when saline was available. Further, the dose–effect curves were relatively flat, leading the authors to conclude that nicotine functioned as a 'marginal reinforcer'.

In all of the above described studies using fixed-ratio schedules of reinforcement, the rate of responding by the animal directly determined the total amount of nicotine delivered. Thus, increasing rates of responding may have led to accumulation of nicotine in the body, which may subsequently have had effects on the continued responding. Generally with continuous reinforcement schedules, the rate of responding for nicotine is very low and often very unstable. However, it was shown in a study by Risner and Goldberg (1983), that when using a higher fixed-ratio requirement such that 15 lever presses were required per infusion (a fixed-ratio 15 schedule of reinforcement) and a time-out period of 4 min after each nicotine delivery, beagle dogs would self-administer intravenous nicotine (0.003–0.3 mg kg^{-1}), yielding a dose–response curve qualitatively similar to that generated by cocaine (0.003–0.3 mg kg^{-1}) under the same schedule requirements. The peak rates of responding were very high, averaging 0.3 and 0.7 responses per second for nicotine and cocaine respectively, (see Fig. 2.3). When saline was substituted for nicotine or cocaine responding decreased to low levels. Also, pre-session treatment with intravenous mecamylamine (1.0 mg kg^{-1}) markedly decreased responding maintained by nicotine, but not that maintained by cocaine. The results obtained by Risner and Goldberg (1983) suggest that limiting the access to

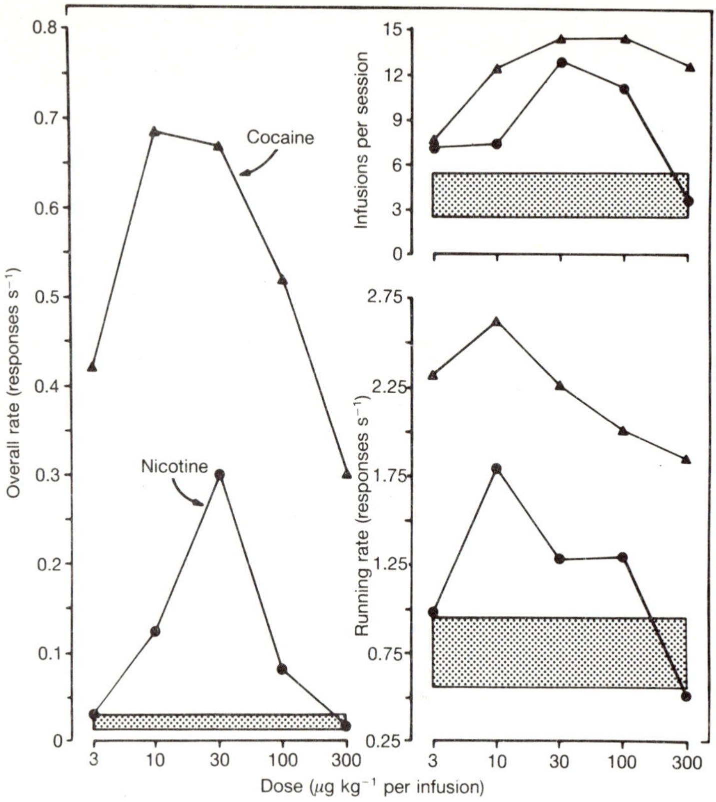

Fig. 2.3. Rates of responding maintained in beagle dogs under a fixed-ratio 15 schedule of intravenous drug infusions with a 4-min timeout after each infusion. The ordinates show doses of nicotine and cocaine tested. The shaded areas at the bottom of each section show the effects of saline. Each point represents the mean of four beagle dogs. The left section shows overall rates of responding. The upper right section shows number of infusions self-administered per session. The lower right section shows running rates (the rate during the last 12 responses of the ratio). (From Risner and Goldberg 1983, with permission.)

nicotine by the use of higher response requirements, and by interposing time-out periods during which no infusions can be obtained, increases the probability of nicotine maintaining self-administration behaviour.

However, the range of response requirements and access limitations suitable for demonstrating self-administration may be more limited with nicotine than with other drugs such as cocaine. For example, Griffiths *et al.* (1979) were unable to maintain self-administration behaviour in baboons required to emit 160 lever presses (a fixed-ratio 160 schedule) to produce an infusion of nicotine (0.001–3.2 mg kg^{-1}) with time-out periods of 3 hr after each infusion, although behaviour was well maintained under these conditions by cocaine. In the study of Risner and Goldberg (1983), as well as in the study of Griffiths *et al.* (1979), the animals had first been trained to self-administer cocaine before being exposed to nicotine. In the Risner and Goldberg (1983) study each dose of nicotine was tested for seven consecutive sessions, while Griffiths *et al.* (1979) tested each dose for 12 days. It is plausible that the more extended period of exposure to nicotine alone, the very high fixed-ratio responding requirement, and the exceedingly long time-out periods in the study of Griffiths *et al.* (1979) as compared to the study of Risner and Goldberg (1983), may have resulted in extinction of self-administration behaviour. Thus the reinforcing effects of nicotine may be more influenced by the conditions under which it is available than is the case with other drugs such as cocaine (for example, Ator and Griffiths 1983; Henningfield and Goldberg 1983).

2.2.2.2 *Progressive-ratio schedules* In attempting to evaluate the relative reinforcing efficacies of drugs, investigators have employed procedures in which the response requirement on a fixed-ratio schedule may be continually increased between self-administration sessions, until a 'break point' is reached where the animal will cease to respond for the drug. When drugs of similar actions (for example, psychomotor stimulants) are compared using a progressive-ratio procedure, those which maintain higher breaking points often prove to be more readily abused by humans (Griffiths *et al.* 1979).

Risner and Goldberg (1983), using beagle dogs, found that the maximum fixed-ratio value at which responding was maintained was:

(1) dose dependent;

(2) higher for nicotine than for saline;

(3) higher for cocaine than for nicotine (see Fig. 2.4).

Similarly, Yanagita *et al.* (1974), using rhesus monkeys responding under a progressive-ratio schedule in which the response requirement was doubled every 16 h, found that nicotine maintained lower breaking points than did cocaine.

2.2.2.3 *Interval schedules of reinforcement* To prevent accumulation of drug in the body, a schedule of reinforcement may be used which limits the

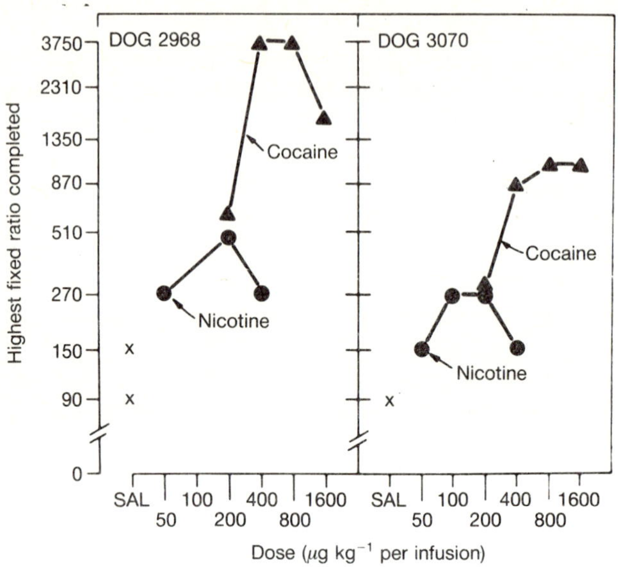

Fig. 2.4. Highest fixed-ratio completed by two beagle dogs responding under a progressive-ratio schedule of intravenous infusions of nicotine and cocaine. The abscissa shows the highest ratio reached. The ordinate shows the doses of the drugs tested. SAL represents results obtained when saline was substituted for the drugs. (From Risner and Goldberg 1983, with permission.)

availability of the drug, by interposing time-out periods between successive opportunities to self-administer the drug (for example, Risner and Goldberg 1983). Another approach is to use a fixed-interval schedule, in which the first response to occur after a fixed interval of time since the previous infusion occurred, produces an infusion. Dougherty *et al.* (1981) used a procedure in which 4 lever presses produced an intravenous infusion of nicotine (0.01–0.27 mg kg^{-1}) after 60 s had elapsed (a fixed-ratio 4 fixed-interval 60-s schedule of reinforcement). Dougherty *et al.* (1981) reported that 2–9 months of daily training was required before three of four rhesus monkeys would self-administer nicotine above saline levels on this schedule.

Spealman and Goldberg (1982), using squirrel monkeys and a procedure in which the first response to occur after 5 min had elapsed (a 5 min fixed-interval schedule) found that intravenous infusions of nicotine (0.003–0.56 mg kg^{-1}) and cocaine (0.003–0.3 mg kg^{-1}) maintained similar rates of lever press responding in two monkeys, whereas in a third monkey cocaine rates were substantially higher; further, with increasing doses of nicotine and cocaine, rates of responding first increased and then decreased (Fig. 2.5). In all monkeys rates of responding were much higher for nicotine than for saline; responding maintained by nicotine but not by cocaine was blocked by

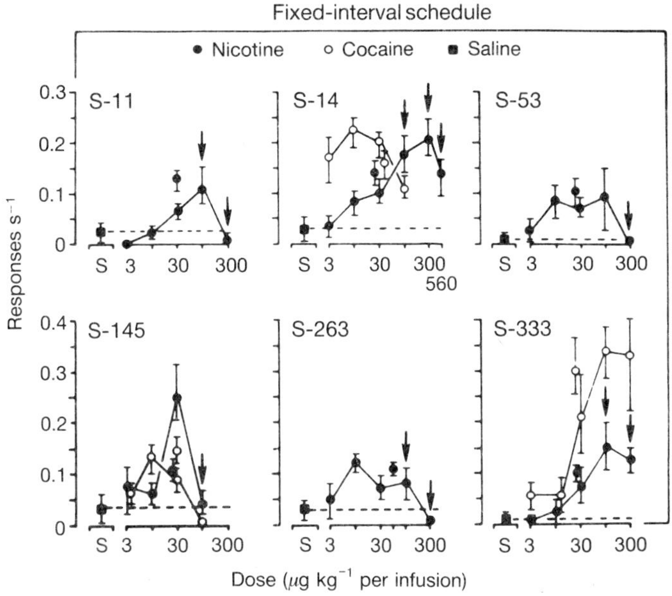

Fig. 2.5. Rates of responding maintained in six squirrel monkeys responding under a 5-min fixed-interval schedule of intravenous nicotine and cocaine infusions. The abscissa shows rate of responding. The ordinates show doses tested. S and dashed horizontal lines represent data obtained when saline was substituted for the drugs. Arrows show doses of nicotine producing vomiting during or shortly after the session. (From Spealman and Goldberg 1982, with permission.)

pre-session treatment with intramuscular injections of 1.0 mg kg^{-1} of mecamylamine.

In a study designed to compare the effectiveness of a fixed-ratio schedule with that of a fixed-interval schedule in maintaining nicotine self-administration in baboons, Ator and Griffiths (1983) compared a procedure in which two responses (a fixed-ratio 2 schedule of reinforcement) produced an infusion of nicotine (0.01–0.32 mg kg^{-1}) with one in which the first response to occur 5 min after the previously reinforced response (a fixed-interval 5 min schedule of reinforcement) produced a nicotine infusion. They found that under the fixed-ratio 2 schedule the first five sessions generated a dose–response curve for nicotine. However, with continued testing the dose–response function flattened out to become no different from saline. With the fixed-interval 5 min schedule, very low rates of responding were maintained and no effect of changes in dose were observed. Ator and Griffiths (1983) therefore concluded that nicotine had only marginally reinforcing effects in the baboon.

2.2.2.4 *Concurrent schedules* When the availability of a reinforcing event such as food delivery is governed by certain schedules of reinforcement, behaviour leading to presentation of a second event such as water delivery, or a drug infusion, is often strengthened during periods of non-availability of the first reinforcing event; this class of behaviour is referred to as adjunctive behaviour or schedule-induced behaviour since its heightened probability occurs as an adjunct to an ongoing schedule of reinforcement. It was first reported by Falk (1961) that food-deprived rats responding on certain schedules of reinforcement for food would drink immediately following each food presentation leading to consumption of excessive amounts of water, even though the rats were not water-deprived. Falk (1971) defined adjunctive behaviour as 'behavior maintained at high probability by stimuli whose reinforcing properties in the situation are derived primarily as a function of schedule parameters governing the availability of another class of reinforcers' (p. 586). This procedure and variations thereof has since been used to establish oral alcohol consumption in rats (George 1987; Lester 1961; Meisch 1975), and subsequently self-administration of a variety of drugs (see review by Sanger 1986).

In one study, rats at 80 per cent of their free-feeding weights, but not food-satiated rats, initiated self-administration of intravenous nicotine; when a free food pellet was delivered every minute to the rats at 80 per cent body weight even higher rates of nicotine self-administration were observed (Lang *et al.* 1977). Once self-administration behaviour had developed, rats receiving the food pellet once per min continued to self-administer nicotine even when they received free food in their home cages (Singer *et al.* 1978). Further, rats kept at 80 per cent body weight continued to self-administer nicotine when the free pellets were discontinued (Smith and Lang 1980). These studies show:

(1) that free-feeding rats did not initiate self-administration of intravenous nicotine;

(2) that reduction of their body weights facilitated initiation of intravenous nicotine self-administration;

(3) that the intermittent presentation of food pellets to rats reduced to 80 per cent body weight enhanced the acquisition of intravenous nicotine self-administration behaviour.

To compare the ability of a concurrent schedule to a fixed-interval schedule of reinforcement in maintaining nicotine self-administration in the same animals, rhesus monkeys were trained in an operant chamber equipped with two response levers, to respond on the right lever such that the first response to occur after 5 min had elapsed since the last food presentation produced a food pellet (a fixed-interval 5 min schedule), whereas each response (fixed-ratio 1) on the left lever produced an infusion of nicotine (Slifer 1983; Slifer and Balster 1985). These studies showed that when both levers were active, more infusions of low doses (0.001 and 0.003 mg kg^{-1}) of nicotine were self-administered than when the drug lever alone was active. Higher doses of nicotine (0.03 and 0.1 mg kg^{-1}) were administered less when both levers were active. Slifer (1983) and Slifer and Balster (1985) suggested that the increased self-administration of low doses of nicotine when both levers were active was a result of schedule-induction of behaviour such that the reinforcing properties of these low doses were enhanced by the food delivery schedule; thus only by occurring in conjunction with the food reinforcement would the weak reinforcing properties of low doses of nicotine maintain responding. The lack of increased self-administration of higher doses of nicotine was attributed to the reinforcing properties of these doses being stronger, and thus not affected by schedule-induction procedures. The finding that self-administration of saline and the lowest dose (0.0001 mg kg^{-1}) of nicotine were not affected by the schedule-induction procedure showed that pharmacological factors were responsible for the increased self-administration (Slifer and Balster 1985).

2.2.2.5 *Second-order schedules* A second-order schedule is composed of components of simpler schedules such as fixed-ratio or fixed-interval schedules. For example, in a second-order schedule each fixed-ratio of 10 lever presses may produce a brief illumination of a light bulb in the operant chamber, but without any food presentation. After completion of 9 fixed-ratio components that resulted in the brief light only, the tenth component will, in addition to presenting the brief stimulus light, also produce a food pellet. In this way the brief stimulus light may acquire a reinforcing effect of its own, since its occurrence signals the advent of the food pellet. Under a second-order drug delivery schedule, the completion of a predetermined amount of behaviour, for example, 10 lever presses, results in an environmental event, such as a brief illumination of a light bulb. Intermittently, the environmental event will also be paired with a drug infusion (Goldberg *et al.* 1975).

Further demonstrations of the reinforcing effects of nicotine, using second-order schedules of reinforcement, were provided by Goldberg *et al.* (1981) and Spealman and Goldberg (1982). Goldberg *et al.* (1981) demonstrated that nicotine maintained persistent and high rates of responding in squirrel monkeys by using a procedure in which every 10 lever presses (fixed-ratio 10) within a 1- or 2- min fixed interval of time would produce a brief light stimulus; the first ten lever presses completed after the interval would produce both the visual stimulus and an intravenous infusion of 0.03 mg kg^{-1} nicotine. A 3 min time-out period followed each infusion. When saline was infused instead of nicotine, and when the centrally acting nicotine antagonist mecamylamine (1.0 mg kg^{-1}) was injected intramuscularly before the session, rates of responding were markedly reduced. Rates of responding were also substantially lower when the brief visual stimulus was omitted, while nicotine continued to be infused (see Fig. 2.6); even though nicotine was the primary reinforcing event, the brief visual stimulus clearly acquired reinforcing

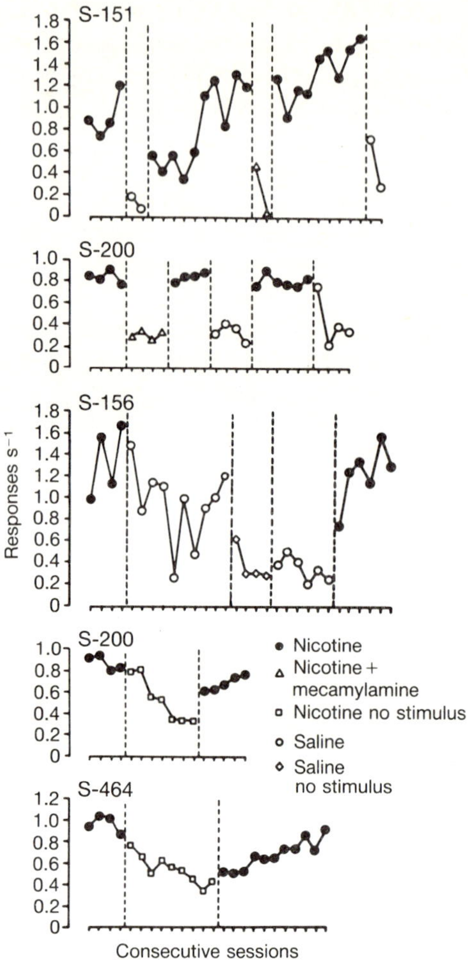

Fig. 2.6. Rates of responding under a second-order schedule of intravenous nicotine infusions in five squirrel monkeys. The abscissas show rates of responding. The ordinates show consecutive sessions. Filled circles represent responding under the second-order schedule with brief stimulus presentations intermittently paired with intravenous infusions of nicotine. Open circles represent the effects of substituting saline infusions for nicotine. Open triangles represent the effects of pretreating the monkeys with intramuscular injections of mecamylamine (1.0 mg kg^{-1}). Open squares represent the effects of omitting the brief light stimulus. Open diamonds represent the effects of substituting saline for the nicotine and omitting the brief stimulus. (From Goldberg et al. 1981, with permission.)

properties as well, since its removal caused rates of responding to decrease.

In a subsequent study, comparing nicotine and cocaine in squirrel monkeys utilizing the same second-order scheduling as in Goldberg *et al.* (1981), Spealman and Goldberg (1982) found qualitatively similar changes in rates of responding for nicotine or cocaine when the dose was varied from 0.003 to 0.1 mg kg^{-1} per infusion, although cocaine sometimes maintained higher maximal rates of responding than did nicotine. By arranging the nicotine infusions this way, the experimenter can control the amount of nicotine delivered and also maintain a stable rate of responding. Also, by analyzing the effects of environmental events upon nicotine intake by animals, a better understanding of the interaction of pharmacological and non-pharmacological events on the induction and maintenance of human smoking behaviour may be achieved.

2.2.2.6 *Punishing effects* The behavioural effects of a drug may be determined by a multitude of events, such as the behavioural history of the subject, the ongoing behaviour of the subject, and the conditions under which the drug is available. For example, Spealman (1979) demonstrated that lever pressing behaviour on one of two levers could be maintained in squirrel monkeys by the postponement of cocaine injections, and that lever pressing on the opposite lever could be simultaneously maintained by infusions of cocaine. The ability of nicotine to suppress behaviour that leads to its administration was studied by Goldberg and Spealman (1983). They demonstrated that nicotine, in addition to maintaining behaviour that will lead to its administration, can also function as a punishing event in a manner similar to electric shock. Squirrel monkeys were trained to press a lever in order to obtain food. When responding intermittently produced either intravenous infusions of nicotine (0.003–0.3 mg kg^{-1}) or brief electric shocks, rates of responding for food were markedly decreased. When the centrally acting nicotine antagonist mecamylamine (0.1–0.3 mg kg^{-1}) was injected intramuscularly prior to the session, the rate-depressant effect of nicotine, but not that of electric shocks, was reversed. In contrast, when the anxiolytic compound chlordiazepoxide (5.6–10.0 mg kg^{-1}) was given prior to the session, the rate-suppressant effects of both nicotine and electric shocks were reversed (see Fig. 2.7). When nicotine or cocaine infusions were intermittently presented to squirrel monkeys responding for food, nicotine but not cocaine suppressed behaviour in a dose-related fashion (Takada *et al.* 1986).

2.2.3 Summary and conclusions

The stability of nicotine self-administration behaviour, and the speed with which it develops in experimental animals, differed from that of self-administration studies using opiates and psychomotor stimulants. Results of earlier animal studies, in which only weak reinforcing effects of nicotine could be demonstrated, were difficult to reconcile with the very persistent

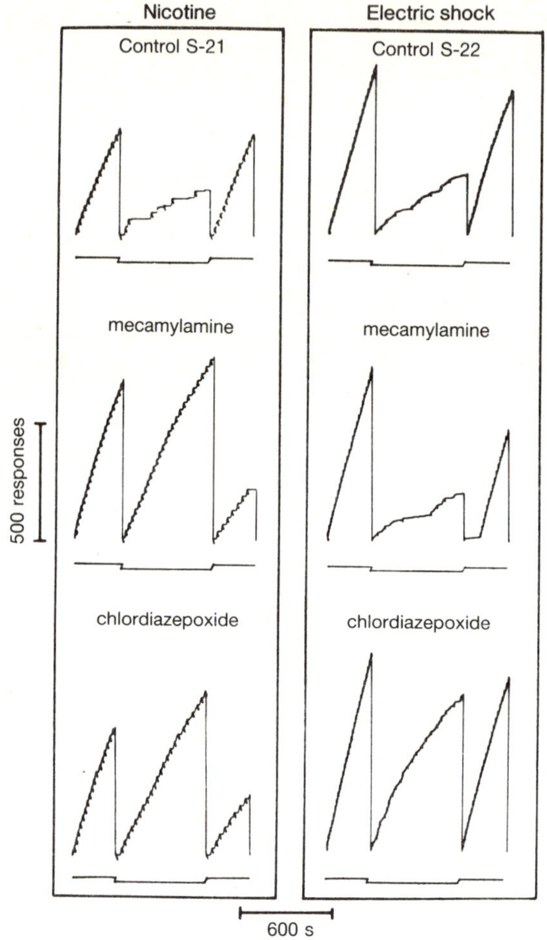

Fig. 2.7. Cumulative records of two squirrel monkeys responding under a two-component fixed-ratio 30 schedule for food presentation in which the first response in every other component produced an electric shock (5 mA) or an intravenous infusion of nicotine (0.01 mg kg^{-1} per infusion). The left section shows responding when nicotine infusions were produced by the first response in every other component. The right section shows responding when electric shocks were produced by the first response in every other component. The two top records show responding when nicotine (left) or electric shocks (right) were produced by the first response in every other component. The two middle records show the effects of giving mecamylamine (0.3 mg kg^{-1}) intramuscularly prior to the session. The bottom records show the effects of administering chlordiazepoxide (10.0 mg kg^{-1}) intramuscularly prior to the session. (From Goldberg and Spealman 1983, with permission.)

behavioural effects of nicotine seen in smoking humans. In fact, such data suggested to some reviewers that the reinforcing effects of nicotine were less robust than those of other drugs (for example, Griffiths *et al.* 1979, 1980), and that the primary function of nicotine in the control of smoking behaviour may be to set an upper limit on smoke intake (Russell 1979; see Russell, Chapter 11, this volume).

However, the utilization of second-order schedules of reinforcement in nicotine self-administration experiments (Goldberg *et al.* 1981; Spealman and Goldberg 1982) provided new insights in the understanding of this pervasive form of human drug self-administration. These experiments helped resolve the controversy between previous animal studies and human data. The results obtained with the second-order schedule suggested that the reinforcing effects of nicotine were related to how the injections were arranged within a session, such that nicotine was more effective in maintaining responding when presented intermittently than when continuous availability prevailed. Subsequent findings with fixed-interval and fixed-ratio schedules of reinforcement (in which the schedule parameters were arranged to provide similar spacing of injections to those under second-order schedules) further demonstrated that the rate of self-administration of nicotine was dependent on scheduling the conditions under which it was available (Risner and Goldberg 1983; Spealman and Goldberg 1982). These findings are in concord with findings in humans that delivering nicotine intermittently by use of tobacco products seemed to maximize its reinforcing efficacy (cf., Henningfield 1984). Another important finding by Goldberg *et al.* (1981) was that environmental stimuli may play an important role in the development and maintenance of responding for nicotine. This finding may have important implications for the understanding of human smoking behaviour since smoking is accompanied by a multitude of stimuli such as visual, olfactory, tactile, and taste stimuli, which may gain control over smoking behaviour.

The resolution of this controversy highlighted the importance of studying the interactions between pharmacological, behavioural, and environmental factors, as determinants in the process of shaping and maintaining drug self-administration behaviour in animals. In a review of the nicotine self-administration literature it was concluded that the conditions under which nicotine is reinforcing are more restricted than for other drugs (Henningfield and Goldberg 1983). By employing second-order schedules of reinforcement when studying the ability of nicotine to initiate and maintain human smoking behaviour, the importance of environmental events supporting this behaviour may be assessed and analyses of interactions between behavioural, environmental, and pharmacological variables may be possible. Such analyses would enhance our understanding of the dependence process as it relates specifically to nicotine in humans. For example, smoking occurring in conjunction with pleasurable environmental events, such as after a meal, with coffee, or when engaging in social conversation, may cause these events to become associated with the effects of nicotine, and thereby acquire the power to elicit smoking behaviour on future occasions. Similarly, pleasurable effects of smoking may, when used in stressful situations, have stress-alleviating effects; thereby stressful situations may then also come to gain control over smoking behaviour. In this way smoking behaviour may be maintained and strengthened by external events over extended periods of time.

The finding that nicotine can also serve to suppress responding in animals (Goldberg and Spealman 1983) further emphasizes the importance of the arrangement of drug injections in a session, relative to the organism's ongoing behaviour, as a critical factor in determining the resultant effects of nicotine upon the behaviour of the animal. This finding may suggest a role of aversive effects of nicotine in the development and control of smoking behaviour in humans as discussed elsewhere (Russell, this volume, chapter 11; Stolerman, this volume, chapter 8). The possibility of utilizing aversive effects of nicotine in smoking termination programs has also been described (Merbaum *et al.* 1979).

2.3 Tolerance development to behavioural effects of nicotine

Tolerance to the effects of a drug is defined as a decreased response to a certain dose of the drug as a result of its repeated administration over time, or as the need for a larger dose of the drug to maintain the effect originally obtained (for example, Jaffe 1985). Tolerance is rarely total, and often develops differentially, depending on the response measured. Tolerance has also been observed to occur as a result of a single administration of a drug, in which case it is referred to as tachyphylaxis or acute tolerance, as opposed to chronic tolerance as defined above. It has been known since the turn of the twentieth century from the proliferation of studies by Langley and others that tolerance may develop to the effects of nicotine. As described by Langley (1905) in studies of nicotine induced muscle contractions in the fowl:

> When the obvious [muscle contraction] has ceased, a second small dose of nicotine will again cause contraction, the contraction being of rather less height and duration, and the injection can be repeated many times with progressive decrease in effect. When a small dose of nicotine has no or only slight effect, a considerable one can still be obtained by injecting a large dose [. . . .], and when this becomes ineffective, a still larger amount [. . . .], will cause more or less contraction. When a given dose ceases to produce an effect, it will usually produce one if an interval of about half-an-hour to an hour is allowed before repeating the injection, (pp. 383–385).

Further, Dixon and Lee (1912) demonstrated tolerance to nicotine in rabbits (see below).

2.3.1 *Mechanisms of tolerance*

It is useful to differentiate three mechanisms by which tolerance may develop (for example, Jaffe 1985). None of these, however, is necessarily by itself the sole determinant of tolerance in any given case. It is also possible that one mechanism may play a larger role in the initiation of tolerance, whereas another may assume a larger role in the maintenance of tolerance. *Pharmacokinetic tolerance* refers to an increased metabolism of a drug through enzyme induction as an effect of prolonged exposure. As a result, a given dose of a drug will be degraded faster by the body and will consequently have a shorter duration of action than in a naive individual. This will lead to the need for a larger dose to maintain a certain level of effect. In an early study Dixon and Lee (1912) measured the effects on blood pressure of decerebrate cats of nicotine-incubated liver extracts from rabbits chronically treated with nicotine, or from control rabbits. Dixon and Lee (1912) found a diminished effect after treatment with the extract from the tolerant rabbits (Fig. 2.8).

The experiment by Dixon and Lee (1912) would suggest that metabolic tolerance to nicotine may develop in rabbits. However, other studies have

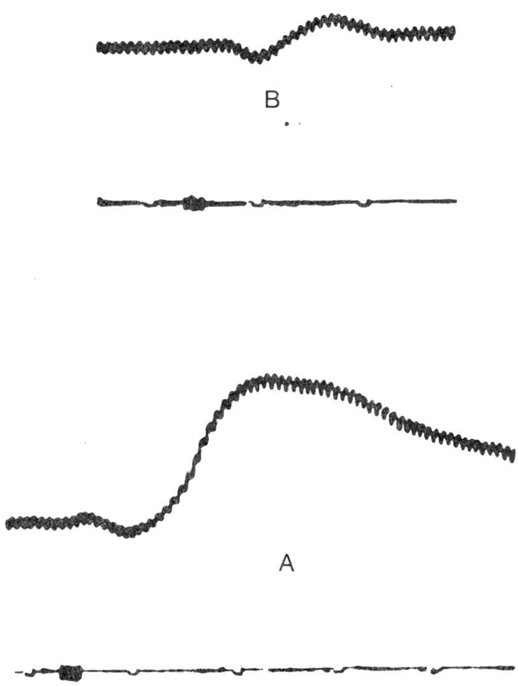

Fig. 2.8. Effects of nicotine on blood pressure of decerebrate cats. Panel A shows the effects of injecting nicotine (10.0 mg) incubated liver extract from a non-pretreated rabbit. Panel B shows the effects of injecting nicotine (10.0 mg) incubated liver extract from a rabbit pretreated with 0.5 cc of a 1 per cent nicotine solution every other day for a month (i.e. a nicotine tolerant rabbit). Blood pressure increases (mmHg) were 40 and 8 in panels A and B, respectively. Time of measurement was 30 s. (Modified from Dixon and Lee 1912, with permission.)

failed to show that chronic nicotine administration can enhance its metabolism in rats (Takeuchi *et al.* 1954; see Schievelbein 1984 for further references) or in mice (Hatchell and Collins 1977; Marks *et al.* 1983). *Behavioural tolerance* refers to the decreased effect of a given dose of a drug as a result of behavioural adaptation to effects of the drug that may interfere with the ability of the individual to interact successfully with his environment (LeBlanc *et al.* 1978; Schuster *et al.* 1966; see Goudie and Demellweek 1986 for a review of additional explanations of behavioural tolerance). Thus behavioural tolerance may occur in situations where it is essential for the organism to adapt to the disruptive effects of a drug. *Pharmacodynamic tolerance* to the effects of a drug refers to an adaptation by cellular mechanisms to the presence of the drug in the body, which will, over time, cause a decrease in response to a given concentration of the drug. Pharmacodynamic tolerance is usually inferred by employing controls to eliminate the possibility of pharmacokinetic or behavioural tolerance.

The contribution of behavioural versus pharmacodynamic factors to the development of tolerance to the effects of nicotine may be very difficult to assess. For example, it has been reported that habituation to the test apparatus (behavioural tolerance) may (Schlatter and Battig 1978) or may not (Clarke and Kumar 1983*a*) influence the development of tolerance to nicotine in rats. Others found inconclusive evidence (Keenan and Johnson 1972), or no evidence (Hatchell and Collins 1977; Hendry and Rosecrans 1982) in favour of the idea that tolerance to nicotine may result from behavioural adaptation to its disruptive effects. An example of procedures that have been employed successfully to assess the contributions of behavioural and pharmacodynamic mechanisms of tolerance is provided by studies of Hendry and Rosecrans (1982). They studied the development of tolerance to nicotine in two groups of mice responding on a fixed-ratio 25 schedule for sweetened milk rewards, using a procedure developed by Chen (1968) and Carlton and Wolgin (1971) to distinguish behavioural from pharmacodynamic tolerance. Initially, before the mice received any differential treatment, the effects of a range of doses (0.2–1.6 mg kg^{-1}) of nicotine on fixed-ratio responding was established in the two groups. Next, during 30 days of chronic treatment, one group received subcutaneous nicotine injections (1.2 mg kg^{-1}) prior to and saline injections after the sessions, whereas the other group received the reverse treatment, i.e. saline injections prior to and nicotine injections (1.2 mg kg^{-1}) after the sessions. Lastly, after the chronic treatment phase, a dose–effect curve for nicotine (0.2–1.6 mg kg^{-1}) was redetermined. Any tolerance development would be seen as a shift either upwards or to the right of the dose–response curve. If tolerance was observed in both groups, pharmacological mechanisms would be postulated as responsible for its development. If, however, tolerance was observed only in the group receiving its injections before the session, then behavioural mechanisms would be implicated such that the animals had learned to adapt behaviourally to the effects of the drug in order to

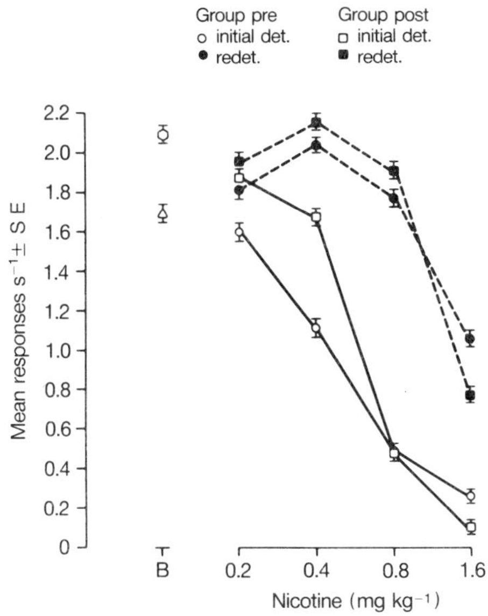

Fig. 2.9. Dose-response curves of nicotine in mice responding on a fixed-ratio 25 schedule for sweetened milk rewards prior to (solid lines) and after chronic nicotine (1.2 mg kg^{-1}) treatment (dashed lines). Circles represent mice given chronic nicotine treatment prior to and saline after each session, and squares represent animals receiving the reverse treatment, i.e. saline before and nicotine after each session. The abscissa represents rates of responding. The ordinate represents doses of nictone and baseline responding (B) during vehicle sessions. (From Hendry and Rosecrans 1982, with permission.)

compensate for its disruptive effects while earning their rewards in the test cage. Hendry and Rosecrans (1982) found that the two groups were very similar in their responding to nicotine (0.2–1.6 mg kg^{-1}) whether compared before or after the 30 days of chronic treatment with 1.2 mg kg^{-1} of nicotine. However, both groups showed an upward shift in the nicotine dose–response curve after the chronic treatment period. There was a significant difference between the determination that occurred before and after chronic treatment such that after chronic treatment only the highest dose in the dose–response curve (1.6 mg kg^{-1}) exerted a depressant action on rate of responding, whereas before chronic administration a dose of 0.4 mg kg^{-1} had clear response depressant effects. Figure 9 shows the results of pre- versus post-session chronic nicotine treatment in the study of Hendry and Rosecrans (1982).

Dissipation of tolerance to nicotine was already evident by the first week after discontinuation of chronic administration, and it was very pronounced by the third week at which point acute tests with the chronic treatment dose of nicotine, (1.2 mg kg^{-1}), severely depressed rates of responding. Hendry and

Rosecrans (1982) concluded that pharmacological but not behavioural mechanisms were responsible for the development of tolerance to nicotine.

A substantial amount of data suggest that pharmacodynamic tolerance best explains most of the instances in which tolerance to the effects of nicotine have been found (see below). In addition, the development of tolerance to nicotine has also been reported to be influenced by genetic and sex differences (for example, Hatchell and Collins 1977, 1980; Marks *et al.* 1986).

2.3.2 *Tolerance development in unlearned behaviours*

Tolerance was reported to the depressant effects of nicotine on rearing frequency in rats after 6 days of subcutaneous administration of 0.5 mg kg^{-1} of nicotine (Keenan and Johnson 1972). Measuring the depressant effects of nicotine on spontaneous locomotor activity in rats in a *Y*-maze, Stolerman *et al.* (1973) reported that intraperitoneal injections of nicotine (0.02–0.65 mg kg^{-1}) dose-dependently decreased locomotor activity in naive rats. Comparing the effects of different routes of nicotine pretreatment, intraperitoneal injections of nicotine (0.33 mg kg^{-1} three times per day for three days), three-day implants with nicotine-containing reservoirs (4.5 mg kg^{-1} per day), and nicotine mixed in the drinking water for seven days (2.1 and 3.9, but not 6.5 mg kg^{-1} per day), produced tolerance to the locomotor depressant effects of 0.33 mg kg^{-1} of nicotine. Further, studying development of acute tolerance to nicotine, Stolerman *et al.* (1973) pretreated rats with a single intraperitoneal injection of 0.33 mg kg^{-1} of nicotine, and gave challenge doses of nicotine (0.33 mg kg^{-1}) at various time intervals. Tolerance to the challenge dose (0.33 mg kg^{-1}) increased as the time interval between pretreatment and challenge doses increased from 0.5 to 2 h. After 8 h tolerance had disappeared. When chronically treating rats with 0.33 mg kg^{-1} of nicotine intraperitoneally three times daily for eight days, tolerance was evident 90 days after the last injection (Stolerman *et al.* 1973).

Using the same procedure as above, Stolerman *et al.* (1974) found that after pretreating rats with a single intraperitoneal injection of 0.24 mg kg^{-1} of nicotine 2 h before challenge doses of nicotine (0.08–1.30 mg kg^{-1}) were given, there was a shift in the nicotine dose–response curve to the right of that for saline treated rats, indicating that acute tolerance to the locomotor depressant effects of nicotine had developed. In the nicotine pretreated rats a 2.4 times higher challenge dose of nicotine was required to achieve the same degree of depression of locomotor activity as in the saline pretreated rats. When varying the acute pretreatment dose (0.04–0.33 mg kg^{-1}) with the challenge dose (0.33 mg kg^{-1}) held constant, tolerance was evident with the 0.08 and 0.16 mg kg^{-1} pretreatments. When the same dose-range (0.04–0.33 mg kg^{-1}) was given chronically (three times per day for 8 days), tolerance was again evident when a 0.65 mg kg^{-1} challenge dose was given 24 h after the last pretreatment, and when tested 7 days later with a 0.33 mg kg^{-1} challenge dose.

Comparing the shapes of the nicotine dose–effect curves after acute (0.04–0.33 mg kg^{-1}) and chronic (0.04–0.33 mg kg^{-1}) nicotine pretreatment, striking similarities were found when nicotine challenge (0.33 mg kg^{-1}) occurred 2 h after pretreatment in the acute group, and when it occurred 7 days after discontinuation of chronic exposure in the chronic group, leading the authors to suggest that similar mechanisms were involved in acute and chronic tolerance. When a pretreatment of 0.33 mg kg^{-1} of nicotine was given every third day for 30 days, tolerance was seen to 0.33 mg kg^{-1} of nicotine 24 h after the last pretreatment, and 7 days later with no treatment in the meantime, tolerance could still be detected.

Tolerance to the ability of nicotine to produce a prostration-immobilization syndrome was reported by Abood *et al.* (1979) in rats given intraventricular administrations of 0.01 mg of nicotine for 3 days consecutively, but not in rats given nicotine every other day. Further, Abood *et al.* (1984), again using the prostration syndrome, found that 0.0008 mg of nicotine injected into the fourth ventricle twice daily for 4 days induced 70 per cent tolerance on day 4, whereas after an additional 3 days (day 7) without the daily nicotine, 10 per cent of the rats were tolerant to the same nicotine dose (0.0008 mg). However, when nicotine was continuously infused into the lateral ventricle at 0.008 mg h^{-1} for 10 days, no tolerance to 0.0008 mg injected into the fourth ventricle was evident on the last day of infusion. After intraperitoneal injections of 1.95 mg kg^{-1} twice daily for 14 days, only a slight tolerance effect was seen when 0.0008 mg of nicotine was injected into the fourth ventricle 12 h after the last intraperitoneal injection. Abood *et al.* (1984) concluded that tolerance to nicotine's ability to induce the prostration syndrome is a transient phenomenon that diminishes with time.

Tolerance to the depressant effects of 0.3 mg kg^{-1} of nicotine on locomotor and exploratory activity was investigated in rats given nicotine (50 mg nicotine base l^{-1}) in drinking water for 28 days, inducing a daily consumption of approximately 4 mg kg^{-1} (Falkeborn *et al.* 1981). At 24 h after withdrawal of the nicotine, these rats showed tolerance to the 0.3 mg kg^{-1} nicotine challenge dose given intraperitoneally, whereas a second group of rats that had received plain water for 28 days showed depression of locomotor behaviour (i.e. no tolerance). A third group of rats that had received plain water for 28 days, and subsequently were tested with saline, were not different from the rats that had received 28 days exposure to nicotine. When tested 31 days after withdrawal no tolerance was seen in any of the groups.

Tolerance was reported to the locomotor suppressant effects of 0.45 mg kg^{-1} of nicotine in rats that had been pretreated with 0.45 mg kg^{-1} of nicotine subcutaneously for a period of 14 days (Larsson *et al.* 1986). Rats given 14 days of saline treatment showed reduced activity after 0.45 mg kg^{-1} of nicotine, whereas rats receiving saline after 14 days of saline treatment were not different from the chronic nicotine group receiving nicotine. The same kind of effects were also seen on body temperature; i.e. the chronic saline rats receiving nicotine had lower body temperature than the two other groups (Larsson *et al.* 1986).

In contrast to development of tolerance to the depressant effects of nicotine, it has been reported that chronic exposure to nicotine may increase sensitivity to its stimulatory effects (Clarke and Kumar 1983*a*, *b*; Ksir *et al.* 1985, 1987; Morrison and Stephenson 1972), an effect not found to be subject to tolerance (Morrison and Stephenson 1972). Clarke and Kumar (1983*b*) found that subcutaneous nicotine (0.1–0.4 mg kg^{-1}) depressed locomotor activity in non-tolerant rats during the first 20 min of the 80 min observation period, whereas in rats made tolerant by daily or even only weekly injections of nicotine (0.4 mg kg^{-1}) the initial depressant activity was replaced by a stimulant action on locomotor behaviour which lasted throughout the session. Mecamylamine (0.5, 1.0 mg kg^{-1}) prevented the initial depressant action of nicotine in non-tolerant rats, and mecamylamine (0.1, 0.32, 1.0 mg kg^{-1}) also prevented the initial stimulant action in tolerant rats. Hexamethonium (0.2, 1.0, 5.0 mg kg^{-1}) did not have these effects, leading the authors to suggest that the effects of nicotine were of central origin, and probably caused by stimulation rather than by blockade of central nicotine receptors.

Another example of development of increased sensitivity to the stimulatory effects of nicotine was provided in a study by Ksir *et al.* (1985). These authors reported increased sensitivity to occur after 5 days of repeated injections of 0.1, 0.2 or 0.4 mg kg^{-1} of nicotine in three groups of rats. Another group of rats was given 0.2 mg kg^{-1} of nicotine for 5 days and then saline for 7 days, and was first tested with 0.2 mg kg^{-1} on day 8, at which point behavioural stimulation was observed. Other rats were pretreated for 5 days with 0.2 mg kg^{-1} of nicotine followed by 20 days of saline injections; at 21 days after the last nicotine injection no stimulant effect of 0.2 mg kg^{-1} of nicotine was seen. Ksir *et al.* (1987) reported an upward shift in the locomotor activity dose–response curve (0.1–1.6 mg kg^{-1}) in rats after 5 days of exposure to 0.2 mg kg^{-1} of nicotine (Fig. 2.10). When pretreatment doses of 0.01–0.3 mg kg^{-1} of nicotine were given, the behavioural response to 0.2 mg kg^{-1} of nicotine was saline-like at the lowest pretreatment dose (0.01 mg kg^{-1}), and plateaued at a 50 per cent increase thereafter. However, rats pretreated with 1.6 mg kg^{-1} of nicotine did not show behavioural stimulation when tested with 0.2 mg kg^{-1} of nicotine (Ksir *et al.* 1987).

Using a battery of tests consisting of respiratory rate, acoustic startle response, locomotor activity, heart rate, and body temperature, Marks *et al.* (1985) studied the development of tolerance to 0.75 mg kg^{-1} of nicotine in mice given continuous infusions of nicotine (4.0 mg $kg^{-1}h^{-1}$). The mice were challenged with 0.75 mg kg^{-1} of nicotine intraperitoneally at different intervals (0–12 days) after onset of the continuous infusion. Marks *et al.* (1985) reported that tolerance did not develop to the effects on respiratory rate and startle response. However, maximal tolerance to the suppressant effects of

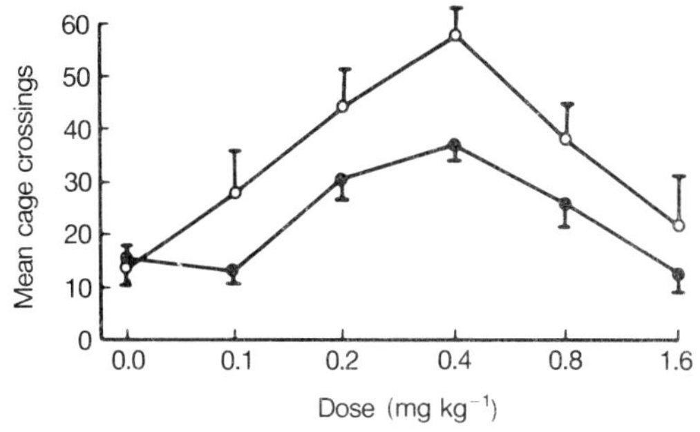

Fig. 2.10. Dose–response curves of nicotine on locomotor activity in rats before (solid circles) and after (open circles) 5 days of chronic exposure to nicotine (0.2 mg kg^{-1}). The abscissa represents the mean number of cage crossings. The ordinate represents doses of nicotine. (From Ksir *et al.* 1987, with permission.)

0.75 mg kg^{-1} on locomotor activity in a *Y*-maze, heart rate and body temperature was seen after 4 days of continuous nicotine infusion, and it was accompanied by increases in number of brain nicotine receptors (see Wonnacott, Fig. 7.5, this volume). Tolerance offset was measured in mice given continuous infusions of nicotine (7.0 mg $kg^{-1}h^{-1}$). Again, with the exception of startle response and respiration rate which were not affected, tolerance was still evident in the other measures 4 days after discontinuation of infusion. After 8 days locomotor activity had returned to control levels; body temperature reached control levels in 12–16 days, whereas heart rate was still not back to control level 20 days after discontinuation of chronic infusion. The offset of tolerance was accompanied by a decrease in the number of nicotine receptors in the brain. Correlating results from [^{3}H]nicotine and [^{125}I]αbungarotoxin binding studies with those from the behavioural measures, high correlations were seen with *Y*-maze locomotor activity and body temperature. The increased number of receptors in the brain reported by Marks *et al.* (1985) as a result of chronic drug administration is unusual since numerous studies have shown that chronic stimulation of receptors will result in a down-regulation rather than in an up-regulation of receptors (Creese and Sibley 1981). However, these findings are supported by other studies on the effects of chronic nicotine administration on brain nicotinic receptors. The possible reasons for this unexpected phenomenon will be addressed elsewhere in this volume (Wonnacott, chapter 7, this volume).

To summarize, tolerance to locomotor depressant effects of nicotine was demonstrated in experiments reviewed above. Tolerance was reported to occur even when nicotine was administered only once every three days (Stolerman *et al.* 1974), or even once a week (Clarke and Kumar 1983*b*). Furthermore, some investigators reported increased sensitivity to stimulatory effects of nicotine after chronic treatment. In contrast to the generally observed tolerance to the behaviourally depressant effects of nicotine, tolerance to the stimulant effects has not been reported. Evidence has also been presented that chronic nicotine administration results in an increase in the number of nicotine binding sites in the brain.

2.3.3 *Tolerance development in learned behaviours*

Rapid development of tolerance to the response rate suppressant effects of 0.25 mg kg^{-1} of nicotine was reported by Domino and Lutz (1973) when rats were administered 0.25 mg kg^{-1} of nicotine intraperitoneally twice daily, before and after the animals were allowed to respond on a fixed-ratio 15 schedule for water reward. A weakness of this study is the absence of a dose–response curve, since only a single dose was used, both as a pretreatment and as challenge dose.

Nicotine (0.2 mg kg^{-1} intraperitoneally) was administered 1 min prior to two daily 30 min sessions to rats responding on a fixed-ratio 50 schedule for water rewards (Dougherty *et al.* 1981.) An interval of 1 h separated the two injections. Initially, the first injection depressed behaviour whereas the second injection had a much smaller effect. By the ninth day, however, neither injection of 0.2 mg kg^{-1} of nicotine had a rate-depressant effect. When the nicotine dose was increased from 0.2 to 0.35, 0.5 and 0.65 mg kg^{-1}, respectively, the difference between the first and the second injection reappeared even though at the same time the overall rate-depressant effect increased dose-dependently. However, complete tolerance to the rate-depressant effects of the higher doses did not develop.

In the experiment of Hendry and Rosecrans (1982) described above (cf., Fig. 2.9), tolerance development in mice responding on a fixed-ratio 25 schedule for sweetened-milk rewards was studied. A range (0.2–1.6 mg kg^{-1}) of subcutaneous doses of nicotine was first administered to determine the effects of dose on the rate-suppressant effects of nicotine in naive mice. After 30 days of chronic nicotine (1.2 mg kg^{-1}) administration a significant upward shift in the nicotine (0.2–1.6 mg kg^{-1}) dose–response curve was observed. Before chronic administration, 0.4 mg kg^{-1} of nicotine exerted response-rate suppressant effects, whereas in the post-chronic determination only the highest dose (1.6 mg kg^{-1}) was effective. Tolerance was wearing off by the first week after chronic treatment was discontinued, and it seemed to be absent by the third week, at which point acute tests with nicotine (1.2 mg kg^{-1}) severely depressed rates of responding.

To summarize, over a variety of treatment regimens tolerance has been reported to occur to the effects of chronic administration of nicotine on learned behaviours.

2.3.4 *Tolerance to consummatory and physiological effects*

Body weights of rats receiving 0.8 mg kg^{-1} of nicotine intraperitoneally, 2 or 3 times per day for 5 weeks, were significantly lower than those of rats in a control group receiving saline, whereas the effects of $0.4\,mg\,kg^{-1}$ were not significantly different from controls (Schechter and Cook 1976). The lower body weights were not paralleled by a significant decrease in food consumption. Neither the centrally acting nicotinic antagonist mecamylamine, nor the peripherally acting nicotinic antagonist hexamethonium, antagonized the weight decreasing effects of nicotine, suggesting that the effects of nicotine on body weight were not mediated by those nicotinic receptors upon which mecamylamine acts. Only after discontinuation of chronic nicotine ($0.8\,mg\,kg^{-1}$) administration did the rats' weights return to the levels of the controls. These findings suggested that tolerance did not develop under these conditions.

The effects on body weight of different routes of chronic administration of nicotine in female rats was investigated by Morgan and Ellison (1987). One group of rats had subcutaneous implants of continuous nicotine release pellets with a diffusion rate of 3.4 mg per day, whereas the two other groups received subcutaneous nicotine injections of either 0.8 or $5.6\,mg\,kg^{-1}$ twice daily. The rats in the implant group lost weight continuously over the first 9 days, after which recovery (i.e. tolerance) was evident up until the point at which the pellet was removed (day 17). The low dose ($0.8\,mg\,kg^{-1}$) injection group also lost weight, but at a slower rate than the pellet implant group, and did not show recovery. The high dose ($5.6\,mg\,kg^{-1}$) injection group, however, did not lose but gained weight continuously up until the time the injections were discontinued, at which point they started losing weight. Thus the loss of body weight and subsequent regain of weight (tolerance) during chronic nicotine treatment seemed to be a function of dose. However, the high-dose nicotine injection group received the same daily amount of nicotine as the implant group but showed an unexpected weight increase during chronic treatment, suggesting that route of administration may be important for tolerance to be demonstrated. This dose ($5.6\,mg\,kg^{-1}$) also caused seizures in all rats during the chronic dosing period; the increase in weight that occurred has yet to be explained.

The effects of chronic nicotine exposure on body weight, and food and water consumption, was studied by Levin *et al.* (1987) using three groups of female rats. Two groups had subcutaneously implanted nicotine pellets diffusing 3.4 mg of nicotine per day, whereas the third group served as controls and had empty pellets implanted. Measuring weight gains in the three groups of rats, it was found that the two nicotine groups lost weight during the first 9 days after the implant of the nicotine pellet. At days 11 and 12 both nicotine groups started to regain weight (i.e. to develop tolerance). At day 14 one group of rats (the 'short-term nicotine group') had their nicotine pellets removed and started to regain weight at an even faster rate, and by day 21 their weights had passed those of the control group (the 'non-nicotine group') as well as those of the group with their pellets still intact (the 'long-term nicotine group'). The 'long-term nicotine group' weighed less than the two other groups at all times from day 14 and on (see Fig. 2.11). Comparing the eating

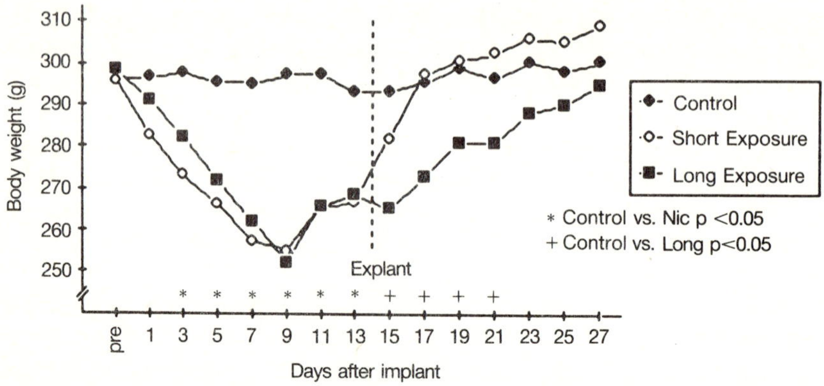

Fig. 2.11. Effects on body weight of chronic nicotine treatment (3.4 mg per rat per day). The abscissa represents body weight. The ordinate represents days after implant of subcutaneous nicotine pellets. Solid circles represent control animals (the 'non-nicotine group') having empty pellets implanted. Open circles represent the 'short-term nicotine group' which had their pellets removed on day 14. Solid squares represent the 'long-term nicotine group'. (From Levin *et al.* 1987, with permission.)

and drinking patterns of the rats it was found that the decrease in weight in the nicotine groups could be attributed to hypophagia and hypodipsia, and that the subsequent increased body weight after nicotine removal was the result of hyperphagia and hyperdipsia.

Measuring heart rate and blood pressure in unanaesthetized rats, Aceto *et al.* (1986) found no tolerance to intravenous bolus injections of 0.005–0.04 mg of nicotine 30 min post-injection; i.e. nicotine produced a dose related increase in blood pressure and a decrease in heart rate. However, when measuring at 1 min intervals tolerance did develop, suggesting that development of tolerance is related to frequency of administration and dose (Aceto *et al.* 1986). Tolerance to the hypothermic effects of nicotine has also been reported (Horstmann 1984; Larsson *et al.* 1986; Marks *et al.* 1983, 1985, 1986, 1987; Marks and Collins 1985).

To summarize, even though some of the experiments reviewed above were not primarily designed to study development of tolerance, the chronic treatment regimens employed do allow for tentative conclusions to be drawn regarding tolerance to weight decreasing effects of nicotine. When given 2 or 3 times daily for 17 days (Morgan and Ellison 1987) or up to five weeks (Schechter and Cook 1976), 0.8 mg kg^{-1} of nicotine did not induce tolerance (i.e. body weights remained lowered during the period of treatment). In rats implanted with nicotine releasing pellets (3.4 mg per day) for 17 days (Morgan and Ellison 1987), 14 days or 27 days (Levin *et al.* 1987), the initial decrease in body weight was reversed with continued exposure (i.e. tolerance). However, rats receiving a convulsant dose of 5.6 mg kg^{-1} of nicotine intraperitoneally (Morgan and Ellison 1987) did not lose weight either initially or after continued exposure, but only after discontinuation of nicotine administration. It would seem, then, that an appropriate dose of nicotine may induce tolerance and that the route of administration may be important. The finding that neither the centrally acting antagonist mecamylamine, nor the peripherally acting antagonist hexamethonium, antagonized the weight-decreasing effects of nicotine may suggest that these effects are not mediated through nicotinic receptors. However, this question needs further research before a conclusion can be reached.

2.3.5 *Withdrawal effects after chronic nicotine administration*

In contrast to the extensive amount of research on nicotine withdrawal effects in humans (Henningfield 1984; Hughes and Hatsukami 1986; see Jaffe, chapter 1, this volume; Russell, chapter 11, this volume), there has been little systematic study of the possibility of a withdrawal syndrome in animals. None the less, more or less systematic observations of withdrawal-like effects have been reported. Wenzel and Azmeh (1970) reported a hypertensive response in rats after withdrawal from a 10-day exposure to nicotine in drinking water. Keenan and Johnson (1972; see also above) reported tolerance to the depressant effects of nicotine on rearing frequency in rats, and a subsequent rebound increase in rearing activity after cessation of administration of 0.5 mg kg^{-1} of nicotine subcutaneously for 10 days, and suggested that these effects were due to compensatory behavioural mechanisms. Stolerman *et al.* (1973), however, found no signs of withdrawal in spontaneous locomotor activity of rats treated with 0.33 mg kg^{-1} of nicotine three times daily for 8 days. In a study by Morgan and Ellison (1987; see also above) rats receiving a convulsive dose of nicotine (5.6 mg kg^{-1} intraperitoneally) gained weight during treatment, and returned to the level of untreated control rats after discontinuation of chronic nicotine administration, whereas rats with nicotine pellet implants that had lost weight during chronic treatment regained weight when the pellets were removed, and the rats also slightly increased their weights to a level above the controls. In the study by Levin *et al.* (1987; see also above), rats that had had nicotine pellets implanted increased their weights up to the level of controls after removal of the pellets; these rats regained weight at a faster rate than did the rats with their nicotine pellets still intact, suggesting possible withdrawal effects. Regarding nicotine's effects on body weight, the general finding is an increase in body weight, up to and sometimes slightly above the level of untreated rats, after discontinuation of chronic treatment. These findings are consistent with human data reporting suppression of appetite in smokers, and weight gain in humans who quit smoking (Grunberg 1982; Wack and Rodin 1982), and with interpretations of withdrawal effects as rebound phenomena in the direction opposite to the drug's effect in a naive organism.

Withdrawal effects can also be studied in a situation where an animal is trained to avoid electric shocks. The animal is typically trained to avoid shocks in an operant chamber by pressing a response lever. A signal, usually visual or auditory, may or may not come on shortly before the shock is due. By pressing the lever the animal can postpone the shock for a certain period of time. An impairment of shock-avoidance behaviour is regarded as an effect of withdrawal symptoms. To study the effects of nicotine on signalled and non-signalled shock-avoidance behaviour, respectively, Morrison (1974) trained rats to emit 10 lever presses to avoid shocks. The results showed that, without the signal, rats receiving 0.4 mg kg^{-1} of nicotine prior to each daily session were impaired in their performance when saline was substituted for the nicotine, as compared with rats trained after saline injections and tested after saline injection, or with nicotine-trained animals receiving nicotine on the test day. When the signal preceeded the shocks, however, the detrimental effects of giving saline to nicotine-trained animals was substantially diminished. However, it is not clear that pharmacological rather than other effects of nicotine were responsible for this effect. It could well be, for example, that avoidance training which took place immediately after injections of 0.4 mg kg^{-1} of nicotine became associated with the presence of nicotine, and that when nicotine was no longer present during the saline test, the rats behaviour became disrupted due to the novelty of the situation. Such an effect would be an instance of state-dependent learning (for example Overton 1984). Further, the degree of exposure to nicotine, 0.4 mg kg^{-1} for four days per week, may have been suboptimal for inducing withdrawal effects.

A more recent attempt to shed light on possible withdrawal effects of nicotine utilized the drug pentylenetetrazol, commonly used to produce animal models of anxiety (for example, Emmett-Oglesby *et al.* 1983) in the drug discrimination procedure (for example, Overton 1984). Harris *et al.* (1986) trained rats in an operant conditioning chamber with two response levers and a food pellet dispenser. After receiving 20 mg kg^{-1} of pentylenetetrazol intraperitoneally the rats had to press a lever 10 times (a fixed-ratio 10 schedule of reinforcement) to obtain food, whereas after they had been injected with saline they had to press the other lever under the same schedule conditions to produce a food pellet. After the rats had learned to discriminate between pentylenetetrazol and saline, training was discontinued and the rats were chronically treated with 1.25 mg kg^{-1} of nicotine for 21 days. At 8, 24 and 48 h after discontinuation of the chronic treatment regimen, rats were again allowed to respond in the operant chamber. The degree to which the rats responded on the pentylenetetrazol lever would indicate the degree of similarity with the anxiogenic drug pentylenetetrazol. At the 48 h post-nicotine interval 50 per cent of the rats responded on the pentylenetetrazol lever, whereas at the 8 and 24 h intervals between 30 and 40 per cent of the rats did so. Harris *et al.* (1986) concluded, therefore, that withdrawal from nicotine produces effects similar to those produced by the anxiogenic drug pentylenetetrazol. The fact that only 50 per cent of the rats responded 'pentylenetetrazol-like' indicates that the effect may be dependent on the doses of pentylenetetrazol and nicotine utilized, and/or that the effect is not very robust. In the absence of a dose–effect curve with additional doses of nicotine used for pretreatment, a firm conclusion cannot yet be reached; however, the approach chosen by Harris *et al.* (1986) to study the possibility of a nicotine withdrawal phenomenon is very promising, and certainly has the potential to provide more conclusive evidence on this issue.

2.3.6 *Summary and conclusions*

It seems that the mechanisms by which nicotine induces tolerance in rats and mice are primarily pharmacodynamic in character, since neither increased metabolism nor behavioural tolerance to nicotine has been demonstrated. Tolerance develops to both learned and unlearned behaviours in a variety of behavioural tests, and after various nicotine treatment conditions. In some studies, the initial behaviourally depressant effects not only disappeared as tolerance developed, but a stimulatory effect was reported to replace it. If this effect was the result of selectively developing tolerance to nicotine's behaviourally depressant effects, and thereby revealing its stimulatory properties as was suggested, it needs to be subjected to further investigation. At the neuronal level chronic nicotine has been demonstrated to result in an up-regulation rather than a down-regulation of the receptor population. This is contrary to expectations, since evidence from other neuronal systems indicate that chronic exposure would lead to a down-regulation of the receptors.

Tolerance did not develop to weight-decreasing effects of a lower dose of nicotine ($0.8\,mg\,kg^{-1}$), neither were these effects reversible by nicotinic antagonists. However, tolerance did develop to higher doses, such that after an initial decrease in body weight an increase in weight occurred during continued nicotine administration. Thus, it seems that tolerance may develop to the weight-decreasing effects of appropriate doses of nicotine. However, more research is needed to discern more carefully under what conditions tolerance does develop to these effects.

The question of whether withdrawal symptoms develop in animals when chronic nicotine administration is discontinued has not been widely studied. However, some observations indicate that withdrawal signs may develop, whereas other observations suggest that they may not. The most convincing evidence to date is the finding that the discriminative effects of the anxiogenic drug pentylenetetrazol shares similarities with the effects induced by the discontinuation of chronic nicotine administration in rats. However, in contrast to clinical symptoms allowing for the definition of a nicotine withdrawal syndrome in humans, the lack of a defined nicotine withdrawal syndrome in animals makes reports on withdrawal effects in animals hard to interpret at this time.

2.4 General conclusions

Animal studies support the characterization of nicotine as a dependence-producing drug. The defining characteristic of drug dependence is the development and subsequent persistence of behaviour directed at securing the drug for self-administration (see section 2.1). Until the 1980's experiments with animals indicated either very weak or no reinforcing efficacy of nicotine, and some researchers had concluded that nicotine's reinforcing effects were dissimilar in animal and man.

This controversy was resolved when Goldberg and colleagues showed that animals would self-administer nicotine at consistent and high rates when a second-order schedule of self-administration was used. Subsequent studies by these investigators utilizing fixed-interval and fixed-ratio schedules, which limited the frequency of nicotine infusions, provided further evidence that under appropriate conditions nicotine was effective in maintaining behaviour leading to its delivery. These studies have shown that the scheduling of nicotine availability and the presence of environmental stimuli are of vital importance for the development and maintenance of nicotine self-administration in animals, and that under appropriate conditions nicotine can be highly efficacious in maintaining drug taking behaviour.

One concomitant to the definition of drug dependence is that tolerance may develop over time, such that increasing doses are required to achieve a desired effect. Tolerance development to various behavioural effects of nicotine has been reported, such as tolerance to behaviourally depressant effects of nicotine.

A second concomitant to drug dependence is that withdrawal symptoms may develop upon discontinuation of chronic drug administration. At this point, only a limited amount of data suggests that animals may show signs of withdrawal after discontinuation of chronic nicotine administration. An orderly dose-related nicotine withdrawal syndrome has yet to be demonstrated in animals.

A recurring issue is whether an animal will self-administer a drug for its direct reinforcing effects, or in order to avoid withdrawal effects that may develop when chronic drug administration is discontinued. It has been repeatedly demonstrated with, for example, opiates and barbiturates, both known to produce severe withdrawal syndromes at certain doses, that for self-administration to occur the alleviation of withdrawal symptoms is not a prerequisite (see Goldberg 1976 for a review). The experiments reporting successful maintenance of nicotine self-administration behaviour all limited maximal daily intake to between 0.3–0.6 mg kg^{-1}, whereas the treatment used in the study in which withdrawal effects were reported (Harris *et al.* 1986) used a daily dose of 3.75 mg kg^{-1}. Thus the doses maintaining self-administration behaviour were between 6 and 12 times lower than that used to study withdrawal effects. It may be concluded that with nicotine, as with other self-administered drugs, the alleviation of withdrawal symptoms caused by discontinuation of chronic drug administration is not a necessary condition for self-administration to occur.

To summarize, nicotine may under appropriate circumstances have powerful reinforcing effects, as shown by its ability to maintain high rates of responding leading to its administration. Nicotine may also induce tolerance to its depressant effects on locomotor behaviour. A withdrawal syndrome for nicotine comparable to those of alcohol or the opiates has not been shown in animals. Some studies seem to suggest, though, that nicotine may produce withdrawal effects of less severity in rats; the evidence for this effect is scarce and further research is needed to answer this question in a more conclusive manner.

References

Abood, L.G., Lowy, K., and Booth, H. (1979). Acute and chronic effects of nicotine in rats and evidence for a noncholinergic site of action. In *Cigarette smoking as a dependence process*, Vol. 23 (N.A. Krasnegor, ed.), pp. 70–90. National Institute on Drug Abuse Research Monograph, US Government Printing Office, Washington, DC.

Abood, L.G., Grassi, S., Constanzo, M., and Junig, J. (1984). Behavioral and biochemical studies in rats after chronic exposure to nicotine. In *Mechanisms of tolerance and dependence*, Vol. 54 (C.W. Sharp, ed.), pp. 348–55. National Institute on Drug Abuse Research Monograph, US Government Printing Office, Washington, DC.

Aceto, M.D., Tucker, S.M., Ferguson, G.S., and Hinson, J.R. (1986). Rapid and brief tolerance to (+) – and (–) nicotine in unanesthetized rats. *European Journal of Pharmacology* **132**, 213–8.

Ando, K. and Yanagita, T. (1981). Cigarette smoking in Rhesus monkeys. *Psychopharmacology* **72**, 117–27.

Ator, N.A. and Griffiths, R.R. (1983). Nicotine self-administration in baboons. *Pharmacology, Biochemistry and Behavior* **19**, 993–1003.

Auerbach, O., Hammond, E.C., Kirman, D., and Garfinkel, L. (1977). Emphysema produced in dogs by cigarette smoking. In *Smoking and health: proceedings of the 3rd world conference on smoking and health* (J. Steinfield, W. Griffiths, K. Ball, and R.M. Taylor, ed.), pp. 139–52. DHEW Publication Number (NIH) 77-1413, US Government Printing Office, Washington, DC.

Brady, J.V. and Lukas, S.E. (1984). Testing drugs for physical dependence potential and abuse liability. *National Institute on Drug Abuse Research Monograph*, Vol. 52, US Government Printing Office, Washington, DC.

Carlton, P.L. and Wolgin, D.L. (1971). Contingent tolerance to anorexigenic effects of amphetamine. *Physiology and Behavior* **7**, 221–3.

Chen, C.S. (1968). A study of the alcohol-tolerance effect and an introduction of a new behavioral technique. *Psychopharmacologia* **12**, 433–40.

Clarke, P.B.S. and Kumar, R. (1983*a*). Characterization of the locomotor stimulant action of nicotine in tolerant rats. *British Journal of Pharmacology* **80**, 587–94.

Clarke, P.B.S. and Kumar, R. (1983*b*). The effects of nicotine on locomotor activity in non-tolerant and tolerant rats. *British Journal of Pharmacology* **78**, 329–37.

Cox, B.M., Goldstein, A. and Nelson, W.T. (1984). Nicotine self-administration in rats. *British Journal of Pharmacology* **83**, 49–55.

Creese, I. and Sibley, D.R. (1981). Receptor adaptations to centrally acting drugs. *Annual Review of Pharmacology and Toxicology* **21**, 357–91.

Deneau, G.A. and Inoki, R. (1967). Nicotine self-administration in monkeys. *New York Academy of Sciences* **142**, 277–9.

Dixon, W.E. and Lee W.E. (1912). Tolerance to nicotine. *Quarterly Journal of Experimental Physiology (London)* **5**, 373–83.

Domino, E.F. and Lutz, M.P. (1973). Tolerance to the effects of daily nicotine on rat bar pressing behavior for water reinforcement. *Pharmacology, Biochemistry and Behavior* **1**, 445–8.

Dougherty, J., Miller, D., Todd, G., and Kostenbauder, H.B. (1981). Reinforcing and other behavioral effects of nicotine. *Neuroscience & Biobehavioral Reviews* **5**, 487–95.

Edwards, G., Arif, A., and Hodgson, R. (1982). Nomenclature and classification of drug and alcohol-related problems: a shortened version of a WHO memorandum. *British Journal of Addiction* **77**, 3–20.

Emmett-Oglesby, M.W., Spencer, D.G. Jr., Lewis, M., Elmesallamy, F., and Lal, H. (1983). Anxiogenic aspects of diazepam withdrawal can be detected in animals. *European Journal of Pharmacology* **92**, 127–30.

Falk, J.L. (1961). Production of polydipsia in normal rats by an intermittent food schedule. *Science* **133**, 195–6.

Falk, J.L. (1971). The nature and determinants of adjunctive behavior. *Physiology and Behavior* **6**, 577–88.

Falkeborn, Y., Larsson, C., and Nordberg, A. (1981). Chronic nicotine exposure in the rat: a behavioral and biochemical study of tolerance. *Drug and Alcohol Dependence* **8**, 51–60.

Ferster, C.B. and Skinner, B.F. (1957). *Schedules of Reinforcement*. Appleton-Century-Crofts, New York.

George, F.R. (1987). Genetic and environmental factors in ethanol self-administration. *Pharmacology, Biochemistry and Behavior* **27**, 379–84.

Glick, S.D., Jarvik, M.E., and Nakamura, R.K. (1970). Inhibition by drugs of smoking behaviour in monkeys. *Nature* **227**, 969–71.

Glick, S.D., Zimmerberg, B., and Jarvik, M.E. (1971). Titration of oral nicotine intake with smoking behaviour in monkeys. *Nature* **233**, 207–8.

Goldberg, S.R. (1976). The behavioral analysis of drug addiction. In *Behavioral pharmacology*, (S.D. Glick and J. Goldfarb ed.) pp. 283–316. Mosby, St Louis.

Goldberg, S.R. and Spealman, R.D. (1982). Maintenance and suppression of behavior by intravenous nicotine injections in squirrel monkeys. *Federation Proceedings* **41**, 216–20.

Goldberg, S.R. and Spealman, R.D. (1983). Suppression of behavior by intravenous injections of nicotine or by electric shock in squirrel monkeys: effects of chlordiazepoxide and mecamylamine. *Journal of Pharmacology and Experimental Therapeutics* **224**, 334–40.

Goldberg, S.R., Kelleher, R.T., and Morse, W.H. (1975). Second-order schedules of drug injection. *Federation Proceedings* **34**, 1771–6.

Goldberg, S.R., Spealman, R.D., and Goldberg, D.M. (1981). Persistent behavior at high rates maintained by intravenous self-administration of nicotine. *Science* **214**, 573–5.

Goudie, A.J. and Demellweek, C. (1986). Conditioning factors in drug tolerance. In *Behavioral analysis of drug dependence*, (S.R. Goldberg and I.P. Stolerman, ed.) pp. 225–85. Academic Press, New York.

Griffiths, R.R., Bigelow, G.E., and Henningfield, J.E. (1980). Similarities in animal and human drug-taking behavior. In *Advances in substance abuse* Vol. 1 (N.K. Mello, ed.), pp. 1–90. JAI Press, Greenwich, Connecticut.

Griffiths, R.R., Brady, J.V., and Bradford, L.D. (1979). Predicting the abuse liability of drugs with animal drug self-administration procedures: psychomotorstimulants and hallucinogens. In *Advances in behavioral pharmacology*, Vol. 2 (T. Thompson and P.B. Dews, ed.), pp. 162–208. Academic Press, New York.

Grunberg, N.E. (1982). The effects of nicotine and cigarette smoking on food consumption and taste preferences. *Addictive Behaviors* **7**, 317–31.

Hammond, E.C., Auerbach, O., Kirman, D., and Garfinkel, L. (1971). Effects of cigarette smoking on dogs. *American Cancer Society,* **21**, 78–94.

Hanson, H.M., Ivester, C.A., and Morton, B.R. (1979). Nicotine self-administration in rats. In *Cigarette smoking as a dependence process*, Vol. 23 (N.A. Krasnegor, ed.), pp. 70–90. National Institute on Drug Abuse Research Monograph, US Government Printing Office, Washington, DC.

Harris, C.M., Emmett-Oglesby, M.W., Robinson, N.G., and Lal, H. (1986). Withdrawal from chronic nicotine substitutes partially for the interoceptive stimulus produced by pentylenetetrazol (PTZ). *Psychopharmacology* **90**, 85–9.

Hatchell, P.C. and Collins, A.C. (1977). Influences of genotype and sex on behavioral tolerance to nicotine in mice. *Pharmacology, Biochemistry and Behavior* **6**, 25–30.

Hatchell, P.C. and Collins, A.C. (1980). The influence of genotype and sex on behavioral sensitivity to nicotine in mice. Psychopharmacology **71**, 45–9.

Hendry, J.S. and Rosecrans, J.A. (1982). The development of pharmacological tolerance to the effect of nicotine on schedule-controlled responding in mice. *Psychopharmacology* **77**, 339–43.

Henningfield, J.E. (1984). Behavioral pharmacology of cigarette smoking. In *Advances in behavioral pharmacology*, Vol. 4 (T. Thompson, P.B. Dews, and J.E. Barrett, ed.), pp. 131–210. Academic Press, New York.

Henningfield, J.E. and Goldberg, S.R. (1983). Nicotine as a reinforcer in human subjects and laboratory animals. *Pharmacology, Biochemistry and Behavior* **19**, 989–92.

Horstmann, M. (1984). Influence of mecamylamine and atropine on tolerance development to nicotine hypothermia in rats. *Journal of Pharmacy and Pharmacology* **36**, 770–1.

Hughes, J.R. and Hatsukami, D. (1986). Signs and symptoms of tobacco withdrawal. *Archives of General Psychiatry* **43**, 289–94.

Jaffe, J.H. (1985). Drug addiction and drug abuse. In *Goodman and Gilman's The pharmacological basis of therapeutics* (7th edn), (A.G. Gilman, L.S. Goodman, T.W. Rall, and F. Murad, ed.), pp. 532–81. MacMillan, New York.

Jarvik, M.E. (1967). Tobacco smoking in monkeys. *Annals of the New York Academy of Sciences* **142**, 280–94.

Keenan, A. and Johnson, F.N. (1972). Development of behavioural tolerance to nicotine in the rat. *Experientia* **28**, 428–9.

Ksir, C., Hakan, R., Hall Jr, D.P., and Kellar, K.J. (1985). Exposure to nicotine enhances the behavioral stimulant effect of nicotine and increases binding of [^{3}H]acetylcholine to nicotinic receptors. *Neuropharmacology* **24**, 527–31.

Ksir, C., Hakan, R.L., and Kellar, K.J. (1987). Chronic nicotine and locomotor activity: influences of exposure dose and test dose. *Psychopharmacology* **92**, 25–9.

Lang, W.J., Latiff, A.A., McQueen, A., and Singer, G. (1977). Self-administration of nicotine with and without a food delivery schedule. *Pharmacology, Biochemistry and Behavior* **7**, 65–70.

Langley, J.N. (1905). On the reaction of cells and of nerve-endings to certain poisons, chiefly as regards the reaction of striated muscle to nicotine and to curari. *Journal of Physiology (London)* **31**, 374–413.

Larsson, C., Nilsson, L., Halen, A., and Nordberg, A. (1986). Subchronic treatment of rats with nicotine: effects on tolerance and on [^{3}H]acetylcholine and [^{3}H]nicotine binding in the brain. *Drug and Alcohol Dependence* **17**, 37–45.

LeBlanc, A.E., Poulos, C.X., and Cappell, H.D. (1978). Tolerance as a behavioral phenomenon: evidence from two experimental paradigms. In *Behavioral tolerance: research and treatment implications*, Vol. 18 (N.A. Krasnegor, ed.), pp. 72–89. National Institute on Drug Abuse Research Monograph, US Government Printing Office, Washington, DC.

Lester, D. (1961). Self-maintenance of intoxication in the rat. *Quaterly Journal of Studies on Alcohol* **22**, 223–31.

Levin, E.D., Morgan, M.M., Galvez, C., and Ellison, G.D. (1987). Chronic nicotine and withdrawal effects on body weight and food and water consumption in female rats. *Physiology and Behavior* **39**, 441–4.

Marks, M.J., Burch, J.B., and Collins, A.C. (1983). Effects of chronic nicotine infusion on tolerance development and nicotinic receptors. *Journal of Pharmacology and Experimental Therapeutics* **226**, 817–25.

Marks, M.J. and Collins, A.C. (1985). Tolerance, cross-tolerance, and receptors after chronic nicotine or oxotremorine. *Pharmacology, Biochemistry and Behavior* **22**, 283–91.

Marks, M.J., Romm, E., Gaffney, D.K., and Collins, A.C. (1986). Nicotine-induced tolerance and receptor changes in four mouse strains. *Journal of Pharmacology and Experimental Therapeutics* **237**, 809–19.

Marks, M.J., Stitzel, J.A., and Collins, A.C. (1985). Time course study of the effects of chronic nicotine infusion on drug response and brain receptors. *Journal of Pharmacology and Experimental Therapeutics* **235**, 619–28.

Marks, M.J., Stitzel, J.A., and Collins, A.C. (1987). Influence of kinetics of nicotine administration on tolerance development and receptor levels. *Pharmacology, Biochemistry and Behavior* **27**, 505–12.

Meisch, R.A. (1975). The function of schedule-induced polydipsia in establishing ethanol as a positive reinforcer. *Pharmacological Reviews* **27**, 465–73.

Meisch, R.A. and Thompson, T. (1974). Ethanol as a reinforcer: an operant analysis of ethanol dependence. In *Drug addiction: neurobiology and influences on behavior*, Vol. 3 (J.M. Singh and H. Lal, ed.), pp. 117–33. Stratton, New York.

Merbaum, M., Avimier, R., and Goldberg, J. (1979). The relationship between aversion, group training and vomiting in the reduction of smoking behavior. *Addictive Behaviors* **4**, 279–85

Morgan, M.M. and Ellison, G. (1987). Different effects of chronic nicotine treatment regimens on body weight and tolerance in the rat. *Psychopharmacology* **91**, 236–8.

Morrison, C.F. (1974). Effects of nicotine and its withdrawal on the performance of rats on signalled and unsignalled avoidance schedules. *Psychopharmacology* **38**, 25–35.

Morrison, C.F. and Stephenson, J.A. (1972). The occurrence of tolerance to a central depressant effect of nicotine. *British Journal of Pharmacology* **46**, 151–6.

Nettesheim, P., Guerin, M., Kendrick, J., Caton, J., Dalbey, W., and Greissmer, R. (1977). Acute and chronic smoke inhalation in rats. In *Smoking and health: proceedings of the 3rd world conference on smoking and health*, (J. Steinfield, W. Griffiths, K. Ball, and R.M. Taylor, ed.), pp. 153–60. DHEW Publication Number (NIH) 77-1413, US Government Printing Office.

Overton, D.A. (1984). State dependent learning and drug discriminations. In *Handbook of psychopharmacology*, Vol. 18 (L.L. Iversen, S.D. Iversen, and S.H. Snyder, ed.), pp.60–127. Plenum Press, New York.

Risner, M.E. and Goldberg, S.R. (1983). A comparison of nicotine and cocaine self-administration in the dog: fixed-ratio and progressive-ratio schedules of intravenous drug infusion. *Journal of Pharmacology and Experimental Therapeutics* **224**, 319–26.

Rogers, W.R., Bass, R.L., Johnson, D.E., Kruski, A.W., McMahan, C.A., Montiel, M.M., Mott, G.E., Wilbur, R.L., and McGill, Jr., H.C. (1980). Atherosclerosis-related responses to cigarette smoking in the baboon. *Circulation* **61**, 1188–93.

Rogers, W.R., McCullough, B., and Caton, J.E. (1981). Cigarette smoking by baboons: *in vivo* assessment of particulate inhalation using bronchoalveolar lavage to recover [^{14}C]dotriacontane. *Toxicology* **20**, 309–21.

Russell, M.A.H. (1979). Tobacco dependence: is nicotine rewarding or aversive? In *Cigarette smoking as a dependence process*, Vol. 33 (N.A. Krasnegor, ed.), pp. 100–22. National Institute on Drug Abuse Research Monograph, US Government Printing Office, Washington, DC.

Sanger, D.J. (1986). Drug taking as adjunctive behavior. In *Behavioral Analysis of Drug Dependence*, (S.R. Goldberg and I.P. Stolerman, ed.), pp. 123–60. Academic Press, New York.

Schechter, M.D. and Cook, P.G. (1976). Nicotine-induced weight loss in rats without an effect on appetite. *European Journal of Pharmacology* **38**, 63–9.

Schievelbein, H. (1984). Nicotine, resorption and fate. In *Nicotine and the tobacco smoking habit* (2nd edn), (D.J.K. Balfour, ed.), pp. 1–15. Pergamon Press, New York.

Schlatter, J. and Battig, K. (1978). Effects of maze habituation and habituation to nicotine and amphetamine on exploratory locomotion patterns in rats. In *International workshop on behavioral effects of nicotine* (K. Battig, ed.), pp. 55–69. Karger, Basel.

Schuster, C.R. and Johanson, C.E. (1974). The use of animal models for the study of drug abuse. In *Research advances in alcohol and drug problems*, Vol. 1 (R.J. Gibbins, Y. Israel, H. Kalant, R.E. Popham, W. Schmidt, and R.G. Smart, ed.), pp. 1–31. Wiley, New York.

Schuster, C.R. and Thompson, T. (1969). Self-administration of and behavioral dependence of drugs. *Annual Review of Pharmacology* **9**, 483–502.

Schuster, C.R., Dockens, W.S., and Woods, J.H. (1966). Behavioral variables affecting the development of amphetamine tolerance. *Psychopharmacologia* **9**, 170–82.

Singer, G., Simpson, F., and Lang, W.J. (1978). Schedule induced self injections of nicotine with recovered body weight. *Pharmacology, Biochemistry and Behavior* **9**, 387–9.

Skinner, B.F. (1938). The behavior of organisms. Appleton - Century - Crofts, New York.

Slifer, B.L. (1983). Schedule-induction of nicotine self-administration. *Pharmacology, Biochemistry and Behavior* **19**, 1005–9.

Slifer, B.L. and Balster, R.L. (1985). Intravenous self-administration of nicotine: with and without schedule-induction. *Pharmacology, Biochemistry and Behavior* **22**, 61–9.

Smith, L.A. and Lang, W.J. (1980). Changes occurring in self administration of nicotine by rats over a 28-day period. *Pharmacology, Biochemistry and Behavior* **13**, 215–20.

Spealman, R.D. (1979). Behavior maintained by termination of a schedule of self-administered cocaine. *Science* **204**, 1231–3.

Spealman, R.D. and Goldberg, S.R. (1978). Drug self-administration by laboratory animals: control by schedules of reinforcement. *Annual Review of Pharmacology and Toxicology* **18**, 313–39.

Spealman, R.D. and Goldberg, S.R. (1982). Maintenance of schedule-controlled behavior by intravenous injections of nicotine in squirrel monkeys. *Journal of Pharmacology and Experimental Therapeutics* **223**, 402–8.

Stolerman, I.P., Fink, R., and Jarvik, M.E. (1973). Acute and chronic tolerance to nicotine measured by activity in rats. *Psychopharmacologia* **30**, 329–42.

Stolerman, I.P., Bunker, P., and Jarvik, M.E. (1974). Nicotine tolerance in rats: role of dose and dose interval. *Psychopharmacologia*, **34**, 317–24.

Takada, K., Katz, J.L., Barrett, J.E., Larscheid, P., Cook, J.M., and Goldberg, S.R. (1986). Aversive effects of drugs: drugs as punishers in squirrel monkeys. *Psychopharmacology* **89**, S35.

Takeuchi, M., Kurogochi, Y., and Yamaoka, M. (1954). Experiments on the repeated injection of nicotine into albino rats. *Folia Pharmacologia Japonica* **50**, 66–9.

Wack, J. and Rodin, J. (1982). Smoking and its effects on body weight and the systems of caloric regulation. *American Journal of Clinical Nutrition* **35**, 366–80.

Wenzel, D.G. and Azmeh, N. (1970). Chronically administered nicotine and the blood pressure of normotensive and hypertensive rats. *Archives Internationales de Pharmacodynamie et de Therapie* **187**, 367–76.

Yanagita, T., Ando, K., Oinuma, N., and Ishida, K. (1974). Intravenous self-administration of nicotine and an attempt to produce smoking behavior in monkeys. *Report of the thirty-sixth annual scientific meeting, Committee on Problems of Drug Dependence*, pp. 567–78. National Academy of Sciences, Washington, DC.

3 Psychopharmacological aspects of nicotine

David M. Warburton

3.1 Introduction

Any author who writes about the psychopharmacology of nicotine faces a dilemma. If the review is restricted to studies in which nicotine alone has been used, the chapter will be very short. If the review is extended to include the effects of cigarette smoking on behaviour, then the question arises of whether the effects described are due to nicotine alone or whether they are the consequence of some other smoke constituent. Given that there are well over 2000 compounds besides nicotine in cigarette smoke, that possibility can never be ruled out. However, nicotine is pharmacologically the most potent agent in cigarette smoke at smoking-doses, and carbon monoxide has little or no psychological effects at smoking-doses (Guillerm *et al.* 1978). Consequently, this chapter includes studies of smoking as well as those using nicotine itself, and it discusses evidence of the extent to which nicotine was the essential agent in the behavioural changes.

An important issue which is also covered in this chapter is the extent to which a decreased response to nicotine develops with repeated use, and the extent to which the effects of nicotine can be explained by relief of withdrawal symptoms. It is often found that the effect of a drug gradually diminishes when it is given continuously or repeatedly. The term *tolerance* is conventionally used to describe a more gradual decrease in responsiveness to a drug, taking days or weeks to develop. *Desensitization* and *tachyphylaxis* are terms used to describe decreased responsiveness when it develops rapidly, often as quickly as a few minutes. Many different mechanisms can give rise to this type of phenomenon and they are rather poorly understood (Rang and Dale 1987).

Tolerance to some drugs occurs partly because repeated administration of the same dose produces a progressively lower plasma concentration due to increased metabolism. Reduction of a drug's effect may occur because it is nullified by physiological adaptation to the drug's effects. For example, prolonged exposure to some agonists results in a gradual reduction in the number of receptors as measured by drug binding studies. In other cases, desensitization is associated with depletion of essential substances, like precursors or transmitters.

The consequence of some types of tolerance is that there will be withdrawal or abstinence symptoms when drug use is stopped, especially when physiological adaptation has occurred. According to DSM-III-R (American Psychiatric Association 1987), the essential feature of a withdrawal syndrome is the development of a substance-specific syndrome that follows the cessation of, or reduction in, intake of a psychoactive substance that the person previously used regularly. The syndrome that develops varies according to the psychoactive substance the person was using. The nicotine withdrawal syndrome includes craving for nicotine, irritability, frustration or anger, anxiety, difficulty concentrating, restlessness, decreased heart rate, and increased appetite or weight gain (see Jaffe, chapter 1, this volume). It is said that changes in performance and mood can be detected after two or three hours of abstinence in some smokers. But as is pointed out in DSM-III-R (American Psychiatric Association 1987):

> In any given case it is difficult to distinguish a withdrawal effect from the emergence of psychological traits that were suppressed, controlled or altered by the effects of nicotine or from a behavioral reaction (e.g. frustration) to the loss of a reinforcer (p. 150).

In this chapter, the question arises about whether the enhancement of performance and mood produced by smoking or nicotine in smokers are due to relief of withdrawal symptoms, or are due to a true effect of nicotine. Crucial evidence comes from studies of nicotine in non-smokers, and studies in smokers after short periods of abstinence (minimal deprivation). In this chapter, minimal deprivation refers to less than two or three hours abstinence, i.e. before abstaining smokers report any changes in mood and performance.

3.2 Information processing

3.2.1 *Information input*

In this section the small amount of evidence on the effects of nicotine on the detection of simple stimuli will be discussed. From subjective reports, there is little evidence that nicotine has any effect on perception. While some research has suggested that smoking raises visual detection thresholds slightly (Larsson *et al.* 1961), there is no evidence to relate this effect to nicotine. Certainly, smoking doses of carbon monoxide increase visual thresholds (Halperin *et al.* 1959) and so it is thought that the effects of smoking are due to carbon monoxide rather than to nicotine.

In contrast, there is evidence that nicotine can change thresholds in the critical flicker-fusion test. This task requires the subject to discriminate between an intermittent light which either appears to flicker or be steady. The dependent variable is the frequency at which the change in appearance occurs from flicker to steady. A rise in threshold indicates better sensory processing. In a study using a dose of 0.1 mg of nicotine, which was held in the mouth for buccal absorption, there was an elevation of the flicker-fusion threshold, and the two-flash threshold, by about seven per cent in both smokers and non-smokers (Warwick and Eysenck 1963).

Although this result is surprising given the very small dose, it has been found that smoking increases the threshold and that there was a dose-response effect, with a greater effect from using higher nicotine cigarettes (Larsson *et al.* 1950; Tong and Leigh 1975; Caille and Bassano 1976). For example, Waller and Levander (1980) found an initial sharp increase in threshold after the first puff by non-deprived smokers with a maximum 8 min later, and the threshold remained significantly higher for up to 20 min after the last puff. It would seem that the amount of nicotine from a single puff was sufficient to improve flash resolution, and data which are discussed later support the conclusion that a single puff can improve performance.

The flash-resolution mechanism that underlies the flicker-fusion phenomenon is mediated at the cortical level (Bobon *et al.* 1982), supporting Warwick and Eysenck's (1963) claim that nicotine improves cortical reactivity. Further evidence for nicotine's action on cortical function comes in later sections.

3.2.2 *Response output*

The immediate effect of smoking one cigarette on the simple reaction time was investigated after non-deprived subjects smoked a cigarette containing 29 mg tar and 1.5 mg nicotine (Cotten *et al.* 1971). Reaction time was measured by the elapsed time between the onset of a light and a press on the response button. The mean reaction times immediately following smoking, and five minutes after, were significantly slower than all other test intervals. However, the reaction time at 40 and 55 min was significantly faster than the reaction time before smoking. While this study found that cigarette smoking slows simple reaction time for a short period, perhaps due to some subjects having the test cigarette too soon after another, or too strong a cigarette as Morgan and Pickens (1982) suggest, it is more commonly found that smoking speeds up reaction time immediately after smoking. For example, Morgan and Pickens (1982) tested smokers (whether they were deprived or not is not stated) on a reaction-time task immediately after smoking: reaction time was significantly faster when (1) smoking a standard cigarette within a prescribed puff pattern or (2) smoking a standard cigarette ad libitum, but not when (3) smoking their own cigarette ad lib. The pattern of change was exactly the same in the latter case.

The smokers were separated for statistical analysis depending upon whether or not they typically smoked while they worked. The hypothesis was that those subjects who normally would not smoke while working would be affected differently by smoking than those who normally smoke during work. On both simple and complex reaction time scores the pattern of performance differed as a function of whether the subjects typically smoked while working; people who normally smoked while working performed better than others.

This result is in accord with the effects of smoking on sustained reaction-time performance of 12-h-deprived smokers (Frankenhaeuser *et al.* (1971).

The experimental sessions lasted 80 min, during which time subjects continually performed a simple visual reaction-time test. In the non-smoking condition, the speed of reaction decreased over time, whereas there was little change over the session when subjects had a cigarette every 20 min. In a related study of reaction times, Myrsten and Andersson (1978) compared the effects of smoking in both simple and complex reaction time tasks using 12-h-deprived smokers. In the simple reaction-time testing periods, smoking prevented the significant increase in reaction time which occurred over time in the non-smoking condition. In the complex reaction-time periods, smoking significantly reduced reaction time, whereas reaction times increased slightly over time in the non-smoking condition.

Similar data have been obtained in a choice reaction-time task in 12-h-deprived smokers (Lyon *et al.* 1975; Smith *et al.* 1977). In these studies, the authors demonstrated that heart rate was increased, which provides some evidence that the subjects were receiving nicotine because nicotine is the only known constituent of tobacco smoke which increases heart rate. In both studies, decision-time scores were significantly decreased by smoking, and the high-nicotine cigarette had the greatest effect, which also implies that nicotine is the agent responsible for the effects.

A nicotine study on pure response output, as opposed to reaction to a stimulus, argues directly for a nicotine effect on response output. In this study, West and Jarvis (1986) examined the effects of nasal nicotine solution on a pure motor task, with little sensory processing involved. It consisted of tapping with one finger on a key as fast as possible. The subjects were non-smokers because the aim was to determine the effects of nicotine in subjects who would not have any tolerance to the effects of nicotine.

The first experiment tested whether 4 mg of nasal nicotine solution would increase tapping rate by comparison with a pepper placebo. Tapping rate increased by an average of 4 per cent of baseline performance after the nasal nicotine solution in all four subjects, whereas after the inactive nasal solution there was no change. This was confirmed in a second study in which the increase after 2 mg nasal nicotine solution averaged 5 per cent over baseline, and was significantly greater than after either 0.15 mg nasal nicotine solution or a placebo solution.

A third experiment followed the time course of the effect of 4 mg of nasal nicotine solution on tapping rate in a single subject. The aim of this study was to assess the sensitivity and consistency of the tapping measure, and the duration of the nicotine effect. Tapping rate rose from 400 min^{-1} before nasal nicotine solution to 437 min^{-1} at the eleventh minute after the dose. It remained consistently elevated for a further 20 min and then began to decrease, although it was still above baseline levels after 1 h. This time course fitted very well with the plasma nicotine concentrations, suggesting that the effect was due to some pharmacological action of nicotine.

A fourth experiment tested this hypothesis by examining whether a small dose of the nicotinic antagonist, mecamylamine, would reduce the effect of nicotine on tapping rate. As before, the tapping rate was higher than baseline after taking the nasal nicotine solution. The tapping rate after taking the nasal nicotine was lower in every case when subjects had mecamylamine than when they were on placebo. This study demonstrates that it was a nicotinic effect, although it does not indicate whether it was mediated in the brain, spinal cord, or muscle.

The fifth experiment determined whether successive dosing of nasal nicotine solution on an hourly schedule for 6 h would have the same effects, or whether desensitization would occur after a few doses. West and Jarvis (1986) found that the nasal nicotine solution increased tapping rate after every dose in both subjects studied. The mean increase in tapping rate was 3.2 per cent over baseline in one subject, and 4.1 per cent in the other. There was no sign of desensitization with repeated doses.

This is an important study which shows that motor performance is reliably enhanced by nicotine by about 5 per cent in non-smokers. An increase could be produced by a single 2 mg dose of nicotine. The increase in tapping rate after nasal nicotine solution was shown to last for at least 30 min and there was no evidence of desensitization with repeated dosing on an hourly schedule.

3.2.3 *Information selection*

In the processing of information, an essential aspect is the selection of the relevant stimuli from all the possible inputs, and maintenance of that selection. This cognitive mechanism is often called attention or concentration. Surveys have indicated that the majority of smokers (86 per cent in a smoking clinic population, 59 per cent among hospital workers, and 86 per cent of students) believe that smoking helps them to concentrate (Russell *et al.* 1974; Warburton and Wesnes 1978), and the question arises whether smoking does improve attention, and if so whether nicotine is responsible for these effects of smoking.

Attention tests involve either focused or divided attention. Focused-attention tasks require subjects to attend to one source of information to the exclusion of others. Divided-attention tasks require subjects to divide their monitoring between two or more sources of information.

3.2.3.1 *Focused attention* Vigilance is one paradigm for examining focused attention. In these tasks attention has to be directed to an input for long periods of time, and the subject is required to detect and respond to brief, infrequent changes in the input. Performance in vigilance situations is often assessed in terms of the detection rate, i.e. the proportion of signals correctly detected. During a typical vigilance session, the detection rate decreases, a change called the vigilance decrement.

In a study of smoking and visual vigilance, the Mackworth Clock (Mackworth 1950) was used because it produces a reliable vigilance decrement.

Smoking cigarettes at 20 min intervals helped smokers, who were previously deprived for 10 h, maintain their detection of experimental targets in the 80 min vigilance task (Wesnes and Warburton 1978). In contrast, detection dropped for a group of non-smokers from a baseline of 0.76 to 0.70, and from a baseline of 0.82 to 0.65 for a group of deprived smokers who were not allowed to smoke. There was no significant difference between deprived smokers and non-smokers in their performance, i.e. no evidence of a withdrawal effect. This suggests that smoking was not merely reversing deprivation, but was enhancing performance.

Similar results were obtained in auditory vigilance studies. A comparison of non-smokers, 3-h-deprived smokers not smoking, non-deprived and 3-h-deprived smokers smoking, was made on a 60 min auditory vigilance task (Tong *et al.* 1977). They found that while non-smokers detected fewer signals as the test progressed, smokers after smoking increased their number of detections. In contrast, the detection rate of the deprived smokers remained stable over time. In spite of the slight vigilance decrement, the non-smokers performed significantly better than the smokers.

At first sight, the latter finding could be taken as evidence for a withdrawal effect. However, the interpretation is not that simple. The smokers-not-smoking group were minimally deprived for 3 h, while the smokers smoking had one cigarette on arrival at the laboratory, and another after instructions and 5 min practice on the task, i.e. about 10 min later. Smokers who had smoked two cigarettes differed from the minimally-deprived smokers only on the first block, and at that point they were worse, not better (perhaps because of too much nicotine). Thus there was no evidence of smoking reversing a withdrawal effect, and it certainly did not bring the performance of smokers any closer to that of the non-smokers. It could even be argued that there was not even an enhancement by smoking, but merely recovery from the effects of two cigarettes in 10 min.

In another auditory vigilance test, it was found that 10-h-deprived smokers maintained their initial level of detection of signals embedded in noise over an 80 min vigilance session when they smoked cigarettes at 20 min intervals. In contrast, their detection rate decreased over time from 0.92 to 0.83 while smoking herbal, nicotine-free cigarettes (Wesnes and Warburton 1978). A similar study used smokers who were deprived for only 2 h (Mangan 1982). The test involved detecting changes in sound intensity which occurred on average once per minute and found essentially the same result. Smoking maintained signal detection performance during the 30 min session compared with the minimally-deprived smokers not allowed to smoke.

If nicotine is the essential ingredient of cigarette smoke for producing the effect, then nicotine alone should produce the same effects on visual vigilance performance as smoking. This theory was addressed by using unbuffered nicotine tablets which were held in the mouth for five minutes and then swallowed. From Fig. 3.1, it can be seen that nicotine reduced the vigilance

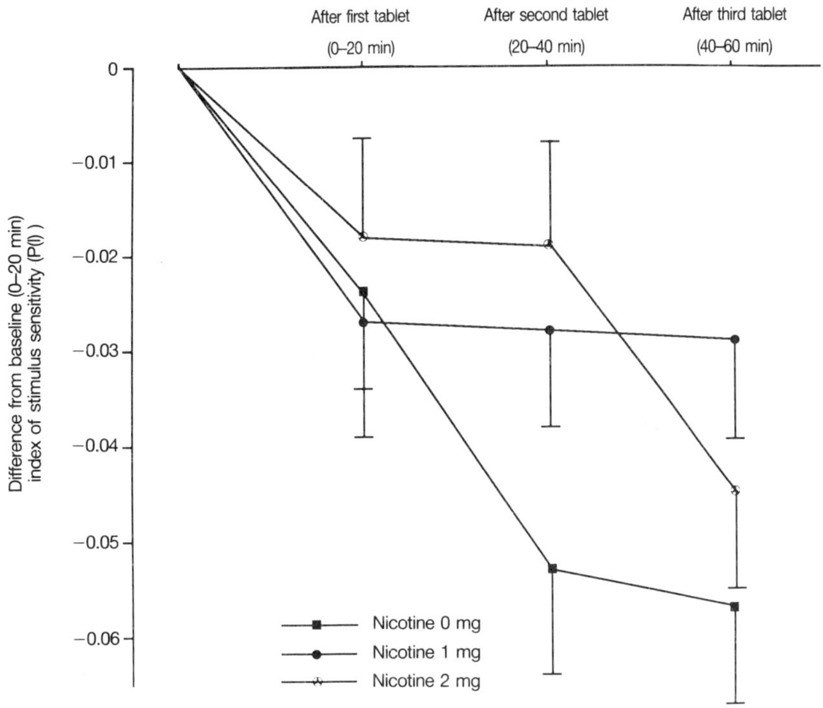

Fig. 3.1. The effects of nicotine tablets (0, 1, and 2 mg) on the difference from baseline stimulus sensitivity scores in a vigilance task during three 20-min periods following baseline. The bars represent standard errors of the mean. Data replotted from Wesnes *et al.* 1983.)

decrement which occurred over time in the placebo condition (Wesnes *et al.* 1983). The nicotine tablets produced the same effects in non-smokers, light smokers, and heavy smokers (see Fig. 3.2). These data have implications for the issue of tolerance to be discussed later.

A factor analysis of a large number of attention tests found that picking out various sequences of numbers or letters from an array was most heavily loaded on an 'attention' or 'mental concentration' factor, (Wittenborn 1943). Williams (1980) tested the effects of smoking on 10 h-deprived smokers with a test of this kind, which involved crossing out each letter E on sheets of randomly ordered letters arranged in lines of 30 letters. A highly significant improvement in performance of the letter cancellation was produced by smoking, in comparison with sham smoking. This finding was replicated in a study with 22 women and 26 men (Williams *et al.* 1984).

A computer version of this test is the Bakan task (Bakan 1959), in which a series of digits is presented at a rapid rate and subjects are required to detect certain specified three-digit sequences. Measures of both the speed and the

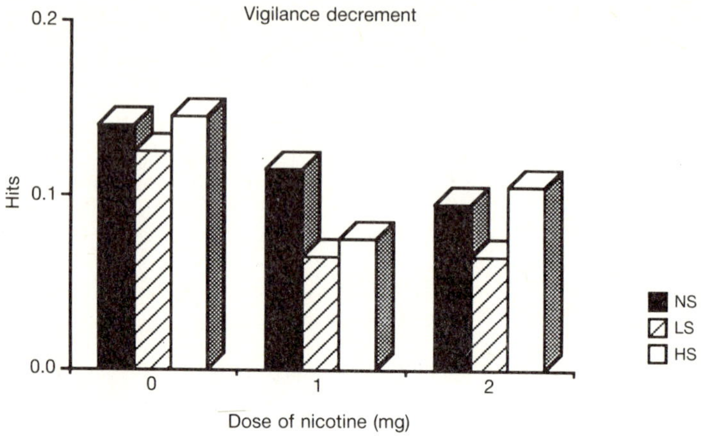

Fig. 3.2. A comparison of the vigilance decrement for non-smokers (NS), light smokers (less than 5 per day; LS) and heavy smokers (more than 15 per day; HS) for 0, 1, and 2 mg nicotine tablets. (Data from Wesnes *et al.* 1983.)

accuracy of detection are made. Performance on this rapid visual information processing task by 10-h-deprived smokers after smoking was improved in both speed and accuracy above baseline levels, whereas either not smoking, or smoking herbal nicotine-free cigarettes, resulted in a decline in speed and accuracy below baseline levels (Wesnes and Warburton 1983, 1984*a*). Analyses of performance while smoking the cigarette show puff-by-puff improvement in both speed and accuracy as each bolus of nicotine enters the brain. The maximum level of improvement was of the order of 15 per cent (see Fig. 3.3). The improvement in both speed and accuracy is important because it shows that there is no speed and accuracy trade-off, i.e. an overall improvement in processing efficiency.

In general, high-nicotine yielding cigarettes (1.5 and 1.7 mg) have improved performance more than low-yield ones (0.9 mg; Wesnes and Warburton 1984*a*), and although the effect of other smoke constituents cannot be ruled out, the most likely hypothesis is that nicotine is the important smoke constituent for these attentional effects. This proposal was supported by studies of 10-h-deprived smokers with cigarettes which differed less in nicotine yield than in tar and in carbon monoxide (1.4 mg and 1.7 mg of nicotine, 11.2 mg and 16.9 mg of tar, and 9.9 mg and 15.1 mg of CO). Cigarettes with similar nicotine yields had the same effect on speed and accuracy (Warburton 1988).

Further support for the hypothesis that nicotine was responsible comes from a study in which tablets containing 0.5 mg, 1.0 mg, or 1.5 mg of nicotine were given to non-smokers for buccal absorption. They then performed the rapid-

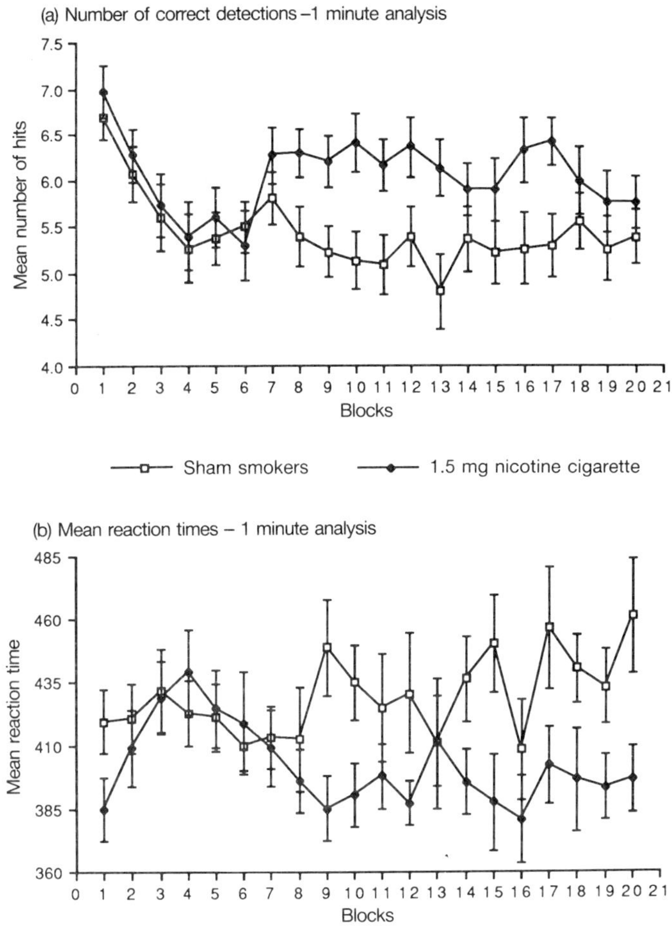

Fig. 3.3. The puff-by-puff effects of smoking on performance in the rapid visual information processing task. Each point represents the difference after each puff and the bar represents the standard error of the mean. Smoking was begun on the sixth minute and finished on the sixteenth minute.

visual information-processing task (Wesnes and Warburton 1984). The taste of the nicotine was masked by capsaicin which was placed on both the active and placebo tablets. The buccal absorption of nicotine was aided by using alkaline tablets. The highest dose of nicotine (1.5 mg) produced a performance improvement in non-smokers which closely resembled that produced by smoking in smokers. The 1.0 mg dose had a lesser effect and 0.5 mg was without effect (see Fig. 3.4). This finding provides strong evidence that

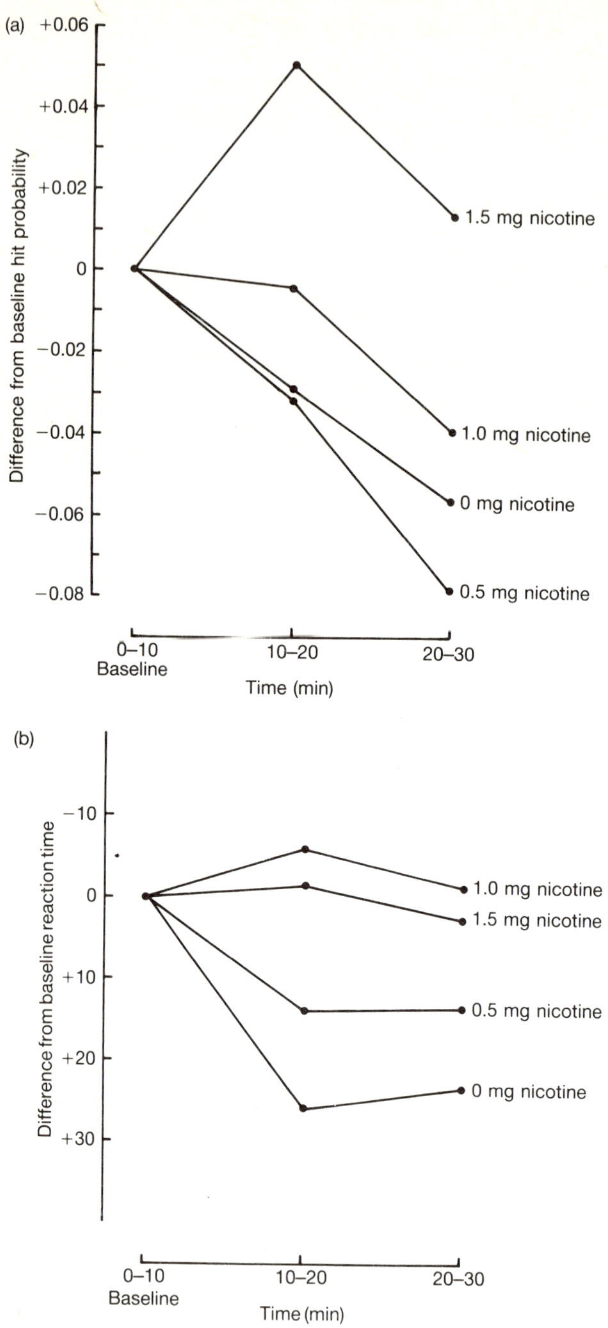

Fig. 3.4. (a) The effects of nicotine tablets on the performance of non-smokers assessed in terms of high probability in the rapid visual information processing tests. (b) The effects of nicotine tablets on the performance of non-smokers assessed in terms of reaction time in the rapid visual information processing tests. (Data from Wesnes and Warburton 1984*b*.)

nicotine plays the major role in the improvements in focused-attention tasks that are produced by smoking. The use of capsaicin as an irritancy control argues against the performance improvements being due to peripherally-mediated arousal. The improvement in the performance of non-smokers argues against the improvement being only a reversal of withdrawal.

Another type of focused attention task has looked at performance under conditions of distraction. A task which has been used in nicotine research to examine distraction effects on attention is the Stroop Test (Stroop 1935). The Stroop Test uses three sets of display, a list of colour words printed in black, a set of colour patches, and a list of colour words with the words printed in incongruent colours (for example, the word GREEN printed in blue). Colour naming the incongruently printed colour words takes much longer than naming the colour patches, and the time difference between the two conditions is the Stroop effect. This score indicates the subject's ability to focus attention on a relevant stimulus dimension of print colour and ignore the irrelevant semantic one.

The effects of doses of nicotine (0, 1.0 and 2.0 mg) administered buccally were studied on the Stroop performance of both 10 h-deprived smokers and non-smokers (Wesnes and Warburton 1978). The buccally-absorbed nicotine increased heart rate by 6 beats min^{-1} for the 2 mg dose. Both doses of nicotine reduced the size of the Stroop effect in this study by 16 per cent. No differences were found between the baselines of the deprived smokers and non-smokers, or in the amount of improvement produced by nicotine at the time of the peak heart rate. Thus there was no evidence of tolerance to the effects of nicotine in smokers for this type of performance.

3.2.3.2 *Divided attention* One test of divided attention in a task designed to simulate driving (Tarriere and Hartemann 1964) combined central guiding with peripheral visual monitoring. The task lasted for two and a half hours, and the measure of performance reported was the percentage of the peripheral visual signals which were missed during the session. Monitoring performance was maintained by smoking, in contrast to the large increase in the percentage of missed signals when not smoking. The performance was similar for non-smokers and smokers who were deprived for 20 h.

In another study of simulated driving and smoking, subjects had to divide their attention between several sources of information. Heimstra *et al.* (1967) measured four aspects of performance; steering efficiency, reaction time, brake light vigilance, and meter vigilance. Although tracking error was significantly greater in the smoker group which had been deprived for an unspecified time, no significant differences were found between non-smokers and smokers smoking. There were no statistically significant differences between groups in reaction times, although reaction times increased significantly over time in the non-smoker and deprived smoker groups, but not for the smoking group. Brake-light vigilance was significantly lower ($p < 0.05$) for the deprived smokers than for the other two groups. In the meter-vigilance task there were no between-group differences, but the non-smokers made significantly more errors as time progressed (a vigilance decrement), whereas performance remained at a constant level for the group allowed to smoke.

In summary, this study showed that smoking maintained performance for reaction time, which is consistent with the data shown in section 3.2.2, and smoking prevented the vigilance decrement in meter vigilance which is in accord with the results that were reported earlier in this section. Some evidence suggested that deprived smokers showed worse performance than non-smokers, but this difference was only significant for steering efficiency and brake-light vigilance.

Another study of divided attention (cited in Warburton and Walters 1988) used a test which was based on the rapid visual information task (Wesnes and Warburton 1984*a*). Subjects were presented with digits at a rate of 50 min^{-1} in both the visual and auditory modality, a different sequence for each modality. The detection of sequences in both modalities improved significantly after 10-h-deprived smokers smoked a cigarette, in comparison with not smoking; an improvement of 7 per cent. Smoking a cigarette also prevented the increase in reaction times that occurred in the non-smoking condition.

In summary, the evidence from this section is consistent. It seems that smoking improves general attentional processing capacity without modifying one specific type. Although nicotine has not been tested in every situation, its effects parallel those of smoking whenever it has been used.

3.2.4 *Information manipulation*

Information is manipulated during thinking and problem solving. Once again, surveys have shown that many smokers believe that smoking helps them to think better (Russell *et al.* 1974; Warburton and Wesnes 1978). Problems come in two broad classes: problems that are well-defined, and those that are ill-defined. A well-defined problem is one that has a clearly stated goal, and so a definite way to tell whether the problem solving is proceeding in the correct direction. Convergent thinking will produce logically correct answers.

A simple example of a well-defined problem is addition. Hull (1924) found a marked increase in the rate of complex mental addition after smoking, but no measurable effect on the accuracy of addition. It has also been found that the reduced efficiency of mental addition that was produced by doing a tracking task was ameliorated by smoking (Kucek 1975). In the latter study, the improvement was manifested most in the most neurotic subjects. We will return to this finding in section 3.5.1.

Ill-defined problems are solved by divergent thinking which leads to inventive solutions to problems. The solutions are not ones which can be seen as logically correct or not. A task which has elements of both convergent and divergent thinking is the Luchins Jar Test (Luchins 1942), in which subjects are asked to solve a number of 'numerical problems' which involve calculating a quantity of water by means of a set of measuring jars. For the first six trials exactly the same solution to the problem can be used, but after trial six both the old formula and a new, easier formula are appropriate. A measure of convergent thinking is performance on the first six trials, while divergent thinking is assessed from the time taken to discover the new, easier solution.

10-h-deprived smokers held 0, 1.0, and 1.5 mg nicotine tablets in their mouths for 5 min. It was found that they performed better on the first half of the test where subjects could use repeatedly the same solution to the problem (convergent thinking). This result is consistent with the findings of Hull (1924) and Kucek (1975) on smoking and arithmetic performance. However, they were slower to switch to a simpler solution when it was available. This result is not easy to interpret. It could be that nicotine was interfering with the ability to think divergently, or it was producing response perseverance (Warburton and Wesnes 1985).

Clearly, much more work using nicotine needs to be done in this complex area of problem solving before strong statements can be made. The fragmentary evidence does not contradict the smoker's belief that smoking and nicotine help some kinds of thinking.

3.2.5 *Neurochemical bases for improved information processing*

The more efficient information processing (in terms of more accurate and faster processing) produced by nicotine or smoking is consistent with our knowledge of the psychophysiology and neuropharmacology of cognition. It is known that a shift from low-frequency high-amplitude activity in the electroencephalogram (8–13 Hz) to high-frequency low-amplitude activity (13–20 Hz) correlates with more efficient information processing (see Clarke, chapter 5, this volume). A number of human studies have found that smoking produces these sorts of shifts (Caille and Bassano 1976; Ulett and Itil 1969; Knott and Venables 1977; Warburton and Wesnes 1979). Animal studies have demonstrated that nicotine produces similar changes (Kawamura and Domino 1969). Smoking also decreases the latency of the P_3, colloquially known as the P300, component of the event-related brain potentials (Edwards *et al.* 1985). This change is shown in Fig. 3.5, and the similarity to the change in reaction time in Fig. 3.3 is striking. The occurrence of the P_3 wave depends on the completion of certain stimulus-evaluation processes (Donchin 1984), and the reduced latency of the P_3 indicates quicker stimulus evaluation i.e. more efficient processing. Cholinergic antagonists like scopolamine have the opposite effect (Callaway *et al.* 1985).

These data argue for nicotine affecting cortical mechanisms involved in information processing (Warburton 1981). Doses of nicotine (for example, $20\,\mu g\,kg^{-1}$ intraperitoneally in the cat) that are thought to be equivalent to those experienced during human smoking produce excitation of cortical cells and release of acetylcholine (Armitage *et al.* 1969). Cortical acetylcholine release and cortical excitation can be produced by electrical stimulation of the

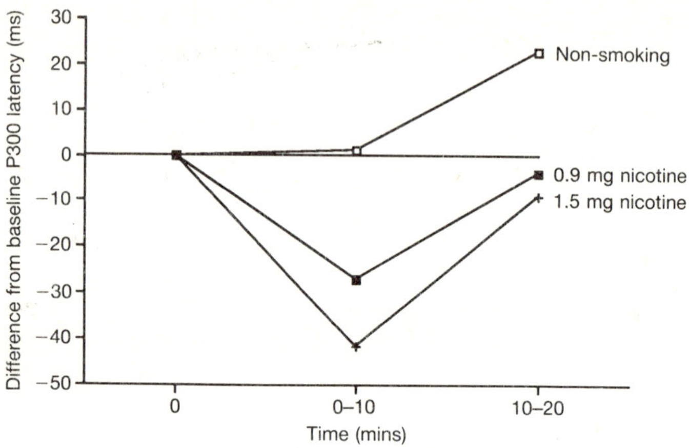

Fig. 3.5. The effects of smoking upon rapid visual information processing: difference from baseline changes in P300 latency. (Data from Edwards *et al.* 1985).

mesencephalic reticular formation, and this phenomenon can be reduced in one hemisphere by destruction of this region ipsilaterally (Celesia and Jasper 1966). Kawamura and Domino (1969) found that 'smoking' doses of nicotine did not produce cortical desynchronization in cats after extensive bilateral lesions in the tegmental region of the midbrain (an area occupying the tip of the midbrain to the central portion of midbrain near the pons). The tegmental region in the mesencephalic reticular formation activates the cholinergic pathway which terminates on cells at the sensory cortex and produces electro-cortical arousal there (see review by Warburton 1981). Improvements in attentional processing during smoking, and after nicotine tablets, are probably the outcome of changes in activity in the pathways which project to cortical sensory-neurones from the nucleus basalis of Meynert (Robinson 1985).

The outcome of activation of the pathways is release of acetylcholine at the cortex, and cortical desynchronization. Many human studies have shown that smoking increases the amount of cortical desynchronization in the form of an upward shift in dominant alpha frequency (see extensive review in Edwards and Warburton 1983). This cortical desynchronization is indistinguishable from the alert pattern in a person who is concentrating. This system can be seen as providing a state appropriate for information processing as state models of cognitive function describe (Hockey and Hamilton 1983; Warburton 1986). The brain is not fixed in a 'state', but the states are varying as shown by the fluctuations in EEG pattern. Given this processing state system, how does nicotine improve attentional processing? The answer may be that nicotine maintains the cortex in a desynchronized state by driving the cholinergic pathways ascending from the nucleus basalis and from the tegmental area. Thus it reduces fluctuations in cortical arousal and the concomitant variations in information processing that occur in the undrugged state.

The hypothesis that nicotine improves information processing by releasing acetylcholine and maintaining cortical desynchronization was tested by examining the effects of scopolamine. Scopolamine produces impairments in attentional performance (Wesnes and Warburton 1984*b*), and increases the P_3 latency (Callaway *et al.* 1985). Scopolamine blocks muscarinic acetylcholine receptor sites at the cortex and decreases electrocortical arousal (Ostfeld and Aruguette 1962), and so it should antagonize the effects of the cortical acetylcholine released by nicotine's action on the brain stem.

Pharmacological antagonism was tested using the rapid-visual information-processing task, and the Stroop Test. It was found that a buccal dose of 1.5 mg of nicotine completely counteracted the impairment in performance produced by an oral dose of 1.2 mg of scopolamine, on both the rapid information processing task and on the Stroop Test (Wesnes and Revell 1984).

In summary, the last section has provided evidence of nicotine acting on cholinergic pathways which ascend to the cortex and release acetylcholine, and which in turn produces cortical desynchronization. This state of cortical arousal is found when a person is processing information efficiently. The beneficial effects of nicotine on information processing can be explained by it producing and maintaining that cortical state.

3.3 Information storage

Information storage consists of input of the information, registration of the information in immediate, shorter-term memory, and consolidation of the information in longer-term memory. However, information storage can only be demonstrated by the effective retrieval of the information from storage. Thus it is important to distinguish a drug's effect on storage from an effect on retrieval. One method of doing this is by using a state-dependent design which will be discussed later. Firstly, the effects of smoking, and by implication, of nicotine, on immediate memory will be considered. Relevant studies on information processing and storage are summarised in Table 3.1.

3.3.1 *Immediate memory*

The first study showing that smoking modified immediate memory came from Andersson and Hockey (1977). Eight words were presented in different positions on a computer screen, and in one condition (List 1) subjects had to remember the words in presentation order, and so by implication location was irrelevant. There were no differences between the performance of smokers after smoking a 2.3 mg nicotine cigarette and smokers deprived on the morning of the test session with respect to the percentage of words that were recalled in the correct order, or the percentage of words that were recalled correctly, regardless of word order. However, the irrelevant information, position on the screen, was recalled significantly less well after smoking when the subjects were not told to attend to location, i.e. incidental memory.

In the second part of the study subjects were told to attend to all three aspects of the material on List 2 and a different picture emerged. The percentage of correctly recalled words, irrespective of order, decreased from the first to the second list in both the smoking and the non-smoking group, although the decline was significant only for the former. When order accuracy was taken into account, the reverse pattern was seen, performance improved from List 1 to List 2, and this difference approached significance ($p<0.10$) for the smoking group. As one would expect, location recall scores were increased in both conditions from the first to the second list, although the improvement was again significant only for the smoking group.

This study suggests that smoking may enhance storage of information, but only of information that is thought to be relevant by the subjects. One interpretation of the finding is an attentional one; the subjects who smoke select out irrelevant information. In the study, heart rate was increased in the smokers who smoked, and so it can be inferred that nicotine was absorbed and was the agent that was responsible for the changes.

Another study on immediate memory (Houston *et al.* 1978) used smokers who were deprived of cigarettes for 3 h. They were matched on a free recall test prior to smoking. One group smoked a 1.5 mg nicotine cigarette rapidly, at the rate of one puff every 25 s for 12 puffs with mandatory inhalation, and the other group smoked a non-nicotine cigarette. The subjects were then given three lists with a free recall test immediately after each one. The immediate recall scores showed that the nicotine-cigarette group had significantly poorer recall than the non-nicotine-cigarette group. The rapid smoking rate with forced inhalation may have resulted in too much nicotine for some subjects.

In a third study of immediate memory (Williams 1980), 10-h-deprived subjects were tested 3 min after smoking one cigarette on three different occasions, using 0.6, 1.3, and 1.8 mg nicotine cigarettes. They were given lists of nine numbers to memorize and then immediately asked to recall them in the correct sequence (constrained recall). Immediate memory errors progressively increased with cigarette yield when pre-smoking performance was controlled.

Peeke and Peeke (1984) studied the effects of smoking on immediate memory in 2-h-deprived smokers. Recall of a 50-word list was tested immediately, and again after intervals of 10 and 45 min. Pre-trial smoking of their own brand resulted in improved recall immediately after learning. Another study compared a low-(0.40 mg) and high-(1.38 mg) nicotine cigarette in light and heavy smokers using pretrial smoking. The high-nicotine cigarette resulted in improved immediate recall. The low-nicotine cigarette was less effective.

In agreement with the last study, when testing was given once just after the input, we found evidence of improved immediate memory (Warburton *et al.* 1986). After smoking a 1.4 mg nicotine cigarette at their own pace (nine puffs at 38 s intervals, on average), the subjects were shown a list of 48 nouns and immediately asked to write down as many as they could. There was better immediate recall after smoking in comparison with not smoking.

3.3.2 *Delayed recall*

The effects of smoking a low-(0.7 mg) and a middle-(1.3 mg) nicotine-yield cigarette on learning to associate pairs of words, and on retention, was assessed under conditions of high and low intra-list interference, and on learning lists of words and retention (Mangan 1983). Cigarettes improved retention in the paired-associate learning, with task difficulty apparently having little relevance. Smoking impeded learning under low-interference conditions, but facilitated learning of high-interference sets. Performance during learning suggests that improvement is produced by smoking on longer-term memory rather than shorter-term memory.

Mangan and Golding (1983) studied the effects on memory of 1 h-deprived smokers, and those smoking a single cigarette after one hour of deprivation but immediately after acquisition, on a paired-associate learning task. Subjects were tested for retention of the memorized material at intervals of 30 min, 1 day, 1 week and 1 month. Deprived smokers showed superior recall compared with all smokers at the 30 min retest. After one month subjects who had smoked a 0.8 mg and 1.3 mg nicotine cigarette showed better recall than those who had not smoked, and those who had smoked a 2 mg nicotine cigarette. The authors offered the following interpretation:

> The possibility of nicotine overdose in smokers smoking the high-nicotine cigarette was suggested by smokers' self-reports. Thus, many of the high-nicotine smokers spontaneously reported that they found this cigarette 'very strong' or 'unpleasant to smoke'. No smoker in the low- or middle-nicotine groups reported this effect. The results for butt nicotine analyses were consistent with this possibility, although caution must be exercised in interpreting such data. (Mangan and Golding 1983, p. 74).

In a careful series of studies, Peeke and Peeke (1984) tested the effects of smoking one cigarette on verbal memory and attention in four experiments, using smokers who were deprived for 2 h, i.e. minimally deprived. In one study, subjects were tested when they smoked their own brand of cigarette prior to learning, after learning, and not smoking. Recall of a 50-word list was tested immediately and after intervals of 10 and 45 min. Pre-learning smoking resulted in improved recall but there was no effect of smoking after learning, as Mangan and Golding (1983) had found. In another experiment, a high-nicotine cigarette (1.38 mg) did produce improved recall on both immediate and delayed recall tests, but a low-nicotine (0.4 mg) cigarette was less effective.

3.3.3 *Information input and retrieval effects of nicotine*

The possibility of nicotine producing state-dependent learning in human subjects is suggested by the numerous animal studies which have given evidence of state dependency with cholinergic drugs (see review in Warburton 1977). In a state-dependent study, one group of subjects learns after a dose of compound while a second group learns after a placebo, or nothing. For the recall test, both groups are divided so that one half of each group are tested with the agent of learning, and half are switched to the other condition. If the recall scores are better for those groups which learned in the same chemical state, then state-dependent learning is said to have occurred. This state-dependency hypothesis with nicotine has been investigated in several smoking studies in order to differentiate nicotine's effects on information input and information retrieval.

Four studies have found evidence for state-dependent learning with smoking or nicotine (Peters and McGee 1982; Kusendorf and Wigner 1985; Warburton *et al.* 1986, two studies). The first study (Peters and McGee 1982) used the state-dependent design to test nicotine's effect on recall and recognition memory. After smoking a 1.4 mg nicotine cigarette subjects were shown a list of nouns, and immediately asked to write downn as many as they could. There was no evidence of any difference on immediate recall in agreement with Andersson and Hockey (1977) and Houston *et al.* (1978). However, the recognition test on the next day showed a clear state-dependent effect, but no difference between the same-state groups. Thus there was no evidence that nicotine modified storage or retrieval.

More recently, state-dependent recall with smoking an unspecified cigarette was examined using text material (Kusendorf and Wigner 1985). Each group spent 15 min studying a 550-word article on education, and answered six factual questions based on the article after a 10 min break. The treatment conditions were smoking versus no smoking during the study period, and these two groups were then spilt to make four groups in total with smoking and no smoking during testing. When studying and testing were conducted in the same state (either smoking or no smoking) memory was better than when study and testing were conducted in different states. However, there was no evidence of smoking improving or impairing information input or information retrieval.

In a recognition study (Warburton *et al.* 1986), smokers who were deprived of cigarettes for over 10 h were given a 1.4 mg-nicotine cigarette or nothing immediately before serial presentation of a set of Chinese ideograms. The subjects were then divided into four groups: a quarter who did not smoke prior to learning or recall, a quarter who did not smoke prior to learning but had a cigarette prior to recall, those who had a cigarette prior to learning and recall, and those who had a cigarette prior to learning but none prior to recall. Subjects who smoked prior to learning had significantly better recognition scores than the subjects who did not smoke in the first part of the experiment.

There was no effect of smoking on recall performance itself. A significant interaction term indicated that changing the chemical state interfered with recognition, i.e. state dependency.

A fourth study used 1.5 mg nicotine tablets in the state-dependent design in which smokers were deprived for over 10 h (Warburton *et al.* 1986). After the tablet, the subjects listened to 48 words and then did successive subtractions for 1 min to prevent rehearsal. As we have said earlier, immediate recall was improved. One hour later, the subjects were given either nicotine or placebo tablets, depending on their group. They were asked to recall as many of the words as they could in another 10 min free-recall test. Long-term recall was significantly better when subjects had taken nicotine prior to learning, but not when taken prior to recall. A significant interaction term gave evidence for a state-dependent effect of nicotine, and showed that nicotine was facilitating the input of information to storage but had no direct effect on retrieval.

In summary, the data on the effects of smoking and nicotine on memory are mixed. It is clear that nicotine does not facilitate or impair retrieval, but there is a state-dependent effect. Smoking and nicotine do improve longer-term storage of information. The results of the immediate memory studies are not clear cut, but some studies suggest that it can be improved in some circumstances. The reason for a positive effect of nicotine on immediate memory in some studies (for example, Peeke and Peeke 1984; Warburton *et al.* 1986) but not others (for example, Houston *et al.* 1978; Williams 1980) is unclear.

One explanation of this discrepancy may be that some memory tasks had an attentional component. In a typical memory task, attention to the words is controlled by instruction. The list is short and so attention does not play a part. However, the list in the Warburton *et al.* (1986) study was 48-items long, and that in the Peeke and Peeke (1984) study was 50-words long, in comparison with the eight-word lists in the Andersson and Hockey (1977) study, and the nine digit list of Williams (1980), for example. In this regard it is interesting that nicotine did produce a slight improvement in the Andersson and Hockey study when subjects had to remember words, word order, and also location on the computer screen i.e. 24 items.

3.3.4 *Neurochemical bases for information storage effects*

In contrast to the human data, animal studies have demonstrated that information storage is improved in animals given a 'smoking dose' of nicotine, approximately 0.1 mg kg^{-1} (Pratt *et al.* 1983). At these sorts of doses, nicotine facilitated the acquisition of a shuttle-box avoidance, maze learning, and two-choice visual discrimination behaviour, when administered prior to training (for example, Robustelli 1963; Bovet *et al.* 1966; Bovet-Nitti 1969).

Post-trial nicotine treatment has been shown to result in facilitated retention of a variety of tasks (Garg and Holland 1968, 1969; Garg 1969; Bättig 1970; Evangelista *et al.* 1970; Erickson 1971). Flood *et al.* (1981) showed that an intraventricular injection of the nicotine agonist dimethylphenylpiperazinium

(DMPP) given to mice after a single training session, improved retention for an appetitive task upon testing one week later (see Stolerman, Fig 8.4, this volume), and also demonstrated dose-dependent impairments following post-trial administration of the nicotine antagonist mecamylamine.

Similar results to those obtained with nicotine have been obtained with physostigmine when administered pre- and post-trial. Physostigmine is an anticholinesterase and produces increases in functional acetylcholine. As mentioned earlier, (section 3.2.5) nicotine increases the release of acetylcholine. The analogous behavioural and neurochemical actions of the two compounds argues for a common neurochemical action, i.e. via the cholinergic system. Brain cholinergic systems have been implicated in memory storage for many years (see Hunter *et al.* 1977 for a review). More recent research has linked Alzheimer's Disease, in which memory loss is a cardinal symptom, with damage to the ascending cholinergic pathways from the nucleus basalis of Meynert to the cortex (Robinson 1985; Kellar & Wonnacott, chapter 10, this volume).

In summary, the finding of post-trial facilitation in animals confirms the conclusion from the human studies, that it is memory storage processes that are affected by nicotine. Studies with other drugs suggest a cholinergic mechanism.

3.4 Mood

3.4.1 *Mood effects of smoking and nicotine*

The calming effects of smoking seem to be the day-by-day experience of many smokers (McKennell 1970, 1973; Tomkins 1966; 1968). Surveys have found that 80 per cent of smokers say that they smoke more when they are worried, 75 per cent say that they light-up a cigarette when they are angry, and 60 per cent feel that smoking cheers them up (Russell *et al.* 1974; Warburton and Wesnes 1978). These opinions suggest that smokers feel that smoking improves their mood and that this is an important reason for smoking (Ashton and Stepney 1982).

A standard way of producing a psychological and physiological stress response is by making subjects watch a mutilation film. This stressful experience increases heart rate and decreases skin conductance. Smoking a cigarette decreases both the subjective stress-response and smooths out the body's stress response (Gilbert and Hagen 1980). Further evidence has come from a study which found that smoking reduced the subjective annoyance to loud noise, as well as the physiological stress response to noise (Woodson *et al.* 1986). Smoking before testing enabled smokers to stand higher levels of electric shock than when they were deprived (Silverstein 1977; also cited in part by Schachter 1978). In a study of smoking during gambling, smoking helped smokers to be less angry when they knew that they were being cheated by another person (Cherek 1981).

Table 3.1. Summary of studies of the effects of smoking and nicotine on information processing and storage in humans

Studies	Deprived	Nicotine	Results
Critical flicker fusion			
Warwick & Eysenck (1963)	N I	Tablet	Progressive elevation of C F F threshold with nicotine dose
Larson *et al.* (1950)	N I	Cigarette	
Tong & Leigh (1975)	N I	Cigarette	
Caille & Bassano (1976)	36–40 h	Cigarette	
Waller & Levender (1980)	Non Dep	Cigarette	
Reaction time			
Cotten *et al.* (1971)	Non Dep	Cigarette	R T reduced immediately following smoking but increased 40–55 min later
Morgan & Pickens (1982)	N I	Cigarette	Increase in R T following smoking
Frankenhauser *et al.* (1971)	12 h	Cigarette	Smoking maintained R T performance
Myrsten & Andersson (1978)	12 h	Cigarette	Smoking prevented fall in simple R T and reduced complex R T
Lyon *et al.* (1975)	12 h	Cigarette	Choice R T reduced by smoking
Smith *et al.* (1977)	12 h	Cigarette	Decision times reduced by smoking
West & Jarvis (1986)	N S	Nasal	Nicotine increased tapping rate
Focused attention			
Wesnes & Warburton (1978)	10 h	Cigarette	Smoking helped maintain vigilance. No difference was found between deprived and non-deprived Ss.
Tong *et al.* (1977)	3 h	Cigarette	Improvement in detections when smoking
Mangan (1982)	10 h	Cigarette	Level of detection maintained by smoking
Wesnes *et al.* (1983)	N S	Tablets	Nicotine reduced the decrement which occured in the placebo condition
Williams (1980)	10 h	Cigarette	Smoking improved performance on letter cancellation task
Williams *et al.* (1984)	10 h	Cigarette	
Wesnes & Warburton (1978)	10 h	Tablets	Nicotine improved performance on R V I P
Wesnes & Warburton (1984)	10 h	Cigarette	Improved performance on R V I P when smoking
Warburton (1988)	10 h	Cigarette	Cigarettes with similar nicotine levels produced the same effects on R V I P
Divided attention			
Tarriere & Hartemann (1964)	20 h	Cigarette	Monitoring performance maintained by smoking
Heimstra *et al.* (1967)	1–6 h	Cigarette	Smoking maintained R T and prevented vigilance decrement
Information manipulation			
Hull (1924)	N I	Cigarette	Increase in rate of mental addition when smoking
Kucek (1975)	Non Dep	Cigarette	Smoking prevented the decrement in the addition task produced by an interfering tracking task
Wesnes & Warburton (1985)	10 h	Tablets	Nicotine produced response perseveration
Immediate memory			
Andersson & Hockey (1977)	N I	Cigarette	Recall of incidental information worse after smoking
Housten *et al.* (1978)	3 h	Cigarette	Smoking group showed poorer recall
Williams (1980)	10 h	Cigarette	Errors increased with nicotine yield of cigarette
Peeke & Peeke (1984)	2 h	Cigarette	Pre-trial smoking of own brand improved recall
Warburton *et al.* (1986)	10 h	Cigarette	Improved recall after smoking at own pace
Delayed recall			
Mangan (1983)	3 h	Cigarette	Smoking improved retention in paired associate learning
Mangan & Golding (1983)	1 h	Cigarette	Deprived smokers showed superior recall at 30 min. After 1 month non-deprived smokers performed better
Peeke & Peeke (1984)	2 h	Cigarette	Pre-learning smoking resulted in improved recall but there was no effect of post-learning smoking

Key: N I = No indication given: N S = Only non-smokers used: Non Dep = Subjects were not deprived

In a recent paper, Pomerleau *et al.* (1984) found that anxiety, as measured by the Profile of Mood States Questionnaire, could be generated using anagrams. In all subjects the anxiety was markedly decreased, more after smoking their own nicotine-containing cigarettes than after non-nicotine cigarettes. The consistency and magnitude of the reduction supports previous observations on the beneficial effects of cigarette smoking on mood. An important part of the study was that subjects were minimally deprived of cigarettes for half an hour prior to the study.

Pomerleau *et al.* (1984) also found that smoking a nicotine-containing cigarette nearly doubled the pain awareness thresholds in the cold-pressor test. Pain tolerance thresholds were also elevated, although the increase was not as pronounced. In three of the five subjects, the effects of smoking on pain awareness and tolerance thresholds were paralleled by changes in perceived pain, as assessed by the McGill Pain Questionnaire. The amount of nicotine intake is not clear from this study because plasma nicotine levels were not measured.

These findings of Pomerleau *et al.* (1984) extend the studies by Nesbitt (1973) and Silverstein (1977, 1982) and argue for a revision of their conclusions. Silverstein found that pain tolerance thresholds for smokers smoking high-nicotine cigarettes were similar to those for non-smokers who had not received nicotine. However, both Nesbitt (1973) and Silverstein (1977) demonstrated that smokers smoking high-nicotine cigarettes exhibited significantly greater pain awareness thresholds than non-smokers who had not had nicotine. Pomerleau *et al.* (1984) argue that pain awareness thresholds may be a more accurate measure of the antinociceptive effects of nicotine than pain tolerance. From the results of both experiments, Pomerleau *et al.* conclude that nicotine from smoking can produce mood changes without withdrawal.

Schachter (1978) assumed that nicotine withdrawal caused anxiety, which was manifested by a lowering of pain tolerance thresholds. Thus he interpreted the Nesbitt (1973) and Silverstein (1977) data as indicating that the consequence of smoking was not anxiety-reduction and pain-reduction *per se*, but rather relief from nicotine withdrawal. The finding of Pomerleau *et al.* (1984) of reduced pain awareness and anxiety reduction in minimally-deprived smokers who were allowed to smoke supports the hypothesis that nicotine from smoking can reduce anxiety and pain without abstinence from smoking.

Studies of the mood of smokers during the smoking of a cigarette show that the smokers feel calmer, more relaxed, more contented, friendlier, and happier after a cigarette, and are in a much better mood in general when smoking (Mangan and Golding 1978). In our own studies of the puff-by-puff mood effects of cigarettes (Warburton *et al.* 1988) subjects completed a set of Bond-Lader mood scales (Bond and Lader 1974) after each puff. For all cigarettes there were significant increases in calmness on the calm-excited scale, when measured after successive puffs. Subjects also reported that they became more tranquil, more sociable, more friendly, more contented, more relaxed, and

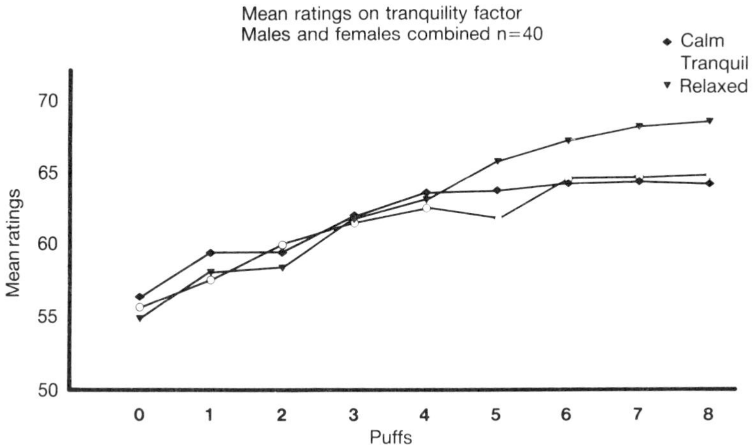

Fig. 3.6. The puff-by-puff effects of smoking on the scales of the tranquillity factor of the Bond-Lader Mood Scales. Non-nicotine cigarettes produced no improvement. (Warburton *et al.* 1988.)

happier over successive puffs and these changes were highly significant (Fig. 3.6).

These improvements in mood follow the same pattern as the increases in plasma nicotine during smoking. Non-nicotine cigarettes produced no improvements. When a second cigarette was given 30 min later (i.e. minimal deprivation) we found that the baseline for the second cigarette was lower than the peak at the end of the first cigarette, but had not returned to pre-smoking levels. While smoking the second cigarette, the mood changes increased until they were above the level that was achieved at the end of the first cigarette. Thus we have done these studies with smokers deprived for 10 h and with smokers minimally deprived for 0.5 h, and have found improvements under both conditions, as Pomerleau *et al.* (1984) found.

From these studies, it seems that cigarette smoking is a mood modifier for smokers, calming them, and reducing their feelings of anxiety and anger. Certainly, people increase their smoking when under stress (Schachter 1978; Ashton and Stepney 1982; Warburton 1987). It is not surprising that people miss these benefits when they stop smoking. Increases in anxiety and anger are two of the common complaints of people who cut down their smoking or give up (Guildford 1966; Schechter and Rand 1974; Shiffman 1979; Shiffman and Jarvik 1976).

The question that still remains is whether nicotine *per se* is producing the mood changes or whether it is the smoking ritual or oral gratification that is important. Murphree (1979) performed a study to determine whether smokers have significant tolerance to some of the effects of nicotine. The subjects were non-smokers and smokers who smoked at least 10 cigarettes per day, but who had been deprived of cigarettes since the previous evening. Nicotine was infused until the heart rate increased by 15 beats per min, i.e. equivalent to a smoking dose. Subjects were asked to fill in a mood adjective checklist before and after each infusion. There was no evidence of any differences in the responses of non-smokers and smokers. Both groups showed similar shifts with nicotine on a number of items. Unfortunately, no statistical analyses were performed on any of the data. Both smokers and non-smokers felt more active, more lighthearted, more energetic, more elated, and calmer. They felt less resentful, less washed out, less tense, and less clutched-up. They also tended to be more forgiving and less suspicious but, as Murphree (1979) points out, these are merely shifts that would be expected when any intravenous infusion test is over, regardless of the drug. These mood shifts are similar to those obtained from smoking, and give some support for the idea that nicotine is the agent responsible for the mood changes. It is also significant that the changes occurred in non-smokers who had not been deprived of nicotine, i.e. it was not withdrawal relief from nicotine deprivation.

The other major evidencc that nicotine is the agent in cigarette smoke which is responsible for mood changes comes from nicotine replacement in smoking cessation studies. At least four studies have examined whether 2 mg nicotine gum can replace smoking in terms of mood changes (Jarvis *et al.* 1982; Hughes *et al.* 1984; West *et al.* 1984; Schneider 1986). Unfortunately, comparison across the studies is made difficult by the fact that different mood measures were used. The ones which fit most closely to the mood states discussed above are anxiety, tenseness, annoyance, hostility, irritability, depression, sociability, and more globally, mood fluctuations.

All of these aspects of mood have been changed significantly in abstaining smokers by nicotine gum, in comparison with placebo gum in at least one study (Hughes *et al.* 1984). Abstinent smokers felt less anxious and less tense when given 2 mg nicotine gum than when they were given placebo gum (Hughes *et al.* 1984). On the anger scales, nicotine substitution resulted in abstinent smokers feeling less annoyed than when they were given placebo (Schneider 1986), also less hostile (Schneider 1986), and less irritable (Jarvis *et al.* 1982; Hughes *et al.* 1984; Schneider 1986; West *et al.* 1984). On other dimensions of mood, 2 mg nicotine gum reduced depression (Jarvis *et al.* 1982; West *et al.* 1984), and increased sociability (West *et al.* 1984).

In summary, smokers report that smoking improves their mood, and laboratory studies substantiate these reports. The effects occur in minimally-deprived smokers. Infused nicotine produces mood changes in non-smokers. Nicotine gum produces mood changes in abstaining smokers. Thus it seems likely that nicotine is the agent which is producing the reduction in anxiety and anger, and the general improvement in mood. The fact that smoking produces mood changes in minimally-deprived smokers and nicotine induces mood changes in non-smokers argues against a withdrawal interpretation.

3.4.2 *Neurochemical bases for mood changes*

The neurochemical basis for the mood changes produced by smoking and by nicotine is poorly understood as yet. Certainly it is clear that nicotine does not act like a conventional euphoriant, or conventional anxiolytic (Warburton 1988). Possible mechanisms are discussed in chapter 8 (Stolerman, this volume).

3.5 Factors affecting nicotine responses

3.5.1 *Personality*

The aspect of individual differences that has been explored most, with respect to behaviour, is the dimension of introversion–extraversion. Eysenck (1957) suggested that individuals who score highly on extraversion should show a greater decrement in performance over time in monotonous tasks than individuals who have a low degree of extraversion. We have examined the relation between smoking, personality, and vigilance performance (Warburton and Wesnes 1978). The subjects in this experiment were non-smokers, light smokers and heavy smokers. They were given nicotine tablets of varying doses (0, 1 and 2 mg), or smoked cigarettes at 20, 40, and 60 min after the start of testing. Among smokers, there was no correlation between performance and extraversion. When all the data from smokers and non-smokers were combined, and then subdivided into a 'high'-extraversion group and 'low'-extraversion (i.e. 'high'-introversion) group about the mean, nicotine reduced the larger vigilance decrement in extraverts that occurred in the placebo condition ($p<0.05$), and had a small and insignificant effect on introverts ($p<0.1$). Given that smokers as a group are more extraverted than average, it is plausible to argue that some smokers use cigarettes to maintain performance in monotonous situations.

Effects on performance were also related to neuroticism. Prior to smoking, subjects with a high degree of neuroticism performed less well than emotionally stable subjects. When performance during the 20 min after smoking was examined, improvement on the task was correlated significantly with the neuroticism scores i.e. smoking was producing greater improvement in subjects with a higher degree of neuroticism (Warburton and Wesnes 1978). Results consistent with these findings were obtained by Kucek (1975), where subjects were tested under conditions of information overload. Kucek (1975) also found that smoking had a beneficial effect on the performance of neurotic subjects who were allowed to smoke. Similar results were also found using nicotine tablets (Warburton and Wesnes 1978), with subjects having higher degrees of neuroticism being improved most by nicotine.

One interpretation of this improvement (Warburton and Wesnes 1978) is that the attentional effects of nicotine enabled the filtering out of distracting thoughts which interfered with performance. However, a number of studies of anxiolytic drugs have found that subjects scoring high on anxiety (which correlates with neuroticism) are improved in performance tasks by doses of anxiolytics (for example, Desai *et al.* 1983). Given the previous evidence for an anxiolytic action of nicotine, our results would be consistent with these findings.

In summary, attempts to relate the psychopharmacological effects of smoking and nicotine to individual differences have rarely been successful. A major problem is that personality questionnaires assess self-reported lifestyle. Although it is believed that a person's lifestyle is based on underlying constitution, it is modulated by cultural factors.

3.5.2 *Desensitization and tolerance*

It is appropriate to return here to the issue of desensitization and tolerance, as defined on the first page of this chapter. Human studies (Benowitz *et al.* 1982) have shown desensitization when nicotine is infused intravenously. Heart rate increased rapidly and reached a maximum at 5 min after the beginning of the infusion, and it remained constant despite continued infusion. When infusion was terminated, the plasma nicotine concentration decreased, but a given concentration was associated with a lower heart rate during the elimination phase than during the infusion (see Benowitz *et al.*, Fig 4.14, this volume). This finding is consistent with the development of partial desensitization of the heart rate response to nicotine.

In the same study, Benowitz *et al.* (1982) also measured skin temperature during infusion, and also after infusion was terminated. Skin temperature decreased gradually during infusion, but returned to baseline in a direct relationship to the plasma nicotine concentration after infusion. In other words, there was no evidence of desensitization of the skin temperature response. Given the differences for these two physiological systems, what are the changes in the psychological system with short interval dosing with nicotine, or with chronic dosing?

The evidence against the occurrence of desensitization with nicotine comes from studies with minimal deprivation. In the introduction to this chapter, minimal deprivation was defined as abstinence for 3 h or less, in contrast to the more usual deprivation of 10 h or more in studies of smoking, or nicotine administration, and behaviour. The shorter the deprivation interval, the better the evidence, but obviously zero deprivation is unsuitable because of the consequent chances of a nicotine 'overdose'.

Performance improvements have been found in vigilance tests with deprivation intervals of 2 h (Mangan 1982), and 3 h (Tong *et al.* 1977; see section 3.2.3.1). On the mood tests, improvements have been found with 0.5 h deprivation (Pomerleau *et al.* 1984; Warburton *et al.* 1988; see section 3.4.1). Also relevant to this discussion is the study of West and Jarvis (1986). They tested subjects with nasal nicotine solution administered at 1 h intervals for 6 h. At 1 h plasma nicotine levels would only have fallen to 60 per cent of their peak values. Yet on each occasion the nasal nicotine solution produced similar increments in finger tapping.

Tolerance to the behavioural effects of nicotine has been discussed by Jarvik (1979). He points out that tolerance does occur to some effects of nicotine in animals, with daily dosing, for example, locomotor activity described in chapter 8 (see Stolerman, this volume).

In order to investigate the importance of tolerance, we designed an experiment (Wesnes *et al.* 1983; see section 3.2.3.1) which studied the effects of nicotine tablets (0, 1 or 2 mg) on non-smokers, light smokers (less than 5 per day), and heavy smokers (more than 15 per day), who were all undergraduates. During the task the subjects took nicotine or placebo tablets at 20, 40, and 60 min. There was no difference in the effects of nicotine on smokers and non-smokers, and no difference in the effect of nicotine on performance of light and heavy smokers (see Fig. 3.2).

In another study, Wesnes and Warburton (1976) compared the sensitivity of heavy (more than 15 cigarettes per day) smokers and non-smokers to nicotine, using the Stroop test (section 3.2.3.1). Using doses of 0, 0.1, and 2.0 mg of nicotine tablets, a significant dose-dependent reduction in the Stroop effect was found, but there was no difference in the effects of nicotine on smokers and non-smokers. It is clear that the improvements in non-smokers could not have been due to nicotine withdrawal. In addition, there was no difference between smokers who had heavy (15 cigarettes or more per day) daily exposure to nicotine or light (5 cigarettes or less per day) exposure. It is worth noting that West and Jarvis (1986) also obtained improvements in the finger-tapping rate of non-smokers given nicotine.

In the study of Murphree (1979; section 3.4.1), nicotine was infused in smokers and non-smokers until the heart rate increased by 15 beats min^{-1}, i.e. equivalent to a smoking dose. The EEG effect of nicotine on alpha activity, at 10.0 cycles s^{-1}, showed a fairly clear cut alpha blockade, and the alpha activity remained suppressed until the end of the experiment, 20 min later. The effect of nicotine on 1.0 cycle s^{-1} EEG activity (delta activity) in deprived smokers and non-smokers was small. Murphree (1979) states:

> 'The conclusion from this work is that by several physiologic measures, smokers and nonsmokers responded in closely similar ways to intravenous nicotine. This suggests that smokers as a group do not have significant pharmacologic tolerance to nicotine.' (p. 237).

In summary, some physiological responses, like heart rate, show desensitization to the effects of nicotine, but others, like skin conductance, do not (Benowitz *et al.* 1982). Similarly, increased activity in rats shows tolerance to nicotine. However, there is no evidence from the studies discussed in this section for desensitization and tolerance for the mood changes and performance improvements.

These findings of differences in tolerance to the same compound are not unique to nicotine. For example, tolerance is found for most of the psychological effects of morphine, including the euphoria, and so chronic users escalate their dose to obtain the desired effects. Yet at 50 times the analgesic dose, users still show marked constipation and pupillary constriction (Rang and Dale 1987). We must be aware that demonstrations of tolerance for one biological system do not necessarily have implications for other systems.

3.5.3 *Withdrawal symptoms*

Mangan and Golding (1984) described what they call the addiction models for smoking which propose that exposure to nicotine from cigarette smoke has induced a change in the smoker's body so that the person needs nicotine to function normally. This need can only be satisfied by further cigarette use, and failure to maintain 'nicotine homeostasis' will result in withdrawal symptoms. This addiction model for smoking has been stated most clearly by Frith (1971), and Schachter (1978). Schachter (1978, pp. 212–3) has written:

> One can, obviously, account for the generally debilitating effects of no or low nicotine by assuming that the deprived smoker is in withdrawal but this assumption alone cannot account for the effects of stress on smoking rate unless one assumes that stress, in some fashion, depletes the available supply of nicotine. And this hypothesis, of course, can account for this pattern of data only if it is the case that the smoker, an addict, is smoking to keep nicotine at a constant level.
>
> Another way of phrasing this same conclusion is that the heavy smoker gets nothing out of smoking. He smokes only to prevent withdrawal. I freely admit that this is a perverse conclusion to reach about a habit that is quite as costly and universally pervasive as smoking but the existing data for humans don't encourage any other conclusion.

This is obviously a recasting of the traditional model of opiate use in which exposure has produced tolerance, and abstinence results in a withdrawal syndrome. The consensus view of the American Medical Association Task Force was that a withdrawal syndrome could be defined as follows: 'The onset of a predictable constellation of signs and symptoms involving altered activity of the central nervous system after the abrupt discontinuation of, or rapid decrease in, dosage of drug' (Rinaldi *et al.* 1988). The question raised by the data on performance and mood improvements is whether exposure to nicotine has altered the neural system, so that a deprived smoker suffers withdrawal symptoms which result in poorer psychological function. Consequently, when a smoker receives nicotine from the cigarette, it is only restoring normal function.

Some support for this position comes from studies which have demonstrated differences in the electrocortical activity of deprived smokers and non-smokers (for example, Ulett and Itil 1969; Knott and Venables 1977). There are also two reports of poorer performance of deprived smokers compared with non-smokers (Heimstra *et al.* 1980; Knott 1984), but other studies which were cited earlier found no differences (see also Jaffe, chapter 1, this volume).

In an attempt to examine the issue further for this chapter I have compared the performance of non-smokers with smokers after 10 h deprivation, a common abstinence interval in the smoking and performance studies. This comparison of 100 young (mean age 19 years) smokers and 100 similar non-smokers revealed no difference in correct-detection probability (smokers' mean 0.86 and non-smokers' mean 0.84, $p > 0.58$). Thus, we have no reason to believe that the performance improvements which are obtained after the deprived smokers smoke a cigarette are only a result of a return to the performance level of non-smokers who have had little exposure to nicotine.

Persuasive evidence against the withdrawal hypothesis comes from the results of animal studies with nicotine, reviewed by Clarke (1987). It has been found that attentional performance (for example, Nelsen and Goldstein 1972), discrimination performance (for example, Geller *et al.* 1971), and memory are improved by nicotine (for example, Battig 1970; see section 3). Obviously, animals have had little exposure to nicotine and the improvements in their performance could not have been due to the reversal of nicotine withdrawal.

Reports of poorer intellectual function and negative mood changes after smoking cessation have been reported, but it is not an invariable consequence of cessation (Shiffman 1979). An alternative interpretation could be that any differences which do occur may merely represent a return to an original constitutional level of functioning (Warburton and Wesnes 1978; Mangan and Golding 1984). As DSM-III (American Psychiatric Association 1987) points out, it is difficult to know whether these represent true withdrawal or a return to a previous state. In fairness, Schachter (1978) did acknowledge this possibility:

> Rather than indicating withdrawal, it is conceivable that people who become smokers are by nature more frightened of shock, more irritated by noise and worse drivers than people who never become smokers, and that for such people smoking is indeed calming and does improve psychomotor performance. (p. 210).

3.6 Conclusions

In the absence of any strong evidence for physiological adaptation to nicotine, a more useful model for understanding nicotine use is a functional model. The functional model of nicotine (Warburton 1987, 1988) views a person's nicotine use as a control of their psychological state.

Nicotine is a unique substance in that it combines performance enhancement with mood-changing properties. Individual nicotine users may have different motives and may have more than one motive for nicotine use. Thus nicotine use may have different functions for different individuals on different occasions. In the functional view, the reasons for nicotine use are purposive, a coping activity for everyday needs. Nicotine use is maintained because it satisfies these needs. The functional model argues that nicotine use is not an irrational behaviour but is adopted because it enables the user to control their psychological state effectively. It can be seen as a resource that the person uses. The functional model explains why individuals experience abstinence symptoms when they stop nicotine use, and why the symptoms vary between individuals depending on the function that it served for that individual.

Acknowledgements

My special thanks to Clifford Arnall, Maureen Kew, Peggy Oberheim, and Amanda Todd, for manuscript preparation and table collation.

References

APA (American Psychiatric Association) (1987). *Diagnostic and Statistical Manual of Mental Disorders*. American Psychiatric Association, Washington, DC.

Andersson, K., and Hockey, G.R.J. (1977). Effects of cigarette smoking on incidental memory. *Psychopharmacologia* **52**, 223–6.

Armitage, A.K., Hall, G.G., and Sellers, C.M. (1969). Effects of nicotine on electrocortical activity and acetylcholine release from the cat cerebral cortex. *British Journal of Pharmacology* **35**, 152–60.

Ashton, H., and Stepney, R. (1982). *Smoking: psychology and pharmacology*. Cambridge University Press, Cambridge.

Bakan, P. (1959). Extraversion–introversion and improvement in an auditory vigilance task. *British Journal of Psychology* **50**, 325–32.

Bättig, K. (1970). The effect of pre- and post- trial application of nicotine on the 12 problems of the Hebb-Williams test in the rat. *Psychopharmacologia* **18**, 68–76.

Benowitz, N.L., Jacob, P., Jones, R.T., and Rosenberg, J. (1982). Interindividual variability in the metabolism and cardiovascular effects of nicotine in man. *Journal of Pharmacology and Experimental Therapeutics* **221**, 368–72.

Bobon, D.P., Ott, H., and Holmberg, H. (1982). *Critical flicker fusion frequency in man: validity, reliability and sensitivity in psychopharmacology*. Stuttgart: Georg Thieme Verlag, Stuttgart. (Also in *Pharmacopsychiatria*, **15**, 1–62).

Bond, A., and Lader, M. (1974). The use of analogue scales in rating subjective feelings. *British Journal of Medical Psychology* **47**, 211–8.

Bovet, D., Bovet-Nitti, F., and Oliverio, A. (1966). Effects of nicotine on avoidance conditioning of inbred strains of mice. *Psychopharmacologia* **10**, 1–5.

Bovet-Nitti, F. (1969). Facilitation of simultaneous visual discrimination by nicotine in four inbred strains of mice. *Psychopharmacologia* **14**, 193–9.

Callaway, E., Halliday, R., Naylor, H., and Schechter G. (1985). Effects of oral scopolamine on human stimulus evaluation. *Psychopharmacology* **85**, 133–8.

Caille, E.J., and Bassano, J-L (1976). La cigarette dans le champ cerebrale. *Psychologie Medicale* **8**, 893–908.

Celesia, G.G., and Jasper, H.H. (1966). Acetylcholine released from the cerebral cortex in relation to state of activation. *Neurology* **16**, 1053–63.

Cherek, D.R. (1981). Effects of smoking different doses of nicotine on human aggressive behavior. *Psychopharmacology* **75**, 339–45.

Clarke, P.B.S. (1987). Nicotine and smoking: a perspective from animal studies. *Psychopharmacology* **92**, 135–43.

Cotten, D.J., Thomas, J.R., and Stewart, D. (1971). Immediate effects of cigarette smoking on simple reaction time of college male smokers. *Perceptual and Motor Skills* **33**, 336.

Desai, N., Taylor-Davies, A., and Barnett, D. (1983). The effects of diazepam and oxyprenolol on short term memory in individuals of high and low state anxiety. *British Journal of Clinical Pharmacology* **15**, 197–202.

Domino, E.F. (1965). Some behavioral actions of nicotine. In *Tobacco alkaloids and related compounds*. (U.S. Von Euler. ed.), pp. 145–62. Pergamon Press, Oxford.

Donchin, E. (1984). Dissociation between electrophysiology and behavior – a disaster or a challenge? In: Donchin, E. (ed.) *Cognitive psychophysiology* Lawrence Erlbaum, Hillsdale, New Jersey.

Edwards, J., and Warburton, D.M. (1983). Smoking, nicotine and electrocortical activity. *Pharmacology and Therapeutics* **19**, 147–64.

Edwards, J.A., Wesnes, K., Warburton, D.M., and Gale, A. (1985). Evidence of more rapid stimulus evaluation following cigarette smoking. Addictive Behaviors **10**, 113–26.

Erickson, C.K. (1971). Studies on the mechanism of avoidance facilitation by nicotine. *Psychopharmacologia* **22**, 357–68.

Evangelista, A.M., Gattoni, R.C., and Izquierdo, I. (1970). Effect of amphetamine, nicotine and hexamethonium on performance of a conditioned response during acquisition and retention trials. *Pharmacology* **3**, 91–6.

Eysenck H.J. (1957). *The dynamics of anxiety and hysteria*. Praeger, New York.

Flood, J.F., Landry, D.W., and Jarvik, M.E. (1981). Cholinergic receptor interactions and their effects on long term memory processing. *Brain Research* **215**, 177–85.

Frankenhaeuser, M., Myrsten, A.L., Post, B., and Johansson, G. (1971). Behavioral and physiological effects of cigarette smoking in a monotonous situation. *Psychopharmacologia* **22**, 1–7.

Frith, C.D. (1971). Smoking behavior and its relationship to the smoker's immediate experience. *British Journal of Social and Clinical Psychology* **10**, 73–8.

Garg, M. (1969). The effect of nicotine on two different types of learning. *Psychopharmacologia* **15**, 408–14.

Garg, M., and Holland, H.C. (1968). Consolidation and maze learning. A further study of post-trial injections of a stimulant drug (nicotine). *Neuropharmacology* **7**, 55–9.

Garg, M., and Holland, H.C. (1969). Consolidation and maze learning: a study of some strain/drug interactions. *Psychopharmacologia* **14**, 426–31.

Geller, I., Hartmann, R., and Blum, K. (1971). Effects of nicotine, nicotine monomethiodide, lobeline, chlordiazepoxide, meprobamate and caffeine on a discrimination task in laboratory rats. *Psychopharmacologia* **20**, 355–65.

Gilbert, D.G., and Hagen, R.L. (1980). The effects of nicotine and extraversion on self-report, skin conductance, electromyographic, and heart responses to emotional stimuli. *Addictive Behaviors* **5**, 247–57.

Guilford, J.S. (1966). *Factors related to successful abstinence from smoking*. American Institutes for Research, Pittsburgh.

Guillerm, R., Radziszewski, E., and Caille, J.E. (1978). Effects of carbon monoxide on performance in a vigilance task (automobile driving). In *Smoking behaviour: physiological and psychological influences* (R.E. Thornton, ed.) Churchill Livingstone, Edinburgh.

Halperin, M.H., McFarland, R.A., Niven, J.J., and Roughton, R.J.W. (1959). The time course of the effects of carbon monoxide on visual thresholds. *Journal of Physiology (London)* **146**, 583–93.

Heimstra, N.W., Fallensen, J.J., Kinsley, S.A., and Warner, N.W. (1980). The effects of deprivation of cigarette smoking on psychomotor performance. *Ergonomics* **23**, 1047–55.

Heimstra, N.W., Bancroft, N.R., and Dekok, A.R. (1967). Effects of smoking upon sustained performance in a simulated driving task. *Annals of the New York Academy of Science* **142**, 295–307.

Hockey, G.R.J., and Hamilton P. (1983). The cognitive patterning of stress states. In *Stress and fatigue in human performance* (G.R. Hockey, ed. Chichester, Wiley.

Houston, J.P., Schneider, N.G., and Jarvik, M.E. (1978). Effects of smoking on free-recall and organization in short and long-term memory. *American Journal of Psychiatry* **135**, 220–2.

Hughes, J.R., Hatsukami, D.K., Pickens, R.W., Krahn, D., Malin, S., and Luknic, G. (1984). Effect of nicotine on the tobacco withdrawal syndrome. *Psychopharmacology* **83**, 82–7.

Hull, C. (1924). The influence of tobacco smoking on mental and motor efficiency. *Psychology Monographs* **33**, (150).

Hunter, B., Zornetzer, S.F., Jarvik, M.E., and McGaugh, J.L. (1977). Modulation of learning and memory. In Handbook of Psychopharmacology, Vol. 8 L.L. Iversen, S.D. Iversen, and S.H. Snyder, ed.), pp. 531–77. Plenum Press, New York.

Jarvik, M.E. (1979). Tolerance to the effects of tobacco. In *Cigarette Smoking as a Dependence Process* (N.A. Krasnegor, ed.) National Institute for Drug Abuse, Washington, D.C.

Jarvis, M.J., Raw, M., Russell, M.A.H., and Feyerabend, C. (1982). Randomised controlled trial of nicotine chewing gum. *British Medical Journal* **285**, (6341), 537–40.

Kawamura, H., and Domino, E.F. (1969). Differential actions of m and n cholinergic agonists on the brainstem activating system. *International Journal of Neuropharmacology* **8**, 105–15.

Knott, V.J. (1984). Noise and task induced distraction effects on information processing: sex differences in smokers and non-smokers. *Addictive Behaviors* **9**, 79–84.

Knott, V.J., and Venables P.H. (1977). EEG alpha correlates of non-smokers, smoking and smoking deprivation. *Psychophysiology* **14**, 150–6.

Kucek, P. (1975). Effect of smoking on performance under load. *Studia Psychologica* **17**, 204–12.

Kunsendorf, R., and Wigner, L. (1985). Smoking and memory: state-specific effects. *Perceptual and Motor Skills* **61**, 558.

Larsson, P.S., Haag, H.B., and Silvette, H. (1961). *Tobacco: experimental and clinical studies*, Williams & Wilkins Co, Baltimore.

Larsson, P.S., Finnegan, J., and Haag, H.B. (1950). Observations on the effect of cigarette smoking on the fusion frequency of flicker. *Journal of Clinical Investigation* **29**, 483–6.

Luchins, R. (1942). Mechanization in problem solving. *Psychological Monographs*, 54.

Lyon, R.J., Tong, J.E., Leigh, G., and Clare, G. (1975). The influence of alcohol and tobacco on the components of choice reaction time. *Journal of Studies on Alcohol* **36**, 587–96.

Mackworth, N.H. (1950). *Researches on the measurement of human performance*. Medical Research Council Special Report, Number 268. HMSO, London.

Mangan, G.L. (1982). The effects of cigarette smoking on vigilance performance. *Journal of General Psychology* **106**, 77–83.

Mangan, G.L. (1983). The effects of cigarette smoking on verbal learning and retention. *Journal of General Psychology* **108**, 203–10.

Mangan, G., and Golding, J. (1978). An 'enhancement' model of smoking maintenance? In *Smoking behaviour: physiological and psychological influences* (R.E. Thornton, ed.) pp. 115–26. Churchill-Livingstone, Edinburgh.

Mangan, G.L., and Golding, J.F. (1983). The effects of smoking on memory consolidation. *Journal of Psychology* **115**, 65–77.

Mangan, G.L., and Golding, J.F. (1984). *The psychopharmacology of smoking*. Cambridge University Press, Cambridge.

McKennell, A.C. (1973). *A comparison of two smoking typologies*. Research paper number 12. Tobacco Research Council, London.

McKennell, A.C. (1970). Smoking motivation factors. *British Journal of Social and Clinical Psychology* **9**, 8–22.

Morgan, S.F., and Pickens, R.W. (1982). Reaction time performance as a function of cigarette smoking procedure. *Psychopharmacology* **77**, 383–6.

Murphree, H.B. (1979). EEG effects in humans of nicotine, tobacco smoking, withdrawal from smoking and possible surrogates. In *Electrophysiological effects of nicotine* (A. Remond and C. Izard, ed.), pp. 227–44. Elsevier - North Holland Biomedical Press, Amsterdam.

Myrsten, A-L., and Andersson, K. (1978). Effects of cigarette smoking on human performance. In *Smoking behavior: physiological and psychological influences* (R.E. Thornton, ed.) Churchill Livingstone, Edinburgh.

Nelsen, J.M., and Goldstein L. (1972). Improvement of performance on an attention task with chronic nicotine treatment in rats. *Psychopharmacologia* **26**, 347–60.

Nesbitt, P.D. (1973). Smoking, physiological arousal, and emotional response. *Journal of Personality and Social Psychology* **25**(1), 137–44.

Ostfeld, A.M., and Aruguette, A. (1962). Central nervous system effects of hyoscine in man. *Journal of Pharmacology and Experimental Therapeutics* **137**, 133–9.

Peeke, S.C., and Peeke, H.V.S. (1984). Attention, memory and cigarette smoking. *Psychopharmacology* **84**(2), 205–16.

Peters, R., and McGee, R. (1982). Cigarette smoking and state-dependent memory. *Psychopharmacology* **76**, 232–5.

Pomerleau, O.F., Turk, D.C., and Fertig, J.B. (1984). The effects of cigarette smoking on pain and anxiety. *Addictive Behaviors*, **9**, 265–271.

Pratt, J.A., Stolerman, I.P., Garcha, H.S., Giardini, V., and Feyerabend, C. (1983). Discriminative stimulus properties of nicotine: further evidence for mediation at a cholinergic receptor. *Psychopharmacology* **81**, 54–60.

Rang, H.P., and Dale, M.M. (1987). *Pharmacology*. Churchill Livingstone, Edinburgh.

Rinaldi, R.C., Steindler, E.M., Wilford, B.B., and Goodwin, D. (1988). Clarification and standardization of substance abuse terminology. *Journal of American Medical Association* **259**, 555–7.

Robinson, S.E. (1985). Cholinergic pathways in the brain. In *Central cholinergic mechanisms and adaptive dysfunctions* (M.M. Singh, D.M. Warburton and H. Lal, ed.). Plenum Press, New York.

Robustelli, F. (1963). Azione delle nicotina sull'apprendimento dell ratto nel labirinto. *Rendiconti Accademia Nazionale Dei Lincei Classe di Scienze fisiche, matematic e naturali, VIII* **34**, 703–9

Russell, M.A.H., Peto, J., and Patel, U.A. (1974). The classification of smoking by factorial structure of motives. *Journal of the Royal Statistical Society A* **137**, 313–33.

Schachter, S. (1978). Pharmacological and psychological determinants of smoking. In: Thornton, R.E. (Editor). *Smoking behaviour* (R.E. Thornton, ed.), pp. 208–28.

Schechter, M.D. and Rand, M.J. (1974). Effect of acute deprivation of smoking on aggression and hostility. *Psychopharmacology* **35**, 19–28.

Schneider, N.G. (1986). Use of 2 mg and 4 mg gum in an individual treatment trial. In *The pharmacologic treatment of tobacco dependence: proceedings of the world congress, November 4–5, 1985*, (J.K. Ockene, ed.), (pp. 233–48). Institute for the Study of Smoking Behavior and Policy, Cambridge, MA.

Shiffman, S.M. (1979). The tobacco withdrawal syndrome. In *Cigarette smoking as a dependence process* (N.A. Kransnegor, ed.). National Institute for Drug Abuse, Washington DC.

Shiffman, S.M. and Jarvik, M.E. (1976). Smoking withdrawal symptoms in two weeks of abstinence. *Psychopharmacology* **50**, 35–9.

Silverstein, B. (1977). An addition explanation of cigarette-induced relaxation. *Dissertation Abstracts International* **37**, 76–17, 863, Ann Arbor, Michigan.)

Silverstein, B. (1982). Cigarette smoking, nicotine addiction, and relaxation. *Journal of Personality and Social Psychology*, **42**, 946–950.

Smith, D.L., Tong, J.E., and Leigh, G. (1977). Combined effects of tobacco and caffeine on the components of choice reaction-time; heart rate, and hand steadiness. *Perceptual and Motor Skills* **45**, 635–9.

Stroop, J.R. (1935). Studies of interference in serial verbal reactions. *Journal of Experimental Psychology* **18**, 643–61.

Tarriere, C. and Hartemann, F. (1964). Investigation into the effects of tobacco smoke on a visual vigilance task. *Ergonomics* **525**, 530.

Tomkins, S.S. (1966). Psychological model of smoking behaviour. *American Journal of Public Health*, **56**, 17–20.

Tomkins, S.S. (1968). A modified model of smoking behaviour. In *Smoking, health and behaviour, 165–6* (E. Borgatta and R. Evans, ed.). Aldine, Chicago.

Tong, J.E. and Leigh, G. (1975). A comment on critical flicker fusion as an objective measure of alcohol tolerance by alcoholics. *Journal of Nervous and Mental Disease* **161**, 343–4.

Tong, J.E., Leigh, G., Campbell, J., and Smith, D. (1977). Tobacco smoking, personality and sex factors in auditory vigilance performance. *British Journal of Psychology* **68**, 365–70.

Ulett, J.F. and Itil, T.M. (1969). Quantitative electroencephalogram in smoking and smoking deprivation. *Science* **164**, 969–70.

Waller, D. and Levander, S. (1980). Smoking and vigilance. The effects of tobacco smoking on CPF as related to personality and smoking habits. *Psychopharmacology* **70**, 131–6.

Warburton, D.M. (1977). Stimulus selection and behavioural inhibition. In *Handbook of psychopharmacology* (L.L. Iversen, S.D. Iversen and S.H. Snyder, ed.), pp. 385–431. Plenum Press, New York.

Warburton, D.M. (1981). Neurochemical bases of behaviour. *British Medical Bulletin* **37**, 121–5.

Warburton, D.M. (1986). A state model for mental effort. In *Energetics and human information processing* (G.R.J. Hockey, A.W.K. Gaillard, and M.G.H. Coles, ed.), pp. 217–32. Martinus Nijhof, Dordrecht. 232.

Warburton, D.M. (1987). The functions of smoking. In *Tobacco smoke and nicotine: a neurobiological approach* (W.R. Martin, G.R. Van Loon, E.T. Iwanmoto, and D.L. Davis, ed.), pp. 51–62. Plenum Press, New York.

Warburton, D.M. (1988). The puzzle of nicotine use. In *The psychopharmacology of addiction* (M. Lader, ed.). Oxford University Press, Oxford.

Warburton, D.M. and Walters, A.C. (1988). Attentional processing. In *Smoking and human behaviour* (A. Ney and A. Gale, ed.). Wiley, Chichester.

Warburton, D.M., Revell, A., and Walters, A.C. (1988). Nicotine as a resource. In *International symposium on nicotine* (M.J.R and K. Thurau, ed.), pp. 359–373.

International Council of Scientific Unions, Press, USA.

Warburton, D.M. and Wesnes, K. (1978). Individual differences in smoking and attentional performance. In *Smoking behaviour* (R.E. Thornton, ed.), pp. 19-43. Curchill-Livingstone, Edinburgh.

Warburton, D.M. and Wesnes, K. (1979). The role of electrocortical arousal in the smoking habit. In *Electrophysiological effects of nicotine* (A. Remond and C. Izard, ed.), pp. 183–200. Elsevier - North Holland Biomedical Press.

Warburton, D.M. and Wesnes, K. (1982). Cholinergic mechanisms and attentional dysfunction. In *Biological Psychiatry* (C. Perris, G. struwe, and B. Jansson, ed.), pp. 778–81. Elsevier, Amsterdam.

Warburton, D.M. and Wesnes, K. (1985). Acetylcholine and attentional disorder. In *Central cholinergic mechanisms and adaptive dysfunctions* (M.M. Singh, D.M. Warburton and H.H. Lal, ed.), pp. 223–46. Pleonum Press, New York.

Warburton, D.M., Wesnes, K., Shergold, K., and James, M. (1986). Facilitation of learning and state dependency with nicotine. *Psychopharmacology* **89**, 55–9.

Warwick, K.M. and Eysenck H.J. (1963). The effects of smoking on the CFF threshold. *Life Sciences* **4**, 219–25.

Wesnes, K. and Revell, A. (1984). The separate and combined effects of scopolamine and nicotine on human information processing. *Psychopharmacology* **84**, 5–11.

Wesnes, K. and Warburton, D.M. (1978). The effect of cigarette smoking and nicotine tablets upon human attention. In *Smoking behaviour: physiological and psychological Influences* (R.E. Thornton, ed.), pp. 131–47. Churchill-Livingstone, Edinburgh.

Wesnes, K. and Warburton, D.M. (1983). Smoking, nicotine and human information processing performance. *Neuropsychobiology* **9**, 223–9.

Wesnes, K. and Warburton, D.M. (1984*a*). The effects of cigarettes of varying yield on rapid information processing performance. *Psychopharmacology* **82**, 338–42.

Wesnes, K. and Warburton, D.M. (1984*b*). Effects of scopolamine and nicotine on human rapid information processing performance. *Psychopharmacology* **82**, 147–50.

Wesnes, K., Warburton, D.M., and Matz, B. (1983). The effects of nicotine on stimulus sensitivity and response bias in a visual vigilance task. *Neuropsychobiology* **9** 41–4.

West, R. J., Jarvis, M. J., Russell, M. A. H., Carruthers, M., and Feyerabend, C. (1984). Effect of nicotine replacement on the cigarette withdrawal syndrome. *British Journal of Addiction* **79**(2), 215–9.

West, R.J. and Jarvis, M.J. (1986). Effects of nicotine on finger tapping rate in non-smokers. *Pharmacology, Biochemistry and Behavior* **25**, 727–31.

Williams, D. (1980). Effects of cigarette smoking on immediate memory and performance in different kinds of smoker. *British Journal of Psychology* **71**, 83–90.

Williams, D., Tata, P. R., and Miskella, J. (1984). Different types of cigarette smokers received similar effects from smoking. *Addictive Behaviours* **9**, 207–10.

Wittenborn, J.R. (1943). Factorial equations for tests of attention. *Psychometrica* **8**, 19–43.

Woodson, P.P., Buzzi, R., Nil, R., and Battig, K. (1986). Effects of smoking on vegetative reactivity to noise in women. *Psychophysiology* **23**(3), 272–82.

4 Pharmacokinetics, metabolism, and pharmacodynamics of nicotine

Neal L. Benowitz, Herve Porchet, and Peyton Jacob, III

4.1 Introduction

This chapter will review current knowledge about the pharmacokinetics, metabolism, and pharmacodynamics of nicotine, primarily in humans, and discuss the implications of these processes in determining dependence on nicotine and disease in humans.

4.2 Nicotine in various tobacco products

Nicotine is a tertiary amine composed of a pyridine and a pyrrolidine ring. Nicotine may exist as one of two stereoisomers. Tobacco contains only (S)-nicotine (also called (−) or *l*-nicotine), which is the most pharmacologically active form. Tobacco smoke also contains the less potent (R)-nicotine (also called (+) or *d*-nicotine) in quantities up to 10 per cent of the total nicotine present (Pool *et al.* 1985). Presumably, some racemization occurs in the combustion process.

The content of nicotine in tobacco products is the ultimate limiting step in self-dosing. Many other factors such as pyrolysis, filtration, and smoking behaviour determine the fraction of nicotine in cigarette tobacco which is delivered to the smoker. The content of nicotine in cigarette or other tobacco products is not specified by manufacturers or governmental agencies. Because tobacco is a plant product, there are differences in content of nicotine among and within different strains of tobacco, including variations in different parts of the plant. The nature of the soil in which the plant is grown may also influence the alkaloid content.

We measured the amount of nicotine in the tobacco of 15 American brands of differing machine-determined yields (Benowitz *et al.* 1983*a*). On average, tobacco contained 1.5 per cent nicotine by weight, and lower-yield cigarettes tended to have higher concentrations than higher-yield cigarettes. However, lower-yield cigarettes contained less tobacco per cigarette, so the total content of nicotine per cigarette, averaging 8.4 mg, was similar in different brands. Thus low-yield cigarettes are not low-yield because they contain less nicotine per cigarette, but because of characteristics of the cigarette which remove tar and nicotine by filtration and/or dilution of smoke with air.

Concentrations of nicotine in other tobacco products are summarized in

Table 4.1. Nicotine content of various tobacco products

Product	Number of brands tested	Concentration of nicotine (mg g^{-1} tobacco)	Typical single dose (g tobacco)	Nicotine in single dose (mg)	Nicotine in dose typically consumed in a day
Cigarettes[a]	15	15.7(13.3–26.9)*	0.54	8.4	168 mg per 20 cigs
Moist snuff[b,c]	8	10.5(6.1–16.6)	1.4	14.5	157 mg per 15 gm
Chewing tobacco[c,d,e]	2	16.8(9.1–24.5)	7.9	133	1176 mg per 70 gm
Pipe tobacco[e]	1	11.3	—	—	—
Cigar tobacco[e]	1	13.9	—	—	—

* – Range
a – Benowitz *et al.* 1983*a*
b – Kozlowski *et al.* 1982
c – Gritz *et al.* 1981
d – Benowitz *et al.* 1988
e – Benowitz and Jacob, unpublished

Table 4.1. Nicotine gum (Nicorette) is a pharmaceutical product in which nicotine is bound to an ion-exchange resin in a gum base. Nicotine is released as the gum is chewed. The gum is available in 2 mg and 4 mg preparations in Canada and Europe, and 2 mg in the United States. Although the major alkaloid in tobacco is nicotine, there are several alkaloids which may be pharmacologically important. These include nornicotine, anabasine, myosmene, nicotyrine and anatabine. These make up 8–12 per cent of the total alkaloid content of tobacco products (Piade and Hoffmann 1980). In some varieties of tobacco, nornicotine concentrations exceed those of nicotine (Schmeltz and Hoffmann 1977). Nornicotine and anabasine have pharmacological activity qualitatively similar to that of nicotine, with potencies of 20–75 per cent depending on the test system and the animal (Clark *et al.* 1965; see Stolerman, Table 8.1, this volume). In addition to direct activity, some of the minor alkaloids may interact with effects of nicotine. For example, nicotyrine inhibits the metabolism of nicotine (Stalhandski and Slanina 1982) in animals. Little is known about the importance of these alkaloids in determining the effects of tobacco consumption in people.

4.3 Absorption of nicotine

Nicotine is distilled from burning tobacco and is carried proximally on tar droplets (mass median diameter 0.3–0.5 μm) which are inhaled. Absorption of nicotine across biological membranes depends on pH (Armitage and Turner 1970; Schievelbein *et al.* 1973). Nicotine is a weak base with a pKa of 8.0 (aqueous solution, 25°C). In its ionized state, such as in acidic environments, nicotine does not rapidly cross membranes. The pH of smoke from flue-cured tobaccos, found in most cigarettes, is acidic (pH 5.5). At this pH, the nicotine is primarily ionized. As a consequence, there is little buccal absorption of nicotine from cigarette smoke, even when it is held in the mouth (Gori *et al.* 1986). The pH of smoke from air-cured tobaccos, the predominant tobacco in pipes, cigars, and in a few European cigarettes, is alkaline (pH 6.5 or higher), and nicotine is primarily un-ionized. Smoke from these products is well-absorbed through the mouth (Armitage *et al.* 1978; Russel *et al.* 1980*b*).

When tobacco smoke reaches the small airways and alveoli of the lung, the nicotine is rapidly absorbed independent of pH of the smoke. Blood concentrations of nicotine rise quickly during and peak at the completion of cigarette smoking (Fig. 4.1). Armitage and co-workers (1975), measuring exhalation of radiolabeled nicotine, found that 82–92 per cent of nicotine in mainsteam smoke was absorbed by four habitual smokers, 29 per cent by another habitual smoker who was presumed to be a non-inhaler, and 30–66 per cent by three non-smokers (who were instructed to smoke as deeply as possible). The rapid absorption of nicotine from cigarette smoke through the lung is presumably because of the huge surface area of the alveoli and small airways, and dissolution of nicotine into fluid of pH in the human physiologic range, which facilitates transfer across cell membranes.

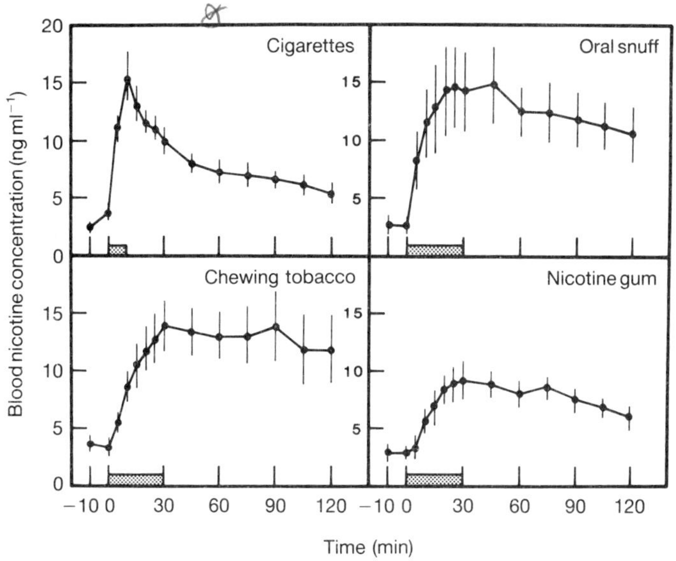

Fig. 4.1. Blood nicotine concentrations during and after cigarette smoking for 9 min (1-1/3 cigarettes), oral snuff (2.5 g), chewing tobacco (average 7.9 g), and nicotine gum (two 2 mg pieces). Average values for 10 subjects (± S E M). Shaded bars above time axis indicate period of tobacco or gum exposure.

Chewing tobacco, snuff, and nicotine gum are buffered to alkaline pH to facilitate absorption of nicotine through mucous membranes. Concentrations of nicotine in the blood rise gradually with use of smokeless tobacco, and tend to plateau at about 30 min with levels persisting and declining only slowly over 2 h or more (Fig. 4.1).

Rates and time course of absorption have been studied in more detail by analysing blood concentration data using the point area deconvolution technique (Vaughan and Dennis 1978). This method applies a unit disposition function (which describes the concentration of drug over time following an intravenous bolus injection) to average blood concentrations, to generate an input rate (absorption) function. With cigarette smoking, absorption of nicotine occurs only during the 10 min of smoking as expected (Fig. 4.2). With use of moist oral snuff or chewing tobacco, the absorption of nicotine accelerates quickly in the first 10 min, but then declines over 30 min despite the continued presence of tobacco in the mouth. Absorption continues at a slower rate for up to 60 min after the tobacco has been removed from the mouth. Possibly, continued absorption of nicotine is due to release of nicotine from mucous membranes and/or absorption of nicotine which had been swallowed. Similar blood level time courses have been described for the use of smokeless tobacco in sachets (such as Skoal Bandits) (Russell *et al.* 1985). Finely ground nasal snuff is sniffed into the nose, from which absorption may be more rapid

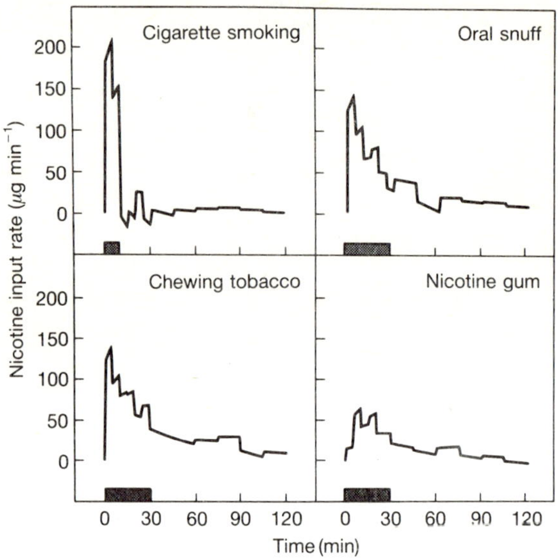

Fig. 4.2. Nicotine absorption rate profiles, estimated by the point-area deconvolution method, during and after cigarette smoking, oral snuff, chewing tobacco, and nicotine gum. Shaded bars above time axis indicate period of tobacco or nicotine gum exposure.

than through buccal mucosa, with resultant blood levels that more closely resemble those after cigarette smoking (Russell *et al.* 1981; see Russell, Fig. 11.2, this volume).

Absorption of nicotine from nicotine gum is gradual with evidence of continued release throughout the entire period of chewing (Figs. 4.1 and 4.2). Blood levels of nicotine are considerably lower, even when chewing two pieces of 2 mg gum, than with the use of tobacco. Extraction of nicotine from nicotine gum is incomplete, averaging 53 per cent and 72 per cent for 2 mg and 4 mg gum, respectively, and highly variable (more than twofold) among individuals (Benowitz *et al.* 1987*a*). In addition, considerable nicotine from the gum is swallowed. For these reasons the systemic dose of nicotine is much less than the amount of nicotine contained in the gum.

Nicotine is poorly absorbed from the stomach due to the acidity of gastric fluid (Travell 1960), but is well absorbed in the small intestine (Jenner *et al.* 1973), which has a more alkaline pH and a large surface area. Following the administration of nicotine in capsules, peak concentrations are reached in just over 1 h (Fig. 4.3). The oral bioavailability of nicotine is about 45 per cent. Bioavailability is incomplete because of first-pass metabolism. Nicotine base can be absorbed through the skin and there have been cases of poisoning after skin contact with pesticides containing nicotine (Benowitz *et al.* 1987*b*; Faulkner 1933). Likewise, there is evidence of cutaneous absorption of nicotine in tobacco field workers (Gehlback *et al.* 1975).

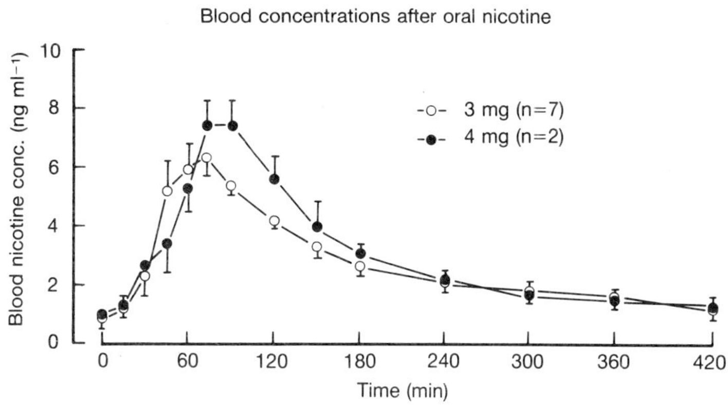

Fig. 4.3. Blood nicotine concentrations (± S E M) following ingestion of capsules containing 3 or 4 mg nicotine base (given as bitartrate salt).

Because of the complexity of the processes of cigarette smoking and use of smokeless tobacco products, the dose of nicotine cannot be predicted from the nicotine content of the tobacco or its absorption characteristics. To determine the dose, one needs to measure blood levels, and also to know how fast the individual eliminates nicotine. This topic, that is, estimation of systemic doses of nicotine consumed from various tobacco products, will be considered in a later section after discussion of relevant pharmacokinetic issues.

4.4 Distribution of nicotine in body tissues

After absorption, nicotine enters the bloodstream where, at pH 7.4, it is about 69 per cent ionized and 31 per cent un-ionized. Binding to plasma proteins is less than 5 per cent (Benowitz *et al.* 1982*a*). The drug is distributed extensively to body tissues with a steady state volume of distribution averaging 2.6 times body weight (Table 4.2). This means on average, at steady state, the concentrations of nicotine in body tissues is 2.6 times that of blood. The

Table 4.2. Human pharmacokinetics of nicotine and cotinine

	Nicotine	Cotinine
Half-life	120 min	18 h
Volume of Distribution	180 l	88 l
Total Clearance	1300 ml min^{-1}	72 ml min^{-1}
Renal Clearance	200 ml min^{-1} (acid urine)	12 ml min^{-1}
Nonrenal clearance	1100 ml min^{-1}	60 ml min^{-1}

Average values based on data from Benowitz *et al.* 1982*a*, and Benowitz *et al.* 1983*b*.

Table 4.3. Steady State Distribution of Nicotine[a]

Tissue	Tissue : Blood ratio	
	Rabbit	Rat
Blood	1.0	1.0
Brain	3.0	2.8
Heart	3.7	1.9
Muscle	2.0	1.6
Adipose	0.5	0.5
Kidney	21.6	13.3
Liver	3.7	3.5
Lung	2.0	2.1
Gastrointestinal	3.5	2.2

a Tissue to blood nicotine concentration ratios based on 24-h constant intravenous infusion of nicotine in rabbits (2.5 μg kg^{-1} min^{-1}, n = 4), and 72-h constant subcutaneous infusion via osmotic minipumps in rats (2.8 μg kg^{-1} min^{-1}, n = 3; 5.6 μg kg^{-1} min^{-1}, n = 5). Steady-state concentrations of nicotine in blood averaged 63 ng ml^{-1} in rabbits, and 112 ng ml^{-1} and 361 ng ml^{-1} in rats with low and high infusion rates, respectively

pattern of tissue uptake cannot be studied in humans, but we have, by measuring concentrations of nicotine in various tissues after infusion of nicotine to steady state, examined tissue uptake in rats and rabbits (Table 4.3). Liver, lungs, and brain have high, and adipose relatively low, affinity for nicotine.

After rapid intravascular injection, concentrations of nicotine decline rapidly due to tissue uptake of the drug. Simulation of tissue levels of nicotine after injection into the lung (simulating smoking) is possible using a perfusion model with partition coefficients derived from animal studies (Fig. 4.4). Immediately after intrapulmonary injection, concentrations in arterial blood, lung, and brain are high, while concentrations in tissues such as muscle and adipose (major storage tissues at steady state) are low (Fig. 4.5). The consequence of this distribution pattern is that uptake into the brain is rapid, occuring within 1 or 2 min of peak arterial concentrations, and blood levels fall due to peripheral tissue uptake for 20 or 30 min after administration. Thereafter, blood concentrations decline more slowly, the rate determined by rates of elimination and rates of distribution out of storage tissues.

Fig. 4.5. Perfusion model simulation of the distribution of nicotine in various tissues following infusion of 1.5 mg of nicotine into the lung compartment over 10 min. The dose, route, and dosing regimen were intended to mimic smoking a cigarette. The perfusion model used is shown in Fig. 4.4. Typical human organ weights and blood flows, and partition coefficients from rabbits, were used for the simulation.

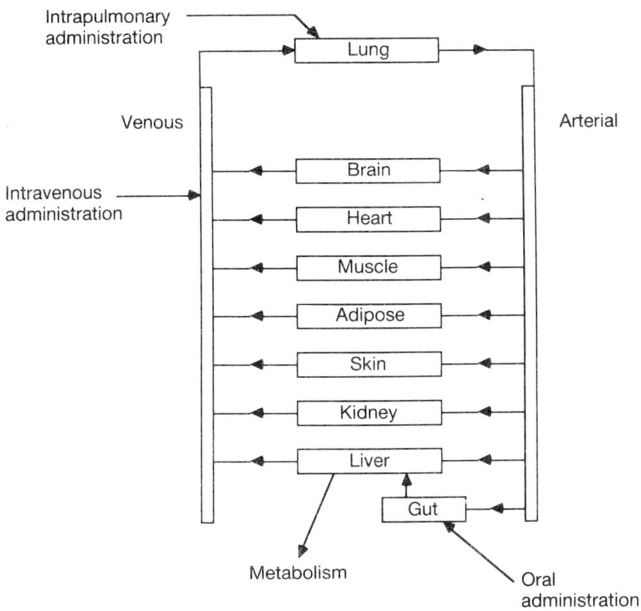

Fig. 4.4. A blood perfusion model used to simulate the disposition kinetics of nicotine in humans. Boxes represent tissue capacitance, the product of mass and partition coefficients. Arrows represent blood flows to individual organs. For details of the modelling procedure, see Benowitz *et al* (1974).

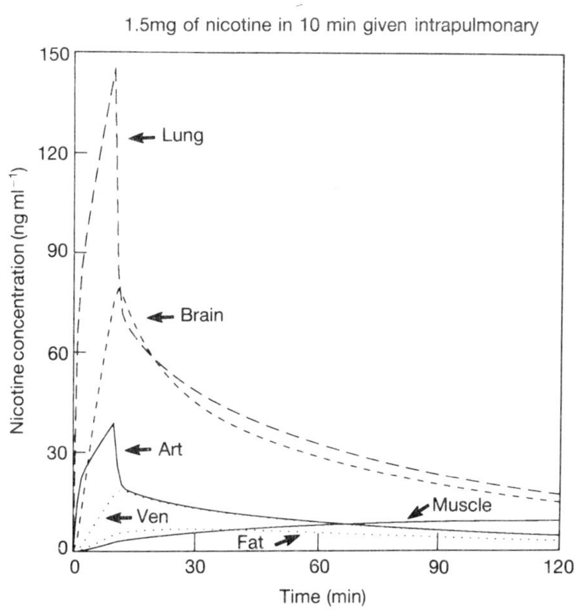

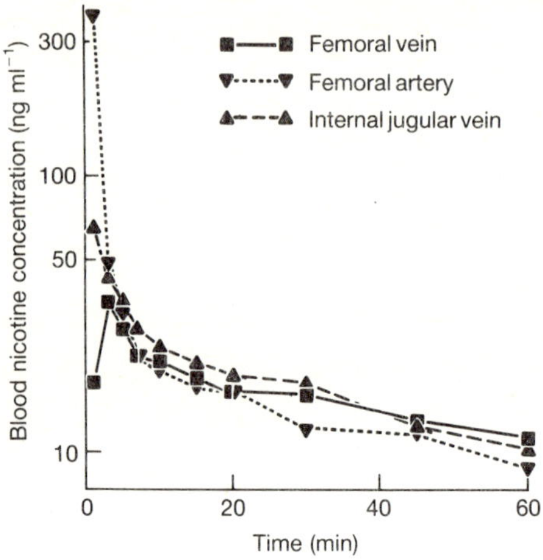

Fig. 4.6. Average blood nicotine concentrations in 6 rabbits following an intravenous bolus of 25 μg kg^{-1} nicotine. The arterial and internal jugular nicotine concentrations are much higher than those in the femoral venous blood during the first 5 min. The internal jugular concentration (reflecting brain concentration)–time curve crosses the arterial concentration–time curve at 4 min, and the femoral venous curve at 8 min, indicating that equilibration between brain and arterial blood is much faster than between peripheral tissue (venous blood) and arterial blood (from Porchet *et al.* 1987).

Detailed analysis of the time course of nicotine concentrations in arterial, peripheral venous, and jugular venous blood (the latter reflecting brain concentrations) shows that levels of nicotine are initially extremely high in arterial blood, with rapid equilibration with brain tissues but slower equilibration with systemic venous blood (Fig. 4.6) (Porchet *et al.* 1987). This is important because most pharmacodynamic studies relate venous nicotine levels to effects. However, when considering nicotine actions on the brain, venous levels may peak later and at a lower level compared to those in the brain. Disequilibrium between venous and brain levels explains, at least in part, acute development of tolerance to actions of nicotine after intravenous injection (see section 4.10 for further discusson of acute tolerance).

Smoking represents an exposure similar to that of rapid intravenous injection, except that the point of entry into the circulation is through the pulmonary rather than the systemic venous circulation. As a consequence of delivery into the lung, peak levels of nicotine may be higher, and lag time between smoking and entry into the brain shorter than after intravenous injection. After smoking, the action of nicotine upon the brain would be expected to occur quickly. Rapid onset of effects after a puff is believed to provide an optimal reinforcement schedule for the development of drug dependence. The effect of nicotine declines as nicotine is distributed to other tissues, as discussed above. The distribution half-life, about 8 min, rather than the elimination half-life, about 2 h, determines the time course of central nervous system actions of nicotine after smoking a single cigarette. With regular cigarette smoking throughout the day, the elimination half-life may be a determinant of the kinetics of tolerance regression.

Nicotine is secreted into saliva (Russell and Feyerabend 1978). Passage of saliva containing nicotine into the stomach, combined with a process of trapping of nicotine in the acidic gastric fluid and reabsorption from the small bowel, provides a potential route for enteric recirculation of nicotine. This recirculation has been speculated to account for some of the oscillations in the terminal decline phase after intravenous infusion of nicotine or cessation of smoking (Russell 1976).

Nicotine is excreted into urine and may, depending on urinary pH, be reabsorbed through the walls of the urinary bladder. When the urine is acidic (pH is less than 6), as is usually the case, nicotine is ionized and there is little reabsorption (Borzelleca 1963). From alkaline urine, there may be significant reabsorption of nicotine, particularly where there is urine stasis. Alkaline urine as well as stasis may be seen in the presence of urinary-tract infections with certain urea-splitting bacteria.

Nicotine freely crosses the placenta and has been found in amniotic fluid and in umbilical cord blood of neonates (Hibberd *et al.* 1978; Luck *et al.* 1982; Van Vunakis *et al.* 1974). The amniotic fluid provides a reservoir for continued delivery of nicotine to the foetus, even when maternal levels are low (Inokuchi *et al.* 1989). Prolonged exposure of the foetus to nicotine might contribute to slowed growth of the foetus *in utero* and resultant small birth weight, which is well documented in babies of smokers. Nicotine is found in breast milk and breast fluid of non-lactating women (Petrakis *et al.* 1978; Hill and Wynder 1979). The concentration of nicotine in breast milk is such that the dose of nicotine consumed by the infant is small and unlikely to have physiologic consequence. Relative concentrations of nicotine in various body fluids are shown in Table 4.4.

4.5 Elimination of nicotine

Nicotine is extensively metabolized, primarily in the liver, but also to a small extent in the lung and kidney (Turner *et al.* 1975). Renal excretion depends on urinary pH and urine flow, typically accounting for 5–10 per cent of total elimination.

4.5.1 *Metabolism*

4.5.1.1 *Pathways of metabolism* The primary metabolites of nicotine are cotinine and nicotine-*N*-oxide (Fig. 4.7). Cotinine is formed in a two-step

Table 4.4. Nicotine and cotinine concentrations in body fluids of smokers

Body Fluid	N	Nicotine concentration ($ng\ ml^{-1}$)	Nicotine ratio body fluid to plasma or blood	Cotinine concentration ($ng\ ml^{-1}$)	Cotinine ratio body fluid to plasma or blood
Saliva					
Feyerabend *et al.* (1982)	82	567[a] (2–2000)	—	—	—
Haley *et al.* (1983)	12	—	—	361 (± 80)	1.47
Jarvis *et al.* (1984)	94	672	45.4	310	1.13
Urine					
Feyerabend *et al.* (1982)		1278 (5–8000)	—	—	—
Jarvis *et al.* (1984)	94	1750	118.2	1391	5.05
Wald *et al.* (1984*a*)	131	—	—	1645 (537–3326)[d]	—
Amniotic Fluid					
Luck and Nau (1984)[b]	29	(1.5–23)	1.5 (± 0.27)	(5–188)	0.72 (± 0.15)
Van Vunakis *et al.* (1974)[c]	37	(ND–31)	—	(ND–129)	—
Umbilical Venous Blood					
Luck and Nau (1984)	27	(0.5–25)	1.12 (± 0.30)	(15–233)	1.05 (± 0.27)

Breast Milk					
Luck and Nau (1984)	23	(2–62)	2.9 (± 1.1)	(12–222)	0.8 (± 0.21)
Luck and Nau (1987)	10 (1–10 CPD)[e]	18 (± 16)	—	76 (± 33)	—
	13 (21–40 CPD)	48 (± 25)	—	230 (± 62)	—
Breast Milk, non-lactating					
Petrakis *et al.* (1978)	4	(46–195)	—	—	—
Hill & Wynder (1979)	5	40.6 (20–56)	1.05 (± 0.31)	169 (111–271)	0.46 (± 0.16)
Cervical Fluid					
Sasson *et al.* (1985)	10	740 (66–2620)	67.2 (± 91)	316 (161–773)	1.41 (± 0.82)

a — Where available, mean values are given; numbers in parentheses are range or SD.
b — 16–24 weeks gestation
c — 14–20 weeks gestation; ND = not detected
d — (10–90 percentile)
e — CPD-Number of cigarettes smoked per day

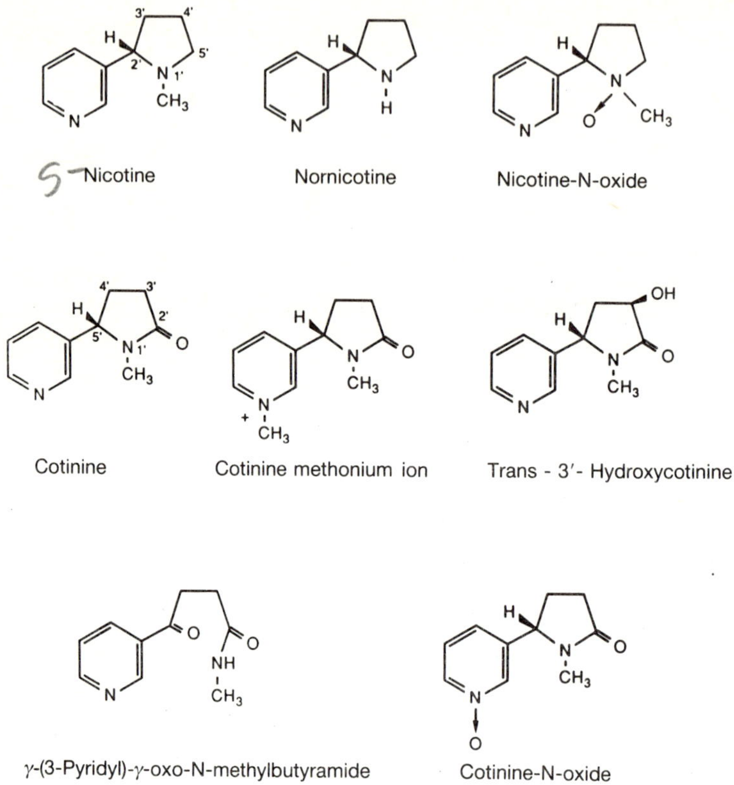

Fig. 4.7. Chemical structures of nicotine and its major metabolites.

process, the first of which involves oxidation of the five position of the pyrrolidine ring in a cytochrome P-450 mediated process, to nicotine-Δ1′(5′)-iminium ion (Peterson *et al.* 1987). The latter is then metabolized by a cytoplasmic aldehyde oxidase to cotinine (Hibberd and Gorrod 1983).

Cotinine itself is also extensively metabolized, with only about 17 per cent excreted unchanged in the urine (Benowitz *et al.* 1983*b*). Several metabolites of cotinine have been reported, including *trans*-3′-hydroxycotinine (McKennis *et al.* 1963*b*), 5′-hydroxycotinine (Bowman and McKennis 1962), cotinine-*N*-oxide (Shulgin *et al.* 1987), and cotinine methonium ion (McKennis *et al.* 1963*a*). Little is known about the quantitative importance of these metabolites. *Trans*-3′-hydroxycotinine appears to be a major metabolite (Jacob *et al.* 1988*b*; Neurath *et al.* 1987), with urinary concentrations exceeding cotinine concentrations by 2–3 fold. We have found that continine *N*-oxide is a minor metabolite in humans, accounting for approximately 3 per cent of ingested nicotine (Shulgin *et al.* 1987). Further oxidative degradation of the pyrrolidine ring gives rise to 3-pyridyl-acetic acid. This compound has been indentified in human urine (McKennis *et al.* 1964), but no quantitative data are available.

Nicotine-1′-*N*-oxide is a minor metabolite of nicotine. Oxidation of the nitrogen atom of the pyrrolidine ring is dependent upon a microsomal flavoprotein system, and produces a mixture of the two diasteriomers, 1′-(R)-2′-(S)-*cis*- and 1′-(S)-2′-(S)-*trans*-nicotine-1′-*N*-oxide (Booth and Boyland 1970). After intravenous injection, 100 per cent of nicotine-*N*-oxide is excreted unchanged in the urine, indicating no further metabolism (Beckett *et al.* 1971*b*). However, after oral administration only 30 per cent is recovered in the urine as nicotine-*N*-oxide; the remainder is recovered as nicotine and its metabolites. After rectal administration, less than 100 per cent is recovered as nicotine-*N*-oxide (Beckett *et al.* 1970). These findings indicate reduction of nicotine-*N*-oxide back to nicotine within the gastrointestinal tract, believed to be a consequence of bacterial action.

Although this review considers primarily the human pharmacology of nicotine, there are important species differences in the metabolism of nicotine which should be noted. Jenner and Gorrod (1973) compared the metabolism of nicotine *in vitro* in several rodent species using liver 10 000 × g hepatic supernatant preparations. Liver homogenates from rabbits and hamsters produced more cotinine than nicotine *N*-oxide. About equal amounts of cotinine and nicotine *N*-oxide were produced from incubation of nicotine with mouse liver preparations, but twice as much nicotine *N*-oxide as cotinine was formed in incubations of rat liver homogenates. Since cotinine formation is quantitatively more important than nicotine *N*-oxide formation in humans, rabbits and hamsters would be more suitable than mice and rats for studies designed to study the metabolic profile of nicotine as it might relate to humans.

Quantitative aspects of the conversion of nicotine to its metabolites have not been well-defined. Studies of excretion of cotinine in urine, collected for 24 h after intravenous injection of nicotine, indicate a range from less than 10 per cent in non-smokers to an average of 25 per cent in smokers (Beckett *et al.* 1971*a*). Another study, comparing 24 h urinary excretion of cotinine with the nicotine content of cigarette butts after smoking, indicated 46 per cent recovery as cotinine (Schievelbein 1982). However, both of these studies underestimate the conversion of nicotine to cotinine because the urine collection period, 24 h, was too short. Cotinine has a half-life averaging 20 h (Benowitz *et al.* 1983*a*), so that in 24 h only a little more than half of cotinine will be recovered. Urine collection for at least 72 h is necessary to recover more than 90 per cent of cotinine in most subjects. In addition, since only 17 per cent of cotinine is excreted unchanged (Benowitz *et al.* 1983*a*), urinary recovery underestimates cotinine generation rate.

To investigate quantitative aspects of conversion of nicotine to its metabolites, we measured urinary excretion of nicotine and its metabolites, cotinine,

Table 4.5. Urinary excretion of cotinine-*N*-oxide and other nicotine metabolites in 25 cigarette smokers[a]

	Nicotine (μg per 24 h)	Nicotine *N*-oxide (μg per 24 h)	Cotinine (μg per 24 h)	Cotinine *N*-oxide (μg per 24 h)
Mean	2522	1078	2883	716
S D	1827	686	1539	370
Range	212–5786	135–3192	441–6196	68–1536

a From Shulgin *et al.* 1987

nicotine-*N*-oxide, and cotinine-*N*-oxide, in 24 h urine collections in habitual smokers (Table 4.5). Assuming steady state, the rate of excretion of metabolites reflects generation rate. 100 per cent of nicotine-*N*-oxide but only 17 per cent of cotinine is excreted unchanged in the urine. Based on the ratio for cotinine to nicotine-*N*-oxide of 2.9, the relative generation rate of cotinine compared to nicotine-*N*-oxide is 17 to 1. We and Beckett *et al.* (1971*b*) have found that about 4 per cent of nicotine is excreted as nicotine-*N*-oxide. Therefore, about 70 per cent of nicotine appears to be converted to cotinine. Our current estimate of quantitative disposition of nicotine is shown in Fig. 4.8. Of note is that 17 per cent of nicotine is not accounted for by metabolism to cotinine or nicotine-*N*-oxide. Other metabolites, such as nicotine isomethonium ion and nornicotine, which may have pharmacological activity, need to be investigated as candidates for the missing 17 per cent.

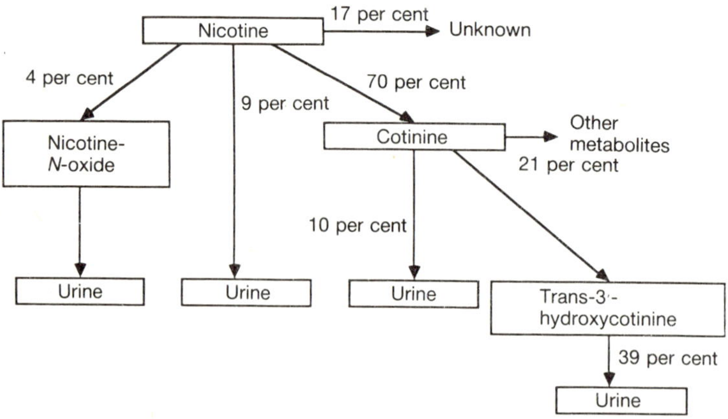

Fig. 4.8. Quantitative disposition of nicotine, based on analysis of 24-h urine collections in 25 habitual cigarette smokers.

4.5.1.2 *Quantitative aspects of nicotine metabolism* The rate of metabolism of nicotine can be determined by measuring blood levels after administration of a known dose of nicotine. We studied habitual cigarette smokers given intravenous infusions of nicotine for 30–60 min (Benowitz *et al.* 1982*a*). Total and renal clearance were computed directly, and the non-renal or metabolic clearance computed as the difference between total and renal clearance. Total clearance averaged 1292 ml min^{-1}. Non-renal clearance averaged 1089 ml min^{-1} (Table 4.2). Non-renal clearance represents about 70 per cent of liver blood flow. Assuming most nicotine is metabolized by the liver (data in animals indicate only a small contribution by the lung) (Turner *et al.* 1975), this means about 70 per cent of the drug is extracted from the blood in each pass through the liver.

The metabolism of cotinine is much slower than that of nicotine (Table 4.2). Total clearance averages 72 ml min^{-1} and metabolic clearance 60 ml min^{-1} (Benowitz *et al.* 1983*a*). As it is slowly metabolized, the rate of elimination of cotinine is predicted not to be substantially influenced by changes in liver blood flow. Pharmacokinetic data from animal species are summarized in Table 4.6.

4.5.1.3 *Factors influencing the rate of nicotine metabolism* The level of any drug in the body during repeated dosing depends on the rate of intake and the rate of elimination of the drug. There is considerable individual variability in rate of elimination of nicotine among individuals (Benowitz *et al.* 1982*a*). In pharmacokinetic terms, clearance is a number which predicts steady-state blood level of a drug achieved at a particular level of intake. The relationship is expressed mathematically as $C_{ss} = R/Cl$, where C_{ss} is the steady state blood concentration, R is the rate of intake and Cl is clearance. Since there may be fourfold inter-individual differences in clearance of nicotine, a person with an extremely high clearance will have blood and body nicotine levels one-fourth as high as a person consuming the same amount of nicotine but having a low clearance. If people smoke to achieve a particular body level of nicotine, then clearance could influence self-determined rate of cigaratte smoking. Differences in clearance among people could result in different exposures to and risks from other combustion products. For this reason, factors which influence the rate of metabolism of nicotine may be important determinants of cigarette smoking.

Cigarette smoking itself may influence the rate of metabolism of nicotine. Cigarette smoking is known to accelerate the metabolism of many drugs (Dawson and Vestal 1982). One report indicates that clearance of nicotine is slower in non-smokers compared to habitual cigarette smokers (Kyerematen *et al.* 1982), which is consistent with the expected effects of cigarette smoking to accelerate drug metabolism. However, we have found that in habitual smokers, brief (1–2 weeks) abstinence is associated with an increased rather

Table 4.6. Pharmacokinetics of S-nicotine in animals

Species	Body weight (kg)	Dose (mg kg^{-1})	Terminal half-life (h)	Volume of Distribution	Total clearance
Rat (Miller *et al.* 1977)	0.17–0.34	0.4	0.92	5.2 l kg^{-1a}	65 ml min^{-1} kg^{-1}
Rabbit (Brady *et al.* 1979)	3.1–4.1	3.58	0.90	6.9 l^{b}	112 ml min^{-1}
Rabbit (Jacob *et al.* 1987)	3–4	0.18 (infused over 60 min)	1.3	2.9 l kg^{-1a}	57 ml min^{-1} kg^{-1}
Dog (Monji and Castro 1979)	12.2 (average)	0.1	1.2	9.5 l^{b}	183 ml min^{-1}
Dog (Jacob *et al.* 1988*a*)	9.0–11.6	0.1	0.92	3.0 l kg^{-1b}	36 ml min^{-1} kg^{-1}

a = $Vd\beta$ = apparent volume of distribution
b = Vd_{ss} = steady state volume of distribution

than decreased clearance of nicotine (Lee *et al.* 1987*a*). Possible explanations for this observation include inhibition of metabolism by a metabolite of nicotine, or inhibition by another component of tobacco smoke, such as carbon monoxide or hydrogen cyanide. For an individual smoker, the rate of metabolism of nicotine appears to be determined by opposing factors, of general acceleration by chronic tobacco smoke exposure, and inhibition by some component of tobacco smoke.

Gender differences in drug metabolism have been reported for some drugs (Wilson 1984). In a small group of habitual heavy cigarette smokers, we found that clearance of nicotine corrected for body weight was significantly greater in men than women (Benowitz and Jacob 1984). In a study of 330 smokers, blood levels of nicotine were similar on average in men and women, while men smoked more cigarettes than women (Russell *et al.* 1980*c*). This finding is consistent with the prediction that men, because of more rapid clearance, need to consume more nicotine to achieve similar levels and effects compared to women.

An implication of the high degree of hepatic extraction is that clearance of nicotine should be dependent upon liver blood flow. That is, about 70 per cent of the volume of blood perfusing the liver should be cleared of nicotine. Thus, physiological events such as meals, posture, exercise, or other drugs, which perturb hepatic blood flow, are predicted to have significant effects on the rate of nicotine metabolism. Feyerabend and Russell (1978) noted a decline in plasma nicotine concentrations after lunch in one subject when smoking or chewing nicotine gum. We have recently found that a high protein meal consumed during a steady-state infusion of nicotine results in a small (12 per cent) but consistent decline in nicotine concentrations (Lee *et al.* 1989), the maximal effect seen 30 min after the end of the meal. The biological importance of this finding remains to be determined.

4.5.2 *Renal excretion*

Nicotine is excreted by glomerular filtration and tubular secretion, with variable reabsorption depending on urinary pH (Table 4.7). In acid urine, nicotine is mostly ionized and tubular reabsorption is minimized; renal clearance may be as high as 600 ml min^{-1} (urinary pH 4.4), depending on urine flow rate (Benowitz *et al.* 1983*a*; Rosenberg *et al.* 1980). In alkaline urine, a larger fraction of nicotine is unionized, allowing net tubular reabsorption with renal clearances as low as 17 ml min^{-1}(urine pH 7.0). With uncontrolled urine pH, renal clearance averages about 100 ml min^{-1}, accounting for the elimination of 10–15 per cent of the daily intake of nicotine. Because urinary pH is a determinant of elimination rate and, hence, body levels of nicotine, it has been suggested that urine pH may influence self-determined nicotine intake (Schachter 1978). This issue will be addressed in section 4.8.

Renal clearance of cotinine is much less than glomerular filtration rate (Benowitz *et al.* 1983*a*). Since cotinine is not appreciably protein bound, this

Table 4.7. Influence of urinary pH on renal clearance of cotinine and nicotine[a]

	Baseline[b] (mean ± SD)	Acidified urine (mean ± SD)	Alkalinized urine (mean ± SD)
Urine pH	5.8 ± 0.3	4.4 ± 0.4[c]	6.7 ± 0.3
Urine volume (ml)	2350 ± 720	2738 ± 514	2466 ± 725
Cotinine renal clearance (ml min^{-1})	12.3 ± 5.9	18.6 ± 10.0[c]	14.0 ± 10.4
Nicotine renal clearance (ml min^{-})	102 ± 41	598 ± 238[c]	43 ± 66[c]

a From Benowitz *et al.* (1983*b*)
b Baseline smoking condition when urine pH was uncontrolled
c Significantly different from baseline, $p < 0.05$

indicates extensive tubular reabsorption. Renal clearance of cotinine can be enhanced by up to 50 per cent with extreme urinary acidification (Table 4.7). Cotinine excretion is less influenced by urinary pH than nicotine because it is less basic and, therefore, is primarily in the unionized form within the physiological pH range. As for nicotine, the rate of excretion of cotinine is influenced by urinary flow rate. Renal excretion of cotinine is a minor route of elimination (averaging 17 per cent of the total clearance); most cotinine is metabolized. In contrast, 100 per cent of nicotine-*N*-oxide appears to be excreted unchanged in the urine (Beckett *et al.* 1971*b*). Since blood levels have not been determined in those studies, the clearance of nicotine-*N*-oxide cannot be computed.

4.6 Nicotine and cotinine blood levels during tobacco use

4.6.1 *Nicotine levels*

Blood or plasma nicotine concentrations sampled in the afternoon in smokers generally range from 10–50 ng ml^{-1}, although we have observed one patient with a level of over 100 ng ml^{-1}. The increment in venous blood nicotine concentration after smoking a single cigarette ranges from 5 to 30 ng ml^{-1}, depending on how the cigarette is smoked (Armitage *et al.* 1975; Herning *et al.* 1983; Isaac and Rand 1972). Blood levels peak at the end of smoking a cigarette and decline rapidly over the next 20 min due to tissue distribution (Fig. 4.1).

Peak blood levels of nicotine are similar, although the rate of rise of nicotine is slower, for cigar smokers, and users of snuff and chewing tobacco, compared to cigarette smokers (Armitage *et al.* 1978; Turner *et al.* 1977; Gritz *et al.* 1981; Russell *et al.* 1980*b*, 1981). Pipe smokers, particularly those who have previously smoked cigarettes, may have blood and urine levels of nicotine as high as cigarette smokers (McCusker *et al.* 1982; Turner *et al.* 1977; Wald *et al.* 1984*b*). Others who are primary pipe smokers, who have not previously smoked cigarettes, tend to have lower nicotine levels. Likewise, cigar smokers who have previously smoked cigarettes may inhale more deeply and achieve higher blood levels of nicotine than primary cigar smokers (Armitage *et al.* 1978), although on average, based on urinary cotinine levels, daily nicotine intake appears to be less for cigar as compared to cigarette or pipe smokers (Wald *et al.* 1981).

The earliest published studies of the elimination kinetics of nicotine reported half-lives of 20–40 min (Armitage *et al.* 1975; Isaac and Rand 1972). In those

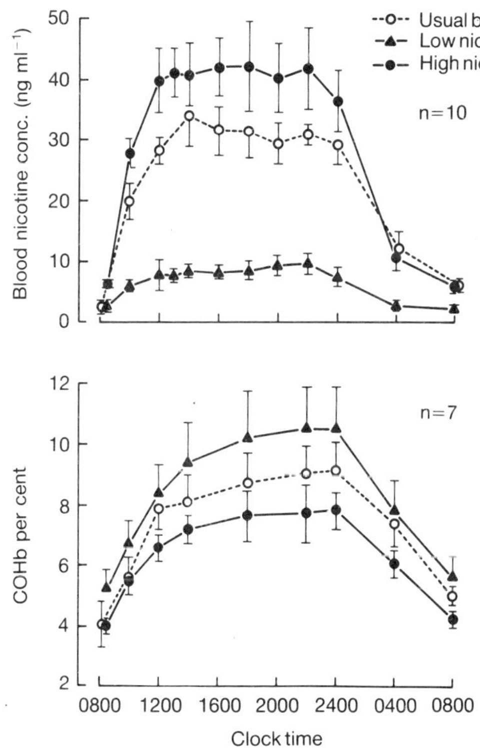

Fig. 4.9. Blood nicotine and carboxyhemoglobin concentrations (± SEM) in subjects smoking high-nicotine (2.5 mg), and low-nicotine (0.4 mg) Kentucky Reference cigarettes, and their usual brand (average nicotine yield 1.2 mg) of cigarettes. Subjects smoked on a fixed schedule of one cigarette every 0.5 h from 8.30 a.m. to 11.00 p.m. for a total of 30 cigarettes per day. Blood samples were collected just before the next scheduled cigarette (from Benowitz *et al.* 1982*b*, with permission).

studies blood levels were followed only for 30 to 60 min, which is not long enough to determine the terminal half-life. Thus, half-lives were based on blood levels which included the distribution phase. When blood levels are followed for several hours after the end of nicotine infusion, there is a log-linear decline phase with a half-life of about 2 h (Benowitz *et al.* 1982*a*).

The half-life of a drug is useful in predicting the rate of accumulation of that drug in the body with repetitive doses, and the time course of decline after cessation of dosing. Based on a half-life of 2 h for nicotine, one would predict accumulation over 6–8 h (3–4 half-lives) of regular smoking, and persistence of significant nicotine levels for 6–8 h after cessation of smoking. If a smoker smokes until bedtime, significant levels should persist all night. Studies of blood levels in regular cigarette smokers confirm these predictions (Fig. 4.9) (Benowitz *et al.* 1982*b*). Peaks and troughs follow each cigarette, but as the day progresses trough levels rise and the influence of peak levels becomes less important. Thus, nicotine is not a drug to which people are exposed inter-

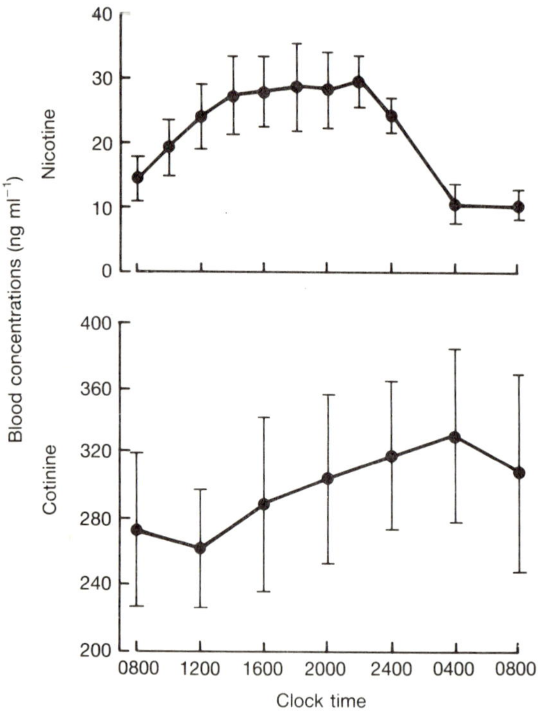

Fig. 4.10. Circadian blood concentrations of nicotine and cotinine during unrestricted cigarette smoking. There is much less fluctuation throughout the day in cotinine compared with nicotine concentrations. Data are shown as mean (± SEM) for 8 subjects (from Benowitz *et al.* 1983*b*, with permission).

mittently and which is eliminated rapidly from the body. To the contrary, smoking represents a multiple-dosing situation with considerable accumulation while smoking and persistent levels for 24 h of each day.

4.6.2 *Cotinine levels*

Cotinine is present in the blood of smokers in much higher concentrations than those in nicotine. Cotinine blood levels average about 250–300 ng ml^{-1} in groups of cigarette smokers (Benowitz *et al.* 1983*a*; Haley *et al.* 1983; Langone and Van Vunakis 1975; Zeidenberg *et al.* 1977). We have seen levels ranging from 10–900 ng ml^{-1}. After stopping smoking, levels decline in a log-linear fashion with a half-life averaging 20 h (range 11–37 h). Because of the long half-life, there is much less fluctuation in cotinine concentrations throughout the day than in nicotine concentrations (Fig. 4.10). As expected, there is a gradual rise in cotinine levels during the day, peaking at the end of smoking, and persisting in high concentrations overnight.

Because some studies have shown pharmacological activity of cotinine in animals, we studied subjective and cardiovascular actions of cotinine in humans (Benowitz *et al.* 1982*b*). Cotinine was infused in abstinent smokers to achieve levels averaging 387 ng ml^{-1}, similar to those found in heavy cigarette smokers. No subjective or cardiovascular effects were noted. At levels to which smokers are generally exposed, therefore, cotinine appears to exert little, if any, direct pharmacological effect.

4.7 Intake of nicotine from various nicotine-containing products

4.7.1 *Cigarette smoking*

Nicotine intake from single cigarettes has been measured by spiking cigarettes with radiolabelled nicotine (Armitage *et al.* 1975). That study of eight subjects, smoking a single filter-tipped cigarette, indicated an intake range of 0.36–2.62 mg. Intake was higher in smokers than in non-smokers. This study was an important first step, but the method is not applicable to measuring nicotine consumption in natural smoking conditions. We estimated daily intake of nicotine by a method similar to that used in drug bioavailability studies (Benowitz and Jacob 1984). Metabolic clearance of nicotine was determined after intravenous injection. Metabolic clearance data were then used in conjunction with blood and urinary concentrations of nicotine measured during a 24 h period of smoking to determine the daily intake of nicotine. This computation is based on the relationship $D = AUC_{\text{nic}} \times CL_{\text{total}}$, where D is the daily dose of nicotine, AUC_{nic} is the area under the blood-nicotine concentration time curve while smoking over 24 h, and CL_{total} is nicotine clearance based on non-renal clearance, determined from prior nicotine infusion, and renal clearance determined on the day of the smoking

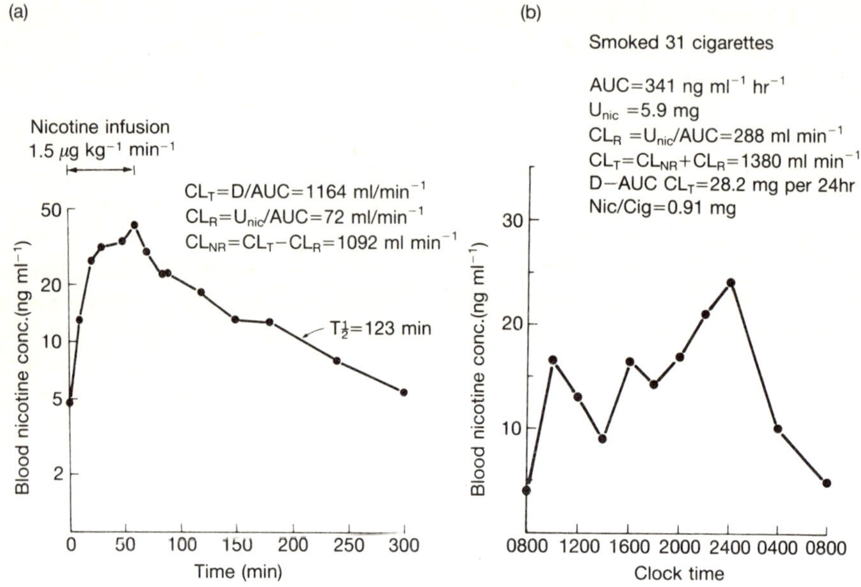

Fig. 4.11. Data and computations used to estimate daily intake of nicotine in one subject. (a) Blood concentrations of nicotine during and after nicotine infusion. (b) Blood concentrations of nicotine measured throughout the day during ad lib cigarette smoking (Benowitz and Jacob 1984, with permission).

study. An example of data and relevant analysis for one subject is shown in Fig. 4.11.

In 22 habitual cigarette smokers, 13 men and 9 women, who smoked an average of 36 (range 20–62) cigarettes per day, we found an average daily intake of 37.6 mg nicotine, with a range from 10.5–78.6 mg. Nicotine intake per cigarette averaged 1.0 mg (range 0.37–1.56 mg). Intake per cigarette did not correlate with smoking-machine predicted yield. However, on average smoking machines predicted a yield of 1 mg which was similar to that observed in the smokers. Using similar methods, Feyerabend *et al.* (1985) found a similar average intake of 1.06 mg per cigarette (range 0.58–1.49) in five subjects smoking six cigarettes over 3 h.

4.7.2 *Smokeless tobacco*

Assuming average disposition kinetics, and using the deconvolution technique described previously, we have estimated average nicotine intake during use of oral snuff and chewing tobacco (Benowitz *et al.* 1988). With 2.5 g of moist oral snuff held in the mouth for 30 min, the average intake was 3.6 mg. With the use of an average of 7.9 g chewing tobacco for 30 min, the average intake was 4.5 mg. Thus, per dose intake of nicotine is substantially (2–4 fold) greater while using smokeless tobacco than while cigarette smoking.

4.7.3 *Nicotine gum*

Using techniques similar to those described for cigarette smoking, daily intake of nicotine has been determined in people chewing nicotine polacrilex gum (Benowitz *et al.* 1987*a*). In subjects chewing 2 and 4 mg gum (7 subjects in each group), the daily systemic doses of nicotine were 10.2 ± 4.9 (SD) and 14.7 ± 5.5 mg while chewing 12 pieces per day. This is equivalent to an intake of 0.85 ± 0.33 and 1.22 ± 0.46 per piece of 2 mg and 4 mg gum, respectively. The systemically-absorbed dose was less than expected based on extraction from nicotine gum. Disproportionately higher levels of cotinine while chewing gum compared to smoking suggested that some nicotine was swallowed and underwent first-pass metabolism. Assuming that the bioavailability of nicotine by the oral route is 30 per cent, and using ratios of nicotine to cotinine while smoking and chewing nicotine gum, we estimated how much nicotine was absorbed through the buccal mucosa and how much was swallowed. Nicotine that could not be accounted for was assumed to be lost by expectoration of saliva (which was observed in many of the subjects chewing 4 mg gum). Of the nicotine extracted from the gum, 22 and 53 per cent from the 2 and 4 mg gum, respectively, were swallowed. 25 per cent of nicotine from the 4 mg gum was expectorated. The greater amount swallowed and expectorated with the 4 mg gum may be due to the greater saliva production with that product. A conception of typical absorption of nicotine from 2 and 4 mg gum is shown in Fig. 4.12.

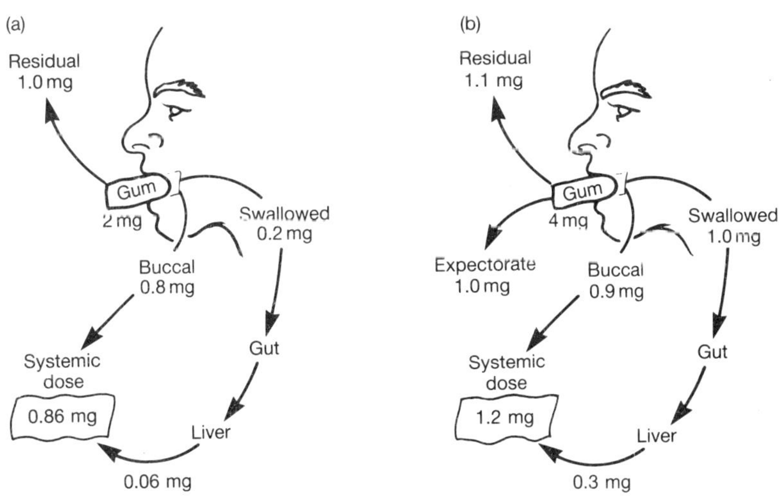

Fig. 4.12. Disposition of nicotine when chewing (a) 2 mg or (b) 4 mg nicotine polacrilex gum. Estimates are based on average values for seven subjects chewing 12 pieces of gum per day (see Benowitz *et al.*, 1987*a*).

4.8 Elimination rate as a determinant of nicotine intake by cigarette smoking

There is considerable evidence that smokers adjust their smoking behaviour to try to regulate or maintain a particular level of nicotine in the body (Gritz 1980; Russell 1976; see Russell, chapter 11, this volume). Techniques for measuring daily intake of nicotine (Benowitz and Jacob 1984) have been applied to study the influence of elimination on nicotine intake.

Rate of renal elimination of nicotine was manipulated by administration of ammonium chloride or sodium bicarbonate to acidify or alkalinize the urine, respectively (Benowitz and Jacob 1985). Compared to daily excretion during placebo treatment (3.9 mg nicotine per day), acid loading increased (to 12 mg per day) and alkaline loading decreased (to 0.9 mg per day) daily excretion of nicotine. Average blood nicotine concentrations were similar in placebo and bicarbonate treatment conditions, but were 15 per cent lower during ammonium chloride treatment. Daily intake of nicotine was 18 per cent higher during acid loading, indicating compensation for increased urinary loss. The compensatory increase in nicotine consumption was only partial, replacing about half the excess urinary nicotine loss.

These results would seem to support the suggestion of Schachter that emotional stress might increase cigarette smoking by resulting in more acidic urine, thereby accelerating nicotine elimination from the body (Schachter 1978). But caution must be exercised in applying our findings to usual smoking situations. Our studies were performed under conditions of extreme urinary acidification or alkalinization, so that the changes in renal clearance would be maximized. Even with extreme differences in urinary pH, differences in overall nicotine elimination rate and smoking behaviour were modest. This is because renal excretion is a minor pathway for elimination of nicotine; most is metabolized. Smaller changes in urinary pH, such as occur spontaneously throughout the day, or that might be related to stressful events, would not be expected to influence substantially nicotine elimination or smoking behaviour. Thus, the Schachter theory, while theoretically reasonable, appears in light of quantitative considerations not to be relevant to human smokers. Alkali therapy has been advocated as a way to slow the rate of nicotine elimination and to reduce smoking (Fix *et al.* 1983). Our data show that there is relatively little excretion of nicotine in baseline conditions (averaging 4 mg per day) and a decrease to 0.9 mg per day with alkali treatment has very little effect on body levels or smoking behaviour. Thus, bicarbonate therapy would be expected to have little, if any, effect on cigarette consumption. This prediction is supported by other experimental studies (Cherek *et al.* 1981).

The primary route of nicotine elimination is metabolism. In preliminary studies, we have found a significant correlation between metabolic clearance and daily intake of nicotine in smokers smoking their customary brands. There was no correlation between average nicotine blood levels and clearance. These observations support the idea that rapid metabolizers generally consume more nicotine (smoke more) than slow metabolizers, while maintaining similar nicotine blood levels and, presumably, similar effect.

4.9 Biochemical markers of nicotine intake

4.9.1 *Comparison of different markers*

Simple measures of nicotine intake are desirable for studies of smoking in the natural environment. Different investigators have used blood or urinary nicotine concentrations, blood, salivary, or urinary cotinine concentrations, expired carbon monoxide, or carboxyhemoglobin concentrations, or plasma or salivary thiocyanate concentrations as measures of tobacco consumption. We examined relationships between daily intake of nicotine, daily exposure to nicotine (that is, blood concentrations of nicotine integrated over 24 h), various parameters of cigarette consumption and different measures of nicotine intake (Benowitz and Jacob 1984).

The number of cigarettes smoked per day best correlated with nicotine intake. Correlations between intake, exposure, and various biochemical markers are shown in Table 4.8. The best marker of nicotine intake and exposure was a random blood nicotine concentration measured at 4 p.m. This level was drawn independently of when the last cigarette was smoked. This finding is consistent with the idea that nicotine levels accumulate throughout the day and plateau in the early afternoon (Fig. 4.9). At steady state, there should be a reasonably good correlation between nicotine concentrations and daily intake. Carboxyhemoglobin concentrations in the afternoon were the next best markers of nicotine intake. Morning (8 a.m.) levels of nicotine and carboxyhemoglobin also correlated with intake, presumably reflecting persistence of nicotine and carboxyhemoglobin in the blood from exposure from the day before.

Table 4.8. Correlation of various markers of tobacco smoke with daily intake of nicotine

Measure	Correlation (r)
Blood nicotine concentration at 1600 h	0.81*
Carboxyhemoglobin level at 1600 h	0.69*
Blood cotinine concentration at 1600 h	0.53*
Urinary cotinine excretion at 24 h collection	0.62*
Urinary nicotine excretion at 24 h collection	0.39

* $p < 0.01$; ** $p < 0.05$.
From Benowitz and Jacob 1984

Although cotinine is a highly specific marker for nicotine exposure, blood levels of cotinine were not as good a quantitative marker of nicotine intake as blood levels of nicotine or carboxyhemoglobin. This is presumably because of individual variability in fractional conversion of nicotine to cotinine, and in the rate of elimination of cotinine itself. However, for longitudinal within-subject studies, cotinine level would be expected to be a good marker of changes in nicotine intake.

As expected by the known variation in renal clearance due to effects of urinary flow and pH, urinary concentrations of nicotine did not correlate well with nicotine intake. In contrast, urinary cotinine is less influenced by urinary flow or pH, it correlates well with plasma concentrations ($r = 0.81$, Jarvis *et al.* 1984) and it is as good a marker as blood cotinine concentration.

Salivary cotinine concentrations also correlate well with blood cotinine concentrations ($r = 0.82$–0.90) (Haley *et al.* 1983; Jarvis *et al.* 1984). Therefore, salivary cotinine concentrations should be as useful as blood levels in indicating nicotine intake. However, cotinine is present in small amounts in tobacco and in tobacco smoke (Piade and Hoffmann 1980). Contamination of the mouth with cotinine from tobacco, particularly when using oral snuff or chewing tobacco, can potentially invalidate relationships between blood and salivary levels.

4.9.2 *Cotinine as a quantitative indicator*

Because cotinine has a long half-life, time of sampling with respect to smoking the last cigarette, or time of day, is less critical than for nicotine. Cotinine levels are also easier for most laboratories to measure. It would be useful to be able to translate cotinine levels, at least on average, into daily intake of nicotine. Such a conversion can be computed using data presented earlier for average clearance of cotinine and fractional conversion of nicotine to cotinine. At steady state, which is assumed in persons with relatively stable daily smoking habits, the daily elimination of cotinine is the same as the daily conversion of nicotine to cotinine:

$$COT_{\text{gen}} = COT_{\text{elim}} = F \times D_{\text{nic}} \qquad \text{(eqn. 4.1)}$$

where COT_{gen} and COT_{elim} are the amounts of cotinine generated and eliminated per day, respectively, F is the fractional conversion of nicotine to cotinine, and D_{nic} is the daily intake of nicotine. Cotinine elimination rate can be estimated as the product of average concentration (C_{cot}) and total body cotinine clearance:

$$COT_{\text{elim}} = \overline{C_{\text{COT}}} \times CL_{\text{COT}} \qquad \text{(eqn. 4.2)}$$

Combining and rearranging eqns 4.1 and 4.2,

$$D_{\text{nic}} = \frac{\overline{C_{\text{COT}}} \times CL_{\text{COT}}}{F} \qquad \text{(eqn. 4.3)}$$

Using our average data for cotinine clearance (72 ml min^{-1}) and $F(0.70)$,

$$D_{\text{nic}}\left(\frac{\text{mg}}{24\,\text{h}}\right) = 0.12 \times \overline{C_{\text{COT}}}\left(\frac{\text{ng}}{\text{ml}}\right) \qquad \text{(eqn. 4.4)}$$

Thus, an average blood concentration of cotinine of 100 ng ml^{-1} corresponds to an average 24 hr consumption of 12 mg nicotine. Galeazzi and colleagues (1985) have measured plasma cotinine concentrations directly during repeated intravenous dosing of nicotine. In six volunteer smokers, they found the relationship: Steady state cotinine concentration (ng ml^{-1}) = 0.783 × nicotine intake (μg kg^{-1} per day). With such data, a 70 kg smoker with a plasma cotinine concentration of 100 ng ml^{-1} would have an estimated intake of 9 mg per day. The typical smoker with a blood concentration of 300 ng ml^{-1} would be predicted to consume 27–36 mg nicotine per day. The latter estimate is in agreement with the 37.6 mg estimated in experimental studies (Benowitz and Jacob 1984).

4.10 Pharmacodynamics of nicotine

4.10.1 *General considerations*

The pharmacology of nicotine is reviewed by Clarke, chapter 5, and Stolerman, chapter 8, this volume. We will focus on the relationship between levels and effects of nicotine (pharmacodynamics), and how pharmacodynamic considerations may explain consequences of cigarette smoking in people. Two issues are particularly relevant in understanding the pharmacodynamics of nicotine.

Firstly, the dose–response relationship for nicotine is complex. In classical pharmacology, nicotine is commonly discussed as an example of a drug which in low doses causes ganglionic stimulation, and in high doses causes ganglionic blockade (following brief stimulation) (Comroe 1960). Dose-response characteristics *in vivo* are often biphasic as well, although the mechanisms are far more complex. For example, at very low doses, similar to those seen during cigarette smoking, cardiovascular effects appear to be mediated by the central nervous system, either via activation of chemoreceptor afferent pathways or via direct effects on the brain stem (Comroe 1960; Su 1982) (Fig. 4.13). The net result is sympathetic neural discharge with an increase in blood pressure and heart rate. At higher doses, there may be direct effects on the peripheral nervous system, such as ganglionic stimulation, with the release of adrenal catecholamines. With extremely high doses, there may be hypotension and heart rate slowing, mediated either by peripheral vagal activation, or direct depressor effects mediated by effects on the brain (Ingenito *et al.* 1972; Porsius and Van Zwieten 1978). Before extrapolating pharmacologic observations from animals to humans, blood concentrations should be measured to ensure that the effects are being studied in a portion of the dose–response curve related to smokers.

A second pharmacologic issue of importance is development of tolerance.

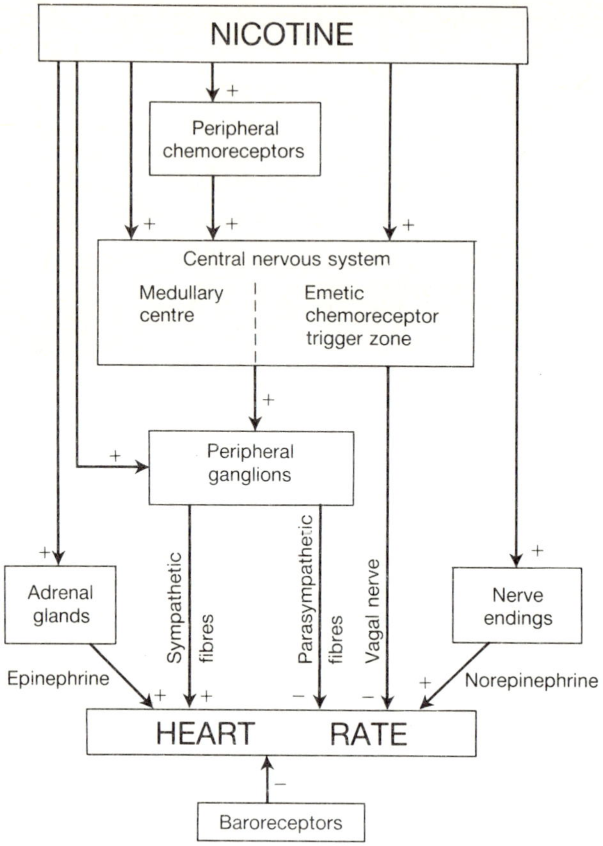

Fig. 4.13. Multiple sites at which nicotine may influence heart rate. Heart rate acceleration is indicated by +, deceleration by −.

Tolerance is usually defined as when, after repeated doses, a given dose of a drug produces less effect, or when increasing doses are required to achieve the effect observed with the first dose. We further define pharmacodynamic tolerance as when a particular concentration of a drug at a receptor site (in humans approximated by the concentration in blood) produces less of an effect than it did after a prior exposure. Pharmacokinetic tolerance refers to accelerated drug metabolism as a mechanism for diminished effect after repeated doses of a drug. Pharmacokinetic factors appear not to be of significance as a mechanism for development of tolerance to nicotine and will not be discussed further. Tolerance in subsequent portions of this text will refer to pharmacodynamic tolerance.

Smokers know that tolerance develops to some effects of smoking. Smoking the first cigarette as a teenager is commonly associated with dizziness, nausea, and/or vomiting, effects to which the habitual smoker rapidly becomes tolerant. In habitual smokers, smoking a cigarette after 24 h cigarette abstinence results in dizziness and other subjective effects, as well as acceleration of heart rate (West and Russell 1987). In contrast, smoking a cigarette in the afternoon of a normal smoking day results in neither subjective effects nor heart-rate acceleration, demonstrating tolerance within a normal day's smoking. Tolerance may even develop to toxic effects—nausea, vomiting, pallor—during the course of nicotine poisoning, despite persistence of nicotine in the blood in extremely high concentrations (200–300 ng ml^{-1}) (Benowitz *et al.* 1987*b*). Likewise, in *in vitro* and *in vivo* pharmacological studies in animals, tolerance to various effects develops rapidly, although tolerance may not be complete (Loffelholz 1970; Marks *et al.* 1985; Steinsland and Furchgott 1975; see Swedberg *et al*, chapter 2, this volume). Studies of the human pharmacology of nicotine necessarily take place in people who have a degree of tolerance prior to dosing with nicotine. Based on studies of other drugs, the extent of tolerance is likely to depend on both the level of nicotine, the duration of exposure to a given level, and the rate of increase of nicotine exposure to a particular organ (Kalant *et al.* 1971). In designing pharmacological studies in man, it is essential to consider recent exposure to nicotine, and the degree of tolerance which may be present at the time nicotine is administered. Individual differences in degree of tolerance may be important determinants of individual differences in pharmacological responses observed in experimental studies.

4.10.2 *Cardiovascular effects*

The cardiovascular effects of cigarette smoking have been extensively studied. Smoking a cigarette activates the sympathetic nervous system and results, in healthy people, in an increase in heart rate and blood pressure (Cryer *et al.* 1976), cardiac stroke volume and output (Irving and Yamamoto 1963), and coronary blood flow (Bargeron *et al.* 1957). Smoking or nicotine causes peripheral vascular changes including cutaneous vasoconstriction (Freund and Ward 1960), associated with a decrease in skin temperature, systemic venoconstriction (Eckstein and Horsley 1960), and increased muscle blood flow (Rottenstein *et al.* 1960). Smoking results in increased circulating concentrations of noradrenalin, consistent with neural adrenergic stimulation, and adrenalin, indicating adrenal medullary stimulation (Cryer *et al.* 1976). Circulating free fatty acids, glycerol, and lactate concentrations increase. Cardiovascular and metabolic effects are prevented by combined alpha and beta adrenergic blockade (Cryer *et al.* 1976), indicating that the cardiovascular effects of cigarette smoking are mediated by activation of the sympathetic nervous system.

Nicotine appears to be the substance in cigarette smoke responsible for activation of the sympathetic nervous system. Evidence for this comes from studies comparing the effects of nicotine-containing versus denicotinized cigarettes, comparing cigarettes of differing nicotine yield, and studying the effects of nicotine *per se*, either as intravenous nicotine or nicotine gum (Aronow *et al.* 1971; Benowitz *et al.* 1982*a*; Nyberg *et al.* 1982; Spohr *et al.* 1979).

4.10.3 *Pharmacodynamics of intravenous nicotine*

Although there have been many studies of the actions of smoking or nicotine *per se*, until recently little has been known about the relationship between

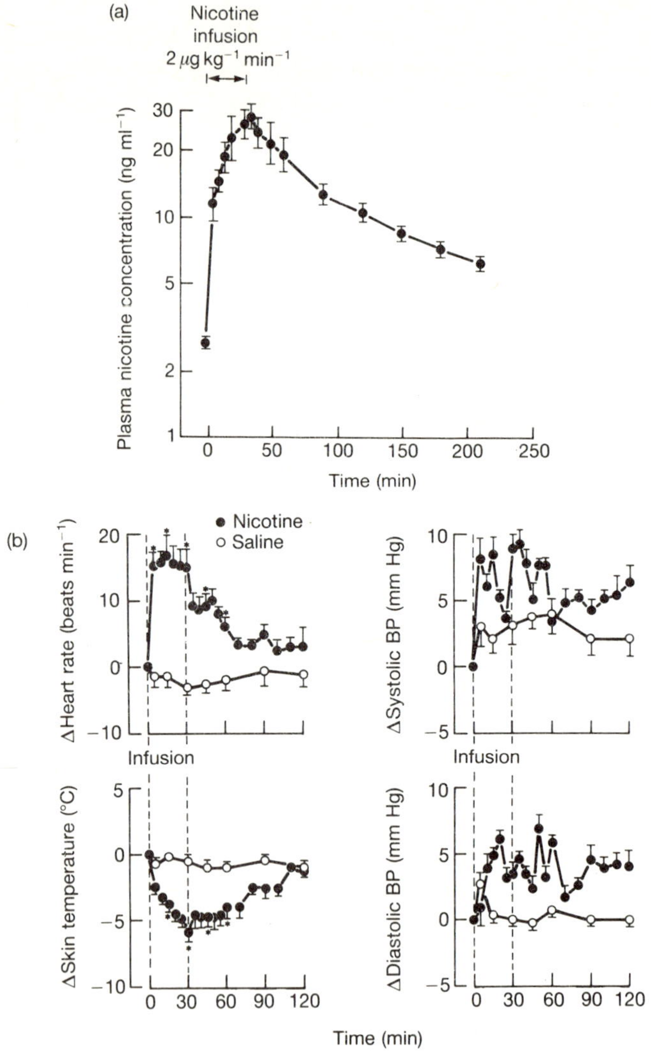

Fig. 4.14. (a) Plasma nicotine concentrations (± S E M) in 5 subjects during and after constant infusion for 30 min (from Benowitz *et al.* 1982*a*, with permission). (b) Cardiovascular responses to nicotine (n = 5) and saline (n = 5) infusions. Asterisks indicate $p < 0.05$, comparing nicotine and saline conditions (Benowitz *et al.* 1982*a*, with permission).

nicotine blood concentrations and cardiovascular effects. Such considerations are important to be able to relate knowledge of blood levels throughout the day while smoking to the expected cardiovascular consequences. We studied the effects of intravenous bolus injections of nicotine, given in a manner to simulate cigarette puffing (Rosenberg *et al.* 1980). A series of injections was given every 30 min for 3 h. Blood levels and cardiovascular effects were similar to those observed after cigarette smoking. Heart rate and blood pressure increased sharply after the first series of nicotine injections. Values remained above baseline, but there was little increment with successive injections, despite rising and falling blood nicotine levels. In contrast, skin temperature fell progressively during the period of nicotine dosing, gradually returning toward baseline at the end of the study. These data indicated rapid development of tolerance to heart rate and blood pressure responses, but tolerance was not complete in that heart rate and blood pressure remained above baseline.

To clarify these pharmacodynamic observations, a second group of subjects was studied with continuous infusion of nicotine. Heart rate and systolic blood pressure increased mostly in the first 5–10 min after infusion despite the progressive increase in blood nicotine levels over 30 min (Fig. 4.14). Skin temperature declined during the infusion and increased gradually after cessation of the infusion. Analysis of the blood concentration–response

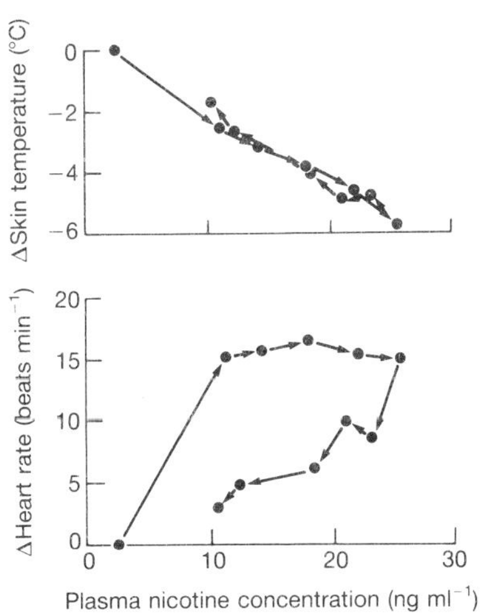

Fig. 4.15. Hysteresis loops for skin temperature and heart rate responses plotted against simultaneous plasma nicotine concentration (mean values n = 5). Arrows indicate progression of time during and after nicotine infusion. Change in skin temperature is directly related to plasma nicotine concentration. The heart rate hysteresis plot demonstrates a maximal response at low nicotine concentrations, and less of a response during the phase of declining as compared to rising nicotine concentrations, suggesting development of tolerance (Benowitz *et al.* 1982*a*, with permission).

characteristics by hysteresis plots (Fig. 4.15) illustrates that heart rate increased at relatively low blood nicotine concentrations, and plateaued despite rising levels. Heart rate was lower for a given blood nicotine concentration in the decline phase, suggesting development of tolerance within the time course of the infusion study. This acute tolerance would be similar to what other authors have called tachyphylaxis. Skin temperature declined and rose in association with changes in blood nicotine concentrations, showing no evidence of tolerance. Similarly, West and Russell (1987) found selective development of

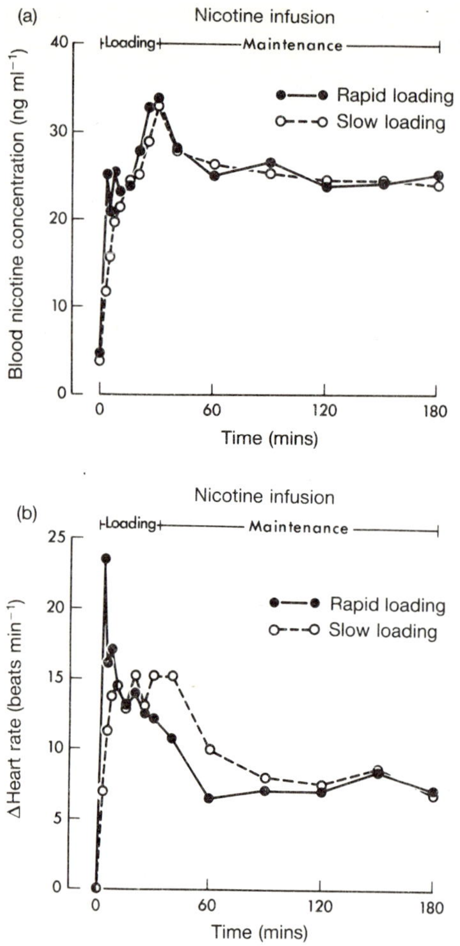

Fig. 4.16. (a) Average blood nicotine concentration and (b) increase in heart rate in 7 subjects. On the rapid loading day, subjects received 17.5 $\mu g\ kg^{-1}\ min^{-1}$ nicotine for 1.5 min, then 1.75 $\mu g\ kg^{-1}\ min^{-1}$ for 30 min; on the slow loading infusion day, 2.5 $\mu g\ kg^{-1}\ min^{-1}$ for 30 min. Maintenance infusion on both days was 0.35 $\mu g\ kg^{-1}\ min^{-1}$ from 30–180 min (Porchet *et al.* 1987, with permission).

tolerance to heart rate acceleration but not to the fall in skin temperature after cigarette smoking in a normal smoking day.

As mentioned previously, however, one must be cautious in using venous blood levels which may not reflect levels at the site of action, to make judgments about pharmacodynamic tolerance. To examine the importance of distribution in the development of acute tolerance to nicotine, we compared effects of rapid- and slow-loading infusions of nicotine, followed by a maintenance infusion (Porchet *et al.* 1987). Concentrations in the brain, a possible site of action of nicotine-mediated cardiovascular effects, were simulated using a perfusion model with blood-brain partition data derived from studies in rabbits. Figure 4.16 shows that rapid and slow infusions of nicotine result in similar peak venous blood levels, but rapid infusion produces a much greater cardiac acceleration as well as greater subjective effects (not shown). Simulation of brain concentration of nicotine (Fig. 4.17) shows that brain levels more closely resemble the heart rate acceleration curve with the two modes of administration. Nicotine concentration versus heart rate acceleration hysteresis plots show apparent tolerance when using venous concentrations, but no tolerance when using simulated brain concentrations. In other words, with a rapid rate of rise of nicotine in the body, as during infusion or inhalation of nicotine, venous nicotine levels underestimate brain levels, and pharmacodynamic analysis based on venous levels incorrectly indicates acute tolerance.

However, as indicated by the repeat infusion studies described previously, pharmacodynamic tolerance to nicotine does develop. To characterize the time course of tolerance development and regression, we conducted a study in

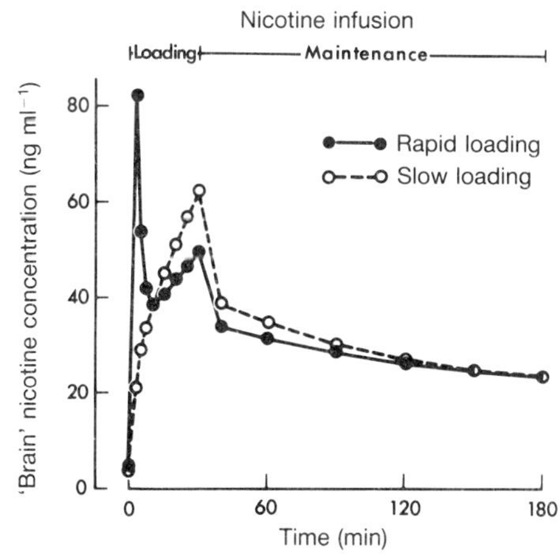

Fig. 4.17. Predicted mean 'brain' concentrations of nicotine over time (based on a physiological knietic model, fit to the rabbit data, and scaled to humans). Infuson rates are as described in Fig. 4.16. Note that 'brain concentrations' peak at the same time as heart rate (see Fig. 4.16) (from Porchet *et al.* 1987, with permission).

which subjects received paired intravenous infusions of nicotine, separated by different time intervals (Porchet *et al.* 1988). Despite higher blood concentrations (Fig. 4.18), heart rate acceleration (Fig. 4.19), and subjective effects, were much less when a second infusion was given at 60 or 120 min after the first infusion. But at 240 min the response was fully restored. The pharmacodynamics of tolerance were modeled using a pharmacokinetic–pharmacodynamic model (Porchet *et al.* 1988). This modeling indicates a half-life of development and regression of tolerance of 35 min. Such modeling suggests that three half-lives, or one and a half hour after a cigarette, nearly full sensitivity should have been regained. The interval at which cigarette smokers smoke cigarettes may be determined to a considerable degree by the kinetics of regression of tolerance.

This modeling also indicates that tolerance to the effect of nicotine is not total. The model predicts that for a mean steady-state concentration of 25 ng ml^{-1} of nicotine (which corresponds to the average level achieved by habitual cigarette smokers), approximately 25 per cent of the pharmacologic effect, as compared with the effect at a similar level of nicotine in the non-tolerant state, is still present. This indicates the persistence of nicotine effect during the day despite the development of acute tolerance.

Our studies have focused on the heart rate and other cardiovascular responses to nicotine. We have not had a reliable continuous marker of nicotine actions directly on the central nervous system. The cardiovascular responses to nicotine at low doses are mediated by the nervous system (Comroe 1960). We speculate that tolerance to central nervous system actions exhibits similar pharmacodynamics as that of cardiovascular effects.

The above studies indicate rapid development of tolerance to some (but not all) actions of nicotine. These studies were performed in habitual smokers after overnight abstinence. But even after overnight abstinence significant quantities of nicotine persist in the body, and there is conceivably some persistence of tolerance. To determine if there is a longer-term tolerance and, if so, its extent, we examined cardiovascular responses to infused nicotine after overnight and after seven days tobacco abstinence (Lee *et al.* 1987*a*). Heart-rate and blood pressure responses were greater after more prolonged abstinence (Fig. 4.19). However, within 60–90 min the blood concentration effect relationship in subjects after prolonged abstinence approximated to that observed after brief abstinence. Thus, we conclude that there is an element of tolerance that persists throughout the daily smoking cycle and which is lost with prolonged abstinence, but that tolerance is rapidly re-established with subsequent exposure.

4.10.4 *Pharmacodynamics of nicotine during cigarette smoking*

Pharmacodynamic observations with intravenous dosing of nicotine explain some observations made in cigarette smokers. That bolus doses (or

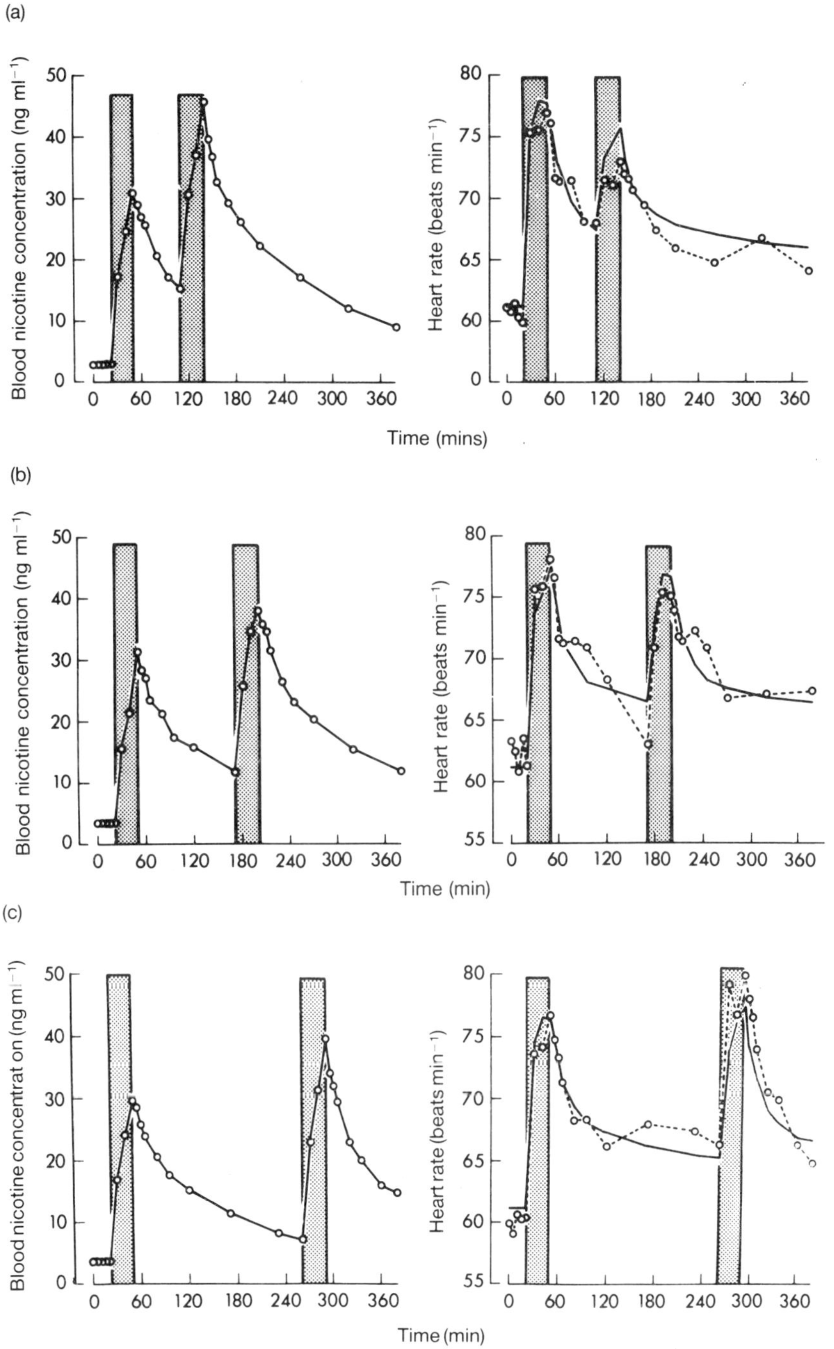

Fig. 4.18. Mean blood concentrations of nicotine (left) and the corresponding mean heart rate (right) in 8 subjects after two 30 min intravenous infusions of 2.5 μg kg^{-1} min^{-1} of nicotine separated by (a) 60 min (b) 120 min, or (c) 210 minutes. The shadowed areas indicate the periods during which nicotine was infused. The solid line shows the fit of the model to the effect data.

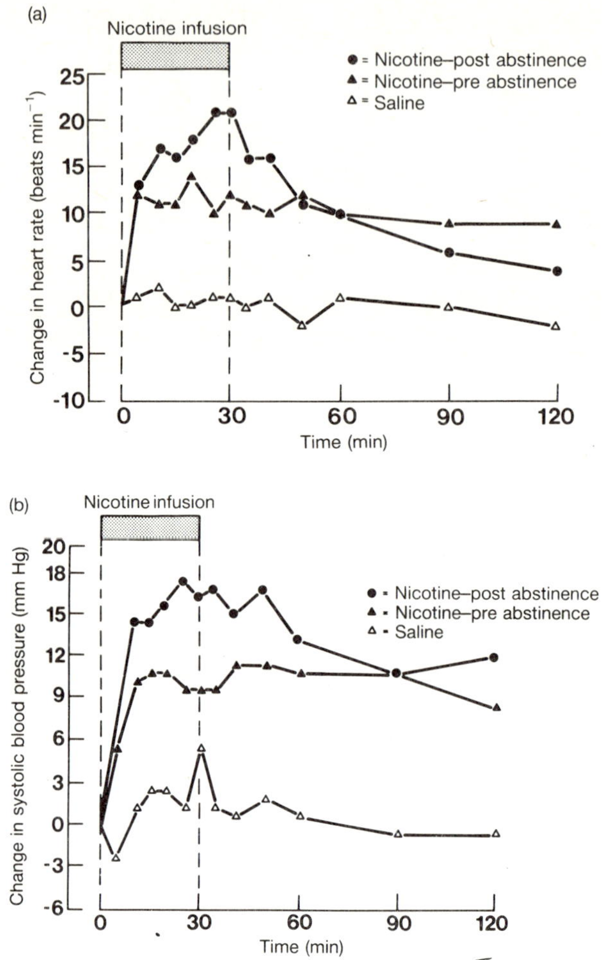

Fig. 4.19. Cardiovascular responses to nicotine (2 μg kg^{-1} min^{-1} for 30 min) and saline infusions after overnight ('pre-abstinence') compared with 7 days of tobacco abstinence ('post-abstinence') (mean of 7 subjects). Saline infusions had no significant effect on cardiovascular indices. Changes in systolic blood pressure and heart rate were significantly greater when nicotine was infused after prolonged compared with brief abstinence (from Lee *et al.* 1987*a*, with permission).

inhalations) of nicotine increase heart rate to a maximum suggests that heart rate will increase most with the first few cigarettes of the day, but subsequently will not vary in relation to the amount of nicotine consumed. That only partial tolerance developed to heart-rate acceleration due to nicotine suggested that effects on heart rate may persist as long as significant levels of nicotine persist, including persistence overnight. These predictions were tested in a study during which healthy volunteers smoked either high- or low-yield research cigarettes, or abstained from smoking (Benowitz *et al.* 1984). The research cigarettes were non-filter cigarettes which were high- or low-yield because they contained more or less nicotine. Thus full compensation for the low-yield cigarette was impossible. Resultant nicotine blood levels were more than fourfold different (Fig. 4.9). As predicted, heart rate, assessed by continuous ambulatory EKG monitoring, increased in the morning, more on smoking than non-smoking days, and the increase occurred with the first few cigarettes of the day (Fig. 4.20). Subsequently, heart rate followed a normal circadian pattern, but was always higher when smoking compared to abstinence. Only at 6–7 a.m., prior to the next day's smoking, were heart rates during smoking similar to those in abstinence. Also, as predicted, heart rate was no different when smoking low-yield compared to high-yield cigarettes, despite the fourfold difference in blood nicotine concentration.

Pharmacodynamic considerations may help us understand the development of smoking behaviour as well as adverse effects of cigarette smoking. For example, the daily smoking cycle can be conceived as follows. The first cigarette of the day produces substantial pharmacologic effects, primarily arousal, but at the same time tolerance begins to develop. A second cigarette

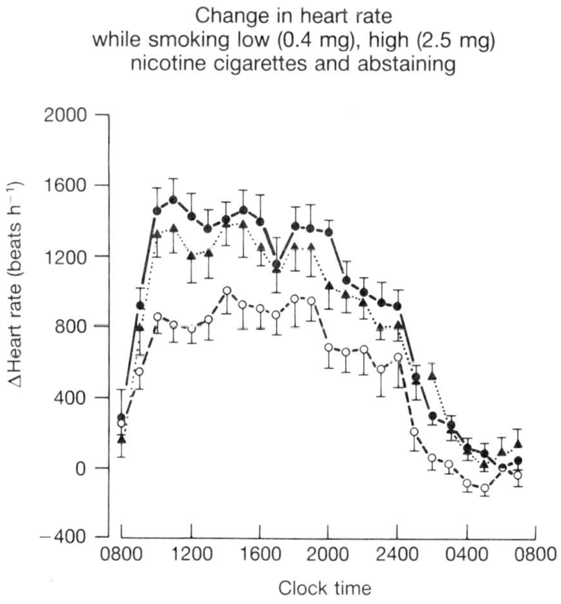

Fig. 4.20. Heart rate acceleration (compared to a 5–6 a.m. baseline value) in 10 subjects smoking low-(triangles) or high-(closed circles) nicotine cigarettes, and while abstaining from cigarettes (open circles) (mean ± S E M). Blood nicotine concentrations are shown in Fig. 4.9 (Benowitz *et al.* 1984).

may be smoked at a later time, at which the smoker has learned that there is some regression of tolerance. With subsequent cigarettes, there is accumulation of nicotine in the body, resulting in a greater level of tolerance. Transiently high brain levels of nicotine following smoking individual cigarettes may partially overcome tolerance. But the effects of individual cigarettes tend to lessen throughout the day. Overnight abstinence allows considerable resensitization to actions of nicotine. However, full resensitization requires days, or longer. Because of dose–response and tolerance characteristics, habitual smokers need to smoke at least 15 cigarettes, and consume 20 to 40 mg nicotine, per day to achieve desired affects of cigarette smoking throughout the day.

Pharmacodynamic aspects of nicotine action may explain how cigarette smoking causes coronary heart disease. As noted before, because of the accumulation and dose–response characteristics, heart rate is increased during cigarette smoking for 24 h per day. We have found that there is increased circulating plasma catecholamine concentrations as well (Benowitz 1986*a*), consistent with the idea that cigarette smoking produces sympathetic activation 24 h per day. Persistent sympathetic activation could affect lipid metabolism, resulting in a more atherogenic lipid profile, promote platelet aggregation and hypercoagulability, induce vasoconstriction and coronary spasm, increase heart rate and myocardial contractility, thereby reducing oxygen delivery while increasing the oxygen demands of the heart, and increase circulating catecholamines, which can promote cardiac arrhythmias. These factors could then accelerate atherosclerosis and could contribute to acute myocardial infarction in a person with pre-existing coronary atherosclerosis (Benowitz 1986*b*). Because of persistent effects of nicotine throughout the day, one would not expect to see a specific correlation between acute coronary events and when a person smokes a cigarette, which appears to be the case.

4.11. Conclusions

Over the past 15 years, owing to advances in analytical technology, much new information on the pharmacokinetics, metabolism, and pharmacodynamics of nicotine has emerged. Such research has enhanced our understanding of why people consume tobacco, why particular tobacco products and patterns of consumption are favoured, and possible reasons for individual differences in tobacco consumption. Most important perhaps are studies that document regulation of particular levels of nicotine in the bodies of habitual smokers, and the idea that factors which influence levels of nicotine in the body may affect smoking behaviour. Pharmacokinetic and metabolic studies provide methods and the rationale for detecting and quantifying human consumption of tobacco. These methods have been of great utility in studying tobacco-consuming behaviour in large populations, in smoking-cessation research, and potentially in individual smoking-cessation therapy. Pharmacodynamic research findings provide clues as to likely contributions of nicotine to the adverse health consequences of tobacco use.

There is much current interest in the development of nicotine as a therapeutic drug (see Kellar and Wonnacott, chapter 10, this volume). At present, therapeutic use is limited to nicotine replacement therapy, as an adjunct to smoking cessation therapy. Pharmacokinetic and pharmacodynamic research has and will continue to contribute to development of optimal dosage forms, and may ultimately be of use in therapeutic drug monitoring to optimize efficacy.

Acknowledgements

Research described in this chapter was supported in part by U.S. Department of Health and Human Services Grants DA02277, CA32389, HL29476, and DA01696. We thank Kaye Welch for preparation of the manuscript.

References

Armitage, A.K. and Turner, D.M. (1970). Absorption of nicotine in cigarette and cigar smoke through the oral mucosa. *Nature* **226**, 1231–2.

Armitage, A.K., Dollery, C.T., George, C.F., Houseman, T.H., Lewis, P.J., and Turner, D.M. (1975). Absorption and metabolism of nicotine from cigarettes. *British Medical Journal* **4**, 313–6.

Armitage, A., Dollery, C., Houseman, T., Kohner, E., Lewis, P.J., and Turner, D. (1978). Absorption of nicotine from small cigars. *Clinical Pharmacology and Therapeutics* **23**, 143–50.

Aronow, W.S., Dendinger, B.S., and Rokaw, S.N. (1971). Heart rate and carbon monoxide level after smoking high, low and non-nicotine cigarettes. *Annals of Internal Medicine* **74**, 697–702.

Bargeron, L.M., Ehmke, D., Gonlubol, F., Castellanos, A., Siegel, A., and Bing, R.J. (1957). Effect of cigarette smoking on coronary blood flow and myocardial metabolism. *Circulation* **15**, 251–7.

Beckett, A.H., Gorrod, J.W., and Jenner, P. (1970). Absorption of (–)-nicotine-1′-*N*-oxide in man and its reduction in the gastrointestinal tract. *Journal of Pharmaceutics and Pharmacology* **22**, 722–3.

Beckett, A.H., Gorrod, J.W., and Jenner, P. (1971*a*). The effect of smoking on nicotine metabolism *in vivo* in man. *Journal of Pharmaceutics and Pharmacology* **23**, 62–7S.

Beckett, A.H., Gorrod, J.W., and Jenner, P. (1971*b*). The analysis of nicotine-1′-*N*-oxide in urine, in the presence of nicotine and cotinine, and its application to the study of *in vivo* nicotine metabolism in man. *Journal of Pharmaceutics and Pharmacology* **23**, 55–61S.

Benowitz, N.L. (1986*a*). Increased 24-hour energy expenditure in cigarette smokers (Ltr), *New England Journal of Medicine* **314**, 1641–2.

Benowitz, N.L. (1986*b*). Clinical pharmacology of nicotine. *Annual Reviews of Medicine* **37**, 21–32.

Benowitz, N.L., Forsyth, R.P., Melmon, K.L., and Rowland, M. (1974). Lidocaine disposition kinetics in monkey and man. I. Prediction by a perfusion model. *Clinical Pharmacology and Therapeutics* **16**, 87–98.

Benowitz, N.L. and Jacob, P. III, (1984). Daily intake of nicotine during cigarette smoking. *Clinical Pharmacology and Therapeutics* **35**, 499–504.

Benowitz, N.L. and Jacob, P. III, (1985). Nicotine renal excretion rate influences nicotine intake during cigarette smoking. *Journal of Pharmacology and Experimental Therapeutics* **234**, 153–5.

Benowitz, N.L., Hall, S.M., Herning, R.I., Jacob, P. III, Jones, R.T., and Osman, A-L. (1983*a*). Smokers of low yield cigarettes do not consume less nicotine. *New England Journal of Medicine* **309**, 139–42.

Benowitz, N.L., Jacob, P. III, Jones, R.T., and Rosenberg, J. (1982*a*). Inter-individual variability in the metabolism and cardiovascular effects of nicotine in man. *Journal of Pharmacology and Experimental Therapeutics* **221**, 368–72.

Benowitz, N.L., Jacob, P. III, and Savanapridi, C. (1987*a*). Determinants of nicotine intake while chewing nicotine polacrilex gum. *Clinical Pharmacology and Therapeutics* **41**, 467–73.

Benowitz, N.L., Kuyt, F., and Jacob, P. III, (1982*b*). Circadian blood nicotine concentrations during cigarette smoking. *Clinical Pharmacology and Therapeutics* **32**, 758–64.

Benowitz, N.L., Kuyt, F., and Jacob, P. III, (1984). Influence of nicotine on cardiovascular and hormonal effects of cigarette smoking. *Clinical Pharmacology and Therapeutics* **36**, 74–81.

Benowitz, N.L., Kuyt, F., Jacob, P. III, Jones, R.T., and Osman, A-L. (1983*b*). Cotinine disposition and effects. *Clinical Pharmacology and Therapeutics* **309**, 139–42.

Benowitz, N.L., Lake, T., Keller, K.H., and Lee, B.L. (1987*b*). Prolonged absorption with development of tolerance to toxic effects following cutaneous exposure to nicotine. *Clinical Pharmacology and Therapeutics* **42**, 119–20.

Benowitz, N.L., Porchet, H., Sheiner, L., and Jacob, P. III. (1988). Nicotine absorption and cardiovascular effects with smokeless tobacco use: comparison to cigarettes and nicotine gum. *Clinical Pharmacology and Therapeutics*, **44**, 23–8.

Booth, J. and Boyland, E. (1970). The metabolism of nicotine into two optically active stereoisomers of nicotine-1′-oxide by animal tissues *in vitro* and by cigarette smokers. *Biochemical Pharmacology* **19**, 733–42.

Borzelleca, J.F. (1963). Drug movement from the isolated urinary bladder of the rabbit (Abstract). *Federation Proceedings* **22**, 661.

Bowman, E.R. and McKennis, N., Jr. (1962). Studies on the metabolism of (−)-cotinine in the human. *Journal of Pharmacology and Experimental Therapeutics* **135**, 307.

Brady, M.E., Ritschel, W.A., Saelinger, D.A., Cacini, W., and Patterson, A.J. (1979). Animal model and pharmacokinetic interpretation of nicotine poisoning in man. *International Journal of Clinical Pharmacology and Biopharmacology* **17**, 12–7.

Cherek, D.R., Lowe, W.C., and Friedman, T.T. (1981). Effects of ammonium chloride on urinary pH and cigarette smoking behavior. *Clinical Pharmacology and Therapeutics* **29**, 762–70.

Clark, M.S.G., Rand, M.J., and Vanov, S. (1965). Comparison of pharmacological activity of nicotine and related alkaloids occurring in cigarette smoke. *Archives of International Pharmacodynamics* 363–79.

Comroe, J.H. (1960). The pharmacological actions of nicotine. *Annals of the New York Academy of Sciences* **90**, 48–51.

Cryer, P.E., Haymond, M.W., Santiago, J.V., and Shah, S.D. (1976).

Norepinephrine and epinephrine release and adrenergic mediation of smoking-associated hemodynamic and metabolic events. *New England Journal of Medicine* **295**, 573–7.

Dawson, G.W. and Vestal, R.E. (1982). Smoking and drug metabolism. *Pharmacology and Therapeutics* **15**, 207–21.

Eckstein, J.W. and Horsley, A.W. (1960). Responses of the peripheral veins in man to the intravenous administration of nicotine. *Annals of the New York Academy of Sciences* **90**, 133–7.

Faulkner, J.M. (1933). Nicotine poisoning by absorption through the skin. *Journal of the American Medical Association* **100**, 1664–5.

Feyerabend, C., Higenbottam, T., and Russell, M.A.H. (1982) Nicotine concentrations in urine and saliva of smokers and non-smokers. *British Medical Journal* **284**, 1002–4.

Feyerabend, C., Ings, R.M.J., and Russell, M.A.H. (1985). Nicotine pharmacokinetics and its application to intake from smoking. *British Journal of Clinical Pharmacology* **19**, 239–47.

Feyerabend, C. and Russell, M.A.H. (1978). Effect of urinary pH and nicotine excretion rate on plasma nicotine during cigarette smoking and chewing nicotine gum. *British Journal of Clinical Pharmacology* **5**, 293–7.

Fix, A.J., Daughton, D., Kass I., Smith, J.L., Wickiser, A., Golden, C.J., and Wass, A.R. (1983). Urinary alkalinization and smoking cessation. *Journal of Clinical Psychology* **39**, 617–23.

Freund, J. and Ward, C. (1960). The acute effect of cigarette smoking on the digital circulation in health and disease. *Annals of the New York Academy of Science* **90**, 85–101.

Galeazzi, R.L., Daenens, P., and Gugger, M. (1985). Steady-state concentrations of cotinine as a measure of nicotine intake by smokers. *European Journal of Clinical Pharmacology* **28**, 301–4.

Gehlbach, S.H., Perry, L.D., Williams, W.A., Freeman, J.I., Langone, J.J., Peta, L.V., and Van Vunakis, H. (1975). Nicotine absorption by workers harvesting green tobacco. *Lancet* **1**, 478–80.

Gori, G.B., Benowitz, N.L., and Lynch, C.J. (1986). Mouth versus deep airways absorption of nicotine in cigarette smokers. *Pharmacology, Biochemistry and Behavior* **25**, 1181–4.

Gritz, E.R. (1980). Smoking behavior and tobacco abuse. In *Advances in substance abuse*, Vol. 1 (N. Mello, ed.), pp. 91–158. JAI Press, Greenwich, CN.

Gritz, E.R., Baer-Weiss, V., Benowitz, N.L., Van Vunakis, H., and Jarvik, M.E. (1981). Plasma nicotine and cotinine concentrations in habitual smokeless tobacco users. *Clinical Pharmacology and Therapeutics* **30**, 201–9.

Haley, N.J., Axelrad, C.M., and Tilton, K.A. (1983). Validation of self-reported smoking behavior: biochemical analyses of cotinine and thiocyanate. *American Journal of Public Health* **73**, 1204–7.

Herning, R.I., Jones, R.T., Benowitz, N.L., and Mines, A.H. (1983). How a cigarette is smoked determines nicotine blood levels. *Clinical Pharmacology and Therapeutics* **33**, 84–90.

Hibberd, A.R. and Gorrod, J.W. (1983). Enzymology of the metabolic pathway from nicotine to cotinine *in vitro*. *European Journal of Drug Metabolism and Pharmacokinetics* **8**, 151–62.

Hibberd, A.R., O'Connor, V., and Gorrod, J.W. (1978). Detection of nicotine, nicotine-1'-*N*-oxide and cotinine in maternal and fetal body fluids. In Biological oxidation of nitrogen (J.W. Gorrod, ed.), pp. 353–61. Elsevier, Amsterdam.

Hill, P. and Wynder, E.L. (1979). Nicotine and cotinine in breast fluid. *Cancer Letters* **6**, 251–4.

Ingenito, A.J., Barrett, J.P., and Procita, L. (1972). Direct central and reflexly mediated effects of nicotine on the peripheral circulation. *European Journal of Pharmacology* **17**, 375–85.

Inokuchi, T., Rudolph, A.M., Benowitz, N.L., Wang, L.H., and Benet, L.Z. (1989). The disposition and pharmacokinetics of nicotine in the sheep maternal–fetal unit. *Journal of Pharmacology and Experimental Therapeutics* submitted.

Irving, D.W. and Yamamoto, T. (1963). Cigarette smoking and cardiac output. *British Heart Journal* **25**, 126–32.

Isaac, P.F. and Rand, M.J. (1972). Cigarette smoking and plasma levels of nicotine. *Nature* **236**, 308–10.

Jacob, P. III, Benowitz, N.L., Copeland, J.R., Risner, M.E., and Cone, E.J. (1988*a*). Disposition kinetics of nicotine and cotinine enantiomers in rabbits and beagle dogs. *Journal of Pharmacological Science*, **77**, 396–400.

Jacob, P. III, Benowitz, N.L., and Shulgin, A.T. (1988*b*). Recent studies of nicotine metabolism in humans. *Pharmacology, Biochemistry and Behavior*, **30**, 249–53.

Jarvis, M., Tunstall-Pedoe, H., Feyerabend, C., Vesey, C., and Salloojee, Y. (1984). Biochemical markers of smoke absorption and self-reported exposure to passive smoking. *Journal Epidemiology and Community Health* **48**, 335–9.

Jenner, P. and Gorrod, J.W. (1973). Comparative *in vitro* hepatic metabolism of some tertiary *N*-methyl tobacco alkaloids in various species. *Research Communications in Chemistry, Pathology and Pharmacology* **6**, 829–43.

Jenner, P., Gorrod, J.W., and Beckett, A.H. (1973). The absorption of nicotine-1′-*N*-oxide and its reduction in the gastrointestinal tract in man. *Xenobiotica* **3**, 341–9.

Kalant, H., LeBlanc, A.E., and Gibbins, R.J. (1971). Tolerance to and dependence on some nonopiate psychotropic drugs. *Pharmacological Reviews* **23**, 135–91.

Kozlowski, L.T., Appel, C.-P., Frecker, R.C., and Khouw, V. (1982). Nicotine, a prescribable drug available without prescription. *Lancet* **1**, 334.

Kyerematen, G.A., Damiano, M.D., Dvorchik, B.H., and Vesell, E.S. (1982). Smoking-induced changes in nicotine disposition: Application of a new HPLC assay for nicotine and its metabolites. *Clinical Pharmacology and Therapeutics* **32**, 769–80.

Langone, J.J. and Van Vunakis, H. (1975). Quantitation of cotinine in sera of smokers. *Research Communications in Chemistry Pathology and Pharmacology* **10**, 21–8.

Lee, B.L., Benowitz, N.L., and Jacob, P. III. (1987*a*). Influence of tobacco abstinence on the disposition kinetics and effects of nicotine. *Clinical Pharmacology and Therapeutics* **41**, 474–9.

Lee, B.L., Benowitz, N.L., Jarvik, M.E., and Jacob, P. III. (1989). Food and nicotine metabolism. *Pharmacology, Biochemistry and Behavior* **9**, 621–5.

Loffelholz, K. (1970). Autoinhibition of nicotinic release of noradrenaline from postganglionic sympathetic nerves. *Naunyn-Schmiedeberg's Archives of Pharmacology* **267**, 49–63.

Luck, W., Hansen, R., Steldinger, R., and Nau, H. (1982). Nicotine and cotinine—two pharmacologically active substances as parameters for the strain on fetuses and babies of mothers who smoke. *Journal of Perinatal Medicine* **10**, 107–8.

Luck, W. and Nau, H. (1984). Nicotine and cotinine concentrations in serum and milk of nursing mothers. *British Journal of Clinical Pharmacology* **18**, 9–15.

Luck, W. and Nau, H. (1987). Nicotine and cotinine concentrations in the milk of smoking mothers: influence of cigarette consumption and diurnal variation. *European Journal of Pediatrics* **146**, 21–6.

Marks, M. J., Stitzel, J. A., and Collins, A. C. (1985). Time course study of the effects of chronic nicotine infusion on drug response and brain receptors. *Journal of Pharmacology and Experimental Therapeutics* **235**, 619–28.

McCusker, K., McNabb, E., and Bone, R. (1982). Plasma nicotine levels in pipe smokers. *Journal of the American Medical Association* **248**, 577–8.

McKennis, H. Jr., Schwartz, S. L., and Bowman, E. R. (1964). Alternate routes in the metabolic degradation of the pyrrolidine ring of nicotine. *Journal of Biological Chemistry* **239**, 3990–6.

McKennis, H. Jr., Turnbull, L. B., and Bowman, E. R. (1963*a*). *N*-methylation of nicotine and cotinine *in vivo*. *Journal of Biological Chemistry* **238**, 719–23.

McKennis, H. Jr., Turnbull, L. B., Bowman, E. R., and Tamaki, E. (1963*b*). The synthesis of hydroxycotinine and studies on its structure. *Journal of Organic Chemistry* **28**, 283–7.

Miller, R. P., Rotenberg, K. S., and Adir, J. (1977). Effect of dose on the pharmacokinetics of intravenous nicotine in the rat, *Drug Metabolism and Disposal* **5**, 436–43.

Monji, N. and Castro, A. (1979). Plasma nicotine pharmacokinetics in dogs after intravenous administration: determination by radioimmunoassay, *Research Communications in Chemistry, Pathology and Pharmacology* **23**, 267–77.

Neurath, G. B., Dunger, M., Orth, D., and Pein, F. G. (1987). Trans-3′-hydroxycotinine as a main metabolite in urine of smokers. *International Archives in Occupational and Environmental Health* **59**, 199–201.

Nyberg, G., Panfilov, V., Sivertsson, R., and Wilhelmsen, L. (1982). Cardiovascular effects of nicotine chewing gum in healthy non-smokers. *European Journal of Clinical Pharmacology* **23**, 303–7.

Peterson, L. A., Trevor, A., and Castagnoli, N. (1987). Stereochemical studies on the cytochrome P-450 catalyzed oxidation of (S)-nicotine to the (S)-nicotine Δ1′(5′)-iminium species. *Journal of Medicinal Chemistry* **30**, 249–54.

Petrakis, N. L., Gruenke, L. D., Beelen, T. C., Castagnoli, N., and Craig, J. C. (1978). Nicotine in breast fluid of nonlactating women. *Science* **199**, 303–5.

Piade, J. J. and Hoffmann, D. (1980). Chemical studies on tobacco smoke LXVII. Quantitative determination of alkaloids in tobacco by liquid chromatography. *Journal of Liquid Chromatography* **3**, 1505–15.

Pool, W. E., Godin, C. S., and Crooks, P. A. (1985). Nicotine racemization during cigarette smoking. *The Toxicologist* **5**, 232.

Porchet, H. C., Benowitz, N. L., and Sheiner, L. B. (1988). A pharmacodynamic model of tolerance: application to nicotine. *Journal of Pharmacology and Experimental Therapeutics* **244**, 231–6.

Porchet, H. C., Benowitz, N. L., Sheiner, L. B., and Copeland, J. R. (1987). Apparent tolerance to the acute effect of nicotine results in part from distribution kinetics. *Journal of Clinical Investigation* **80**, 1466–71.

Porsius, A. J. and Van Zwieten, P. A. (1978). The central actions of nicotine on blood pressure and heart rate after administration via the left vertebral artery of anesthetized cats. *Arzheim Forschein* **28**, 1628–31.

Rosenberg, J., Benowitz, N. L., Jacob, P., and Wilson, K. M. (1980). Disposition kinetics and effects of intravenous nicotine. *Clinical Pharmacology and Therapeutics* **28**, 516–22.

Rottenstein, H., Peirce, G., Russ, E., Felder, D., and Montgomery, H. (1960). Influence of nicotine on the blood flow of resting skeletal muscle and of the digits in normal subjects. *Annals of the New York Academy of Science* **90**, 102–13.

Russell, M.A.H. (1976). Tobacco smoking and nicotine dependence. In *Research advances in alcohol and drug problems*, Vol. 3, (R.J. Gibbins, Y. Israel, H. Kalant, R.E. Popham, W. Schmidt, and R.G. Smart ed.) pp. 1–48, John Wiley, New York.

Russell, M.A.H. and Feyerabend, C. (1978). Cigarette smoking: a dependence on high-nicotine boli. *Drug Metabolism Reviews* **8**, 29–57.

Russell, M.A.H., Jarvis, M.J, Devitt, G., and Feyerabend, C. (1981). Nicotine intake by snuff users. *British Medical Journal* **283**, 814–7.

Russell, M.A.H., Jarvis, M.J., and Feyerabend, C. (1980*a*). A new age for snuff?, *Lancet* **1**, 474–5.

Russell, M.A.H., Jarvis, M., Iyer, R., and Feyerabend, C. (1980*c*). Relation of nicotine yield of cigarettes to blood nicotine concentrations in smokers. *British Medical Journal* **280**, 972–6.

Russell, M.A.H., Jarvis, M.J., West, R.J., and Feyerabend, C. (1985). Buccal absorption of nicotine from smokeless tobacco sachets. *Lancet*, December 14, 1370.

Russell, M.A.H., Raw, M., and Jarvis, M.J. (1980*b*). Clinical use of nicotine chewing gum. *British Medical Journal* **280**, 1599–1602.

Sasson, I.M., Haley, N.J., Hoffmann, D., Wynder, E.L., Hellberg, D., and Nilsson, S. (1985). Cigarette smoking and neoplasia of the uterine cervix: smoke constituents in cervical mucus. *New England Journal of Medicine* **312**, 315–6.

Schachter, S. (1978). Pharmacological and psychological determinants of smoking, *Annals of Internal Medicine* **88**, 104–14.

Schievelbein, H. (1982). Nicotine. Resorption and fate. *Pharmacology and Therapeutics* **18**, 233–48.

Schievelbein, H., Eberhardt, R., Loschenkohl, K., Rahlfs, V., and Bedall, F.K. (1973). Absorption of nicotine through the oral mucosa. 1. Measurement of nicotine concentration in the blood after application of nicotine and total particulate matter. *Agents and Actions* **3/4**, 254–8.

Schmeltz, I. and Hoffmann, D. (1977). Nitrogen-containing compounds in tobacco and tobacco smoke. *Chemical Reviews* **77**, 295–311.

Shulgin, A.T., Jacob, P. III, Benowitz, N.L., and Lau, D. (1987). Identification and quantitative analysis of cotinine-*N*-oxide in human urine. *Journal of Chromatography* **423**, 365–72.

Spohr, U., Hoffmann, K., Steck, W., Harenberg, J., Walter, E., Hengen, N., Augustin, J., Morl, H., Koch, A., Horsch, A., and Weber, E. (1979). Evaluation of smoking-induced effects on sympathetic, hemodynamic and metabolic variables with respect to plasma nicotine and COHb levels. *Atherosclerosis* **33**, 271–83.

Stalhandske, T. and Slanina, P. (1982). Nicotyrine inhibits *in vivo* metabolism of nicotine without increasing its toxicity. *Toxicology and Applied Pharmacology* **65**, 366–72.

Steinsland, O.S. and Furchgott, R.F. (1975). Desensitization of the adrenergic neurons of the isolated rabbit ear artery to nicotinic agonists. *Journal of Pharmacology and Experimental Therapeutics* **193**, 138–48.

Su, C. (1982). Actions of nicotine and smoking on circulation. *Pharmacology and Therapeutics* **17**, 129–41.

Travell, J. (1960). Absorption of nicotine from various sites. *Annals of the New York Academy of Science* **90**, 13–30.

Turner, D.M., Armitage, A.K., Briant, R.H., and Dollery, C.T. (1975). Metabolism of nicotine by the isolated perfused dog lung. *Xenobiotica* **5**, 539–51.

Turner, J. A. M., Sillett, R. W., and McNicol, M. W. (1977). Effect of cigar smoking on carboxyhaemoglobin and plasma nicotine concentrations in primary pipe and cigar smokers and ex-cigarette smokers. *British Medical Journal* **2**, 1387–9.

Van Vunakis, H., Langone, J. J., and Milunsky, A. (1974). Nicotine and cotinine in the amniotic fluid of smokers in the second trimester of pregnancy. *American Journal of Obstetrics and Gynecology* **120**, 64–6.

Vaughan, D. P. and Dennis, M. (1978). Mathematical basis of point-area deconvolution method for determining *in vivo* input functions. *Journal of Pharmaceutical Science* **67**, 663–5.

Wald, N. J., Boreham, J., Bailey, A., Ritchie, C., Haddow, J. E., and Knight, G. (1984*a*). Urinary cotinine as marker of breathing other people's tobacco smoke. *Lancet* **1**, 230–1.

Wald, N. J., Idle, M., Boreham, J., Bailey, A., and Van Vunakis, H. (1981). Serum cotinine levels in pipe smokers: evidence against nicotine as cause of coronary heart disease. *Lancet* **2**, 775–7.

Wald, N. J., Idle, M., Boreham, J., Bailey, A., and Van Vunakis, H. (1984*b*). Urinary nicotine concentrations in cigarette and pipe smokers. *Thorax* **39**, 365–8.

West, R. J. and Russell, M. A. H. (1987). Cardiovascular and subjective effects of smoking before and after 24 h of abstinence from cigarettes. *Psychopharmacology* **92**, 118–21.

Wilson, K. (1984). Sex-related differences in drug disposition in man. *Clinical Pharmacokinetics* **9**, 189–202.

Zeidenberg, P., Jaffe, J. H., Kanzler, M., Levitt, M. D., Langone, J. J., and Van Vunakis, H. (1977). Nicotine: cotinine levels in blood during cessation of smoking. *Comprehensive Psychiatry* **18**, 93–101.

5 The central pharmacology of nicotine: electrophysiological approaches

Paul B.S. Clarke

5.1 Introduction

Although nicotine exerts diverse pharmacological actions in both the peripheral and central nervous systems, it is the central actions of the alkaloid which appear particularly important in reinforcing the tobacco smoking habit (see Clarke 1987*b*). For this reason, this review is concerned mainly with the central nervous system (CNS).

Global electroencephalographic measures are reviewed first. Subsequently, electrophysiological analyses at the level of the single neurone are considered. The main part of this chapter concerns single-unit recording studies, and seeks to explore three questions: what are the pharmacological characteristics of central nicotinic receptors, where in the CNS does nicotine act, and how do the electrophysiological observations mesh with the findings of receptor binding studies? Subsequent sections describe ionic mechanisms which underlie nicotine's actions; the pharmacology of nicotine in relation to cholinergic circuitry, and modulation of nicotinic actions by peptides and other endogenous neuromodulators. Finally, the evidence for possible subtypes of nicotinic receptor is outlined, and the chapter closes with some recommendations concerning 'anti-smoking' drugs.

5.2 Gross electrophysiological measures

5.2.1 *Electroencephalogram (EEG)*

The effects of nicotine and smoking on the EEG have been reviewed in depth (for example, Gilbert 1979; Edwards and Warburton 1983; see Warburton, chapter 3, this volume). The EEG provides a window on CNS activity although it is not known precisely which neural components contribute. One much studied EEG phenomenon is electrocortical desynchrony, which is associated with a reduction of alpha (8–13 Hz) electrical activity, and an increase of low-voltage, high-frequency (13–20 Hz) activity. Electrocortical desynchrony, together with hippocampal theta activity (which can only be studied in animals), have received particular attention, since in some situations it is correlated with behavioural and subjective indices of arousal or alertness. However, in recent years, the significance of this correlation has been seriously questioned (Vanderwolf and Robinson 1981), and more specifically Gilbert

(1979) has argued that the complex effects of nicotine on EEG cannot be adequately explained in terms of a single concept such as cortical or CNS arousal.

Most studies in humans show that tobacco smoking can increase EEG indices traditionally associated with arousal. For example, Knott and Venables (1977) compared deprived and non-deprived smokers with non-smokers, and reported that the smoking of two cigarettes resulted in an increased dominant frequency which persisted for the entire 15 min recording period. Subjects in this experiment were required to puff at 30 s intervals for 10 min, adding a degree of procedural rigour, but possibly denying the 'finger-tip control' of nicotine intake that is offered by cigarette smoking. In this and other studies, smoking abstinence was associated with EEG 'sedation', despite the occurrence of restlessness.

In almost all studies to date, the effects of nicotine or smoking on the EEG have been examined under baseline conditions of low arousal. However, Mangan and Golding (1978) have demonstrated the importance of behavioural context. When subjects were exposed to white noise (a presumed stressor), smoking increased the proportion of alpha activity, possibly reflecting a sedative effect, whereas in conditions of low-sensory stimulation, smoking had the opposite effect. Nicotine intake was not measured in this study, and it is unclear whether a constant dose of the drug would have opposite effects in the two experimental situations. This point is particularly relevant because smoking behaviour is altered in stressful situations, and studies in animals show that the EEG effects of nicotine are sensitive to changes in dose and rate of administration (Armitage *et al.* 1968).

Kumar *et al.* (1977) compared the effects of cigarette smoking with those of comparable intravenous doses of nicotine. In some sessions, subjects were instructed to take 12 deep puffs from normal or nicotine-free cigarettes, each puff separated by 1 min from the next; in other sessions, nicotine (or distilled water) was injected as a series of 10 intravenous pulses given at 1 min intervals. The injected doses of nicotine were approximately 10 and 20 $\mu g\, kg^{-1}$ free base. Intravenous nicotine was found to increase the mean rectified, integrated voltage of the β waveband (13.5–26 Hz) - an indication of electrocortical desynchrony. Smoking normal cigarettes tended to produce the same effect, but this result was not statistically significant.

Thus, the encephalographic effects of smoking in human subjects are most probably due, at least in part, to nicotine. This conclusion is supported indirectly by animal studies. In cats, electrocortical desynchrony has been reported following exposure to tobacco smoke (Hall 1970). Cortical activation produced by exposure to tobacco smoke was reduced, albeit infrequently abolished, by the centally acting nicotinic antagonist, mecamylamine (Hall 1970). However, electrocortical desynchrony was also obtained when subjects pretreated with mecamylamine were administered carbon monoxide in amounts comparable to those obtained in cigarette smoke. These results suggest that both nicotine and carbon monoxide contribute to the EEG effects of tobacco smoking.

In several animal species, administration of 'smoking doses' of nicotine (5–20 $\mu g\ kg^{-1}$ intravenously) rapidly results in electrocortical desynchrony and hippocampal theta activity (for example, Domino 1967; Hall 1970). Pharmacological studies indicate a central site (or sites) of action (Kawamura and Domino 1969). Nicotine-induced cortical desynchrony is prevented by transsection of the brain stem and by electrolytic lesions of the mesencephalic reticular formation in cats (Kawamura and Domino 1969). The precise role of midbrain reticular neurones in the response to nicotine is not known. It cannot be assumed that nicotine acts directly on these neurones.

The EEG-activating effects of nicotine in cats are blocked by mecamylamine and are susceptible to tachyphylaxis (Domino 1967; Hall 1970), and hence appear to be mediated by nicotinic receptors bearing a pharmacological resemblance to those of the ganglionic (C6) type. An extensive series of experiments by Domino and colleagues (Domino 1967; Kawamura and Domino 1969) indicates that both nicotine and muscarinic agonists can induce EEG signs of arousal in cats. Nicotine-induced EEG activation was effectively antagonized by atropine as well as by mecamylamine, whereas similar effects produced by a muscarinic agonist were only sensitive to atropine. Possibly, then, nicotine evokes EEG activation by releasing acetylcholine onto muscarinic receptors. Evidence in favour of this view has been obtained with hemicholinium-3. This drug, which depletes cholinergic terminals of acetylcholine, produced EEG changes which were reversed by a muscarinic agonist, but not by nicotine (Domino 1967). Presumably, nicotine failed to reverse the effects of hemicholinium because intact stores of acetylcholine were required for nicotine to act.

In animals, nicotine-induced electrocortical desynchrony is commonly followed by behavioural and EEG signs of sedation (for example, Domino 1967). This second phase has not been reported in humans. However, there is very little data available from animals chronically treated with nicotine, and the possibility remains that tolerance develops to these secondary effects.

5.2.2 *Event-related potentials (ERPs)*

Evoked potentials and the contingent negative variation (CNV), collectively known as ERPs, are obtained with scalp macroelectrodes, but differ from traditional EEG measures in that they reflect responses to sensory stimuli or to more complex contingencies.

An evoked potential is a complex electrical waveform which occurs in response to a brief stimulus (such as a flash of light, or a brief tone), and reflects the central processing of sensory information. When a stimulus is presented repeatedly, the resulting ERP waveform consists of two components. The so-called exogenous component occurs first (0–250 ms), and is mainly dependent on the physical characteristics of the stimulus. The second, endogenous component reflects the psychological state of the experimental subject. Tobacco smoking can alter the exogenous and endogenous components of ERPs.

The psychological significance of these effects is generally far from clear. However, Edwards and colleagues (see Edwards and Warburton 1983; see Warburton, chapter 3, this volume) have shown that tobacco smoking and nicotine can improve performance on tasks requiring rapid information-processing, and have provided evidence of an association with changes occurring within the endogenous component of the ERP (reduced P300 latency; see Warburton, Fig. 3.5., this volume).

Woodson *et al.* (1982) examined the effects of smoking experimental cigarettes on the averaged visual-evoked responses of human smokers. Cigarettes with higher-nicotine content (1.34 mg per cigarette) enhanced peak-to-peak amplitudes relative to other cigarettes which were considered to contain a pharmacologically negligible amount of nicotine (0.14 mg). The cigarettes did not differ in tar and water content, and it was concluded that the effects of cigarette smoking on visual-evoked potentials are due to nicotine. The authors also argued that the alterations observed in the visual-evoked responses were related to attentional processes, since these changes were not mimicked by procedures designed to increase subjective 'arousal'. As further evidence for some sort of specific effect, it was noted that whereas nicotine typically increases visual-evoked potentials, it depresses auditory-evoked potentials (see Woodson *et al.* 1982).

The CNV comprises a slow negative shift of potential, occurring during the interval following a warning signal and prior to a second expected stimulus. This electrophysiological measure does not directly correspond to any one psychological state or process, although it may bear some relation to the subject's state of alertness. Ashton and coworkers (1974) found that tobacco smoking could increase or decrease the amplitude of the CNV, and the direction of change was consistent within individual subjects. Smoking of an unlit cigarette was without effect. CNV magnitude was enhanced in the more extrovert subjects, and depressed in the more introvert individuals.

Attempts were made to relate the observed changes in CNV to subjective reports following smoking. This analysis was complicated by difficulties of terminology. For example, some smokers claimed that smoking 'relaxed' them, but also commented that their ability to concentrate was improved as a result. Nevertheless, subjects who exhibited a reduction of CNV magnitude tended to report a calming effect, whereas increased CNV was associated with a subjective stimulant effect. From an analysis of butt length, it was concluded that extroverts inhaled less nicotine than introverts (Ashton *et al.* 1974).

Intravenous administration of 'smoking doses' of nicotine (500 or 750 μg) produced CNV alterations of similar direction and magnitude as tobacco smoking itself (Ashton *et al.* 1980). Individuals differed in their sensitivity to nicotine, but in general, low doses of nicotine (12.5–50 μg) increased CNV

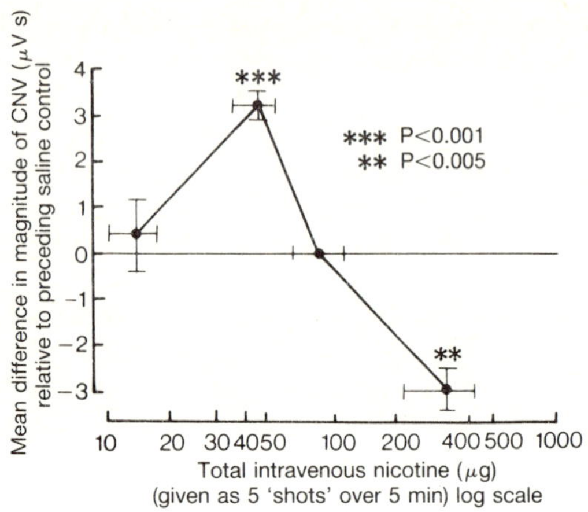

Fig. 5.1. Effect of intravenous nicotine on CNV magnitude. The mean change in CNV magnitude is shown at the lowest dose of nicotine administered, at the doses which in each subject produced the largest increases and decreases in CNV, and the mean dose of nicotine at which the curve crossed the baseline. Means and S E M (n = 8 subjects) are given for dose and response. The nicotine dosage range at zenith is significantly different from that at nadir. From Ashton *et al.* (1980).

magnitude, whereas higher doses (100–800 μg) diminished it (see Fig. 5.1). In this study, each dose of nicotine was divided into 5 equal parts, given at intervals of 1 min, for comparison with cigarette smoking. Since habitual smokers absorb approximately 1 mg of nicotine per cigarette (Benowitz and Jacob 1984; see Benowitz *et al.*, chapter 4, this volume), distributed in 15 puffs or so, it is questionable whether simulated puffs comprising 2.5–10 μg are relevant to smoking. Nevertheless, these findings serve as a reminder that nicotine exerts complex dose-dependent pharmacological actions, and that different actions of the drug may be important in different individuals, as well as in different situations.

5.3 Mapping sites of action of nicotine in the CNS: methodology

The global measures of cortical activity, discussed above, give little indication of where nicotine acts in the brain. This question has been addressed by recording the electrical activity of single neurones ('units'), or from small numbers of neurones using microscopically-fine glass or metal electrodes (see Curtis and Crawford 1969 for review). Such experiments are confined to laboratory animals. In the majority of studies, single units have been recorded extracellularly in anaesthetised subjects, and responses to application of a drug have been measured in terms of changes in firing rate. The drug is typically applied in the vicinity of the neurone, by microiontophoretic ejection. Recently, intracellular recordings have been made *in vitro*; this technique can be used to investigate mechanisms of cell excitability in relation to drug action.

5.3.1 *Interpretational issues*

Single-unit-recording studies, particularly those involving drug iontophoresis, must be interpreted with caution. Some of the more important issues are noted below:

1. Anaesthetics can profoundly alter neuronal response to drugs (Curtis and Crawford 1969).

2. The absolute amount of drug ejected is seldom known. The quantity of drug released is usually expressed in terms of the magnitude and duration of the ejecting current. However, identical ejection parameters do not guarantee similar drug quantities ejected from different micropipettes. It is therefore difficult to compare, for example, the potency of two agonists or antagonists even when they are applied to the same cell.

3. The local tissue concentration of drug is unknown. Generally, the concentration gradient from the tip of the micropipette to the surrounding tissue can only be guessed. It is not usually possible to determine whether the whole neurone is exposed to drug. Typically, the drug is ejected close to the tip of the recording barrel, and hence close to the cell body; if the relevant drug receptors are confined to the distal dendrites, the cell may be unresponsive to iontophoretic application. Conversely, drug application may indirectly alter neuronal firing by affecting a neighbouring cell; this possibility can be tested *in vitro*, and to some extent *in vivo*, by experimentally preventing synaptic transmission.

4. Micropipettes used for extracellular recording are associated with a sample bias towards larger cells which are spontaneously active. Neurones which are normally 'silent' can be induced to fire by iontophoretically-applied excitatory amino acids, but this approach is seldom taken.

5. A given nucleus of the brain typically contains a number of distinct neuronal populations, differing from one another in their anatomical connections, transmitter identity, and function. Studies which treat all neurones within a brain area as equivalent are therefore of less value than those which distinguish between different groups of neurones.

6. Precautions are not always taken to control for the possible effects of current flow and pH changes on neuronal firing.

7. Typically, iontophoretic applications of drugs are of several seconds duration. The pharmacokinetics of iontophoretic and systemic administration are thus quite different. This is particularly relevant in the case of nicotine, because tachyphylaxis can rapidly occur when this drug is applied to the CNS (Curtis and Ryall 1966*a*; Egan and North 1986).

5.3.2 *Selectivity of nicotinic antagonists*

Central nicotinic receptors, as far as they have been characterized, are pharmacologically similar to peripheral receptors of the ganglionic (C6) type (Clarke 1987*a*, 1987*b*; Wonnacott 1987; see Wonnacott, chapter 7, and Stolerman, chapter 8, this volume). Several nicotinic antagonists have been employed in iontophoretic studies of the CNS: of these, α-bungarotoxin and decamethonium are selective for peripheral C10 (muscle endplate) receptors, *d*-tubocurarine and dihydroβerythroidine are potent at both C6 and C10 receptors, whilst mecamylamine, hexamethonium, and chlorisondamine act selectively at C6 receptors.

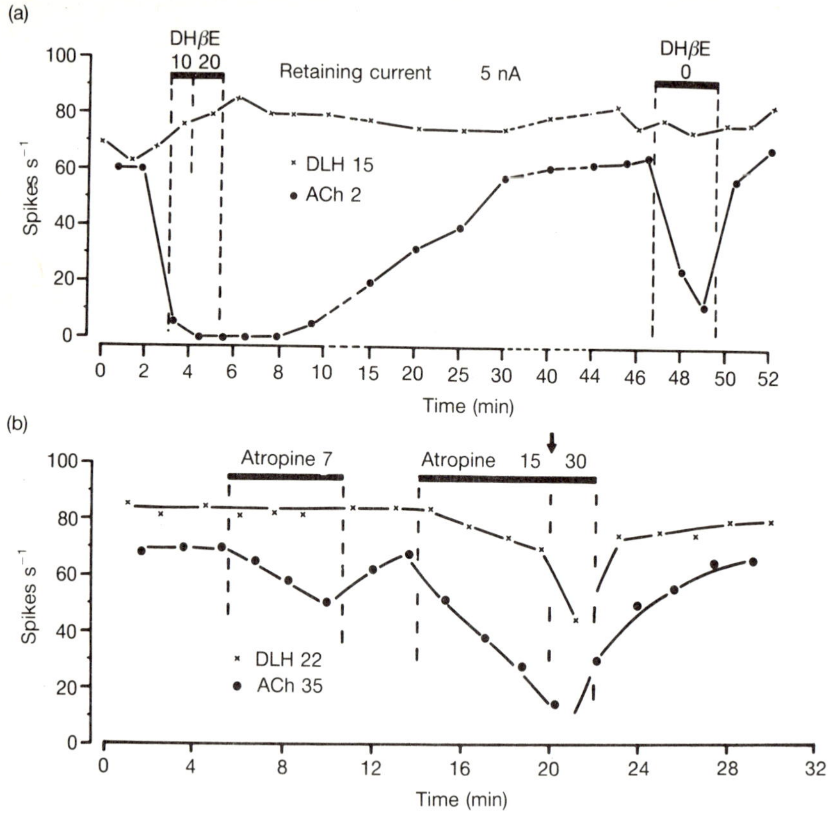

Fig. 5.2. (a) Selective depression of the acetylcholine response of a Renshaw cell by dihydroβerythroidine (D H B E). The symbols plot the maximum firing frequencies induced by an excitatory amino acid (D L-homocysteic acid, x) and acetylcholine (o), both intermittently iontophoresed for 18 s on each occasion. Dihydroβerythroidine iontophoresed (10 and 20 nA current) or simply permitted to diffuse (0 nA) from a third drug micropipette, as indicated. (b) Depression of responses to both acetylcholine and to an excitatory amino acid by atropine. This experiment was similar to the one shown in (a), except that atropine was the antagonist (from Curtis and Ryall 1966*b*).

There are several reports that hexamethonium, decamethonium, and *d*-tubocurarine can excite central neurones when iontophoresed alone (Curtis and Crawford 1969). Such excitation could conceivably result from a blockade of nicotinic cholinergic neurotransmission. Two observations make this unlikely. Firstly, central nicotinic actions are predominantly excitatory. Secondly, dihydroβerythroidine serves as a potent antagonist in the CNS, but shows little or no excitatory activity when applied alone. Excitatory effects of *d*-tubocurarine, in particular, appear to derive from non-cholinergic actions of the drug (Krnjevic 1975).

Mecamylamine is non-selective when given iontophoretically, since it can reduce neuronal responses to excitatory amino acids (Krnjevic and Phillis 1963*b*; McCance *et al.* 1968*b*; MacLeod *et al.* 1984). Dihydroβerythroidine, the most commonly used nicotinic anatagonist in electrophysiological studies, does not share this action (Andersen and Curtis 1964*b*; Curtis and Ryall 1966*b*; Fig. 5.2.). In general, the non-selective actions of nicotinic antagonists, when applied iontophoretically to central neurones, may result from local high concentrations of drug and non-uniform drug distribution, and need not imply that central nicotinic receptors are pharmacologically different from peripheral receptors.

5.3.3 *Central cholinoceptors: nicotinic, muscarinic, or mixed?*

In the periphery, nicotinic cholinergic neurotransmission is a rapid affair. The post-junctional effect of released acetylcholine is complete within a few milliseconds. In contrast, muscarinic responses to endogenous acetylcholine are of slow onset and persist for several seconds or more (Adams and Brown 1982). These temporal characteristics reflect different transduction mechanisms; nicotinic cholinoceptors contain a cation ionophore which opens when nicotinic agonists bind (see Wonnacott, chapter 7, this volume), resulting in rapid depolarisation, whereas muscarinic receptor-channel coupling is less direct (Brown 1983).

In the spinal cord and in several brain regions, nicotinic and muscarinic actions have been differentiated not only pharmacologically, but also by temporal characteristics (Curtis and Eccles 1958; Misgeld *et al.* 1980; Egan and North 1986; McCormick and Prince 1987*b*). In particular, a fast nicotinic response has been identified in the locus coeruleus (Egan and North 1986), interpeduncular nucleus (Brown *et al.* 1983; Takagi 1984), medial habenula (McCormick and Prince 1987*a*), and in the dorsolateral geniculate nucleus (McCormick and Prince 1987*b*). It should be noted, however, that nicotinic and muscarinic effects are defined by the actions of agonists and antagonists, and not by temporal criteria. Hence, the possible occurrence of less rapid nicotinic responses should not be ignored.

As outlined below, many studies provide evidence for distinct populations of nicotinic and muscarinic receptors. Other studies point to the possible existence of a receptor type having a 'mixed' pharmacological profile:

1. Electrophysiological studies. Generally, iontophoretic experiments employing selective nicotinic and muscarinic agonists support the existence of distinct nicotinic and muscarinic receptors. Thus, there are several reports where actions of nicotine are blocked by nicotinic but not by muscarinic antagonists, and where nicotinic antagonists are ineffective against muscarinic agonists (for example, Curtis and Ryall 1966*b*; Barker *et al.* 1971; Bradley and Lucy 1983; McLennan and Hicks 1978; Takagi 1984). It appears that like autonomic ganglion cells (Brown 1983), some central neurones possess both classes of cholinergic receptor (Curtis and Ryall 1966*a*, 1966*b* , 1966*c*; Barker *et al.* 1971), whereas others possess only one class or none at all.

Pharmacological ambiguities arise when mixed nicotinic/muscarinic agonists such as acetylcholine or carbamylcholine are employed: many investigations have reported excitatory actions of acetylcholine which are completely blocked not only by nicotinic antagonists, but also by muscarinic antagonists (Andersen and Curtis 1964*b*; Phillis *et al.* 1967; Bird and Aghajanian 1976; Bradley and Lucy 1983; see Curtis and Crawford 1969). It is unclear whether these results reflect the existence of an 'intermediate' type of nicotinic/muscarinic receptor, as some have suggested. Muscarinic mechanisms controlling neuronal excitability have been elucidated in some detail, and it appears that actions of acetylcholine at muscarinic receptors serve primarily to facilitate other inputs, rather than to trigger action potentials directly (Krnjevic *et al.* 1971; Brown 1983). In short, acetylcholine may potentiate its nicotinic actions by acting at muscarinic receptors. Such a neuromodulatory role may also explain the observation that atropine reduces neuronal excitation induced not only by acetylcholine, but also by excitatory amino acids (for example, Andersen and Curtis 1964*b*; Curtis and Ryall 1966*b*; McCance *et al.* 1968*b*).

In some brain nuclei (interpeduncular nucleus, locus coeruleus, medial habenula), excitatory actions of acetylcholine have been identified which are attenuated by nicotinic antagonists, but minimally affected by muscarinic antagonists (Brown *et al.* 1983; Egan and North 1986; McCormick and Prince 1987*a*, 1987*b*). In the medial habenula, this result is readily explained by a sparsity of muscarinic receptors (McCormick and Prince 1987*a*). In the interpeduncular nucleus, however, the source of antagonist selectivity is puzzling, since atropine-selective responses to selective muscarinic agonists are clearly present (Brown *et al.* 1983). In the locus coeruleus, a fast-depolarizing action of acetylcholine was selectively inhibited by nicotinic antagonists, and this response was distinct from a muscarinic component (Egan and North 1986).

2. Non-electrophysiological studies. Central effects of nicotine which are blocked by nicotinic antagonists, but not by muscarinic antagonists, have been identified in studies of transmitter release (Chesselet 1984), gastric secretion (Osumi *et al.* 1980), cardiovascular function (Kubo and Misu 1981), temperature regulation (Hall 1972), and behaviour (Rosecrans and Meltzer 1981). In other studies, a lack of antagonist selectivity had been found

(Domino 1967; Morrison *et al.* 1969), but since responses of the whole organism were involved, nicotinic and muscarinic blockade need not have occurred at the same receptor, or even on the same neurone. Radioligand-binding studies strongly support the existence of distinct muscarinic and nicotinic receptor populations in the brain, differentiated by drug affinities, regional distributions, and *in vivo* regulation by cholinergic agonists (Marks and Collins 1985; Schwartz 1986; see Wonnacott, chapter 7, this volume).

5.4 Nicotinic actions in the spinal cord

Although the central actions of iontophoretically-applied nicotine have been examined in scores of studies, any account logically begins with the pioneering investigations of the spinal Renshaw cell, conducted by Eccles and colleagues. Renshaw cells are interneurones which receive a cholinergic innervation from the axon colaterals of alpha-motoneurones. They are excited by systemic and iontophoretic application of cholinomimetics; nicotine, and other nicotinic agonists, are particularly potent (Curtis and Eccles 1958; Curtis and Ryall 1966*a*).

When acetylcholine is iontophoresed onto single Renshaw cells, the nicotinic response is seen to have three noteworthy characteristics: rapidity, high sensitivity to acetylcholine, and a propensity to fade with continued application of the transmitter (Curtis and Eccles 1958; Curtis and Ryall 1966*a*). Thus the response was seen as soon as 30 ms after the beginning of the acetylcholine application, and usually ended within 0.5 s when the application was terminated. As little as 10^{-14} moles of acetylcholine was effective. As shown in Fig. 5.3 (Curtis and Ryall 1966*a*), when a constant current of acetylcholine was applied, the neuronal firing rate reached a peak after 1–2 s, and subsequently waned. This 'response fade' is partly attributable to a

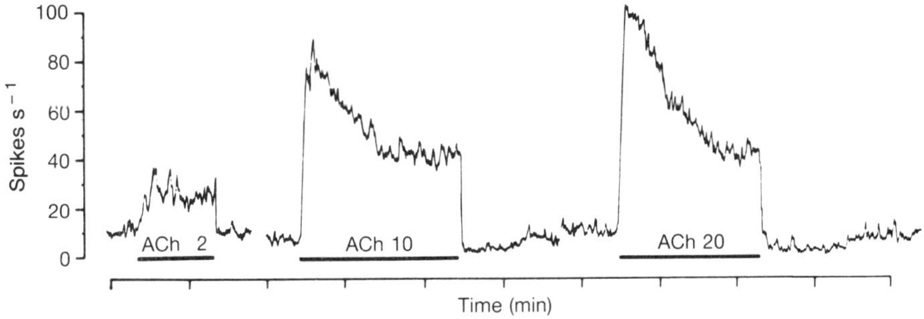

Fig. 5.3. Patterns of firing of a feline Renshaw cell produced by iontophoretic ejections of acetylcholine. The ordinate shows the rate of neuronal firing determined by extracellular recording. The abscissa indicates time in minutes. Iontophoretic ejections (with acetylcholine current in nA) are shown by horizontal bars. With prolonged application, the neuronal response waned (from Curtis and Ryall 1966*a*).

desensitization of nicotinic receptors. Iontophoresed nicotine was also extremely potent; the response was more prolonged, presumably because, unlike acetylcholine, nicotine is not rapidly hydrolyzed.

Renshaw cells can be driven synaptically by electrical stimulation of the ventral roots of the spinal cord. The resulting excitation is antagonized by mecamylamine (Ueki *et al.* 1961), indicating possible mediation by a ganglionic (C6) type of receptor. Subsequent iontophoretic experiments have confirmed the predominantly nicotinic nature of the response, and have in addition revealed the co-existence of muscarinic and nicotinic receptors on individual Renshaw cells (Curtis and Ryall 1966*a*, 1966*b*). The synaptic response to ventral-root stimulation includes a fast-onset, high-frequency discharge mediated by nicotinic receptors, and a subsequent late excitation mediated by muscarinic receptors (Curtis and Ryall 1966*c*). The fast nicotinic response and the slower muscarinic excitation were mimicked by iontophoretic application of nicotine and acetyl-β-methylcholine, respectively, and selectively reduced by nicotinic and muscarinic antagonists (Curtis and Ryall 1966*b*). These experiments show that acetylcholine excites individual Renshaw cells by two distinctly different mechanisms, mediated respectively by nicotinic and muscarinic receptors.

5.5 Nicotinic actions in the brain

Individual neurones in the CNS vary markedly in their responsiveness to iontophoresed nicotine, and an important factor in this respect is the density of receptors on the cell membrane. Recently, autoradiographic maps have been produced which show the relative density of putative nicotinic receptors in different brain areas (Clarke *et al.* 1984, 1985*b*; London *et al.* 1985; Schwartz 1986). Tritiated nicotine and [^{3}H]acetylcholine ([^{3}H]ACh) (with muscarinic receptors occluded) label a single population of sites (Clarke *et al.* 1985*b*; Marks *et al.* 1986; Martino-Barrows and Kellar 1987) which represent pharmacological sites of action for nicotine (see Clarke 1987*a*); in contrast, [^{125}I]αbungarotoxin labels a different population whose functional significance is unclear (Clarke 1987*a*; see Wonnacott, chapter 7, this volume). The one-to-one correspondence of [^{3}H]nicotine and 'nicotinic' [^{3}H]ACh binding sites finds a counterpart in electrophysiology: nicotinic actions of acetylcholine (i.e. actions blocked by nicotinic cholinergic antagonists) are almost always mimicked by nicotine, and vice versa.

Whilst autoradiographic maps do not permit an analysis of receptor density on individual neurones, there is nevertheless a rough correlation between [^{3}H]nicotine (and nicotinic [^{3}H]ACh) labelling and neuronal responsiveness to nicotine, as illustrated below. Areas possessing high [^{3}H]nicotine-binding density are discussed first.

5.5.1 *Interpeduncular nucleus (IPN)*

This nucleus has been extensively explored from a neuroanatomical standpoint, but its function remains unclear. It possesses the highest density of

$[^3H]$nicotine and $[^3H]$ACh labelling of any brain area (Clarke *et al.* 1984, 1985; London *et al.* 1985; Schwartz 1986). Although labelling within the IPN is not uniform, all subnuclei are densely labelled (Hamill *et al.* 1986). Binding of $[^{125}I]$αbungarotoxin is dense in certain subnuclei and sparse in others (Hamill *et al.* 1986).

Many IPN neurones are sensitive to the application of cholinergic agonists (Lake 1973; Sastry 1978; Brown *et al.* 1983), including nicotine itself (Takagi 1984). In an extensive investigation employing an *in vitro* slice preparation, the majority of single units responded to acetylcholine and carbamylcholine (Brown *et al.* 1983). The action of carbamylcholine was shown to be post-synaptic, since it persisted when the overall concentration of Ca^{2+} was lowered in order to prevent synaptic transmission. Actions of acetylcholine and

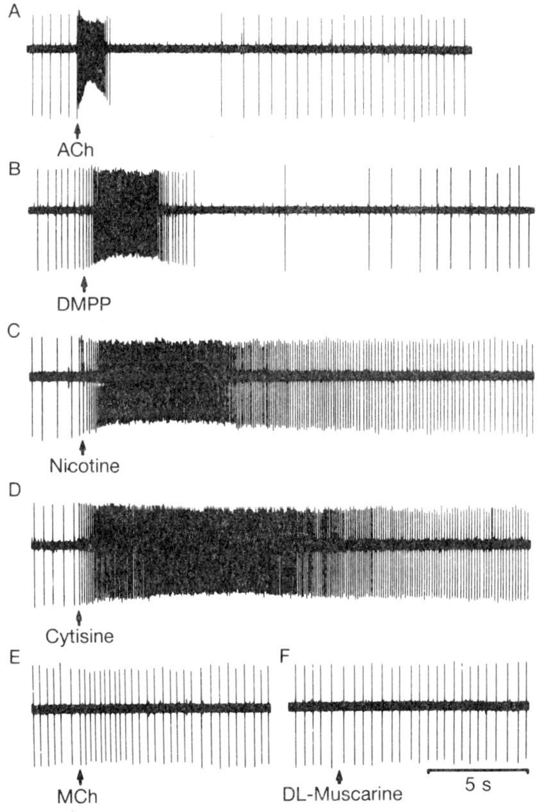

Fig. 5.4. Pharmacological characterization of receptors mediating responses to acetylcholine (ACh) in the guinea pig medial habenula. All data are from the same neurone. Application of acetylcholine (A), and DMPP (B), caused an excitation followed by an inhibition. In contrast, application of the nicotinic agonists cytisine (C), and nicotine (D), caused prolonged excitation with little or no inhibition. Muscarinic agonists induced a slight excitation (acetyl-β-methylcholine: MCh, E) or no response (DL-muscarine, F). (From McCormick and Prince 1987*a*).

carbamylcholine were greatly reduced or blocked by hexamethonium, *d*-tubocurarine, and mecamylamine, whereas atropine was much less effective, indicating that the neuronal response was mediated predominantly by nicotinic receptors. αBungarotoxin was ineffective.

5.5.2 *Medial habenula (mHb)*

Like the IPN to which it projects, the mHb is as yet a structure without a clear biological role. This nucleus is also densely labelled with [^{3}H]nicotine and [^{3}H]ACh, but not by [^{125}I]αbungarotoxin (Clarke *et al.* 1985*b*; London *et al.* 1985; Schwartz 1986). In addition, techniques of molecular biology suggest that cell bodies within this nucleus synthesize nicotinic receptors (Boulter *et al.* 1986; see Wonnacott, Fig. 7.8, this volume). The actions of nicotine and other cholinergic agonists have been examined using single unit recording techniques in an *in vitro* slice preparation of guinea pig brain (McCormick and Prince 1987*a*). Local application of the nicotinic agonists, nicotine, cytisine, and dimethylphenylpiperazinium (DMPP) caused long-lasting excitation; for reasons presently unclear, only DMPP produced inhibitory responses (see Fig. 5.4). Acetylcholine induced a rapid excitation, followed by inhibition. When Mg^{2+} ions were substituted for Ca^{2+} ions in the bathing solution, synaptic transmission was shown to be blocked; the actions of acetylcholine were unchanged, and were therefore probably post-synaptic. Muscarinic agonists were not consistently effective. Both excitatory and inhibitory actions of acetylcholine were blocked by hexamethonium, and were unaffected by muscarinic antagonists. These electrophysiological results are consistent with data from radioligand binding studies, which indicate a high density of nicotinic but not muscarinic cholinoceptors in this nucleus (Rotter *et al.* 1979; Clarke *et al.* 1984, 1985; London *et al.* 1985; Schwartz 1986).

5.5.3 *Cerebral cortex*

The cerebral cortex is concerned with sensorimotor and higher functions. A laminar distribution of [^{3}H]nicotine and [^{3}H]ACh labelling is apparent in all species examined (rat, cat, rhesus monkey, human; see Kellar and Wonnacott, Fig. 10.4, this volume). In the rat, binding is present in all layers, but it is particularly dense in layers I and III/IV (Clarke *et al.* 1984, 1985); dense labelling has also been reported in layer V (London *et al.* 1985), possibly confined to certain cortical regions. In the rhesus monkey, [^{3}H]nicotine binds predominantly in layer III, especially in the primary sensory areas (Friedman *et al.* 1985). In the cat visual cortex, labelling is concentrated primarily in layer IV of areas 17 and 18; lesion studies indicate that these binding sites are located on the presynaptic terminals of geniculocortical projection neurones (Prusky *et al.* 1987).

Most investigators have found the actions of acetylcholine in cerebral cortex to be muscarinic (for example, Krnjevic and Phillis 1963*b*). Nevertheless, nicotinic actions have been reported. McLennan and Hicks (1978) observed

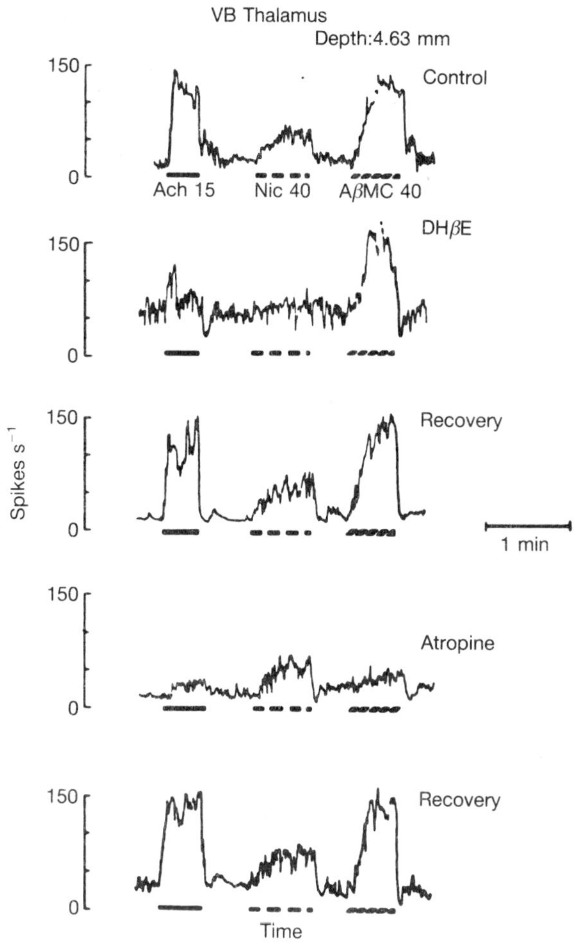

Fig. 5.5. Responses of a single neurone of the ventrobasal thalamus to nicotinic (dihydroβerythroidine; D H β E) and muscarinic (atropine) antagonism of excitations produced by acetylcholine (A Ch) and by cholinomimetics. Responses to selective nicotinic (Nic) or muscarinic (acetyl-β-methylcholine; A βM C) agonists were antagonized by the corresponding blocker. In contrast, acetylcholine-induced excitations were antagonized by either blocker (from McLennan and Hicks 1978).

single units in an unspecified region of rat cerebral cortex, which were responsive to iontophoresed acetylcholine and also to nicotinic and muscarinic agonists (see Fig. 5.5). The predominant action was excitatory, and excitations produced by nicotine, and by the nicotinic agonist DMPP, were usually more rapid in onset and offset than those induced by a muscarinic agonist. Responses to carbamylcholine resembled those to nicotinic agonists, and were consistently antagonized by nicotinic blockers rather than by atropine.

Although the actions of carbamylcholine were predominantly nicotinic, those of acetylcholine were abolished by both nicotinic and muscarinic antagonists.

In the above study, single units responsive to nicotine were encountered at all depths within the cortex. In contrast, Stone (1972) was unable to detect responses to iontophoresed nicotine at more than 0.6 mm below the cortical surface. In an unspecified number of superficial units, these responses were mimicked by acetylcholine and were blocked by *d*-tubocurarine, which did not apparently reduce glutamate-induced excitation. Other nicotinic antagonists were not tested. Application of nicotine was associated with unusually long response latencies, and after-discharges were encountered, quite unlike those associated with the 'classical' nicotinic response of Renshaw cells. It must therefore be questioned whether this action of nicotine was post-synaptic.

Rather different results have been reported by Lamour and coworkers (1982). In rat somatosensory (S1) cortex, neuronal responses to cholinergic agents were almost exclusively excitatory. The actions of acetylcholine and carbamylcholine were almost wholly confined to the deep layers (V and VI), and were much more effectively suppressed by atropine than by mecamylamine. A muscarinic mechanism was further suggested by the temporal characteristics of the response, and by the greater potency of muscarinic agonists compared to nicotinic agonists. The excitatory effect of nicotine was rarely observed. For reasons presently unclear, half of the neurones which were excited by nicotine were not excited by acetylcholine, and it is regrettable that conventional nicotinic antagonists were not tested. Whereas Stone (1972) reported nicotinic excitation confined to the superficial layers of cortex, Lamour *et al.* (1982) detected such excitations in deeper layers as well. The reason for this discrepancy is unclear. Possibly it reflects the propensity of nicotinic acetylcholine receptors to desensitize (Egan and North 1986); it has been suggested that nicotinic agonists may leak from micropipettes in sufficient amounts to cause a preemptive inactivation of receptors (Stone 1972).

The majority of studies suggest that acetylcholine and nicotine exert largely excitatory actions on cortical cells (for example Krnjevic and Phillis 1963*a*; Lamour *et al.* 1982). However, Phillis and York (1968) reported that in cat pericruciate cortex, almost half of all cells tested were depressed by acetylcholine and other cholinomimetics, including nicotine, when these drugs were iontophoresed. Both muscarinic and nicotinic antagonists were effective against acetylcholine-induced inhibition, and also against depressions induced by other cholinomimetics (non-cholinergic inhibitory agents such as GABA were not tested). This latter finding is puzzling, since in studies where the excitatory actions of acetylcholine have been antagonized by both types of blocker, antagonist selectivity has been maintained against selective agonists such as nicotine and muscarine (see section 5.3.3). For these and other reasons (see Stone 1972), the possibility of a nicotinic inhibitory action in this tissue must be viewed with caution.

5.5.4 *Substantia nigra zona compacta (SNC) and ventral tegmental area (VTA)*

These nuclei contain the cells of origin of the nigrostriatal and mesolimbic dopaminergic systems, which are involved, respectively, with motor and motivational functions (see Reavill, chapter 9, this volume). When dopaminergic neurones were selectively destroyed by microinjection of the neurotoxin 6-hydroxydopamine, [^{3}H]nicotine binding was markedly reduced (Clarke and Pert 1985). In anaesthetized rats, dopaminergic cells within the SNC are excited by iontophoresed acetylcholine and nicotine (Lichtensteiger *et al.* 1976), and these actions are blocked by iontophoretically-applied dihydroβerythroidine (see Fig. 5.6; Lichtensteiger *et al.* 1982). These neurones are also excited by fairly high systemic doses of nicotine (Lichtensteiger *et al.* 1976); this effect is prevented by systemic administration of mecamylamine, but not of the quaternary antagonist chlorisondamine (Clarke *et al.* 1985*a*). Because chlorisondamine does not readily cross the blood–brain barrier, these results

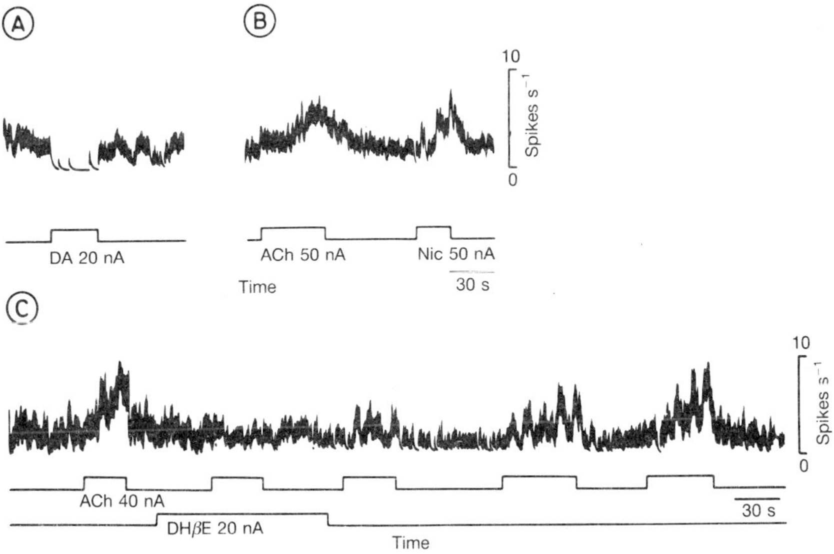

Fig. 5.6. Extracellular recording of a single neurone in rat substantia nigra zona compacta. The neuronal firing rate was depressed by iontophoretic application of dopamine (DA); this action is mediated by autoreceptors and identifies the cell as dopaminergic (A). The cell was excited by iontophoresed acetylcholine and nicotine (B). Dihydroβerythroidine (D H β E) prevented the action of acetylcholine (C). (From Lichtensteiger *et al.* 1982.)

point to a central and presumably direct action. The subjacent pars reticulata of the substantia nigra contains few dopaminergic cells, and has a low density of [^{3}H]nicotine labelling (Clarke and Pert 1985). Unexpectedly, non-dopaminergic cells in this nucleus were potently stimulated by intravenous doses of nicotine; in this case, however, both mecamylamine and chlorisondamine prevented the effect (Clarke *et al.* 1985*a*). This study shows that nicotine need not act centrally in order to exert profound effects on the activity of neurones within the CNS. It is worth noting that this indirect effect was produced by doses as low as $2 \mu g\, kg^{-1}$ intravenously, thought to approximate to the dose received from a single puff of tobacco smoke (Armitage *et al.* 1968).

Two groups have compared the effects of systemic nicotine on the firing rate of dopaminergic cells of the SNC and VTA in rats. Grenhoff *et al.* (1986) reported that in rats anaesthetized with chloral hydrate, systemic administration of nicotine ($0.5\, mg\, kg^{-1}$ intraperitoneally) significantly increased the firing rate of dopaminergic cells in the SNC, but not in the VTA. However, when the temporal pattern of firing was examined, more subtle effects emerged. The basal firing of dopaminergic cells in both SNC and VTA was found to consist of single spikes and bursts. The latter were characterized as series of 2–15 spikes of diminished amplitude, separated by short intervals, and followed by a compensatory pause. Although nicotine significantly increased burst firing in SNC and VTA dopaminergic neurones which had demonstrated this type of neuronal activity during baseline recording, the drug did not induce burst-firing in non-bursting neurones. In addition, no correlation was observed between nicotine-induced changes in firing rate and in burst-firing activity, and it was suggested that the nicotinic effect was not purely excitatory, but possessed a modulatory component as well.

Since general anaesthetics can disrupt the pharmacological responses of these neurones, Yoon *et al.* (1986) carried out experiments in subjects under chloral hydrate (general) anaesthesia, and also in locally-anaesthetized animals. Consistent with previous findings, nicotine ($50–200 \mu g\, kg^{-1}$ intravenously) increased the firing rate of SNC dopaminergic neurones by approximately 25 per cent, regardless of the anaesthetic employed. In sharp contrast, nicotine in this dose range failed to excite dopaminergic cells in the VTA in deeply-anaesthetized rats, but increased neuronal activity by up to 70 per cent in locally-anaesthetized subjects. All the above effects were reversed and prevented by mecamylamine. This study makes an important point: effects of nicotine on central neurones may be masked be general anaesthesia. These results accord with neurochemical findings which show that in freely-moving rats, reasonable systemic doses of nicotine increase dopaminergic activity selectively in the mesolimbic system (Imperato *et al.* 1986; Clarke *et al.* 1988).

5.5.5 *Thalamus*

[^{3}H]Nicotine and [^{3}H]ACh labelling is concentrated in the so-called 'specific' nuclei of the thalamus which relay sensory and motor information to and from the cerebral cortex (Clarke *et al.* 1984, 1985*b*). The following nuclei are particularly densely labelled in the rat: laterodorsal, anteroventral, anteromedial, lateral posterior, and ventroposterior (Clarke *et al.* 1984, 1985*b*; London *et al.* 1985; Schwartz 1986). The dorsolateral and medial geniculate nuclei are also heavily labelled. An analogous distribution has been observed in [^{3}H]nicotine autoradiographs of rhesus monkey brain (Friedman *et al.* 1985). Comparable data are not available for the cat, which is regrettable since may electrophysiological studies have been conducted using this species.

In the thalamus, single unit responses to nicotine have been less thoroughly investigated than those to acetylcholine. Nevertheless, a nicotinic component of action if frequently seen. Neurones sensitive to iontophoretically-applied acetylcholine occur throughout the thalamus, and they are particularly well-represented in the ventrobasal complex, comprising the ventroposterolateral and ventroposteromedial nuclei (McCance *et al.* 1968*a*). Acetylcholine excited almost all neurones tested within this complex (Anderson and Curtis 1964*a*). Thalamocortical relay cells demonstrated a sensitivity to acetylcholine which approached that of Renshaw cells, but the onset and offset of firing were invariably less prompt (Andersen and Curtis 1964 *a*; McCance *et al.* 1968*a*). Acetylcholine exerted a less potent, modulatory effect on interneurones, revealed by increased sensitivity to excitatory amino acids or synaptic excitation (Andersen and Curtis 1964*a*). Both dihydroβerythroidine and atropine antagonized acetylcholine-excitation (Andersen and Curits 1964*b*; McCance *et al.* 1968*b*), but atropine also reduced neuronal sensitivity to an excitatory amino acid (Andersen and Curtis 1964*b*). It has been concluded that thalamic receptors mediating acetylcholine-excitation are of an intermediate nicotinic/muscarinic type. However, in the light of recent findings (see section 5.3.3), these results are probably consistent with a fast-nicotinic action and a slower-modulatory action mediated by muscarinic receptors.

The dorsal lateral geniculate nucleus (dLGN) and medial geniculate nucleus (MGN) contain neurones which relay, respectively, visual and auditory information to the cerebral cortex. Acetylcholine strongly excites feline dLGN cells (for example, Sillito *et al.* 1983). In early work (Phillis *et al.* 1967), acetylcholine-induced excitation was found to be prevented by muscarinic antagonists and not by dihydroβerythroidine; nicotine was consistently less potent than acetylcholine. Despite the apparent preponderance of muscarinic effects, some of the extracellular responses to acetylcholine were extremely rapid in onset and offset (Sillito *et al.* 1983), and a recent *in vitro* intracellular recording study has confirmed the existence of a fast nicotinic component

(McCormick and Prince 1987*b*). Thus in cat thalamic slices, a fast depolarization was found in practically all neurones sampled in the dLGN and MGN, following the local application of acetylcholine, nicotine, or DMPP. This type of response was generated in thalamocortical relay neurones. Depolarization occured with an onset latency (10 ms) comparable to that associated with Renshaw cells, and was susceptible to tachyphylaxis. A fast, presumably nicotinic, deporization was also observed in dLGN slices of guinea-pig and rat, although the proportion of neurones showing this response was much less. Significantly, many neurones demonstrating this fast nicotinic depolarization also displayed more sluggish hyperpolarizing and/or depolarizing responses which were mediated by muscarinic receptors. These experiments demonstrate the power of intracellular recording techniques and clearly reveal that distinct nicotinic and muscarinic receptors mediate different actions of acetylcholine on geniculate neurones in the cat.

5.5.6 *Dentate gyrus*

This structure is intimately connected with the hippocampus and is believed to play a role in learning and memory. [^{3}H]Nicotine labelling is densely represented in the middle third of the molecular layer of the dentate gyrus (Clarke *et al.* 1984), suggesting that nicotinic receptors may be present on the dendrites of dentate granule cells. These neurones are excited by iontophoretic application of acetylcholine (Wheal and Miller 1980). The most common response was of slow onset and prolonged duration, and was antagonized by atropine. However, a significant number of neurones displayed a rapid, transient response to low iontophoretic currents of acetylcholine; the pharmacology of this response was not investigated.

5.5.7 *Neostriatum*

One of the principal components of the extrapyramidal motor system, the neostriatum possesses a moderate density of [^{3}H]nicotine and [^{3}H]ACh binding sites, and a significant proportion of these sites are probably located on dopaminergic and serotonergic nerve terminals in the rat (Schwartz *et al.* 1984; Clarke and Pert 1985; see Wonnacott, chapter 7, this volume). Binding of [^{125}I]αbungarotoxin, in contrast, is very low (for example, Clarke *et al.* 1985*b*). Misgeld and colleagues have employed an *in vitro* slice technique in order to study the receptors which are involved in intrinsic cholinergic excitation within the rat neostriatum. Following electrical stimulation close to the recording site, multi-unit potentials were evoked which could be blocked by anticholinergic drugs and potentiated by acetylcholinesterase inhibitors (Misgeld *et al.* 1980). More specifically, this 'intrinsic excitation' was suppressed by micromolar concentrations of mecamylamine and *d*-tubocurarine, but not by αbungarotoxin or by atropine. Intracellular recordings revealed not only a fast nicotinic excitation, but also a slow muscarinic excitation. The temporal and pharmacological characteristics of these responses were comparable to those of sympathetic ganglia and Renshaw cells. It is interesting that electrically-evoked nicotinic responses were readily observed, whereas muscarinic responses were only demonstrated convincingly in the presence of an acetylcholinesterase inhibitor. In contrast, the results of iontophoretic studies indicate a predominant muscarinic action.

5.5.8 *Inferior colliculus*

This midbrain structure is involved in auditory information processing. The inferior colliculus appears almost devoid of [^{3}H]nicotine and [^{3}H]ACh labelling (Clarke *et al.* 1985*b*; Schwartz 1986), but displays dense [^{125}I]αBungarotoxin labelling (Clarke *et al.* 1985*b*). In anaesthetized rats, most single units driven by auditory stimuli were either excited or depressed by iontophoretic acetylcholine and carbamylcholine (Farley *et al.* 1983). αBungarotoxin did not alter baseline (tone-evoked) firing but blocked agonist effects. Mecamylamine inhibited firing in the presence and absence of agonist, and hence it is not clear whether specific antagonism was involved. Dihydroβerythroidine was ineffective, confirming previous results (Curtis and Koizumi 1961); this is surprising since this antagonist is effective at both C6- and C10-type nicotinic receptors in the periphery.

5.5.9 *Cerebellum*

The cerebellum participates in motor coordination and in classical conditioning. [^{3}H]Nicotine labelling in the cerebellum is concentrated in an unidentified band running parallel with the cortical surface (Clarke *et al.* 1984). Binding of [^{125}I]αbungarotoxin is concentrated in clusters within the vermal regions of folia I, IX, and X (Hunt and Schmidt 1978). Nicotinic responses to acetylcholine have been recorded in the feline cerebellar cortex, particularly in the deeper layers (McCance and Phillis 1968). These responses were susceptible to tachyphylaxis. In the rat cerebellar cortex, locally-applied nicotine altered single unit activity in a manner strictly dependent on cell type (de la Garza *et al.* 1987*a*). Thus nicotine consistently inhibited Purkinje cells, and excited interneurones. The inhibitory effects of nicotine were antagonized by hexamethonium in most cases, whereas curare was rarely effective. Conversely, the excitatory effects of nicotine was consistently blocked by curare, but infrequently by hexamethonium. Acetylcholine mimicked both actions of nicotine, but muscarinic as well as nicotinic effects were observed. Neither action of nicotine was significantly reduced by simultaneous iontophoresis of Mg^{2+}, although administration of Mg^{2+} appeared to reduce the presynaptic release of neurotransmitters; this finding suggests that the inhibitory as well as the excitatory actions of nicotine are directly postsynaptic (de la Garza *et al.* 1987*b*).

In a subsequent study (de la Garza *et al.* 1987*c*), αbungarotoxin blocked the excitatory action of nicotine, leaving the inhibitory action unchanged. These results point to the existence of two subtypes of nicotinic receptor located on different populations of rat cerebellar neurones. The excitatory actions appear to be mediated by a neuromuscular type of receptor; the receptor mediating inhibitory actions is suggested to be ganglionic (C6) in nature, although this does not account for the ineffectiveness of curare.

5.5.10 *Locus coeruleus (LC)*

The LC contains an important group of noradrenergic cell bodies which innervate large parts of the CNS. [^{3}H]nicotine labelling is moderately dense in this nucleus. Nicotine (10–50μg kg^{-1} intravenously) caused a dose-dependent, short-lasting increase in firing rate of LC neurones in anaesthetized rats (Svensson and Engberg 1980). However, on the basis of iontophoretic experiments, these authors concluded that the excitation was indirect. Nevertheless, more recently Egan and North (1986) have identified a direct excitatory action of nicotine. Intracellular recordings were made in tissue slices *in vitro*. Both acetylcholine (in the presence of a muscarinic antagonist) and nicotine depolarized LC neurones and increased the firing rate. A direct action was shown by the persistence of these effects in conditions where synaptic transmission was prevented. Depolarizations evoked by local application of acetylcholine could be resolved into fast and slow components (see Fig. 5.7). The fast component was brief (1–2 s duration), and was reduced by

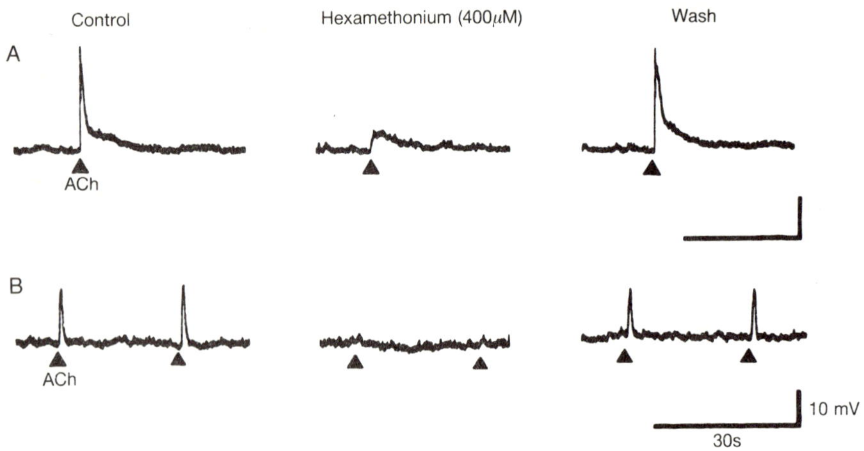

Fig. 5.7. Intracellular recording of rat locus coeruleus neurones *in vitro*. The ordinate shows the membrane potential which, in the resting state, is negative with respect to the extracellular space. Local application of acetylcholine caused cells to depolarize (shown by an upward deflection). (A) The typical response was a biphasic depolarization, the first component of which was blocked by hexamethonium. The residual slow depolarization was blocked by muscarinic antagonists (not shown). The trace on the right shows recovery upon removal of hexamethonium. (B) A similar experiment on an atypical neurone which displayed only the fast component of depolarization (from Egan and North 1986).

hexamethonium but not by muscarinic antagonists. Slow components were selectively blocked by muscarinic antagonists. In other experiments, the membrane potential was maintained at a constant value by voltage clamp, and neuronal responses were measured by membrane current. Dihydroβerythroidine did not completely block nicotinic responses; the residual component of the responses was blocked by hexamethonium. Possibly, then, nicotinic receptors in the LC are pharmacologically heterogeneous.

5.5.11 *Hypothalamus*

The various nuclei of the hypothalamus control the autonomic nervous system and the endocrine system, and serve to organize behaviours which are essential to survival. [^{3}H]nicotine binding is low throughout the hypothalamus, whereas [^{125}I]αbungarotoxin binding is dense (Clarke *et al.* 1985*b*; Schwartz 1986). Acetylcholine can excite neurosecretory cells in the supraoptic nucleus; this action was mimicked by nicotine and antagonized by dihydroβerythroidine, but not by atropine (Barker *et al.* 1971). This nicotinic response was seen in cells which also displayed a depressant response to acetylcholine, mediated by muscarinic receptors. In the paraventricular nucleus, neurosecretory cells were excited by an osmotic stimulus administered intravenously, and this response was reduced by iontophoretic application of hexamethonium (Akaishi and Negoro 1983). In the rabbit, neurosecretory cells in the paraventricular nucleus were identified by antidromic stimulation of the neurohypophysis (Moss *et al.* 1972). These neurones were almost all excited by iontophoretic acetylcholine. Atropine reduced acetylcholine excitations in half of the cells sampled. Approximately 50 per cent of neurosecretory cells were excited by iontophoretic nicotine, and 25 per cent were depressed by this drug. Dihydroβerythroidine blocked excitation produced by nicotine and acetylcholine in 8 out of 12 cells stimulated by both agonists. These results suggest that neurosecretory cells in the paraventricular nucleus possess nicotinic and muscarinic receptors mediating excitatory responses to acetylcholine. Three different types of response were distinguished on temporal grounds. One type of response, presumably nicotinic, was immediate, and waned completely over 15–20 s of continuous acetylcholine current. Nicotine depressed neuronal activity in some cells, and this effect was neither mimicked by acetylcholine nor significantly reduced by dihydroβerythroidine.

5.5.12 *Medulla and pons*

These hindbrain components are concerned with the control of sleep and arousal, relaying information to and from higher centres, and with vegetative functions. Acetylcholine-sensitive neurones are distributed unevenly in the pons and medulla. Excitatory responses to acetylcholine appear to possess mixed nicotinic and muscarinic properties (Bradley *et al.* 1966). In most studies, neurones responsive to iontophoresed acetylcholine have not been physiologically identified. However, Bradley and Lucy (1983) have compared the actions of iontophoresed cholinergic compounds on non-respiratory and respiratory-related neurones of the rat medulla. The two types of neurone responded in a similar fashion. Most neurones were excited by acetylcholine, nicotine, and muscarine. Responses to nicotine and muscarine were respectively blocked by a nicotinic and muscarinic antagonist (dihydroβerythroidine and atropine). Responses to acetylcholine were reduced or blocked by both antagonists, and it was concluded that individual neurones possessed both nicotinic and muscarinic receptors rather than a single population of 'mixed' receptors. The C6-selective antagonist mecamylamine failed to block the excitatory actions of nicotine and acetylcholine.

In the cat, single-unit excitatory responses to acetylcholine were found to be scattered in the medulla (Salmoiraghi and Steiner 1963). Two-thirds of these excitations were reduced or blocked by iontophoresed dihydroβerythroidine or hexamethonium, whilst atropine was ineffective. However, these putative nicotinic responses were of slow onset (2–20 s).

5.5.13 *Hippocampus (excluding dentate gyrus)*

The rat hippocampus possesses a very low density of [^{3}H]nicotine and nicotinic [^{3}H]ACh binding sites as detected by receptor autoradiography (Clarke *et al.* 1984, 1985*b*; London *et al.* 1985; Schwartz 1986). There is less agreement between studies which have examined the density of binding in homogenate preparations of microdissected hippocampal tissue (Wonnacott 1987). Binding of [^{125}I]αbungarotoxin is high in certain regions (Hunt and Schmidt 1978; Clarke *et al.* 1985*b*). Pyramidal cells were readily excited by iontophoresed acetylcholine and carbamylcholine (Bird and Aghajanian 1976; Segal 1978). This action was of slow onset and long duration, and was mimicked by muscarinic agonists but not by nicotine or DMPP, which either depressed neuronal firing or had no effect. The acetylcholine-induced excitation of pyramidal cells was consistently blocked by muscarinic antagonists but not by *d*-tubocurarine, confirming a muscarinic profile of action. However, excitation evoked by acetylcholine was blocked by iontophoretic application of the nicotinic antagonist dihydroβerythroidine which acted selectively, in so far as it did not alter responses to excitatory amino acids (Bird and Aghajanian 1976). This antagonist also completely blocked the excitatory actions of 'pure' muscarinic agonists, an unexpected finding since other investigators have found dihydroβerythroidine to exert little if any effect on muscarinic receptors (Curtis and Ryall 1966*b*).

More recently, Rovira *et al.* (1983) have detected a nicotinic input associated with hippocampal pyramidal cells situated in the same area of hippocampus (CA1), and have observed a clear-cut distinction between nicotinic and muscarinic actions. These authors suggest that the anti-muscarinic effects of dihydroβerythroidine observed by Bird and Aghajanian (1976) may have resulted from inappropriately high iontophoretic doses of the antagonist. Rovira *et al.* (1983) also report that the actions of nicotinic agonists (DMPP, tetramethylammonium) were blocked by curare and dihydroβerythroidine, whereas mecamylamine and hexamethonium were ineffective. These findings point to the possible existence of a non-ganglionic type of nicotinic receptor in the hippocampus. Whether such a receptor would be labelled by [^{125}I]αbungarotoxin remains to be investigated.

5.5.14 *Interim summary*

This section has surveyed neuronal responsiveness to nicotine in different brain areas. It is clear that nicotinic responses are found most convincingly in brain areas which possess appreciable [^{3}H]nicotine labelling. The best electrophysiological evidence for brain nicotine receptors is found in the interpeduncular nucleus, medial habenula, substantia nigra zona compacta, ventral tegmental area, neostriatum, and locus coeruleus. In the cerebral cortex, electrophysiological findings have been diverse and difficult to interpret, despite the presence of dense [^{3}H]nicotine labelling. In other areas of lower [^{3}H]nicotine binding (for example cerebellum, inferior colliculus, pons, and medulla), results have been obtained which indicate the possible existence of a non-ganglionic type of nicotinic receptor (see section 5.9).

5.6 Mechanisms of action

In the periphery and in the CNS, nicotine is primarily excitatory. For example, at the neuromuscular junction, nicotinic transmission is associated with increased cation conductance, permitting K^+ to leave the post-junctional cell and Na^+ to enter. The net effect is depolarization. Ion translocation is directly coupled to agonist binding and thus permits the faithful transfer of frequency-encoded, digital signals. An analogous mechanism underlies nicotinic excitations in sympathetic ganglia (Brown 1983). Nicotinic receptors in the mammalian brain are structurally related to those in the periphery (Goldman *et al.* 1987; see Wonnacott, chapter 7, this volume); biochemical and electrophysiological studies suggest that they also control the opening of cation channels (Rapier *et al.* 1987; Egan and North 1986; McCormick and Prince 1987*a*).

Nicotinic responses exhibited by single units in the brain are frequently slower than the 'classical' nicotinic responses seen in the periphery and in the spinal cord. Conceivably, a sluggish response could reflect drug diffusion to receptors located on distal dendrites. Nevertheless, it is also possible that the brain possesses a subtype of receptor with slower kinetics than found elsewhere. The patch-clamp recording technique would best address this question but has only very recently been applied to brain cells *in situ*.

5.7 Pharmacological actions of nicotine in relation to cholinergic transmission

In the periphery, nicotine mimics certain actions of synaptically-released acetylcholine at ganglionic and neuromuscular sites. In the CNS, almost all studies have shown that neuronal excitation evoked by iontophoresed nicotine can be mimicked by acetylcholine. This is evidence of cholinoceptivity, but does not of itself imply that neurones which are stimulated by nicotine are also excited by *endogenously-released* acetylcholine. Several authors have suggested that nicotine may act at 'non-cholinergic' sites in the brain. In essence, two sorts of non-cholinergic action could readily be envisaged:

(1) at nicotinic cholinoceptors located outside cholinergic synapses

(2) at non-cholinoceptive receptors, i.e. receptors that do not recognize acetylcholine even when this is applied exogenously.

Attempting to test a non-cholinergic mode of action, Rosecrans and Meltzer (1981) employed a combination of drugs (physostigmine, atropine, and hexamethonium) which was intended to increase the concentration of transmitter acting on nicotinic receptors within central cholinergic synapses. When administered to rats, this drug combination failed to mimic the discriminative stimulus (cue) properties of nicotine. This result is scarcely conclusive, however, because there is no clear indication that the drug combination had indeed enhanced nicotinic cholinergic transmission in the CNS.

In a very few single-unit recording studies, actions of nicotine have been reported which were not mimicked by acetylcholine (for example, Segal 1978). However, in these cases, a muscarinic action of acetylcholine appears to have predominated, which may have masked any nicotine-like action of acetylcholine.

Receptor-binding experiments provide positive evidence that a significant proportion of central acetylcholine receptors are found at cholinergic synapses. Thus, several groups have reported a down-regulation (reduced density) of sites that are labelled with [^{3}H]nicotine or [^{3}H]ACh, following chronic *in vivo* treatment with acetylcholinesterase inhibitors in rodents; this down-regulation has been observed in all brain regions examined (Marks *et al.* 1983; Costa and Murphy 1983; Schwartz and Kellar 1985; see Wonnacott, chapter 7, this volume).

The recurrent collateral-Renshaw cell synapse within the spinal cord provides the only clear demonstration of nicotinic-cholinergic neurotransmission in the CNS. For several reasons, the brain has proved much less amenable to study: it is generally not possible to stimulate cholinergic pathways without stimulating non-cholinergic pathways; unequivocal identification of central cholinergic neurones and pathways has only been possible in the last few years, with the introduction of immunohistochemical techniques (for review, see Wainer *et al.* 1984; for detailed map, see Reavill, Fig. 9.1, this volume); the responses of central neurones to acetylcholine frequently have a strong muscarinic component. Despite these technical obstacles, several candidate pathways have been examined.

1. The interpeduncular nucleus (IPN) receives a massive cholinergic projection contained within a bilateral pathway (the fasciculus retroflexus). Cholinergic afferents to the IPN derive from the mediobasal forebrain, the medial habenula, and the dorsal tegmentum (see Clarke *et al.* 1986). The IPN also possesses the highest density of [^{3}H]nicotine labelling of any brain nucleus (Clarke *et al.* 1984, 1985*b*; Schwartz 1986). The majority of IPN neurones are excited by iontophoretic application of acetylcholine *in vitro*, and this action is primarily nicotinic (Brown *et al.* 1983). Almost all IPN neurones were excited by pulsed electrical stimulation of the fasciculus retroflexus, and the response was of short latency (2–10 ms), and time-locked to the stimulation. However, these evoked responses were not appreciably affected by high doses of nicotinic antagonists, nor by atropine, nor were they enhanced by an acetylcholinesterase inhibitor. Further experiments suggested that the evoked responses were mediated by an excitatory amino acid. The involvement of transmitters other then acetylcholine is to be expected, since the fasciculus retroflexus is a mixed nerve tract, but the absence of a detectable cholinergic component is hard to reconcile with the concentration of cholinergic markers within this neural system.

2. Nicotinic-cholinergic neurotransmission has been demonstrated within the neostriatum (Misgeld *et al.* 1980). Cholinergic innervation is exclusively derived from intrinsic neurones which are sparsely scattered throughout this structure, but the identity or function of the innervated cells has yet to be established.

3. Dopaminergic cells in the rat substantia nigra zona compacta (SNC) are excited by exogenous acetylcholine and nicotine (see section 5.5.4). Retrograde-tracing experiments indicate a possible input from cholinergic perikarya located in the pedunculopontine nucleus. Local excitation of neurones in the vicinity of this nucleus results in a stimulation of dopaminergic cell firing which is blocked by systemic administration of the nicotinic antagonist mecamylamine (Clarke *et al.* 1987). It would therefore appear that cholinergic pedunculopontine neurones synapse onto the nicotinic receptors of dopaminergic cells in the SNC. Although cholinergic terminals have been demonstrated immunohistochemically within the SNC, the activity of choline acetyltransferase (the synthesizing enzyme for acetylcholine) is quite low. This is consistent with the exquisite sensitivity of nicotinic receptors for acetylcholine which has been noted in certain regions of the CNS (Curtis and Eccles 1958; McCormick and Prince 1987*a*).

4. Neurones in the medial habenula are directly excited by acetylcholine and this action is purely nicotinic (McCormick and Prince 1987*a*). This nucleus appears to receive a cholinergic projection from the diagonal band and posterior septal region.

5. Dentate granule cells are innervated by acetylcholine, and in addition to a slow, atropine-sensitive response, there is a brief excitation of unexplored pharmacology, whose temporal properties suggest a possible nicotinic action (Wheal and Miler 1980). The dense band of [^{3}H]nicotine binding sites within the dentate gyrus molecular layer (Clarke *et al.* 1984) indicates a possible nicotinic cholinergic input to the dendritic tree of granule cells.

6. In the feline cerebellum, granule layer cells are excited by iontophoresed acetylcholine, and this response is mediated by nicotinic receptors (McCance and Phillis 1968). These cells are also excited by electrical stimulation of the mesencephalic reticular formation/nucleus reticularis oralis complex, an area containing cholinergic cell bodies which presumably project to the cerebellum. Iontophoretic application of dihydroβerythroidine abolished acetylcholine-induced and synaptically-driven increases in discharge frequency.

5.8 Neuromodulation of nicotinic actions

The actions of several classical neurotransmitters are known to be altered by neuromodulatory substances which are stored and released from the same terminals (Lundberg and Hokfelt 1983). Acetylcholine has been found to coexist with vasoactive intestinal peptide (VIP), substance P, or corticotropin releasing factor (CRF) in different neuronal populations (see Fuxe *et al.*, chapter 6, this volume). Several peptides, exogenously applied, have been found to alter nicotinic actions. These include cholecystokinin (Mo and Dun 1986), somatostatin, and substance P (Mizobe *et al.* 1979; see Wonnacott, chapter 7, this volume). Iontophoretic application of substance P is reported to block nicotinic excitations of Renshaw cells, leaving responses to excitatory amino acids unaltered (Ryall and Belcher 1977). In various peripheral tissues, substance P has been found either to enhance or depress nicotinic actions. It is thought that this peptide controls nicotinic receptor function in an allosteric manner. Another potent neuromodulator of possible physiological significance is adenosine; both adenosine and its antagonist caffeine potently alter the sympathomimetic effects of nicotine in rats (von Borstel *et al.* 1986). Possible interactions between adenosine or caffeine and nicotine in the CNS await investigation.

5.9 Nicotinic receptor subtypes

Several lines of evidence point to the existence of nicotinic receptor subtypes in the mammalian CNS:

1. [^{125}I]αbungarotoxin labels a population of nicotinic binding sites which is different from the population of sites labelled by [^{3}H]nicotine and [^{3}H]ACh (Clarke *et al.* 1985*b*; Marks *et al.* 1986; Martino-Barrows and Kellar 1987). Two components of high-affinity [^{3}H]nicotine binding have been distinguished having different regional distributions in rat brain (Yamada *et al.* 1985).

2. Certain behavioural and physiological effects of nicotine differ in their sensitivity to blockade by mecamylamine, consistent with the existence of two or more subtypes of central receptor (Collins *et al.* 1986).

3. Molecular biological experiments have identified a family of genes coding for different putative nicotinic receptor subtypes in rodent brain (Goldman *et al.* 1987; see Wonnacott, chapter 7, this volume).

None of these experimental approaches stands alone. Electrophysiology addresses the issues of anatomical localization and physiological significance, providing a bridge between radioligand-binding and molecular biological studies on the one side, and studies of the whole organism on the other. Despite the uncertainties of the iontophoretic technique and single unit recording in general (see section 5.3.1), the electrophysiological approach has helped to clarify the different actions of nicotine in the CNS. In particular, two hypotheses can be put forward:

1. Central nicotinic receptors labelled by [^{3}H]nicotine and [^{3}H]ACh are pharmacologically similar to ganglionic (C6) receptors. This hypothesis finds strong support outside the domain of electrophysiology (see Clarke 1987*a*). Receptors of this sort should be sensitive to C6-selective antagonists (for example, hexamethonium, mecamylamine), to non-selective C6/C10 antagonists (for example, curare, dihydroβerythroidine), but not to C10-selective blockers (for example, decamethonium, αbungarotoxin). Nicotinic receptors possessing this pharmacological profile have been identified on Renshaw cells (Duggan *et al.* 1976), in the interpeduncular nucleus (Brown *et al.* 1983), locus coeruleus (Egan and North 1986), and in the neostriatum (Misgeld *et al.* 1980). In other areas, C10-selective antagonists have not been tested, but otherwise the same pharmacological profile has been observed (for example, substantia nigra zona compacta, medial habenula). All these examples are associated with appreciable [^{3}H]nicotine and [^{3}H]ACh labelling (see Table 5.1).

2. Central nicotinic receptors found in areas of low [^{3}H]nicotine and [^{3}H]ACh autoradiographic labelling possess a different pharmacological profile (see Table 5.1). Typically, the actions of 'smoking' doses of nicotine are blocked by centrally-active C6 selective antagonists such as mecamylamine (see Clarke 1987*b*). There are, nevertheless, reports of nicotinic actions on single neurons which are not blocked by C6-selective antagonists, and these examples all occur in areas possessing low [^{3}H]agonist labelling, i.e. inferior colliculus (Farley *et al.* 1983), CA1 region of hippocampus (Rovira *et al.* 1983), and medulla (Bradley and Lucy 1983). In the inferior colliculus, [^{125}I]αbungarotoxin labelling is very dense, and αbungarotoxin appears to possess some antagonist potency (Farley *et al.* 1983). An antagonistic action of αbungarotoxin has also been demonstrated electrophysiologically in the cerebellum (de la Garza *et al.* 1987*c*), and here there is evidence of a second subtype of nicotinic receptor which is sensitive to hexamethonium (see section 5.5.9). It remains to be determined whether receptor subtypes which are

Region	^{3}H-Nic	^{125}I-BTX	Agonists	Block	No Block	Receptor
I P N	high	high	Carb, ACh, Nic	Hex, Mec, dT C	Atr	C6
mHb	high	low	ACh	Hex	Atr, Scop	C6
			Nic, Cyt, D M P P	—		
Cerebral	high	high	Carb, Nic	D H B E, Mec, dT C	Atr	C6
cortex			A Ch	D H B E, Mec, dT C, Atr		
			D M P P	—		
S N C, V T A	high	low	A Ch, Nic	D H B E	—	C6
			Nic	Mec	Chloris (iv)	
Thalamus	high	low	ACh	D H B E	Atr	C6
			ACh, Nic, D M P P	—	—	
D G	high	high	ACh	—	—	?
Striatum	mod	low	ACh	Mec, dT C	B T X, Atr	C6
I C	low	high	ACh, Carb	B T X	D H B E	?
Cerebellum	low	high	Nic (inhib)	Hex	dT C, B T X	C6
			Nic (excit)	dTC, B T X	Hex	C10
L C	mod	mod	ACh, Nic	Hex, D H B E(?)	B T X, Atr	C6
Hypothalamus	low	high	ACh, Nic	D H B E	Atr	?
Medulla	low	high	Nic	D H B E	Atr, Mec	C10?
& Pons			ACh	D H B E, Atr	Mec	
			ACh	D H B E, Hex	Atr	C6
Hippocampus	low	high	D M P P, T M A	dT C, D H B E	Mec, Hex	C10?

Notes: Brain regions are listed in the order in which they are discussed in the text (sections 5.5.1 to 5.5.13). Electrophysiological studies giving equivocal results are omitted from the table. The density of radioligand binding refers to autoradiographic mapping studies; where dense binding is localized and not found throughout a structure, binding is nevertheless designated as ‘high’. Agonists are abbreviated as follows: A Ch, acetylcholine; Carb, carbamylcholine; Cyt, cytisine; D M P P, dimethylphenylpiperazinium; Nic, nicotine; T M A, tetramethylammonium. Antagonists are as follows: Atr, atropine; B T X, α-bungarotoxin; dT C, *d*-tubocurarine; D H B E, dihydroβerythroidine; Hex, hexamethonium; Mec, mecamylamine; Scop, scopolamine. Receptor identity is tentatively concluded from the known peripheral actions of antagonists: mecamylamine, hexamethonium, and chlorisondamine are ganglion (C6)-selective, α-bungarotoxin is selective toward the skeletal muscle nicotinic receptor (C10-selective), and *d*-tubocurarine and dihydroβerythroidine act with similar potency at both types of receptor.

insensitive to mecamylamine are activated by doses of nicotine in the smoking range.

5.10 Concluding remarks

In the context of cigarette smoking, electrophysiology has two main applications. Firstly, nicotine exerts several neuropsychopharmacological actions which could contribute to the smoking habit (see Clarke 1987*b*), and single-unit recording studies have identified some of the neuronal populations which are excited (or occasionally inhibited) by this drug. Ultimately, this work should help to reveal where nicotine gains access to central mechanisms of reinforcement, but care will be needed in order to ensure that iontophoretic doses of nicotine are comparable to those associated with tobacco smoking.

Recently, nicotinic antagonists have been proposed as a possible aid to smoking cessation (Stolerman 1986; Clarke 1987*b*). There are indications that central nicotinic receptors are somewhat different in molecular terms from their peripheral counterparts, and there is a need for a nicotinic antagonist which is 'tailor-made' for the CNS. In particular, this drug should act selectively at C6-like central receptors which mediate the reinforcing actions of 'smoking doses' of nicotine (Clarke 1987*b*). This antagonist should have minimal effects on peripheral receptors, to avoid unwanted side-effects. The electrophysiological approach should be of value in the testing of novel nicotinic antagonists.

Acknowledgement

The author wishes to thank the Smokeless Tobacco Research Council Inc (USA) for financial support.

References

Adams, P. R. and Brown, D. A. (1982). Synaptic inhibition of the M-current: slow excitation post-synaptic potential mechanism in bullfrog sympathetic neurones. *Journal Physiology* **332**, 263–72.

Akaishi, T. and Negoro, H. (1983). Effects of microelectrophoretically applied acetylcholine- and angiotensin-antagonists on the paraventricular neurosecretory cells excited by osmotic stimuli. *Neuroscience Letters* **36**, 157–61.

Andersen, P. and Curtis, D. R. (1964*a*). The excitation of thalamic neurones by acetylcholine. *Acta Physiologica Scandinavica* **61**, 85–99.

Andersen, P. and Curtis, D. R. (1964*b*). The pharmacology of the synaptic and acetylcholine-induced excitation of ventrobasal thalamic neurones. *Acta Physiologica Scandinavica* **61**, 100–20.

Armitage, A. K., Hall, G. H., and Morrison, C. F. (1968). Pharmacological basis for the smoking habit. *Nature* **217**, 331–4.

Ashton, H., Marsh, V. R., Millman, J. E., Rawlins, M. D., Telford, R., and Thompson, J. W. (1980). Biphasic dose-related responses of the CNV (contingent negative variation) to I.V. nicotine in man. *British Journal of Clinical Pharmacology* **10**, 579–89.

Ashton, H., Millman, J.E., Telford, R., and Thompson, J.W. (1974). The effect of caffeine, nitrazepam, and cigarette smoking on the contingent negative variation in man. *Electroencephalography and Clinical Neurophysiology* **37**, 59–71.

Barker, J.L., Crayton, J.W., and Nicoll, R.A. (1971). Noradrenaline and acetylcholine responses of supraoptic neurosecretory cells. *Journal of Physiology* **218**, 19–32.

Benowitz, N.L., and Jacob, P. (1984). Daily intake of nicotine during cigarette smoking, *Clinical Pharmacology and Therapeutics* **35**, 499–504.

Bird, S.J., and Aghajanian, G.K. (1976). The cholinergic pharmacology of hippocampal pyramidal cells. *Neuropharmacology* **15**, 273–82.

Boulter, J., Evans, K., Goldman, D., Martin, G., Treco, D., Heinemann, S., and Patrick, J. (1986). Isolation of a cDNA clone coding for a possible neural nicotinic acetylcholine receptor α-subunit. *Nature* **319**, 368–74.

Bradley, P.B., Dhawan, B.N., and Wolstencroft, J.H. (1966). Pharmacological properties of cholinoceptive neurons in the medulla and pons of the cat. *Journal of Physiology* **183**, 658–73.

Bradley, P.B., and Lucy, A.P. (1983). Cholinoceptive properties of respiratory neurones in the rat medulla. *Neuropharmacology* **22**, 853–8.

Brown, D.A. (1983). Slow cholinergic excitation– a mechanism for increasing neuronal excitability. *Trends in Neuroscience* **6**, 302–7.

Brown, D.A., Docherty, R.J., and Halliwell, J.V. (1983). Chemical transmission in the rat interpeduncular nucleus *in vitro*. *Journal of Physiology* **341**, 655–70.

Brown, D.A., Docherty, R.J., and Halliwell, J.V. (1984). The action of cholinomimetic substances on impulse conduction in the habenulointerpeduncular pathway of the rat *in vitro*. *Journal of Physiology* **353**, 101–9.

Chesselet, M.F. (1984). Presynaptic regulation of neurotransmitter release in the brain: facts and hypotheses. *Neuroscience* **12**, 347–75.

Clarke, P.B.S. (1987*a*). Recent progress in identifying nicotinic cholinoceptors in mammalian brain. *Trends in Pharmacological Sciences* **8**, 32–5.

Clarke, P.B.S. (1987*b*). Nicotine and smoking: a perspective from animal studies. *Psychopharmacology* **92**, 135–43.

Clarke, P.B.S., Hamill, G.S., Nadi, N.S., Jacobowitz, D.M., and Pert, A. (1986). ^{3}H-Nicotine- and ^{125}I-α-bungarotoxin-labeled nicotinic receptors in the interpeduncular nucleus of rats. II. Effects of habenular deafferentation. *Journal of Comparative Neurology* **251**, 407–13.

Clarke, P.B.S., Hommer, D.W., Pert, A., and Skirboll, L.R. (1985*a*). Electrophysiological actions of nicotine on substantia nigra single units. *British Journal of Pharmacology* **85**, 827–35.

Clarke, P.B.S., Hommer, D.W., Pert, A., and Skirboll, L.R. (1987). Innervation of substantia nigra neurones by cholinergic afferents from the pedunculopontine nucleus in rats: neuroanatomical and electrophysiological evidence. *Neuroscience* **23**, 1011–20.

Clarke, P.B.S., Pert, C.B., and Pert, A. (1984). Autoradiographic distribution of nicotine receptors in the brain. *Brain Research* **323**, 390–5.

Clarke, P.B.S. and Pert, A. (1985). Autoradiographic evidence for nicotine receptors on nigrostriatal and mesolimbic dopaminergic neurons. *Brain Research* **348**, 355–8.

Clarke, P.B.S., Schwartz, R.D., Paul, S.M., Pert, C.B., and Pert, A. (1985*b*). Nicotinic binding in rat brain: autoradiographic comparison of ^{3}H-nicotine, ^{3}H-ACh and ^{125}I-α-bungarotoxin. *Journal of Neuroscience* **5**, 1307–15.

Clarke, P.B.S., Fu, S.D., Jakubovic, A., and Fibiger, H.C. (1988). Evidence that mesolimbic dopaminergic activation underlies the locomotor stimulant action of nicotine in rats. *Journal of Pharmacology and Experimental Therapeutics* **246**, 701-8.

Collins, A.C., Evans, C.B., Miner, L.L., and Marks, M.J. (1986). Mecamylamine blockade of nicotinic responses: evidence for two brain nicotinic receptors. *Pharmacology, Biochemistry and Behavior* **24**, 1767-73.

Costa, L.G., and Murphy, S.D. (1983). ^{3}H-nicotine binding in rat brain: alteration after chronic acetylcholinesterase inhibition. *Journal of Pharmacology and Experimental Therapeutics* **226**, 392-7.

Curtis, D.R., and Crawford, J.M. (1969). Central synaptic transmission—microelectrophoretic studies. *Annual Review of Pharmacology* **9**, 209-40.

Curtis, D.R. and Eccles, R.M. (1958). The excitation of Renshaw cells by pharmacological agents applied electrophoretically. *Journal of Physiology* **141**, 435-45.

Curtis, D.R. and Koizumi, K. (1961). Chemical transmitter substances in the brain stem of the cat. *Neurophysiology* **24**, 80-90.

Curtis, D.R. and Ryall, R.W. (1966*a*). The excitation of Renshaw cells by cholinomimetics. *Experimental Brain Research* **2**, 49-65.

Curtis, D.R. and Ryall, R.W. (1966*b*). The acetylcholine receptors of Renshaw cells. *Experimental Brain Research* **2**, 66-80.

Curtis, D.R. and Ryall, R.W. (1966*c*). The synaptic excitation of Renshaw cells. *Experimental Brain Research* **2**, 81-96.

de la Garza, R., Bickford-Wimer, P.C., Hoffer, B.J., and Freedman, R. (1987*a*). Heterogeneity of nicotine actions in the rat cerebellum: an *in vivo* electrophysiological study. *Journal of Pharmacology and Experimental Therapeutics* **240**, 689-95.

de la Garza, R., McGuire, T.J., Freedman, R., and Hoffer, B.J. (1987*b*). The electrophysiological effects of nicotine in the rat cerebellum: evidence for direct postsynaptic actions. *Neuroscience Letters* **80**, 303-8.

de la Garza, R., McGuire, T.J., Freedman, R., and Hoffer, B.J. (1987*c*). Selective antagonism of nicotine actions in the rat cerebellum with α-bungarotoxin. *Neuroscience* **23**, 887-91.

Domino, E.F. (1967). Electroencephalographic and behavioural arousal effects of small doses of nicotine. A neuropsychopharmacological study. *Annals of the New York Academy of Science* **142**, 216-44.

Duggan, A.W., Hall, J.G., and Lee, C.Y. (1976). α bungarotoxin, cobra neurotoxin and excitation of Renshaw cells by acetylcholine. *Brain Research* **107**, 166-70.

Edwards, J.A. and Warburton, D.M. (1983). Smoking, nicotine and electrocortical activity. *Pharmacology and Therapeutics* **19**, 147-64.

Egan, T.M and North, R.A. (1986). Actions of acetylcholine and nicotine on rat locus coeruleus neurons *in vitro*. *Neuroscience* **19**, 565-71.

Farley, G.R., Morley, B.J., Javel, E., and Gorga, M.P. (1983). Single-unit responses to cholinergic agents in the rat inferior colliculus. *Hearing Research* **11**, 73-91.

Friedman, D.P., Clarke, P.B.S, O'Neill, J.B., and Pert, A. (1985). Distributions of nicotinic and muscarinic receptors in monkey thalamus. *Society for Neuroscience Abstracts* **11**, 307-3.

Gilbert, D.G. (1979). Paradoxical tranquilizing and emotion-reducing effects of nicotine. *Psychological Bulletin* **86**, 643-61.

Goldman, D., Deneris, E., Luyten, W., Kochhar, A., Patrick, J., and Heinemann, S. (1987). Members of a nicotinic acetylcholine receptor gene family are expressed in different regions of the mammalian central nervous system. *Cell* **48**, 965-73.

Grenhoff, J., Aston-Jones, G., and Svensson, T.H. (1986). Nicotinic effects on the firing pattern of mid brain dopamine neurons. *Acta Physiologica Scandinavica* **128**, 351–8.

Hall, G.H. (1970). Effects of nicotine and tobacco smoke on electrical activity of cerebral cortex and olfactory bulb. *British Journal of Pharmacology* **38**, 271–86.

Hall, G.H. (1972). Changes in body temperature produced by cholinomimetic substances injected into the cerebral ventricles of unanaesthetized cats. *British Journal of Pharmacology* **44**, 634–41.

Hamill, G.S., Clarke, P.B.S., Pert, A., and Jacobowitz, D.M. (1986). ^{3}H-nicotine and ^{125}I-α-bungarotoxin receptors in interpeduncular nucleus of rats. I. Subnuclear distribution. *Journal of Comparative Neurology* **251**, 398–406.

Hunt, S. and Schmidt, J. (1978). Some observations on the binding patterns of alpha-bungarotoxin in the central nervous system of the rat. *Brain Research* **157**, 213–32.

Imperato, A., Mulas, A., and DiChiara, G. (1986). Nicotine preferentially stimulates dopamine release in the limbic system of freely moving rats. *European Journal of Pharmacology* **132**, 337–8.

Kawamura, H. and Domino, E.F. (1969). Differential actions of *m* and *n* cholinergic agonists on the brainstem activating system. *International Journal of Neuropharmacology* **8**, 105–15.

Knott, V.J. and Venables, P.H. (1977). EEG alpha correlates of non-smokers, smoking and smoking deprivation. *Psychophysiology* **14**, 150–6.

Krnjevic, K. (1975). Acetylcholine receptors in vertebrate CNS. In *Handbook of Psychopharmacology*, Vol 6 (L.L. Iversen, S.D. Iversen, and S.H. Synder ed.), pp. 97–126. Plenum, New York.

Krnjevic, K. and Phillis, J.W. (1963*a*). Acetylcholine-sensitive cells in the cerebral cortex. *Journal of Physiology* **166**, 296–327.

Krnjevic, K. and Phillis, J.W. (1963*b*). Pharmacological properties of acetylcholine-sensitive cells in the cerebral cortex. *Journal of Physiology* **166**, 328–50.

Krnjevic, K., Pumain, R., and Renaud, L. (1971). The mechanism of excitation by acetylcholine in the cerebral cortex. *Journal Physiology* **215**, 247–68.

Kubo, T. and Misu, Y. (1981). Changes in arterial blood pressure after microinjections of nicotine into the dorsal area of the medulla oblongata in rats. *Neuropharmacology* **20**, 521–4.

Kumar, R., Cooke, E.C., Lader, M.H., and Russell, M.A.H. (1977). Is nicotine important in tobacco smoking? *Clinical Pharmacology and Therapeutics* **21**, 520–9.

Lake, N. (1973). Studies of the habenulo-interpeduncular pathway in cats. *Experimental Neurology* **41**, 113–32.

Lamour, Y., Dutar, P., and Jobert, A. (1982). Excitatory effect of acetylcholine on different types of neurons in the first somatosensory neocortex of the rat: laminar distribution and pharmacological characteristics. *Neuroscience* **7**, 1483–94.

Lichtensteiger, W., Dominiak, F., Lienhart, R., and Hefti, F. (1976). A quantitative correlation between single unit activity and fluorescence intensity of dopamine neurones in zona compacta of substantia nigra, as demonstrated under the influence of nicotine and physostigmine. *Brain Research* **117**, 85–103.

Lichtensteiger, W., Hefti, F., Felix, D., Huwyler, T., Melamed, E., and Schlumpf, M. (1982). Stimulation of nigrostriatal dopamine neurones by nicotine. *Neuropharmacology* **21**, 963–8.

London, E.D., Waller, S.B., and Walmsley, J.K. (1985). Autoradiographic localisation of ^{3}H-nicotine binding sites in the rat brain. *Neuroscience Letters* **53**, 179–84.

Lundberg, J.M. and Hokfelt, Y. (1983). Coexistence of peptides and classical neurotransmitters. *Trends in Neurosciences* **6**, 325–33.

MacLeod, N.K., James, T.A., and Starr, M.S. (1984). Muscarinic action of acetylcholine in the rat ventromedial thalamic nucleus. *Experimental Brain Research* **55**, 553–61.

Mangan, G.L. and Golding, J. (1978). An 'enhancement' model of smoking maintenance? In *Smoking behavior: physiological and psychological influences* (R.E. Thornton, ed.), pp. 87–114. Churchill Livingstone, Edinburgh.

Marks, M.J. and Collins, A.C. (1985). Tolerance, cross-tolerance, and receptors after chronic nicotine or oxotremorine. *Pharmacology, Biochemistry and Behavior* **22**, 283–91.

Marks, M.J., Stitzel, J.A., Romm, E., Wehner, J.M., and Collins, C. (1986). Nicotinic binding sites in rat and mouse brain: comparison of acetylcholine, nicotine, and α-bungarotoxin. *Molecular Pharmacology* **30**, 427–36.

Martino-Barrows, A.M. and Kellar, K.J. (1987). [^{3}H] acetylcholine and [^{3}H](-)nicotine label the same recognition site in rat brain. *Molecular Pharmacology* **31**, 169–74.

McCance, I. and Phillis, J.W. (1968). Cholinergic mechanisms in the cerebellar cortex. *International Journal of Neuropharmacology* **7**, 447–62.

McCance, I., Phillis, J.W., and Westerman, R.A. (1968*a*). Acetylcholine-sensitivity of thalamic neurones: its relationship to synaptic transmission. *British Journal of Pharmacology and Chemotherapy* **32**, 635–51.

McCance, I., Phillis, J.W., Tebecis, A.K., and Westerman, R.A. (1968*b*). The pharmacology of acetylcholine-excitation of thalamic neurones. *British Journal of Pharmacology and Chemotherapy* **32**, 652–62.

McCormick, D.A. and Prince, D.A. (1987*a*). Acetylcholine causes rapid nicotinic excitation in the medial habenular nucleus of guinea pig, *in vitro*. *Journal of Neuroscience* **7**, 742–52.

McCormick, D.A. and Prince, D.A. (1987*b*). Actions of acetylcholine in the guinea-pig and cat medial and lateral geniculate nuclei, *in vitro*. *Journal of Physiology* **392**, 147–66.

McLennan, H. and Hicks, T.P. (1978). Pharmacological characterization of the excitatory cholinergic receptors of rat central neurones. *Neuropharmacology* **17**, 329–34.

Mizobe, F., Kozousel, V., Dean, D.M., and Livett, B.G. (1979). Pharmacological characterization of adrenal paraneurons: substance P and somatostatin as inhibitory modulators of the nicotinic response. *Brain Research* **178**, 555–66.

Misgeld, J., Weiler, M.H., and Bak, I.J. (1980). Intrinsic cholinergic excitation in the rat neostriatum: nicotinic and muscarinic receptors. *Experimental Brain Research* **39**, 401–9.

Mo, N. and Dun, N.J. (1986). Cholecystokinin octapeptide depolarises guinea pig inferior mesenteric ganglion cells and facilitates nicotinic transmission. *Neuroscience Letters* **64**, 263–8.

Morrison, C.F., Goodyear, J.M., and Sellers, C.M. (1969). Antagonism of anti-muscarinic and ganglion-blocking drugs of some of the behavioural effects of nicotine. *Psychopharmacology* **15**, 341–50.

Moss, R.L., Urban, I., and Cross, B.A. (1972). Microiontophoresis of cholinergic and aminergic drugs on paraventricular neurons. *American Journal of Physiology* **183**, 658–73.

Osumi, Y., Ishikawa, T., Nagasaka, Y., and Fujiwara, T. (1980). Central effects of nicotine on gastric acid secretions in rats. *European Journal of Pharmacology* **68**, 409–15.

Phillis, J.W., Tebecis, A.K., and York, D.H. (1967). A study of cholinoceptive cells in the lateral geniculate nucleus. *Journal of Physiology* **192**, 695–713.

Phillis, J.W. and York, D.H. (1968). Pharmacological studies on a cholinergic inhibition in the cerebral cortex. *Brain Research* **10**, 297–306.

Prusky, G.T., Shaw, C., and Cynader, M.S. (1987). Nicotine receptors are located on lateral geniculate nucleus terminals in cat visual cortex. *Brain Research* **412**, 131–8.

Rapier, C., Wonnacott, S., Lunt, G.G., and Albuquerque, E.X. (1987). The neurotoxin histrionicotoxin interacts with the putative ion channel of the nicotinic acetylcholine receptors in the central nervous system. *FEBS Letters* **212**, 292–6.

Rosecrans, J.A. and Meltzer, C.T. (1981). Central sites and mechanisms of action of nicotine. *Neuroscience and Biobehavioral Reviews* **5**, 497–501.

Rotter, A., Birdsall, N.J.M., Burgen, A.S.V., Field, P.M., Hulme, E.C., and Raisman, G. (1979). Muscarinic receptors in the central nervous system of the rat. I. Techniques for autoradiographic localization of the binding of [^{3}H]propylbenzilylcholine mustard and its distribution in the forebrain. *Brain Research Reviews* **1**, 141–65.

Rovira, C., Ben-Ari, Y., Cherubini, E., Krnjevic, K., and Ropert, N. (1983). Pharmacology of the dendritic action of acetylcholine and further observations on the somatic disinhibition in the rat hippocampus *in situ*. *Neuroscience* **8**, 97–106.

Ryall, R.W., and Belcher, G. (1977). Substance P selectively blocks nicotinic receptors on Renshaw cells: a possible inhibitory mechanism. *Brain Research* **137**, 376–80.

Salmoiraghi, G.C., and Steiner, F.A. (1963). Acetylcholine sensitivity of cat's medullary neurons. *Journal of Neurophysiology* **26**, 581–97.

Sastry, B.R. (1978). Effects of substance P, acetylcholine and stimulation of habenula on rat interpeduncular neuronal activity. *Brain Research* **144**, 404–10.

Schwartz, R.D. (1986). Autoradiographic distribution of high affinity muscarinic and nicotinic cholinergic receptors labeled with [^{3}H]acetylcholine in rat brain. *Life Sciences* **38**, 2111–9.

Schwartz, R.D., and Kellar, K.J. (1985). *In vivo* regulation of [^{3}H]acetylcholine recognition sites in brain by nicotinic cholinergic drugs. *Journal of Neurochemistry* **45**, 427–33.

Schwartz, R.D., Lehmann, J., and Kellar, K.J. (1984). Presynaptic nicotinic cholinergic receptors labelled by ^{3}H-acetylcholine on catecholamine and serotonin axons in brain. *Journal of Neurochemistry* **42**, 1495–8.

Segal, M. (1978). The acetylcholine receptor in the rat hippocampus; nicotinic, muscarinic or both? *Neuropharmacology* **17**, 619–23.

Sillito, A.M., Kemp, J.A., and Berardi, N. (1983). The cholinergic influence on the function of the cat dorsal lateral geniculate nucleus. *Brain Research* **280**, 299–307.

Stolerman, I.P. (1986). Could nicotine antagonists be used in smoking cessation? *British Journal of Addiction* **81**, 47–53.

Stone, T.W. (1972). Cholinergic mechanisms in the rat somatosensory cerebral cortex. *Journal of Physiology* **225**, 485–99.

Svensson, T.H., and Engberg, G. (1980). Effect of nicotine on single cell activity in the noradrenergic nucleus locus coeruleus. *Acta Physiologica Scandinavica* Suppl. **479**, 31–4.

Takagi, M. (1984). Actions of cholinergic drugs on cells in the interpeduncular nucleus. *Experimental Neurology* **84**, 358–63.

Ueki, S., Koketsu, K., and Domino, P.F. (1961). Effects of mecamylamine on the

Golgi recurrent collateral-Renshaw cell synapse in the spinal cord. *Experimental Neurology* **3**, 141–8.

Vanderwolf, C. H. and Robinson, T. E. (1981). Reticulo-cortical activity and behavior: A new critique of the arousal theory and a new synthesis. *Behavioral Brain Sciences* **4**, 459–514.

Von Borstel, R. W., Evoniuk, G. E., and Wurtman, R. J. (1986). Adenosine potentiates sympathomimetic effects of nicotinic agonists *in vivo*. *Journal of Pharmacology and Experimental Therapeutics* **236**, 344–9.

Wainer, B. H., Levey, A. I., Mufson, E. J., and Mesulam, M-M. (1984). Cholinergic systems in mammalian brain identified with antibodies against choline acetyltransferase. *Neurochemistry International* **6**, 163–82.

Wheal, H. V., and Miller, J. J. (1980). Pharmacological identification of acetylcholine and glutamate excitatory systems in the dentate gyrus of the rat. *Brain Research* **182**, 145–55.

Wonnacott, S. (1987). Editorial: Brain nicotine binding sites. *Human Toxicology* **6**, 343–53.

Woodson, P. P., Baettig, K., Etkin, M. W., Kallman, W. M., Harry, G. J., Kallman, M., J., and Rosecrans, J. A. (1982). Effects of nicotine on the visual evoked response. *Pharmacology, Biochemistry and Behavior* **17**, 915–20.

Yamada, S., Isogai, M., Kagawa, Y., Takayanagi, N., Hayashi, E., Tsuji, K., and Kosuge, T. (1985). Brain nicotinic acetylcholine receptors: biochemical characterization by neosurugatoxin. *Molecular Pharmacology* **28**, 120–7.

Yoon, K-W. P., Gessa, G. L., Boi, V., Naes, L., Mereu, G., and Westfall, T. C. (1986). Electrophysiological effects of nicotine on dopamine midbrain neurons. *Society for Neuroscience Abstracts* **12**, 411.9.

6 Effects of nicotine on synaptic transmission in the brain

Kjell Fuxe, Kurt Andersson, Anders Härfstrand, Peter Eneroth, Miguel Perez de la Mora, and Luigi F. Agnati

6.1 Introduction

In humans and in animals nicotine is known to produce an EEG and behavioural arousal, which may partly underly its reinforcing properties (see Jaffe and Jarvik 1978). In addition, nicotine, and exposure to cigarette smoke, have been found to markedly affect the endocrine system via actions within the hypothalamus (see Fuxe *et al.* 1977, 1987). The central effects of nicotine and exposure to cigarette smoke appear to be mediated via central nicotinic-cholinergic receptors, belonging to cholinergic neurons and participating in synaptic transmission in the brain (see Fuxe *et al.* 1977, 1987; Giorguieff-Chesselet *et al.* 1979; Larsson and Nordberg 1985; Larsson 1985; Morley *et al.* 1983; Schwartz *et al.* 1982; Schwartz and Kellar 1983; Westfall 1974). The powerful effects of nicotine on the brain are in part due to the ability of these receptors to control monoaminergic neurons, especially in the cerebral cortex, the basal ganglia, the hypothalamus, the preoptic area, the substantia nigra, and ventral tegmental area. The drug nicotine is highly efficient in performing a modulation of monoamine transmission, since nicotinic-cholinergic receptors probably are located *inter alia* on the nerve cell membrane of catecholamine and 5-hydroxytryptamine (5-HT) axon terminals (Balfour 1982; de Belleroche and Bradford 1978; Clarke and Pert 1985; Härfstrand *et al.* 1987; Rowell and Winkler 1984; Schwartz *et al.* 1984). Thus, in the cerebral cortex, cholinergic nerve terminals linked to nicotinic receptors interact with noradrenaline and 5-HT nerve-terminal networks originating from ascending noradrenergic and 5-HT pathways (see Butcher and Woolf 1984; Fuxe *et al.* 1970; Kimura *et al.* 1984; Lewis and Shute 1978). Also, within the hypothalamus and the preoptic area the interactions with the cholinergic and monoaminergic system probably involve interactions between the cholinergic terminals, linked to nicotinic receptors, and monoamine nerve terminals. There is probably also a direct or indirect control by cholinergic terminals, linked to nicotinic receptors, of the arcuate dopamine cells (Fig. 6.1). Small to medium sized cholinergic nerve cells have been identified within the hypothalamus, especially within the arcuate nucleus (Tago *et al.* 1987; Rao *et al.* 1987). These also participate in the innervation of the hypothalamus including

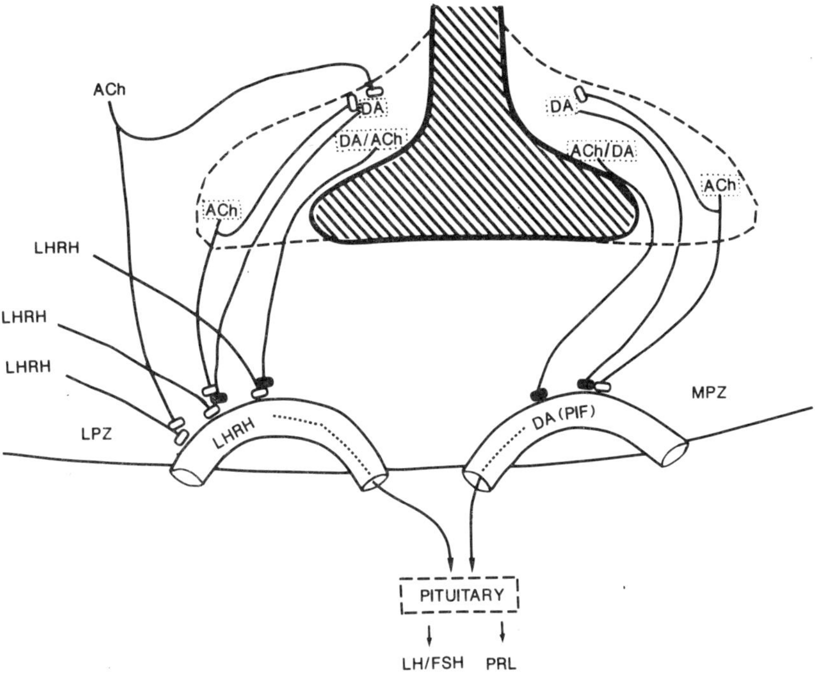

Fig. 6.1. An overall summary of the neuroendocrine functions of the median eminence, with regard to the effects of nicotine and exposure to cigarette smoke on prolactin and L H/F S H secretion. The possible presence of cholinergic synapses in the arcuate nucleus (dashed boundary) and in the external layer of the median eminence is indicated. These synapses may in part operate via nicotinic-cholinergic receptors which are in part located on dopaminergic nerve terminals within the median eminence, but may also control the tubero-infundibular dopaminergic neurons at the arcuate level. The effects of nicotine on the lateral and medial tubero-infundibular dopaminergic systems is suggested to mediate the inhibitory effects of nicotine on L H and prolactin secretion, respectively. It seems likely that the action of nicotine on the tubero-infundibular dopamine neurons is exerted at the cell body level or in the medianosome network. (Open nerve terminals = facilitatory nerve terminals; black nerve terminals = inhibitory nerve terminals; M P Z = medial palisade zone of the median eminence; L P Z = lateral palisade zone of the median eminence.)

the median eminence (Fig. 6.1). We have recently found that some tubero-infundibular dopaminergic neurons show choline acetyltransferase immunoreactivity (Tinner *et al.* 1989), suggesting co-storage of dopamine and acetylcholine in these neurons (Fig. 6.1). In the caudate putamen and in the ventral striatum (tuberculum olfactorium and nucleus accumbens) the dopaminergic nerve terminals are controlled by nicotinic receptors belonging to local cholinergic neurons. Within the ventral tegmental area and the substantia nigra, cholinergic inputs via their nicotinic-cholinergic receptors located on the dopaminergic cell bodies regulate activity within the mesolimbic and mesostriatal dopamine pathways (Clarke and Pert 1985; Grenhoff *et al.* 1986).

The aim of this chapter is to describe the central effects of nicotine and exposure to cigarette smoke (giving nicotine concentrations similar to those found in human smoking) on various transmitter systems. The chapter will focus on the monoamine systems of the brain, and will attempt to show how the effects of nicotine may be related to its rewarding and arousal actions, and especially to its neuroendocrine activities. However, it must be emphasized that nicotine in low doses enhances the release of acetylcholine and monoamines, but in high doses, especially upon repeated treatment, may block transmitter release due to desensitization of the nicotinic receptors, and possibly even to a depolarization-induced inactivation (see Langley and Dickinson 1889; Thesleff 1955). Such phenomena must be born in mind when discussing the complex pharmacology of nicotine.

6.2 Effects of nicotine on central dopamine neurons

6.2.1 *Effects on tubero-infundibular dopamine neurons in relation to neuroendocrine function*

6.2.1.1 *Acute nicotine treatment* Acute intraventricular injections of nicotine and acute intravenous infusion of nicotine have been found to produce a dose-dependent reduction and a dose-dependent increase in dopamine utilization in both the medial and lateral palisade zones (MPZ and LPZ) of the median eminence (Andersson *et al.* 1982*a*). It seems likely that nicotine-induced dopamine release in the LPZ, which in an inhibitory way regulates luteinizing hormone releasing hormone (LHRH) release from the median eminence, at least in part mediates the inhibitory effects of nicotine on LHRH secretion. A D1 dopamine receptor appears to be involved, since a selective D1-receptor antagonist (SCH 23390), but not D2 dopamine receptor antagonists (remoxipride, raclopride), can dose-dependently counteract the inhibitory effects of nicotine on the serum luteinizing hormone (LH) levels (Andersson *et al.* 1988; Fuxe *et al.* 1987). The medial tubero-infundibular dopaminergic pathway, innervating the MPZ, probably releases dopamine as a prolactin inhibitory factor. Therefore it seems likely that the nicotine-induced increase in dopamine release in the MPZ at least in part mediates the inhibitory effects of nicotine on prolactin secretion (Andersson *et al.* 1981*e*). It is also true that the tubero-infundibular dopaminergic neurons in an inhibitory manner control the release of thyrotropin releasing hormone (TRH) from the median eminence (see Andersson and Eneroth 1986; Andersson *et al.* 1985*b*), and that acute nicotine treatment inhibits the secretion of thyroid stimulating hormone (TSH) (Andersson *et al.* 1982*a*). However, this action of

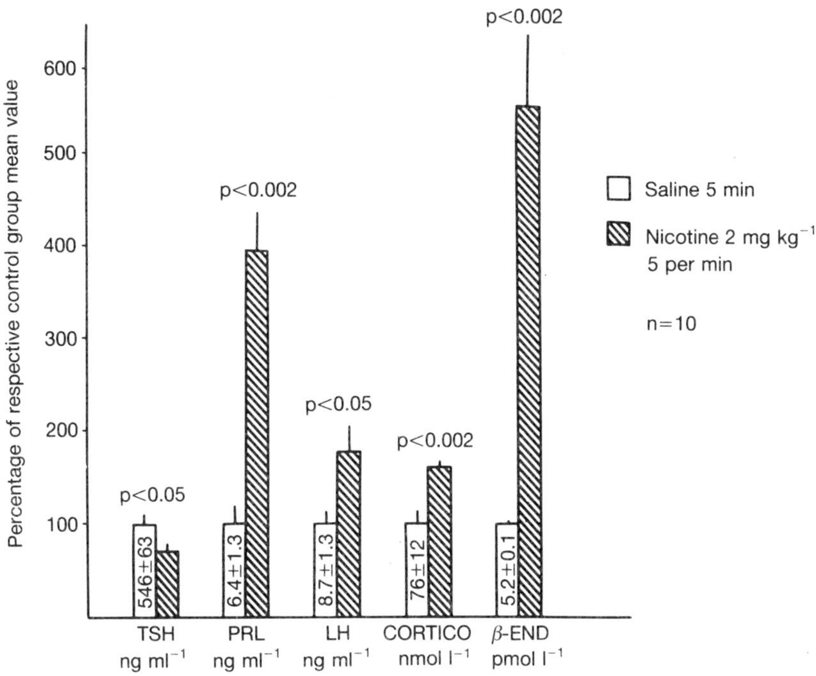

Fig. 6.2. Effects of an acute intraperitoneal injection of nicotine on the serum levels of TSH, LH, prolactin, corticosterone, and β-endorphin. (Mean ± SEM, Mann-Whitney U-test.)

nicotine is not counteracted by D1 or D2 dopamine receptor blocking agents, suggesting that nicotine-induced release of dopamine in the median eminence does not mediate the inhibitory effects of nicotine on TSH secretion (Andersson *et al.* 1988).

All the effects reported above were observed 1 h after acute nicotine treatment. However, 5 min after a single dose (2 mg kg^{-1}) of nicotine, marked increases in serum LH, prolactin, corticosterone and β-endorphin levels have been observed (Fig. 6.2) (Andersson *et al.* 1983*b*; Sharp and Beyer 1986). These marked neuroendocrine effects are not correlated with any changes in dopamine levels in the MPZ and LPZ of the median eminence (Andersson *et al.* 1983*b*). *In vitro*, acetylcholine has also been found to release LHRH from the hypothalamus (Richardson *et al.* 1982). The nicotinic receptors involved in this response appear to desensitize rapidly (Sharp and Beyer 1986), while this is not true for the cholinergic-nicotinic receptors involved in the inhibitory effects of nicotine on gonadotrophin secretion, mediated by the activation of the tubero-infundibular dopaminergic neurons (Andersson *et al.* 1985*c*).

In line with the above view, acute repeated treatment with nicotine (4 × 2 mg kg^{-1} intraperitoneally at 30 min time intervals) in intact male and

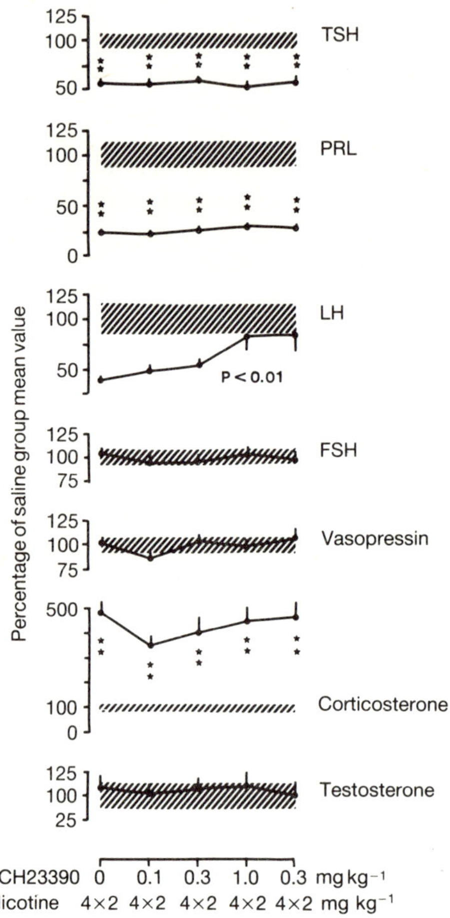

Fig. 6.3. Dose-effects of the D1 dopamine receptor antagonist S C H 23390 on nicotine-induced changes in serum levels of T S H, prolactin, L H, and corticosterone in the normal male rat. Saline or nicotine (2 mg kg^{-1} intraperitoneally) was given 4 times with 30 min intervals, the animals being decapitated 30 min after the last injection. Saline or various doses of S C H 23390 (0.1–3.0 mg kg^{-1} intraperitoneally) was given 5 min before the first saline or nicotine injection. Means ± S E M (n = 6) are given in percentage of saline control group mean values (hatched area represents the means ± S E M of the control group). Statistical analysis according to the Wilcoxon test; one-way classification, comparing all possible pairs of treatments. A dose-related counteraction of the inhibitory effects of nicotine on L H serum levels is seen (Jonckheere-Terpstra test). ** = $p<0.01$.

dioestrous female rats, and in castrated male and female animals, were also found to inhibit the secretion of prolactin, LH, and TSH (Fig. 6.3). These reductions were associated with increases in dopamine utilization in the MPZ and LPZ, and with reductions in dopamine levels in these regions (Andersson 1985; Andersson *et al.* 1980, 1982*b*, 1984*b*; Eneroth *et al.* 1977; Fuxe *et al.* 1977, 1987; see also Lichtensteiger *et al.* 1978). The inhibitory effects on LH secretion can be counteracted by a D1 dopamine receptor antagonist (Fig. 6.3) (Andersson *et al.* 1988). The involvement of a ganglionic type of cholinergic-nicotinic receptor in these inhibitory neuroendocrine actions, and in the increases in dopamine utilization in the median eminence, is indicated by the ability of mecamylamine, a ganglionic blocking agent, to substantially counteract these actions. The major central transmitter system involved in the inhibitory effects of nicotine on TSH secretion is yet unknown.

It seems likely that the tubero-infundibular cholinergic neurons, via their nicotinic receptors, regulate the activity of the tubero-infundibular dopaminergic neurons. Of substantial interest is the recent demonstration of probable dopamine/acetylcholine co-storing neurons in the arcuate nucleus, on which nicotine may act to increase release both of dopamine and acetylcholine (Tinner *et al.* 1989). In view of the latency of onset of the inhibitory effects of nicotine on LH and prolactin secretion, it seems possible that the cholinergic-nicotinic receptors in the median eminence can unmask inhibitory D1 dopamine receptors located on the LHRH nerve terminals (Fig. 6.1) (Andersson *et al.* 1988), via effects on tanycytes (modified glial cells involved in the microregulation of the secretion of releasing hormones in the median eminence; Lichtensteiger *et al.* 1978).

6.2.1.2 *Acute exposure to cigarette smoke* The effects of cigarette smoke have been investigated by placing rats in a Walton horizontal smoking

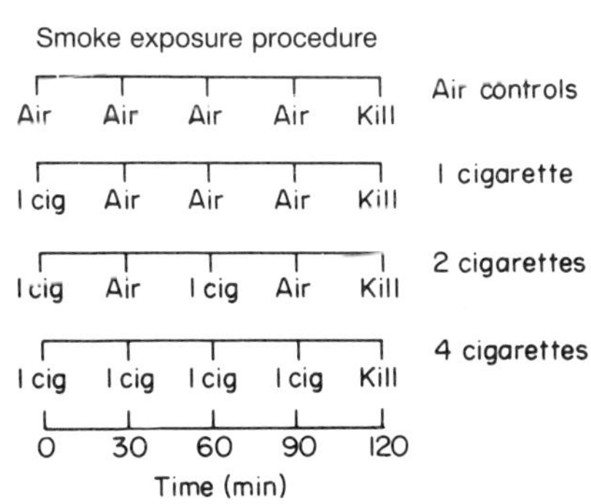

Fig. 6.4. Treatment schedules in the experiments involving acute intermittent exposure of male rats to cigarette smoke using a Walton smoke-exposure machine. '1 cig' means exposure either to filtered or unfiltered cigarette smoke. The control rats were instead exposed to puffs of air according to the schedule indicated. The exposure time for each cigarette is approximately 10 min.

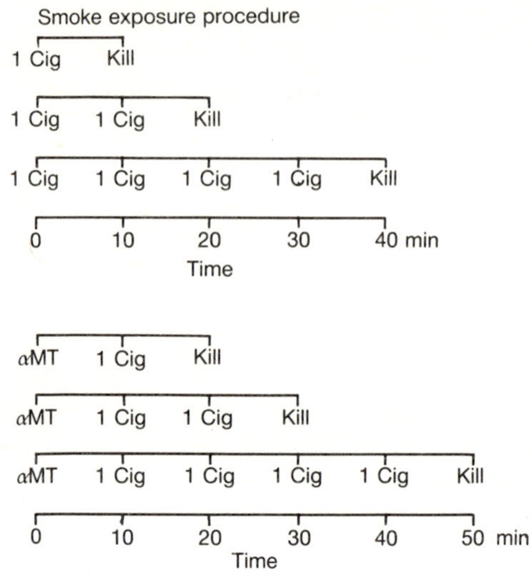

Fig. 6.5. Treatment schedules in the experiments involving acute continuous exposure of male rats to cigarette smoke. '1 cig' means exposure either to filtered or unfiltered cigarette smoke. The control rats were instead exposed to puffs of air according to the schedule indicated. In order to measure changes in catecholamine levels, the tyrosine hydroxylase inhibitor αmethyltyrosine (αMT; 250 mg kg^{-1} intraperitoneally) was given 10 min before onset of exposure to air, or filtered or unfiltered cigarette smoke.

machine (Andersson *et al.* 1985*c*). Cigarettes were of the Kentucky reference IR-1 type (high-nicotine, high-tar type; Tobacco and Health Research Institute, University of Kentucky, Lexington, USA). In control experiments, glass-fiber filters were used to remove nicotine from the smoke. The treatment schedules, in the experiments involving acute-intermittent and acute-continuous exposure of male rats to cigarette smoke, are shown in the Fig. 6.4 and Fig 6.5. The acute-intermittent exposure to cigarette smoke produced dose-dependent increases in the serum nicotine and cotinine levels (see Fig. 6.6). Exposure to air or filtered cigarette smoke did not lead to increases in serum nicotine and cotinine levels.

It was found that acute-intermittent exposure to cigarette smoke increased dopamine utilization in the MPZ and LPZ of the median eminence, changes associated with a marked lowering of serum LH, prolactin, and TSH levels (Figs. 6.7 and 6.8). Filtered cigarette smoke had no effects on catecholamine utilization in the median eminence, nor on neuroendocrine function. Furthermore, mecamylamine counteracted the ability of acute-intermittent exposure to cigarette smoke to increase dopamine utilization in the median eminence and to inhibit the serum LH, prolactin, and TSH levels (Andersson 1985). Thus, as in the case of nicotine administration, nicotinic-cholinergic receptors

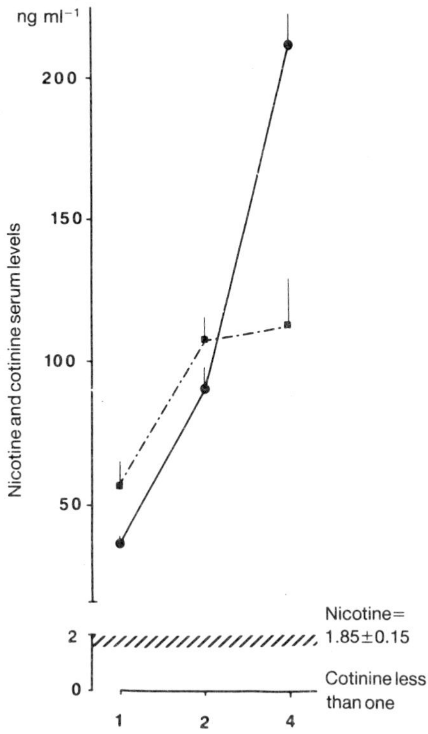

Fig. 6.6. The effects of acute intermittent exposure of male rats to unfiltered cigarette smoke on serum nicotine (solid line) and cotinine (dotted line) levels. For exposure procedure see Fig. 6.4. Means ± S E M (n = 6) are shown after exposure to 1, 2, or 4 cigarettes. The hatched area indicates the levels in the control animals (means ± S E M).

of the ganglionic type appear to be involved in mediating these actions of cigarette smoke. Obviously the results indicate that it is indeed the nicotine component of the cigarette smoke which produces the reported effects upon dopamine utilization and on the neuroendocrine function (see Andersson 1985; Andersson *et al.* 1985*c*).

In the experiments on acute-continuous exposure to cigarette smoke, cigarettes were burnt without any time interval in between successive cigarettes (Fig. 6.5) (See Andersson *et al.* 1987). As in the experiments involving acute intermittent exposure, the exposure time for each cigarette was about 10 min. Serum nicotine levels obtained in these experiments were of a similar magnitude to those obtained following acute intermittent exposure. Also in these experiments a depletion of dopamine stores was observed in the MPZ and LPZ of the median eminence, but in contrast to acute intermittent exposure to

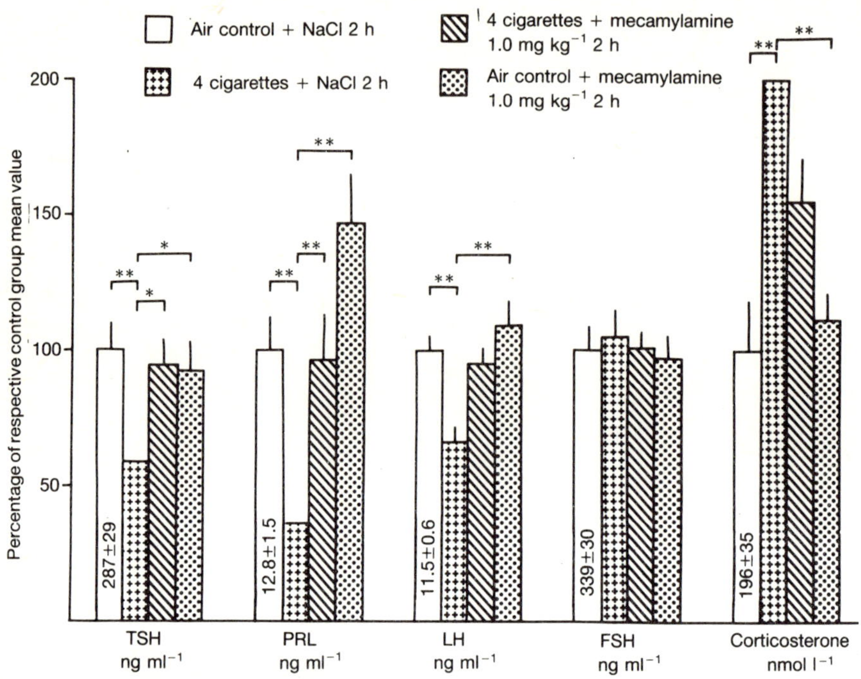

Fig. 6.7. Effects of mecamylamine and acute intermittent exposure to cigarette smoke on serum levels of T S H, prolactin, L H, F S H, and corticosterone in the normal male rat. For exposure procedure see Fig. 6.4. Means ± S E M (n = 12) are shown as percentage of the control (air-treated) group mean values. The serum levels of respective hormones are given (in ng ml^{-1} or nmol L^{-1}) in the open columns representing the saline treated group. Statistical analysis according to the Wilcoxon test: one-way classification, comparing all possible pairs of treatments. * = $p<0.05$; ** = $p<0.01$.

cigarette smoke, no increase in dopamine utilization was observed in this region (Andersson *et al.* 1987). Also, the neuroendocrine effects differed since only serum TSH and prolactin levels were reduced, but not serum LH levels. These results may be explained by a depolarization-induced inactivation and/or a desensitization of the nicotinic receptors after continuous exposure to smoke from 2 and especially 4 cigarettes. Thus, insufficient amounts of dopamine may be released to produce the inhibition of serum LH levels. It should also be noted that the exposure to 1 cigarette (10 min) did not produce an increase in serum LH and prolactin levels as observed in the very acute experiments with nicotine injections (2 mg kg^{-1}) (see also Sharp and Beyer 1986). Taken together, these experiments on acute continuous exposure to cigarette smoke indicate that desensitization of the cholinergic-nicotinic

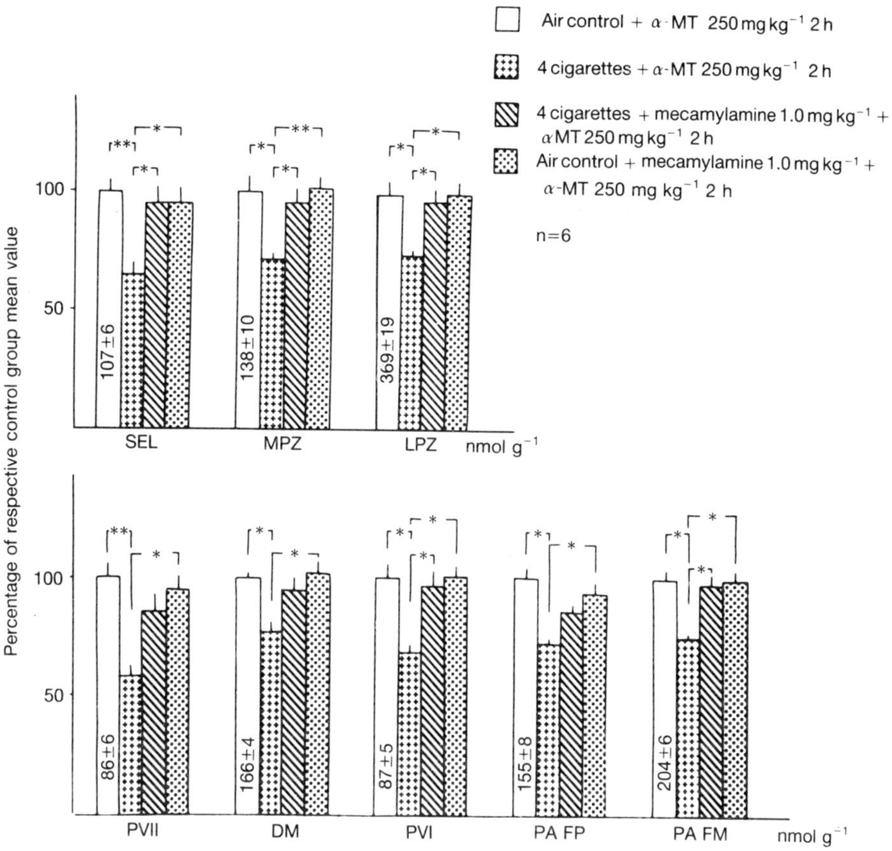

Fig. 6.8. Effects of mecamylamine and acute intermittent exposure to cigarette smoke on dopamine and noradrenaline fluorescence in discrete hypothalamic regions of the normal male rat. For exposure procedure see Fig. 6.4. αMethyltyrosine was given to inhibit catecholamine synthesis. Means ± S E M are shown as a percentage of air-treated group mean values (n = 6 in all groups). The absolute concentrations of catecholanines in the respective areas are given in nmol g^{-1} of tissue wet weight in the open columns representing the saline control group. Statistical analysis according to the Wilcoxon test: one-way classification comparing all possible pairs of treatments. * = $p<0.05$; ** = $p<0.01$. Abbreviations used: S E L = subependymal layer of the median eminence; M P Z = medial palisade zone of the median eminence; L P Z = lateral palisade zone of the median eminence; P V II = posterior periventricular hypothalamic region; D M = dorsomedial hypothalamic nucleus: P V I = anterior periventricular hypothalamic region; P A F P = parvocellular part of the paraventricular hypothalamic nucleus; P A F M = magnocellular part of the paraventricular hypothalamic nucleus; αM T = αmethyltyrosine.

receptors involved in the control of the tubero-infundibular dopaminergic neurons occurs.

6.2.1.3 *Chronic exposure to cigarette smoke* In these studies chronic exposure to unfiltered cigarette smoke for 9 days (2 cigarettes per day) was followed by acute intermittent exposure to unfiltered smoke (1 cigarette every 30 min during 2 h) according to the previously described procedure (see Fig. 6.4) (Andersson *et al.* 1985*a*). Following this chronic exposure, depletion of dopamine stores in the MPZ and LPZ was no longer observed. Instead, an increase in dopamine levels in the LPZ was demonstrated (Andersson *et al.* 1985*a*). These results probably reflect activation of catecholamine synthesis by chronic exposure to cigarette smoke, possibly caused by activation of tyrosine hydroxylase. However, no tolerance had developed with regard to the ability of nicotine to increase dopamine utilization in the median eminence.

It seems likely that also in humans there is an activation of the tubero-infundibular dopaminergic neurons by nicotine, since human studies have demonstrated that smoking subjects have reduced prolactin levels compared with non-smoking controls, and a shorter breast-feeding time has also been reported (Andersen *et al.* 1982, 1984). The work of Blake, Sawyer and collegues, demonstrating that nicotine blocks the suckling-induced rise in circulating prolactin in lactating rats and inhibits the proestrous surge of prolactin (Blake and Sawyer 1972, Blake *et al.* 1973) is pertinent to the human studies. The LH surge in proestrous rats is delayed by nicotine (McLean *et al.* 1977). All the experimental work in the rat indicates that the serum follicle stimulating hormone (FSH) levels are only weakly and inconsistently affected by nicotine treatment and exposure to cigarette smoke (see Fuxe *et al.* 1977, 1987).

6.2.2 *Effects on nigrostriatal dopamine neurons*

The effects of nicotine on this dopaminergic system are dealt with in chapter 9 (see Reavill, this volume). Therefore only a brief summary will be given here.

6.2.2.1 *Acute nicotine treatment* A large number of observations indicate that nicotine can increase the firing rate of dopaminergic neurons. This change is correlated with an increase in the fluorescence intensity of dopamine within the zona compacta of the substantia nigra (see Lichtensteiger *et al.* 1976; Lichtensteiger *et al.* 1982; see Clarke, Fig. 5.6, this volume). Also, an increase in the amount of burst-firing of burst-firing dopamine cells located in the zona compacta (Grenhoff *et al.* 1986) has been observed. As indicated above (section 6.1), there exist nicotinic-cholinergic receptors, both on the dopaminergic perikarya and dendrites in the substantia nigra, as well as on the striatal nerve terminals. *In vitro* experiments on brain slices clearly demonstrate that nicotine produces an increase in dopamine release from the striatum (Giorguieff-Chesselet *et al.* 1979; Takano *et al.* 1983; Westfall 1974; Westfall *et al.* 1983*b*). These effects of nicotine are probably mediated via central nicotinic-cholinergic receptors located on the nigrostriatal dopaminergic axon terminals (Härfstrand *et al.* 1987). *In vivo* experiments involving single intraventricular injections and intravenous infusion of nicotine have been found to reduce, in a dose-dependent way, dopamine stores in both the diffuse and islandic systems of the neostriatum, as well as to increase dopamine utilization in these systems (Andersson *et al.* 1981*c*). Following acute single intraperitoneal injections of nicotine, however, a selective increase in dopamine utilization was observed within striatal dopaminergic terminals located in the medial and caudal part of the neostriatum (Andersson *et al.* 1981*b*). Studies with mecamylamine *in vivo* support a role of central nicotinic receptors in control of striatal dopamine release (Ahtee and Kaakkola 1978).

6.2.2.2 *Acute exposure to cigarette smoke* Acute intermittent exposure to cigarette smoke has not been found to influence dopamine levels or dopamine utilization within the islandic or the diffuse neostriatal dopaminergic nerve terminal systems (Fuxe *et al.* 1986).

A putative anti-Parkinsonian action of nicotine and smoking may be due at least in part to a releasing action of nicotine on dopaminergic nerve terminals in the caudate putamen, and/or an increase in the firing rate of the nigral dopaminergic cells. However, the action is not always revealed in view of the absence of effects in the experiments involving acute intermittent exposure to cigarette smoke (Fuxe *et al.* 1986). It is therefore of substantial interest that nicotine, continuously administered via subcutaneously implanted mini-pumps, can exert protective effects on the nigrostriatal dopaminergic neurons following a partial mes–diencephalic hemitransection, immediately rostral to the substantia nigra. Thus, an increased number of dopaminergic nerve cell bodies seemed to survive (Janson *et al.* 1986, 1987) following nicotine treatment. This is true also for the striatal dopamine-containing terminals, and especially for the nigral dopaminergic dendrites. These results may be due to a reduced degeneration and/or increased regeneration of the dopaminergic cells. It has been hypothesized that these protective effects of nicotine are due to a desensitization of the cholinergic nicotinic receptors located on the dopamine neurons, leading to reduced firing rate, improved ionic homeostasis and thus to reduced energy demands (Janson *et al.* 1986, 1987). Such a mechanism may also help to explain the anti-Parkinsonian action of smoking in man (for review, see Baron 1986; Reavill, chapter 9, this volume).

In view of the above, neostriatal cholinergic interneurons (see for example, Butcher and Woolf 1984) may operate via nicotinic-cholinergic receptors located on the dopaminergic nerve terminals to enhance dopamine release.

6.2.3 *Effects on mesolimbic dopamine neurons*

6.2.3.1 *Acute nicotine treatment* The dopaminergic nerve terminals in the nucleus accumbens and the olfactory tubercule have been analyzed most.

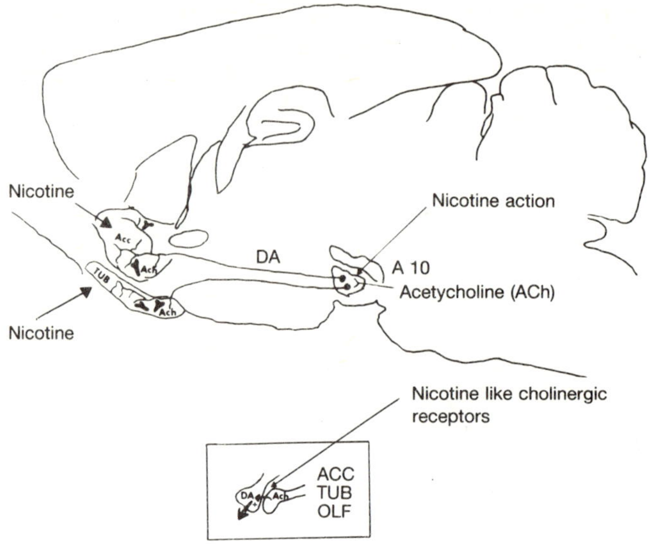

	Striatum		Accumbens		Tuberculum	
	levels	turnover	levels	turnover	levels	turnover
Nicotine	~	↑	↓	↑	↓	↑
Smoking	~	~	↓	↑	~	↑

Fig. 6.9. Schematic illustration of the actions of nicotine on the mesolimbic doparimergic (DA) neurons, and an overall summary of the actions of nicotine and exposure to cigarette smoke on dopamine levels and utilization in the striatum, nucleus accumbens (ACC), and tuberculum olfactorium (TUB OLF).

Within the anterior part of the nucleus accumbens and olfactory tubercule there exist diffuse types of dopaminergic nerve terminals, and in the caudal parts of these regions the dotted type of dopaminergic terminals dominate. The latter also possess cholecystokinin-like immunoreactivity (Fuxe *et al.* 1979; Hökfelt *et al.* 1980). In experiments with single intraventricular injections and intravenous infusion of nicotine, it was possible to demonstrate a dose-dependent reduction in dopamine stores and increases in dopamine utilization in all the various types of discrete dopaminergic nerve-terminal systems of the nucleus accumbens and olfactory tubercule (Andersson *et al.* 1981*c*). However, following acute single intraperitoneal injections of nicotine, differential increases in dopamine utilization were observed within the diffuse types of dopaminergic nerve terminals, which lack cholecystokinin immunoreactivity and which are predominatly present in the anterior parts of the nucleus accumbens and olfactory tubercule (Andersson *et al.* 1981*b*). Recently, nicotine has also been found to increase the firing rate, and mecamylamine to reduce the firing rate of dopaminergic cells of the ventral tegmental area (Grenhoff *et al.* 1986). An increased bursting of the burst-firing cells was also noted after mecamylamine treatment.

6.2.3.2 *Acute exposure to cigarette smoke* It was found that acute intermittent exposure to cigarette smoke induced a dose-related increase in dopamine utilization exclusively within the diffuse types of dopaminergic nerve terminals of the nucleus accumbens and olfactory tubercule (Fig. 6.10) (Fuxe *et al.* 1986).

In view of the possible role of the nucleus accumbens in mechanisms of drug dependence, for example in cocaine-induced euphoria and dependence (see Roberts *et al.* 1979), the present results indicate that the nicotine-induced dopamine release within the anterior parts of the nucleus accumbens may at least in part mediate the nicotine-induced euphoria and dependence (Fig. 6.11). It is also generally considered that mesolimbic dopaminergic

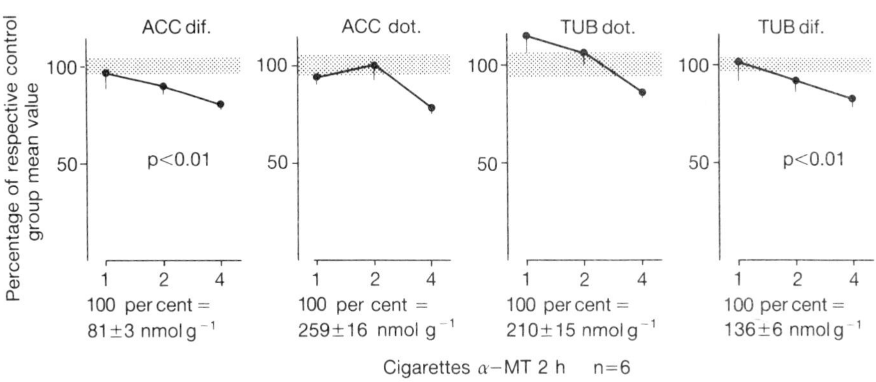

Fig. 6.10. Effects of acute intermittent exposure to cigarette smoke on dopamine fluorescence in various forebrain dopaminergic nerve-terminal systems of the normal male rat. For exposure procedures see Fig. 6.4. All groups were given αmethyltyrosine (αM T; 250 mg kg $^{-1}$ intraperitoneally) immediately before the onset of exposure to inhibit catecholamine synthesis. Means ± S E M are shown as a percentage of air + αM T-treated group mean values (n = 6 rats in all groups except in the 'air controls + αM T' group and the '4 cig + αM T' group where n = 12). Below the respective 100 per cent value, the absolute concentrations of dopamine are given in nmol g^{-1} of tissue wet weight. Statistical analysis according to the Jonckheere-Terpstra test for ordered alternatives. A C C dif. = diffuse type of dopamine fluorescence in the anterior nucleus accumbens; A C C dot. = dotted type of dopamine fluorescence in the posterior nucleus accumbens; T U B dif. = diffuse type of dopamine fluorescence in the lateral posterior olfactory turbercule; T U B dot. = dotted type of dopamine fluorescence in the medial posterior olfactory tubercule.

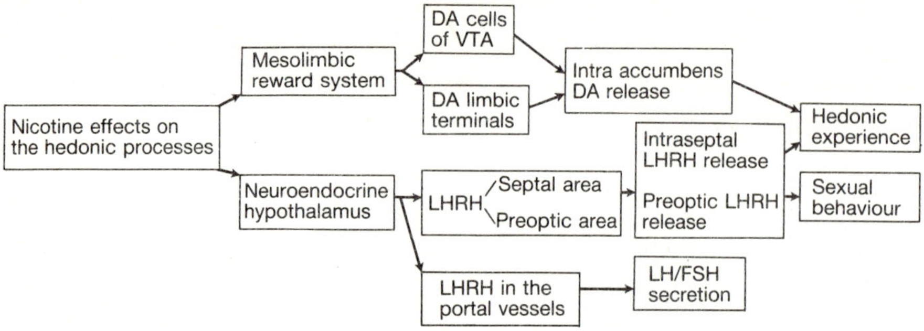

Fig. 6.11. Possible neurochemical mechanisms underlying the effects of nicotine on hedonic processes. The possible involvement of limbic dopamine (D A) and L H R H release is indicated. Preoptic L H R H release may possibly mediate the ability of nicotine to enhance sexual behaviour.

systems are involved in reward mechanisms (see Singer *et al.* 1982; see Reavill, chapter 9, this volume). The acquisition of schedule-induced self injection of nicotine is reduced following destruction of dopaminergic nerve terminals in the nucleus accumbens.

Finally, it should be noted that cortical dopaminergic nerve terminals in the anteromedial frontal cortex were not significantly affected by single intraventricular injections of nicotine, as evaluated from measurements of dopamine levels and utilization in this cortical region (Andersson *et al.* 1981*c*).

The results obtained on the mesolimbic dopaminergic neurons indicate the existence of cholinergic interneurons within the nucleus accumbens and the olfactory tubercule (see for example, Butcher and Woolf 1984), which operate via cholinergic-nicotinic receptors located on the diffuse dopaminergic nerve terminals of these regions to enhance dopamine release.

6.3 Effects of nicotine on central noradrenergic neurons

6.3.1 *Effects on the non-locus-coeruleus noradrenaline systems of the hypothalamus and the preoptic area in relation to neuroendocrine function*

The hypothalamic and preoptic noradrenergic nerve terminal networks mainly arise from axons ascending in the median forebrain bundle from non-locus-coeruleus noradrenaline-containing cell groups of the medulla oblongata and the pons (see Fuxe *et al.* 1970; Ungerstedt 1971). We have especially analyzed the effects of nicotine on the following noradrenergic nerve terminal systems.

1. The preoptic noradrenergic nerve terminal systems, especially those in the medial preoptic necleus which control the activity in the LHRH immunoreactive neurons innervating the median eminence, and which probably also

are involved in regulation of sexual behaviour (Andersson *et al.* 1981*d*). Noradrenaline release in these regions is well known to increase the release of LHRH and enhance sexual activity (see Kalra and Kalra 1983).

2. Noradrenergic nerve terminals have been analyzed in both the magnocellular part and in the parvocellular part of the paraventricular hypothalamic nucleus. Within the magnocellular part they appear to exert an excitatory influence on the vasopressin neuronal systems; in the parvocellular part they appear to exert an excitatory influence on thyrotropin releasing hormone (TRH), and corticotropin releasing factor (CRF) neurons projecting to the median eminence (see Andersson and Eneroth 1987; Andersson *et al.* 1981*d*, 1984*b*, 1985*b*).

3. The noradrenergic nerve terminal systems in the median eminence (for example, in the subependymal layer and in the medial palisade zone) have been analyzed. These terminals appear to exert an inhibitory influence on CRF release, for example (see Andersson *et al.* 1984*b*).

4. Periventricular hypothalamic noradrenergic nerve-terminal systems. In the posterior part they may exert an excitatory influence on prolactin release (Andersson *et al.* 1981*a*).

6.3.1.1 *Acute nicotine treatment* Intravenous injections of nicotine produce within minutes a selective reduction in noradrenaline levels in the subependymal layer of the median eminence and in the posterior periventricular hypothalamic region. At the same time, marked increases in serum levels of adrenocorticotropic hormone (ACTH), vasopressin, and prolactin are observed (Andersson *et al.* 1983*b*). It may be that the prolactin release induced by nicotine is in part mediated via the nicotine-induced release of noradrenaline in the posterior periventricular hypothalamic area. In contrast, the increase in serum ACTH levels is probably caused by a direct activation of the CRF neurons, in view of the probable inhibitory function of noradrenergic nerve terminals in the subependymal layer in the regulation of the CRF secretion (see Andersson *et al.* 1984*b*). Thus cholinergic-nicotinic receptors may be directly located on the soma-dendritic regions of the CRF neurons (see also Cam *et al.* 1979). The same may be true for the vasopressin neuronal systems, as nicotine can induce a release of vasopressin in rat hypothalamic and neurohypophyseal culture systems (Sladek and Joynt 1979). The depletion of noradrenaline in the subependymal layer may therefore be secondary to the release of CRF which, via an ultra-short feedback, causes release of noradrenaline in the subependymal layer and MPZ to inhibit the release of CRF. The marked increase in CRF release by nicotine is probably also responsible for the increase in serum βendorphin levels observed after acute intraperitoneal injections of nicotine, within 5 min of treatment (Fuxe *et al.* 1987; see also Marty *et al.* 1984).

One hour after intraventricular injection of nicotine or intravenous infusion of nicotine, dose-dependent increases in noradrenaline utilization can be observed within all the various hypothalamic and preoptic noradrenergic nerve terminal systems analyzed (Andersson *et al.* 1981*a*, 1982*a*). These results clearly indicate the existence of cholinergic-nicotinic receptors located on the noradrenergic nerve terminal systems of the entire hypothalamus and the preoptic area.

In spite of signs of increased noradrenaline release, which is known to facilitate the release of TSH, prolactin, and LH, reduced serum levels of TSH, prolactin, and LH were observed. This may be due to an activation of an inhibitory dopaminergic mechanism in the median eminence, controlling LH and prolactin release (see section 6.2.1). Also following acute repeated treatment with nicotine in doses of 2 mg kg^{-1} in normal male and castrated animals substantial reductions of noradrenaline levels and increases in noradrenaline utilization have been found in all the various hypothalamic (Fig. 6.8) and preoptic noradrenergic nerve terminal systems. Again, however, these effects on noradrenaline release are associated with an inhibition of LH, TSH, and prolactin serum levels. This is probably in part related to a rapid desensitization of the cholinergic-nicotinic receptors involved in the initial release of these hormones upon nicotine treatment, as well as to the activation of inhibitory mechanisms such as the tubero-infundibular dopaminergic pathway (see section 6.2.1 and also Sharp and Beyer 1986). As indicated above, it seems possible that these nicotinic-cholinergic receptors which rapidly desensitize are located on the somadendritic region of the hypothalamic hormone-containing pathways. In contrast, the cholinergic-nicotinic receptors located on the hypothalamic noradrenergic nerve terminals do not appear to desensitize as rapidly, and those within the parvocellular part of the hypothalamic nucleus may contribute to maintaining the increases in corticosterone serum levels seen upon nicotine treatment, in spite of the rapid desensitization of the acute stimulatory effects of nicotine on rat plasma ACTH levels. It should also be considered that in the case of ACTH secretion, nicotine does not seem to activate any inhibitory mechanism, and therefore no inhibitory effects on ACTH secretion are observed upon various forms of nicotine treatment.

In agreement with the present *in vivo* findings, nicotine can also release noradrenaline from brain slices *in vitro* (see Westfall 1974). Based on lesions with intraventricular injections of the neurotoxin 6-hydroxydopamine, evidence has also been presented that [^{3}H]acetylcholine nicotinic binding sites are located on noradrenergic nerve terminals within the hypothalamus (Schwartz *et al.* 1984). All the effects of nicotine on noradrenaline utilization within the preoptic area and the hypothalamus are counteracted by pretreatment with mecamylamine, indicating that the cholinergic-nicotinic receptors involved are of the ganglionic type. The same appears to be true for all the inhibitory effects of nicotine on neuroendocrine function, but the excitatory effects of nicotine on ACTH and corticosterone secretion are considerably more resistant to the blocking activity of mecamylamine, indicating that the nicotinic-cholinergic receptors controlling CRF pathways may possibly be controlled predominantly by nicotinic-cholinergic receptors similar to those of the 'neuromuscular' type. It will now be of substantial interest to determine if the nicotinic-cholinergic receptors producing the rapid release of LH and prolactin may also be of the 'neuromuscular' type or represent a certain subtype of brain nicotinic receptor.

6.3.1.2 *Acute exposure to cigarette smoke* Acute intermittent exposure to cigarette smoke also leads to widespread dose-dependent reductions in noradrenaline levels, and increases in noradrenaline utilization, in the various hypothalamic and preoptic noradrenergic nerve terminal systems studied (Andersson *et al.* 1985*c*). Increases in noradrenaline utlization and release in the paraventricular hypothalamic nucleus may contribute to the increase in serum corticosterone levels observed (Andersson *et al.* 1985*c*). Mecamylamine blocked all the effects of exposure to cigarette smoke on the hypothalamic and preoptic noradrenergic nerve terminals, underlining the view that a nicotinic-cholinergic receptor of the ganglionic type may be involved (Andersson 1985).

Acute continuous exposure to cigarette smoke has also been performed. The nicotine levels obtained were of a similar magnitude to those found in the experiments with acute intermitted exposure (Andersson *et al.* 1987). Again, a dose-related depletion of noradrenaline stores and increases in noradrenaline utilization were observed in the various types of hypothalamic noradrenergic nerve terminal systems. Moreover, increased plasma ACTH concentrations were again observed, in addition to the previously discussed inhibitory effects of exposure to cigarette smoke on LH, TSH, and prolactin secretion.

6.3.1.3 *Chronic exposure to cigarette smoke* Following chronic exposure to cigarette smoke, increases in noradrenaline utilization in the periventricular noradrenergic nerve-terminal systems were observed (Andersson *et al.* 1985*a*). However, increases in noradrenaline utilization could no longer be observed within the noradrenaline nerve terminals of the paraventricular hypothalamic nucleus or the subependymal layer of the median eminence. These results indicate that a tolerance may have developed in these systems due to a desensitization of the nicotinic-cholinergic receptors involved in their control. With regard to the neuroendocrine actions, tolerance developed only with regard to the ability of exposure to cigarette smoke to increase ACTH and corticosterone secretion. Tolerance development in the paraventricular noradrenergic nerve terminals to nicotine treatment may contribute to these neuroendocrine findings. It should also be added that upon chronic exposure to cigarette smoke, acute intermittent exposure to cigarette smoke could no longer produce wide spread depletions of noradrenaline stores within the hypothalamic nerve terminal systems, but instead small increases in noradrenaline levels were observed in discrete noradrenergic terminal networks, probably related to increases in tyrosine hydroxylase activity upon chronic exposure.

6.3.2 *Effects on the locus coeruleus noradrenergic system*

The locus coeruleus monosynaptically innervates the entire cerebral cortex via ascending fibers in the medial forebrain bundle and, in addition, also innervates the cerebellar cortex and the ventral horns of the spinal cord (see Fuxe *et al.* 1970; Ungerstedt 1971; Lindvall and Björklund 1983). [^{125}I]αBungarotoxin binding sites have been demonstrated in the locus coeruleus (Clarke *et al.* 1985; see Clarke, chapter 5, this volume). Electrophysiologically, it has been found that nicotine, when given intravenously in doses of 10–15 $\mu g\,kg^{-1}$, will increase the firing rate of the locus coeruleus noradrenergic nerve cells (see Svensson and Engberg 1980). Actions of acetylcholine and nicotine have also been studied on the rat locus coeruleus neurons *in vitro*. Acetylcholine (in the presence of a muscarinic antagonist) and nicotine both depolarize the noradrenergic nerve cells and increase the rate of action potential discharge (Egan and North 1986). The acetylcholine-evoked depolarizations were blocked by hexamethonium but not by αbungarotoxin (see Clarke, Fig. 5.7, this volume) These authors also observed that depolarizations induced by nicotine were reduced following superfusion with the agonist for more than 2 min. Also, a reproducable depolarization induced by repeated applications of nicotine was only found when the interval between applications was at least 1 h. These *in vitro* experiments provide evidence for the existence of nicotinic receptors on the locus coeruleus noradrenergic nerve cells, with similar properties to those within the autonomic ganglia (see Clarke, chapter 5, this volume).

It is unclear, however, if there exist any cholinergic-nicotinic receptors located on the noradrenergic terminals in the cerebral cortex arising from the locus coeruleus system, since intraventricular injections of 6-hydroxydopamine producing a degeneration of the cortical noradrenergic nerve terminals did not result in a reduction in the binding of [^{3}H]acetylcholine to nicotinic sites in the cerebral cortex (Schwartz *et al.* 1984). In acute experiments with single intraventricular injections of nicotine, and in intravenous infusion experiments with nicotine, no effects on noradrenaline levels in the anteromedial frontal cortex have been observed; in experiments on noradrenaline utilization only the highest doses of nicotine (100 μg per rat intraventricularly have produced a significant increase in noradrenaline utilization in the anteromedial frontal cortex (Andersson *et al.* 1981*c*). These results indicate that compared with the non-locus-coeruleus noradrenergic system projecting to the hypothalamus and the preoptic area, the cortical noradrenergic nerve terminals are less sensitive to nicotine. The locus coeruleus noradrenergic system is well-known for tonic arousal, and is of importance for selective attention (see Fuxe *et al.* 1970; Jouvet 1972; Fuxe and Lidbrink 1973). In view of the above results it seems possible that nicotine mainly activates the locus coeruleus system via an action at the level of the nerve cell body, an action which may contribute to the ability of nicotine to increase EEG arousal. However, the action on the locus coeruleus noradrenergic cells is short lasting, explaining the lack of increase in cortical noradrenaline turnover with lower doses of nicotine after 1 h. Thus a short-lasting activation of the locus coeruleus may be obtained by nicotine, probably limited by the rapid desensitization of the cholinergic-nicotinic receptors located on the noradrenergic cell bodies of the locus coeruleus (Egan and North 1986).

6.4 Effects of nicotine on central adrenergic neurons

The adrenergic cell bodies are located in the rostral medulla oblongata, and mainly innervate the hypothalamus and cardiovascular centers of the lower brain stem and the spinal cord (Hökfelt *et al.* 1974). Experiments with acute intermittent treatment with nicotine in high doses have shown that nicotine can also deplete adrenaline stores within the hypothalamus, an action which is counteracted by prior treatment with mecamylamine (Andersson *et al.* 1980). These results indicate that acute nicotine treatment also can increase adrenaline utilization within the hypothalamus, possibly related to the existence of a cholinergic-nicotinic receptor located on the adrenaline terminal networks in this region. These results indicate that actions of nicotine on neuroendocrine function may also be mediated via actions on the adrenergic neurons.

6.5 Effects of nicotine on central 5-hydroxytryptamine (5-HT) neurons

The 5-HT nerve-cell bodies are located predominantly in the raphe nuclei of the lower brain stem; the rostral raphe nuclei give rise to large ascending projections into the di- and telencephalon resulting in the diffuse innervation of the entire cortical hemispheres (see Fuxe *et al.* 1970). There exists evidence that binding of [^{3}H]acetylcholine to nicotinic binding sites is reduced in the hypothalamus and in the striatum following intraventricular injections of 5, 7-hydroxytryptamine, leading to degeneration of 5-HT nerve terminals (Schwartz *et al.* 1984). This is consistent with our recent lesion studies on ascending neuronal pathways (Härfstrand *et al.* 1987). In agreement with this view, it has been found that [^{3}H]5-HT in rat hypothalamic and striatal slices can be released by nicotinic receptor activation (Hery *et al.* 1977; Westfall *et al.* 1983*b*). However, the effects of nicotine on spontaneous behaviour in the rat are not altered by selective lesions of the 5-HT axons innervating the hippocampal formation (Balfour *et al.* 1986*a*), although nicotine injections can produce a selective depletion of hippocampal 5-HT stores (Balfour *et al.* 1986*b*). It must be underlined that acute intermittent treatment with nicotine (as well as its main metabolite cotinine) in doses of 2 mg kg^{-1} can produce a reduction of brain 5-HT utilization, effects which are not blocked by pretreatment with mecamylamine (Fuxe *et al.* 1978). These results are important since they indicate that the main metabolite, cotinine, can also have biological activity, and in part mediate the effects of nicotine on central 5-HT systems. In spite of the demonstrated releasing action of nicotine on 5-HT nerve terminals, a reduction of 5-HT turnover was observed. This phenomenon may also be related to the development of a rapid desensitization of the cholinergic-nicotinic receptors located on the 5-HT nerve terminal networks. In view of the above, it seems important to emphasize that 5-HT nerve terminal systems may also be involved in mediating the central actions of nicotine, for example, in modulating neuroendocrine function and mood (Fuxe *et al.* 1970).

6.6 Effects of nicotine on central cholinergic neurons

The various cholinergic systems of the brain are known to operate via large numbers of muscarinic-cholinergic receptors and a small number of nicotinic-cholinergic receptors (see Larson 1985; Rotter 1984). Various subtypes of muscarinic and nicotinic-cholinergic receptors exist. The radioligands $[^3H]$nicotine, and $[^3H]$acetylcholine (in the presence of atropine) appear to label the same high-affinity site in the brain (Larson 1985; Schwartz *et al.* 1982; see Wonnacott, chapter 7, this volume). The nicotinic-cholinergic receptor probably has both a pre- and postsynaptic location at the cholinergic synapse. Presynaptic nicotinic-cholinergic receptors at the cholinergic synapse probably release acetylcholine (Armitage *et al.* 1969; Beani *et al.* 1985; Rowell and Winkler 1984), whereas the postsynaptic receptor may be involved, for example, in the regulation of the release of monoamines via its presence on monoamine axon terminals. Studies with the neurotoxin histrionicotoxin suggest that the ionic channel of the presynaptic nicotinic-acetylcholine receptor of central neurons is similar to that of the peripheral receptor (Rapier *et al.* 1987). There also exists a subpopulation of central nicotinic receptors, labeled by $[^{125}I]\alpha$bungarotoxin, which may represent the neuromuscular type of the central cholinergic-nicotinic receptor. This subtype is especially frequent in some hypothalamic nuclei (Clarke *et al.* 1985; Härfstrand *et al.* 1988). There also exists a low-affinity $[^3H]$nicotine binding site within the brain (Larson and Nordberg 1985), but its relationship to the cholinergic synapses is less clear, since the site is not downregulated by chronic cholinesterase inhibition (Costa and Murphy 1983; see Wonnacott, chapter 7, this volume).

Finally, we know little at the present time about possible pre- and postsynaptic interactions between the muscarinic and nicotinic receptors in the control of cholinergic transmission.

6.7 Effects of nicotine on central GABA neurons

So far it has not been possible to obtain any change in caudate, hippocampal, and cortical GABA levels, or in regional glutamate decarboxylase activity in rat brain upon acute intermittent treatment with high doses (1–3 mg kg^{-1}) of nicotine (Perez de la Mora, and Fuxe, in preparation). However, GABA neurons of the hypothalamus may be implicated in the neuroendocrine actions of nicotine, since repeated injections of nicotine reduce GABA turnover in the hypothalamus (Perez de la Mora, and Fuxe, in preparation).

6.8 Effects of nicotine on central neuropeptide systems

6.8.1 *Effects on luteinizing hormone releasing hormone, somatostatin and enkephalin neurons of the hypothalamus and the preoptic area*

Luteinizing hormone releasing hormone (LHRH), metenkephalin, and somatostatin-like immunoreactivities have been measured in discrete hypothalamic and preoptic nuclei by means of radioimmunoassay procedures. Nicotine (2 mg kg^{-1} intraperitoneally) produced at the 5 min time interval a significant and selective increase in LHRH immunoreactivity in the median eminence, and in the medial preoptic area (Andersson *et al.* 1986). These changes were associated with increases in serum LH levels. These results indicate that the nicotine-induced activation of the LHRH immunoreactive neurons is associated with an enhanced processing of the precursor peptide to LHRH. The studies have also demonstrated the lack of action of nicotine on the metenkephalin and somatostatin-like immunoreactivities in the median eminence and within the paraventricular hypothalamic nucleus. The results underline the importance of nicotinic-cholinergic receptors regulating directly or indirectly the LHRH immunoreactive nerve-cell bodies, and their firing rate within the preoptic area (see Andersson *et al.* 1986). This nicotinic-cholinergic receptor appears to be rapidly desensitized in contrast to the inhibitory effects of nicotine on LHRH activity (see section 6.2.1).

It should also be mentioned that acute nicotine treatment studied at the 5 min time interval failed to change metenkephalin and somatostatin immunoreactivity in the caudate nucleus of the rat (see Fuxe *et al.* 1987).

6.8.2 *Effects on cholecystokinin and substance P neurons in the hypothalamus and in the medial preoptic area*

Cholecystokinin-8 (CCK-8) and substance P-like immunoreactivities have been measured in the median eminence, the arcuate nucleus, the paraventricular hypothalamic nucleus, and in the medial preoptic area (Fuxe *et al.* 1985) by means of radioimmunoassay. Acute continuous exposure to cigarette smoke from 2 cigarettes increased CCK-like immunoreactivity in the paraventricular hypothalamic region, and reduced it within the median emeinence. Both CCK and substance P immunoreactivities were increased within the medial preoptic area. These results indicate that CCK and substance P immunoreactive neurons in certain areas of the brain are influenced by nicotinic-cholinergic receptors. The results obtained in the preoptic area, for example, indicate that the effects of nicotine on LH secretion and sexual behaviour involve substance P and CCK immunoreactive neurons of the medial preoptic area. Since CCK-like immunoreactivities are co-localized in a number of magnocellular oxytocin neurons of the paraventricular hypothalamic nucleus, and since CCK immunoreactivity is also present within the parvocellular CRF immunoreactive neurons of the paraventricular hypothalamic nucleus, it seems possible that the changes observed in CCK immunoreactivity at least in part reflect changes in CCK synthesis and release in paraventricular oxytocin and CRF immunoreactive neurons (see Fuxe *et al.* 1985). Injections of CCK-8 into the paraventricular hypothalamic nucleus and surrounding region are reported to reduce food intake in rats (see Baile and Della-Fera 1985). Therefore, the nicotine-induced increase in the CCK-like immunoreactivity in this area may be involved in the mediation of the anti-appetite effects of nicotine. Central nicotinic-cholinergic receptors in the paraventricular region may thus modulate the multiple functional roles of CCK peptides in this region.

Studies in rat brain slices have indicated a possible role of opiate receptors in the regulation of dopamine release and 5-HT release induced by cholinergic-nicotinic receptors from dopaminergic and 5-HT nerve terminals respectively (Westfall *et al.* 1983*a*, *b*). Thus, certain opioid peptides and morphine can inhibit release of dopamine and 5-HT induced by a nicotinic-cholinergic agonist. These results indicate a possible receptor–receptor interaction between opiate and nicotinic-cholinergic receptors in the nerve-cell membrane of dopaminergic and 5-HT nerve terminals of the neostriatum (see Fuxe and Agnati 1985). Also, evidence has been obtained that presynaptic dopamine D1 receptors located on the hypothalamic noradrenergic terminals are essential for the nicotine-induced release of noradrenaline from these terminals (Anderson *et al.* 1988). Furthermore, in adrenal medullary cells opiate recognition sites modulate nicotinic receptor function, probably by decreasing the gain at which the nicotinic receptors operate. This regulation also appears to be involved in the downregulation of the nicotinic receptor recognition sites (Costa *et al.* 1983).

6.9 Concluding remarks and clinical implications

This chapter summarizes the evidence that central cholinergic nerve terminal networks, via their nicotinic receptors, can regulate the release of monoamines in various types of central monoaminergic neurons via actions at the nerve cell body-dendritic level, and at the local-circuit level where cholinergic and monoaminergic nerve terminals interact via axo–axonic contacts. Furthermore, the available evidence indicates that cholinergic-nicotinic receptors directly influence a number of central peptidergic neurons, such as the LHRH immunoreactive neurons, the CRF immunoreactive neurons, and the vasopressin immunoreactive neurons. It seems possible that:

(1) the locus coeruleus noradrenergic system innervating the cortical hemispheres to some degree is involved in the EEG arousal induced by nicotine treatment (see Knapp and Domino 1962).

(2) the nigrostriatal dopaminergic system at least in part mediates the behavioural arousal seen with nicotine treatment (see Knapp and Domino 1962).

(3) the mesolimbic dopaminergic neurons innervating especially the anterior parts of the nucleus accumbens and olfactory tubercule may be involved in nicotine-induced reward and dependence (Fig. 6.9 and 6.11).

(4) the tubero-infundibular dopaminergic neurons, at least in part, mediate the inhibitory effects of nicotine on LH and prolactin secretion. Of special interest is the demonstration of the co-storage of dopamine and acetylcholine in nerve cells in the arcuate nucleus (Fig. 6.1).

(5) the non-locus coeruleus noradrenergic terminal systems of the hypothalamus and the preoptic area may to some extent be involved in the ability of nicotine to produce an initial and short-lasting release of LH, prolactin, and vasopressin, and a maintained release of ACTH (Fig. 6.1).

The evidence indicates that nicotinic-cholinergic receptors of the ganglionic type are predominantly involved in mediating the effects of nicotine on central monoamine neurons, neuroendocrine function, and behavioural and EEG arousal. However, nicotinic-cholinergic receptors of a special subtype may be involved, especially in the regulation of ACTH secretion, and in initial neuroendocrine events following nicotine administration.

Smokers have been found to have lower sperm counts, associated with the decreased motility and increased percentage of sperm with pathological changes, possibly caused by adrenal gland activation (Biswas *et al.* 1979). Also, reduced development of pubic hair has been reported in early smokers (see Vogel *et al.* 1979). However, no consistent effects of smoking on testosterone levels have been reported in man (see Vogel *et al.* 1979). There also exists a highly significant trend of decreasing fertility with increasing numbers of cigarettes smoked per day (Howe *et al.* 1985; see also Baird and Wilcox 1985). If the results obtained in the rat are relevant to the human smoker, it should be possible to counteract the inhibitory effects of nicotine on LHRH secretion by administration of D1 dopamine receptor antagonists. In this way it seems possible to test if alterations in reproductive function in smokers, especially in fertility, may be related to an activation of the tuberoinfundibular dopaminergic neurons in the median eminence.

Tuberoinfundibular dopaminergic neurons may also exert inhibitory actions on the release of somatostatin from the median eminence, leading to an increase in growth hormone secretion (see Anderson *et al.* 1983*a*). It remains to be determined whether such mechanisms may contribute to the ability of smoking to increase growth hormone secretion (Wilkins *et al.* 1982). These authors have also reported that in chronic smokers nicotine from cigarette smoking can increase the circulating levels of prolactin, indicating that as in the rat, nicotine can also increase prolactin release.

In view of the large number of transmitter-identified neuronal systems which can be influenced by nicotine at the same time, the biological effects of nicotine which mediate the central actions of cigarette smoke are highly complex. The effects observed are dependent on time interval after injection, dose used, and the type of treatment (single or intermittent). An important phenomenon to be considered in this respect is the desensitization of cholinergic-nicotinic receptors, which can be very rapid in some nicotinic-cholinergic populations, while others desensitize more slowly. Also, the development of depolarization-induced blockade in various neuronal systems should be considered upon frequent administration of nicotine. Thus, instead of activation, nicotine will cause an inactivation of the affected neurons. Finally, a marked influence on the endocrine system by nicotine will lead to a hormonal feedback modulation by, for example, corticosterone, of the networks in which the cholinergic, monoaminergic, and peptidergic neurons reciprocally interact with one another. Thus the neurochemical actions of nicotine will differ from moment to moment, and so will its behavioural and neuroendocrine actions, being dependent on the integrated outputs from the neuronal networks controlled *inter alia* by the cholinergic, monoaminergic, and peptidergic neurons.

Acknowledgements

This work has been supported by grants (1223, 1762) from the Council for Tobacco Research, and by a grant from the Svenska Tobaksmonopolet, Stockholm, Sweden. We are grateful to the excellent secreterial assistance of Monia Särd and Anne Edgren.

References

Ahtee, L., and Kaakkola, S. (1978). Effect of mecamylamine on the fate of dopamine in striatal and mesolimbic areas of rat brain; interaction with morphine and haloperidol. *British Journal of Pharmacology* **62**, 213–8.

Andersen, A.N., Lund-Andersen, C., Larsen, J.F., Christensen, N.J., Legros, J.J., Louis, F., Angelo, H., and Molin, J. (1982). Suppressed prolactin but normal neurophysin levels in cigarette smoking breast-feeding women. *Clinical Endocrinology* **17**, 363–8.

Andersen, A.N., Semczuk, M., and Tabor, A. (1984). Prolactin and pituitary-gonadal function in cigarette smoking infertile patients. *Andrologia* **16**, 391.

Andersson, K. (1985). Mecamylamine pretreatment counteracts cigarette smoke induced changes in hypothalamic catecholamine neuron systems and in anterior pituitary function. *Acta Physiologica Scandinavica* **125**, 445–52.

Andersson, K., and Eneroth, P. (1987). Thyroidectomy and central catecholamine neurons of the male rat. *Neuroendocrinology* **45**, 14–27.

Andersson, K., Fuxe, K., Eneroth, P., Gustafsson, J.-Å, and Agnati, L.F. (1980). Mecamylamine induced blockade of nicotine induced inhibition of gonadotrophin and TSH secretion and of nicotine induced increases of catecholamine turnover in the rat hypothalamus. *Acta Physiologica Scandinavica Suppl.* **479**, 27–9.

Andersson, K., Eneroth, P., and Agnati, L.F. (1981*a*). Nicotine-induced increases of noradrenaline turnover in discrete noradrenaline nerve terminal systems of the hypothalamus and the median eminence of the rat and their relationship to changes in the secretion of adenohypophyseal hormones. *Acta Physiologica Scandinavica* **113**, 227–31.

Andersson, K., Fuxe, K., and Agnati, L.F. (1981*b*). Effects of single injections of nicotine on the ascending dopamine pathways in the rat. *Acta Physiologica Scandinavica* **112**, 345–7.

Andersson, K., Fuxe, K., Agnati, L.F., and Eneroth, P. (1981*c*). Effects of acute central and peripheral administration of nicotine on ascending dopamine pathways in the male rat brain. Evidence for nicotine induced increases of dopamine turnover in various telencephalic dopamine nerve terminal systems. *Medical Biology* **59**, 170–6.

Andersson, K., Fuxe, K., Eneroth, P., Blake, C.A., Agnati, L.F., and Gustafsson, J.-Å (1981*d*). Effects of androgenic and adrenocortical steroids on hypothalamic and preoptic catecholamine nerve terminals and on the secretion of anterior pituitary hormones. In *Steroid hormone regulation of the brain* (K. Fuxe, J.-Å. Gustafsson, L. Wetterberg, ed.), pp. 117–33. Pergamon Press, Oxford.

Andersson, K., Fuxe, K., Eneroth, P., Nyberg, P., and Roos, P. (1981*e*). Rat prolactin and hypothalamic catecholamine nerve terminal systems. Evidence for rapid and discrete increases in dopamine and noradrenaline turnover in the hypophysectomized male rat. *European Journal of Pharmacology* **76**, 261–5.

Andersson, K., Fuxe, K., Eneroth, P., and Agnati, L.F. (1982*a*). Effects of acute central and peripheral administration of nicotine on hypothalamic catecholamine nerve terminal systems and on the secretion of adenohypophyseal hormones in the male rat. *Medical Biology* **60**, 98–111.

Andersson, K., Fuxe, K., Eneroth, P., and Agnati, L.F. (1982*b*) Involvement of cholinergic nicotine-like receptors as modulators of amine turnover in various types of hypothalamic dopamine and noradrenaline nerve terminal systems and of prolactin, LH, FSH and TSH secretion in the castrated male rat. *Acta Physiologica Scandinavica* **116**, 41–50.

Andersson, K., Fuxe, K., Eneroth, P., Isaksson, O., Nyberg, F., and Roos, P. (1983*a*). Rat growth hormone and hypothalamic catecholamine nerve terminal systems. Evidence for rapid and discrete reductions in dopamine and noradrenaline turnover in the median eminence of the hypophysectomized male rat. *Eurpoean Journal of Pharmacology* **95**, 271–5.

Anderson, K., Siegel, R., Fuxe, K., and Eneroth, P. (1983*b*). Intravenous injections of nicotine induced very rapid and discrete reductions of hypothalamic catecholamine levels associated with increases of ACTH, vasopressin, and prolactin secretion. *Acta Physiologica Scandinavica* **118**, 35–40.

Andersson, K., Agnati, L.F., Fuxe, K., Eneroth, P., Härfstrand, A., and Benfenati, F. (1984*a*). Corticotropin-releasing factor increases noradrenaline turnover in the median eminence and reduces noradrenaline turnover in the paraventricular region of the hypophysectomized male rat. *Acta Physiologica Scandinavica* **120**, 621–4.

Andersson, K., Fuxe, K., Eneroth P., and Agnati, L.F. (1984*b*). Differential effects of mecamylamine on the nicotine induced changes in amine levels and turnover in hypothalamic dopamine and noradrenaline nerve terminal systems and in the secretion of adenohypophyseal hormones in the castrated female rat. Evidence for involvement of cholinergic nicotine-like receptors. *Acta Physiologica Scandinavica* **120**, 489–98.

Andersson, K., Eneroth, P., Fuxe, K., Mascagni, F., and Agnati, L.F. (1985*a*). Effects of chronic exposure to cigarette smoke on amine levels and turnover in various hypothalamic catecholamine nerve terminal systems and on the secretion of pituitary hormones in the male rat. *Neuroendocrinology* **41**, 462–6.

Andersson, K., Eneroth P., and Roos, P. (1985*b*). Effects of TRH and a rat TSH preparation on discrete hypothalamic and forebrain catecholamine nerve terminal netwroks in the hypophysectomized male rat. *European Journal of Pharmacology* **111**, 295–307.

Andersson, K., Fuxe, K., Eneroth, P., Mascagni, F., and Agnati, L.F. (1985*c*). Effects of acute intermittent exposure to cigarette smoke on catecholamine levels and turnover in various types of hypothalamic DA and NA nerve terminal systems as well as on the secretion of adenohypohypseal hormones and corticosterone. *Acta Physiologica Scandinavica* **124**, 277–85.

Andersson, K, Fuxe, K., Eneroth, P., and Härfstrand, A. (1986). Nicotine-induced increases in brain luteinizing hormone releasing hormone-like immunoreactivity and in serum luteinizing hormone levels of the male rat. *Neuroscience Letters* **71**, 289–91.

Andersson, K., Fuxe, K., Eneroth, P., and Agnati, L.F. (1987). Effects of acute continuous exposure of the rat to cigarette smoke on amine levels and utilization in discrete hypothalamic catecholamine nerve terminal systems and on neuroendocrine function. *Archives of Pharmacology* **335**, 521–8.

Andersson, K., Fuxe, K., Eneroth, P., Härfstrand, A., and Agnati, L.F. (1988). Involvement of D1 dopamine receptors in the nicotine-induced neuroendocrine effects and depletion of diencephalic catecholamine stores in the male rat. *Neuroendocrinology* **48**, 188–200.

Armitage, A.K., Hall, G.H., and Sellers, C.M. (1969). Effects of nicotine on electrocortical activity and acetylcholine release from the cat cerebral cortex. *British Journal of Pharmacology* **35**, 152–60.

Balfour, D. (1982). The effects of nicotine on brain neurotransmitter systems. *Pharmacology and Therapeutics* **16**, 269–82.

Balfour, D.J., Benwell, M.E., Graham, C.A., and Vale, A.L. (1986*a*). Behavioural and adrenocortical responses to nicotine measured in rats with selective lesions of the 5-hydroxytryptaminergic fibres innervating the hippocampus. *British Journal of Pharmacology* **89**, 341–7.

Balfour, D.J., Graham, C.A., and Vale, A.L. (1986*b*). Studies on the possible role of brain 5-HT systems and adrenocortical activity in behavioral responses to nicotine and diazepam in an elevated X-maze. *Psychopharmacology* **90**, 528–32.

Baile, C.A., and Della-Fera, M.A. (1985). Central Cholecystokinin System and the control of feeding. *Annals of the New York Academy of Sciences* **448**, 424–30.

Baird, D.D., and Wilcox, A.J. (1985). Cigarette smoking associated with delayed conception. *Journal of the American Medical Association* **253**, 2979–83.

Baron, J.A. (1986). Cigarette smoking and Parkinson's desease. *Neurology* **36**, 1490–6.

Beani, L., Bianchi, C., Nilsson, L., Nordberg, A., Romanelli, L., and Sivilotti, L. (1985). The effect of nicotine and cytisine on ^{3}H-acetylcholine release from cortical slices of guinea-pig brain. *Archives of Pharmacology* **331**, 293–6.

de Belleroche, J., and Bradford, H.F. (1978). Biochemical evidence for the presence of presynaptic receptors on dopaminergic nerve terminals. *Brain Research* **142**, 52–8.

Biswas, N.M, Patra, P.B., Sanyal, S., and Deb, C. (1979). Adrenogonadal interactions in nicotine-induced alteration of steroidogenesis in male rat. *Andrologia* **11**, 227–33.

Blake, C.A., and Sawyer, C.K. (1972). Nicotine blocks the suckling-induced rise in circulating prolactin in lactating rats. *Science* **177**, 619–21.

Blake, C.A., Norman, R.L., Scaramuzzi, R.J., and Sawyer, C.H. (1973). Inhibition of the prosterous surge of prolactin in the rat by nicotine. *Endocrinology* **92**, ·1334–8.

Butcher, L.L., and Woolf, N.J. (1984). Histochemical distribution of acetylcholinesterase in the central nervous system: Clues to the localization of cholinergic neurons. In *Handbook of chemical neuroanatomy. Vol. 2: Classical transmitters in the CNS*, (A. Björklund, T. Hökfelt, and M.J. Kuhar, ed.), pp. 1–51. Elsevier Science Publishers BV, Amsterdam.

Cam, T.R., Bassett, J.R., and Cairncross, K.D. (1979). The action of nicotine on the pituitary-adrenal cortical axis. *Archives of International Pharmacodynamics and Therapeutics* **237**, 49–66.

Clarke, P., and Pert, A. (1985). Autoradiographic evidence for nicotine receptors on nigrostriatal and mesolimbic dopaminergic neurons. *Brain Research* **348**, 355–8.

Clarke, P.B.S., Schwartz, R.D., Paul, S.M., Pert, C.B., and Pert, A. (1985). Nicotinic binding in rat brain: autoradiogrpahic comparison of [^{3}H] acetylcholine, [^{3}H] nicotine, and [^{125}I]αbungarotoxin. *Journal of Neuroscience* **5**, 1307–15.

Costa, L.G., and Murphy, S.D. (1983). [^{3}H] Nicotine binding in rat brain: alteration after chronic acetylcholinesterase inhibition. *Journal of Pharmacology and Experimental Therapeutics* **226**, 392–7.

Costa, E., Guidotti, A., Hanbauer, I., and Saiani, L. (1983). Modulation of nicotine receptor function by opiate recognition sites highly selective for Met5-enkephalin: Arg6Phe7. *Federation Proceedings* **42**, 2946–52.

Egan, I.M., and North, R.A. (1986). Actions of acetylcholine and nicotine on rat locus coeruleus neurons *in vitro*. Neuroscience **19**, 565–71.

Eneroth, P., Fuxe, K., Gustafsson, J.-Å., Hökfelt, T., Löfström, A., Skett, P., and Agnati, L.F. (1977). The effect of nicotine on central catecholamine neurons and gonadotropin secretion II. Inhibitory influence of nicotine on LH, FSH and prolactin secretion in the ovariectomized female rat and its relation to regional changes in dopamine and noradrenaline levels and turnover. *Medical Biology* **55**, 158–66.

Fuxe, K., and Agnati, L.F. (1985). Receptor–receptor interactions in the central nervous system. A new integrative mechanism in synapses. *Medicinal Research Reviews* **5**, 441–82.

Fuxe, K., and Lidbrink, P. (1973). Biogenic amine aspects of sleep and waking. In *Sleep: physiology, biochemistry, psychology, pharmacology, clinical implications*, (W.P. Koella and P. Levin ed.), pp. 12–26. Karger, Basel.

Fuxe, K., Hökfelt, T., and Ungerstedt, U. (1970). Morphological and functional aspects of central monoamine neurons. *International Reviews of Neurobiology* **13**, 93–126.

Fuxe, K. Agnati, L.F., Eneroth, P., Gustafsson, J.-Å., Hökfelt, T., Löfström, A., Skett, B., and Skett, P. (1977). The effect of nicotine on central catecholamine neurons and gonadotropin secretion I. Studies in the male rat. *Medical Biology* **55**, 148–57.

Fuxe, K., Everitt, B.J., and Hökfelt, T. (1978). On the action of nicotine and cotinine on central 5-hydroxytryptamine neurons. *Pharmacology, Biochemistry and Behavior* **10**, 671–7.

Fuxe, K., Andersson, K., Schwarcz, R., Agnati, L.F., Pérez de la Mora, M., Hökfelt, T., Goldstein, M., Ferland, L., Possani, L., and Tapia, R. (1979). Studies on different types of dopamine nerve terminals in the forebrain and their possible interactions with hormones and with neurons containing GABA, glutamate and opioid peptides. In *Advances in Neurology,* Vol. 24, (L.J. Piorier, T.L. Sourkes, and P.J. Bédard, ed.), pp. 199–214. Raven Press, New York.

Fuxe, K., Siegel, R.A., Andersson, K., Eneroth, P., Mascagni, F., and Agnati, L.F. (1985). Acute continuous exposure to cigarette smoke produces discrete changes in cholecystokinin and substance P levels in the hypothalamus and preoptic area of the male rat. *Acta Physiologica Scandinavica* **125**, 437–43.

Fuxe, K., Andersson, K., Härfstrand, A., and Agnati, L.F. (1986). Increases in dopamine utilization in certain limbic dopamine terminal populations after a short period of intermittent exposure of male rats to cigarette smoke. *Journal of Neural Transmission* **67**, 15–29.

Fuxe, K., Andersson, K., Eneroth, P., Härfstrand, A., and Agnati, L.F. (1987). Effects of nicotine and exposure to cigarette smoke on discrete dopamine and noradrenaline nerve terminal systems of the telencephalon and diencephalon of the rat. Relationship to reward mechanisms and neuroendocrine functions and distribution of nicotinic binding sites in brain. In *Tobacco Smoking and Nicotine,* (W.R. Martin, G.R. Van Loon, E.T. Iwamoto, and L. Davis, ed.), pp. 225–62. Plenum Publishing Corporation, New York.

Giorguieff-Chesselet, M.F., Kemel, M.L., Wandscheer, D., and Glowinski, J. (1979). Regulation of dopamine release by presynaptic nicotinic receptors in rat striatal slices: Effect of nicotine in a low concentration. *Life Sciences* **25**, 1257–62.

Grenhoff, J., Aston-Jones, G., and Svensson, T.H. (1986). Nicotinic effects on the firing pattern of midbrain dopamine neurons. *Acta Physiologica Scandinavica* **128**, 351–8.

Härfstrand, A., Fuxe, K. Andersson, K., Agnati, L.F., Janson, A.M., and Nordberg, A. (1987). Partial di-mesencephalic hemitransections produce disappearance of ^{3}H-nicotine binding in discrete regions of rat brain. *Acta Physiologica Scandinavica* **130**, 161–3.

Härfstrand, A., Adem, A., Fuxe, K. Agnati, L.F., Andersson, K., and Nordberg, A. (1988). Distribution of nicotinic receptors in the rat tel- and diencephalon: a quantitative receptor autoradiographical study using ^{3}H-acetylcholine, ^{125}Iαbungarotoxin and ^{3}H-nicotine. *Acta Physiologica Scandinavica* **132**, 1–14.

Hery, F., Bourgoin, S., Hamon, M., Ternaux, J.P., and Glowinski, J. (1977). Control of the release of newly synthetized ^{3}H-5-hydroxytryptamine by nicotinic and muscarinic receptors in rat hypothalamic slices. *Naunyn Schmiedeberg's Archives of Pharmacology* **296**, 91–7.

Hökfelt, T., Fuxe, K., Goldstein, M., and Johansson, O. (1974). Immunohistochemical evidence for the existence of adrenaline neurons in the rat brain. *Brain Research* **66**, 235–51.

Hökfelt, T., Skirboll, L., Rehfeld, J.F., Goldstein, M., Markey, K., and Dann, O. (1980). A subpopulation of mesencephalic dopamine neurons projecting to limbic areas contains a cholecystokinin-like peptide: evidence from immunohistochemistry combined with retrograde tracing. *Neuroscience* **5**, 2093–124.

Howe, G., Westhoff, C., Vessey, M., and Yeates, D. (1985). Effects of age, cigarette smoking, and other factors on fertility: findings in a large prospective study. *British Medical Journal* **290**, 1697–700.

Jaffe, J.H. and Jarvik, M.E. (1978). Tobacco use and tobacco use disorder. In *Psychopharmocology: a generation of progress*, (M.S. Lipton, A. DiMascio, K.F. Killam, ed.), pp. 1665–77. Raven Press, New York.

Janson, A.M., Fuxe, K., Kitayama, I., Härfstrand, A., and Agnati, L.F. (1986). Morphometric studies on the protective action of nicotine on the substantia nigra dopamine nerve cells after partial hemitransection in the male rat. *Neuroscience Letters* **26**, S88.

Janson, A.M., Agnati, L.F., Kitayama, I., Härfstrand, A., Andersson, K., Goldstein, M., and Fuxe, K. (1988). Chronic nicotine treatment counteracts the disappearance of tyrosine-hydroxylase immunoreactive nerve cell bodies, dendrites and terminals in the meso-striatal dopamine system of the male rat after partial hemitransection. *Brain Research* **455**, 332–45.

Jouvet, M. (1972). The role of monoamines and acetylcholine-containing neurons in the regulation of the sleep-waking cycle. In *Ergebnisse der physiologie (reviews of physiology) neurophysiology and neurochemistry of sleep and wakefulness*, pp. 166–307. Springer, Berlin.

Kalra, S.P. and Kalra, P.S. (1983). Neural regulation of luteinizing hormone secretion in the rat. *Endocrine Reviews USA* **4**, 311–51.

Kimura, H., McGeer, P., and Peng, J.-H. (1984). Choline acetyltransferase-containing neurons in the rat brain. In *Handbook of chemical neuroanatomy. Vol. 2: classical transmitters in the CNS*, (A. Björklund, T. Hökfelt, and M.J. Kuhar, ed.), pp. 51–68. Elsevier Science Publishers BV. North Holland.

Knapp, D.E. and Domino, E.F. (1962). Action of nicotine on the ascending reticular activating system. *International Journal of Neuropharmacology* **1**, 333–51.

Langley, J.N. and Dickinson, W.L. (1889). On the local paralysis of peripheral ganglia, and on the connection of different classes of nerve fibres with them. *Proceedings of the Royal Society* **46**, 423–31.

Larsson, C. (1985). Nicotine receptors in the central nervous system: methodological and functional aspects. *Acta Universitas Upsaliensis, abstracts of Uppsala disserition from the Faculty of Pharmacy* **104**, 4–45.

Larsson, C. and Nordberg, A. (1985). Comparative analysis of nicotine-like ligand receptor interaction in rodent brain homogenate. *Journal of Neurochemistry* **45**, 24–31.

Lewis, P.R. and Shute C.C.D. (1978). Cholinergic pathways in CNS. In *Handbook of psychopharmacology*, Vol. 9 (L.L. Iversen, S.D. Iversen, and S.H. Snyder, ed.), pp.315–55. Plenum Press, New York.

Lichtensteiger, W., Felix, D., Lienhart, R., and Hefti, F. (1976). A quantitative correlation between single unit and fluorescence intensity of dopamine neurons in zona compacta of substantia nigra, as demonstrated under the influence of nicotine and physostigmine. *Brain Research* **117**, 85–103.

Lichtensteiger, W., Richards, R.G., and Kopp, H.G. (1978). Changes in the distribution of non-neuronal elements in rat median eminence and in anterior pituitary hormone secretion after activation of tubero-infundibular dopamine neurones by brain stimulation or nicotine. *Brain Research* **157**, 759–64.

Lichtensteiger, W., Hefti, F., Felix, D. Huwyler, T., Melamed, E., and Schlumpfe, M. (1982). Stimulation of nigrostriatal dopamine neurones by nicotine. *Neuropharmacology* **21**, 963–8.

Lindvall, O. and Björklund, E. (1983). Dopamine- and norepinephrine-containing neurons systems: Their anatomy in the rat brain. In *Chemical Neuroanatomy*, (P.C. Emson, ed.), p. 229–55. Raven Press, New York.

Marty, M.A. Erwin, V.G., Cornell, K, and Zgombick, J.M. (1984). Effects of nicotine on β-endorphin, α-MSH, and ACTH secretion by isolated perfused mouse brains and pituitary glands *in vitro. Pharmacology, Biochemistry and Behavior* **22**, 317–25.

McLean, B.K., Rubel, A., and Nikitovitch-Winer, M.V. (1977). The differential effects of exposure to tobacco smoke on the secretion of luteinizing hormone and prolactin in the proestrous rat. *Endocrinology* **100**, 1566–70.

Morley, B.J., Farley, G.R., and Javel, E. (1983). Nicotinic acetylcholine receptors in mammalian brain.. *Trends in Pharmacological Science* **4**, 225–7.

Rao, Z.R., Yamano, M., Wanaka, A., Tatehata, T., Shiosaka, S., and Tohyama, M. (1987). Distribution of cholinergic neurons and fibers in the hypothalamus of the rat using choline acetyltransferase as a marker. *Neuroscience* **20**, 923–34.

Rapier, C., Wonnacott, S., Lunt, G.G., and Albuquerque, E.X. (1987). The neurotoxin histrinicotoxin interacts with the putative ion channel of the nicotinic acetylcholine receptors in the central nervous system. FEBS Letters **212**, 292–6.

Richardson, S.B., Prasad, J.A., and Hollander, C.S. (1982). Acetylcholine, melatonin, and potassium depolarization stimulate release of luteinizing hormone-releasing hormone from rat hypothalamus in vitro. *Proceedings of the National Academy of Sciences USA* **79**, 2686–9.

Roberts, D.C.S., Cochran, M.E., and Fibiger, H.C. (1979). Recovery of cocaine selfadministration after 6-OHDA lesion of the nuc. accumbens correlates with residual dopamine levels. In *Catecholamines: basic and clinical frontiers*, Vol. 2, (E. Usdin, I.J. Kopin, and F., Barchas, ed.), pp. 1774–6. Pergamon, New York.

Rotter, A. (1984). Cholinergic receptors. In *Handbook of chemical neuroanatomy Vol. 2: classical transmitters in the CNS*, (A. Björklund, T. Hökfelt, and M.J. Kuhar, ed.), pp. 277–304. Elsevier Science Publishers BV. Amsterdam.

Rowell, P.P. and Winkler, D.L. (1984). Nicotinic stimulation of [^{3}H]acetylcholine release from mouse cerebral cortical synaptosomes. *Journal of Neurochemistry* **43**, 1593–8.

Schwartz, R.D. and Kellar, K.J. (1983). Nicotinic cholinergic receptor binding sites in brain: *in vivo* regulation. *Science* **220**, 214–6.

Schwartz, R.D., McGee, R., and Kellar, K.J. (1982). Nicotinic cholinergic receptors labelled by [^{3}H]acetylcholine in rat brain. *Molecular Pharmacology* **22**, 56–62.

Schwartz, R.D., Lehmann, J., and Kellar, K.J. (1984). Presynaptic nicotinic cholinergic receptors labeled by ^{3}H-acetylcholine on catecholamine and serotonin axons in brain. *Journal of Neurochemistry* **42**, 1495–8.

Sharp, B.M. and Beyer, H.S. (1986). Rapid desensitization of the acute stimulatory effects of nicotine on rat plasma adrenocorticotropin and prolactin. *Journal of Pharmacology and Experimental Therapeutics* **238**, 486–91.

Singer, G., Wallace, M., and Hall, R. (1982). Effects of dopaminergic nucleus accumbens lesions on the acquisition of schedule induced self injection of nicotine in the rat. *Pharmacology, Biochemistry and Behavior* **17**, 579–81.

Sladek, C.D. and Joynt, R.J. (1979). Cholinergic involvement in osmotic control of vasopressin release by the organ-cultured rat hypothalamo-neurohypophyseal system. *Endocrinology* **105**, 367–71.

Svensson, T. and Engberg, G. (1980). Effect of nicotine on single cell activity in the noradrenergic nucleus locus coeruleus. *Acta Physiologica Scandinavica* Suppl. **479**, 31–4.

Tago, H., McGeer, P.L., Bruce, G., and Hersh, L.B. (1987). Distribution of choline acetyltransferase-containing neurons of the hypothalamus. *Brain Research* **415**, 49–62.

Takano, Y., Sakurai, Y., Kohjimoto, Y., Honda, K., and Kamiya, H.O. (1983). Presynaptic modulation of the release of dopamine from striatal synaptosomes:

differences in the effects of high K^+ stimulation, methamphetamine and nicotinic drugs. *Brain Research* **279**, 330–4.

Thesleff, S. (1955). The mode of neuromuscular block caused by acetylcholine, nicotine, decamethonium and succinylcholine. *Acta Physiologica Scandinavica* **34**, 218–31.

Tinner, B., Fuxe, K., Köhler, Ch., Hersh, L., Andersson, K., Jansson, A., Goldstein, M., and Agnati, L.F. (1989). Evidence for the existence of a population of arcuate neurons costoring choline acetyltransferase and tyrosine hydroxylase immunoreactivities in the male rat. *Neuroscience Letters* **99**, 44–49.

Ungerstedt, U. (1971). Stereotaxic mapping of the monoamine pathways in the rat brain. *Acta Physiologica Scandinavica* Suppl. **367**, 1–48.

Vogel, W., Broverman, D.M., and Klaiber, E.L. (1979). Gonadal, behavioral and electroencephalographic correlates of smoking. In *Electrophysiological effects of nicotine*, (A. Rémond, and C. Izard, ed.), pp. 201–15. Elsevier/North Holland, Biomedical Press.

Westfall, T.C. (1974). Effect of nicotine and other drugs on the release of ^{3}H-norepinephrine and ^{3}H-dopamine from rat brian slices. *Neuropharmacology* **13**, 639–700.

Westfall, T.C., Grant, H., Naes, L., and Meldrum, M. (1983*a*). The effect of opioid drugs on the release of dopamine and 5-hydroxytryptamine from rat striatum following activation of nicotinic-cholinergic receptors. *European Journal of Pharmacology* **19**, 35–42.

Westfall, T., Grant, H., and Perry, H. (1983*b*). Release of dopamine and 5-hydroxytryptamine from rat striatal slices following activation of nicotinic cholinergic receptors. *General Pharmacology* **14**, 321–5.

Wilkins, J.N., Carlson, H.E., Van Vunakis, H, Hill, M.A., Gritz, E., and Jarvik, M.E. (1982). Nicotine from cigarette smoking increases circulating levels of cortisol, growth hormone, and prolactin in male chronic smokers. *Psychopharmacology* **78**, 305–8.

7 Characterization of nicotine receptor sites in the brain

S. Wonnacott

7.1 Introduction

From the earlier chapters in this volume it is recognized that nicotine exerts its principal behavioural and psychological effects by interacting with the central nervous system. This interaction at the cellular level has been discussed in the previous chapter. The target for nicotine can be further refined to the molecular level, and the present text will consider the interaction of nicotine with its specific recognition sites on neurones. In this context it is fitting that studies with nicotine led to the initial concept of a 'receptive substance' on the cell surface that translates the chemical signal of a neurotransmitter into a cellular response (Langley 1907). Nicotine was instrumental, too, in the subclassification of acetylcholine receptors into muscarinic (sensitive to muscarine) and nicotinic types (Dale 1914).

It has been recognized since these early studies that nicotinic receptors are present on both skeletal muscle cells and neurons, but the molecular characterization of these receptors has commenced only relatively recently. The structure of the muscle nicotinic receptor has been elucidated in detail (thanks to special factors discussed below), but understanding of the molecular targets of nicotine's actions in the central and autonomic nervous systems is just beginning. The initially slow progress in this area reflects the daunting complexity of the brain and the paucity of nicotinic substrates in the central nervous system, where muscarinic responses predominate. Autonomic ganglia, on the other hand, are anatomically simpler and amenable to detailed pharmacological analyses of nicotinic transmission, but lack of tissue frustrates attempts to carry out biochemical studies. This has been circumvented to some extent by using related cell lines (for example PC12 cells, representative of sympathetic neurones), although transformed cells present their own problems such as uncertain phenotype.

Nevertheless, increasing research effort is being directed at the neuronal nicotinic receptors that are now recognized as having significant roles in the brain, not only in sustaining the tobacco smoking habit but also in cognitive processes, and are affected in neurological disorders (see chapters by Warburton, Reavill, Kellar and Wonnacott, this volume). Because the direction of this research has been markedly influenced by the success story of the characterization of the nicotinic-acetylcholine receptor in skeletal muscle, it is appropriate first to outline the current knowledge of the muscle receptor. This has both helped and hampered study of its neuronal counterparts: the availability of plausible research strategies already tried and tested in the peripheral system has been an advantage, but the expectation that the neuronal nicotinic receptor will closely resemble the muscle prototype has been misleading. It is now clear that nicotinic receptors in muscle and neurons are not identical, and differ both structurally and pharmacologically (which is perhaps implicit in their early subclassification into C10 and C6 types, Paton and Zaimis 1949). Furthermore, heterogeneity of receptor structure is emerging, and it may be anticipated that this will be reflected in functional heterogeneity. Understanding this diversity will ultimately help to unravel the complexity of nicotine's actions.

7.2 The muscle nicotinic acetylcholine receptor

The nicotinic receptor at the neuromuscular junction responds rapidly to the release of acetylcholine into the synaptic cleft, by opening an ion channel that allows cations to flow into the muscle cell. This ion flux may sufficiently depolarize the muscle cell to cause muscle contraction. Two important factors have facilitated the study of this nicotinic receptor. Firstly, nature has provided a model tissue in the form of the electric organs (electroplax) of electric ray (*Torpedo* species) and electric eel (*Electrophorus electricus*). Embryologically derived from striated muscle, these tissues represent a concentrated form of neuromuscular junction. Thus synaptic transmission is essentially exclusively nicotinic: the high concentration of nicotinic receptor (2 500 times greater than in mammalian muscle; Dolly and Barnard 1984) has enabled the purification and characterization of this protein to proceed far in advance of any other transmitter or hormone receptor. The second factor crucial in these achievements is the occurrence, in the venom of snakes of the krait and cobra families, of very potent polypeptide toxins that act on the muscle nicotinic receptor. Of these so-called αtoxins, αbungarotoxin from the venom of *Bungarus multicinctus* is the most notable. It is an almost irreversible antagonist of the receptor, and can be radiolabelled to very high specific activity. This makes it an extremely effective high affinity ligand for localizing receptors *in situ*, and for assaying for their presence during purification and experimental manipulation.

The wealth of information on this receptor has been the subject of many excellent reviews (see Hucho 1986), and the detail is outside the scope of this volume. However, the salient features of the protein will be summarized, for comparison with neuronal nicotinic receptors.

7.2.1 *Subunit structure*

The nicotinic-acetylcholine receptor isolated from *Torpedo* electroplax is considered to be composed of five polypeptide chains of four different types,

Table 7.1. Subunit composition of nicotinic receptor candidates

Subunit	Electroplax	Muscle	Brain					
			α bungarotoxin binding protein			[^{3}H]nicotine binding protein		
	Torpedo[1]	chick[2]	chick[3]	rat[4]	rat[5]	rat[5]	chick[6]	bovine[7]
			$M_r \times 10^{-3}$					
α	40*	43*	48	49	45	51	49	51
β	48	50	56*	53	52	79*	59*/75*	74*
γ	60	55	(62)	55*	57			
δ	68	57	69		65			
			(72)					

Constituent subunits of nicotinic receptor proteins were separated by gel electrophoresis under denaturing conditions, and their relative molecular mass (M_r) is given ($\times$ 10^{-3}). Numbers in parenthesis indicate additional subunits that are occasionally present. In chick brain, 2 proteins having been isolated and differ with respect to the size of their larger subunit. The α, β, γ and δ labelling convention is applicable to electroplax and muscle receptors only. *Acetylcholine binding subunit, indicated by affinity labelling experiments.
Data from (1) Hucho (1986) (2) Barnard *et al.* (1986) (3) Conti-Tronconi *et al.* (1985); Norman *et al.* (1982) (4) Kemp *et al.* (1985) (5) Whiting and Lindstrom (1987*a*, *b*) (6) Whiting *et al.* (1987*a*) (7) Whiting and Lindstrom (1988).

designated α, β, γ, δ, with two copies of the smallest (α) subunit. The polypeptides have apparent relative molecular masses, M_r, of α 40 000, β 48 000, γ 60 000, δ 68 000 (Table 7.1; see Fig. 7.4). Electron microscopy of negatively-stained receptor-rich membranes gives an insight into the subunit arrangement in the pentameric protein. The receptor appears as a ring-like structure (diameter 8 nm) with a five-fold axis of pseudo symmetry, probably representing the five subunits, and a densely stained central pit (diameter 2.5 nm). This is interpreted as the circular arrangement of membrane-spanning subunits to create a central pore or ion channel (Fig. 7.1). Evidence to support this has come from studies showing that all the subunits can be recognized on both the intracellular and extracellular faces of the membrane by proteolytic enzymes or monoclonal antibodies, and the amino acid sequence data for the polypeptides reveals that they each contain highly homologous hydrophobic (membrane spanning) sequences (see section 7.2.4). Furthermore, reconstitution of purified receptor into artificial membranes with full recovery of acetylcholine-stimulated cation flux indicates that the ion channel must be an integral feature of the pentameric protein. This is corroborated by the expression of functional receptor from the messenger RNA's for the α, β, γ and δ subunits (see section 7.2.4.3).

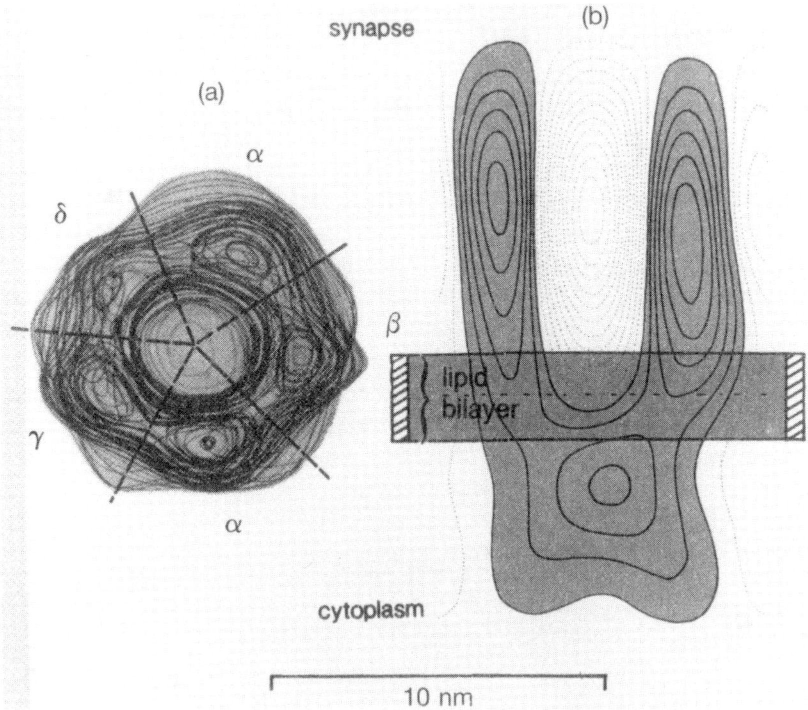

Fig. 7.1. Topology of the nicotinic-acetylcholine receptor. (a) Three dimensional electron-image map of a receptor molecule in crystals grown from *Torpedo* membrane vesicles, viewed as it would appear from the synaptic cleft. Note the five-fold symmetry consistent with five transmembrane subunits forming a central pore or ion-channel. (b) Side view, section through the channel axis, showing the channel opening extending through the membrane. The channel becomes too narrow at the cytoplasmic side to be resolved by present techniques. (From Brisson and Unwin 1985, with permission.)

7.2.2 *The acetylcholine binding site*

The binding site for acetylcholine and for drugs that compete for its binding (both agonists, including nicotine, and antagonists) has been localized on the α subunits, primarily from studies with covalently bound cholinergic ligands. The two major quaternary ammonium alkylating agents used to affinity label the receptor are [^{3}H]MBTA [(4-*N*-maleimido)benzyltrimethylammonium] and [^{3}H]bromacetylcholine (see Karlin 1980). These reagents are potent competitors of acetylcholine binding (thus acting at the same locus) and covalently label the site by interaction with the reduced disulphide group that is very close to the acetylcholine binding site.

7.2.3 *The ion channel*

The ion channel is a water-filled pore selective for cations. Binding of two agonist molecules (presumably one to each of the α subunits) opens the channel by an allosteric mechanism (Changeux *et al.* 1984) allowing Na^+ to flow into the cell, creating a local depolarization. Competitive antagonists (such as αbungarotoxin, *d*-tubocurarine and dihydroβerythroidine) prevent this activation of the channel by binding at or near the same site as acetylcholine, without promoting the allosteric change. Non-competitive antagonists (such as the frog toxin histrionicotoxin, and the neuroleptic chlorpromazine, see below) do not interfere with agonist binding but prevent receptor activation by interacting with the channel itself. Hence these drugs are sometimes referred to as 'channel blockers'.

The prolonged presence of an agonist inactivates (desensitizes) the receptor: agonist binding becomes tighter as the receptor is converted into a high-affinity state but the channel is unresponsive and remains in the closed configuration. Nicotine is a particularly effective desensitizing drug. Thus a minimal three-state model can be proposed:

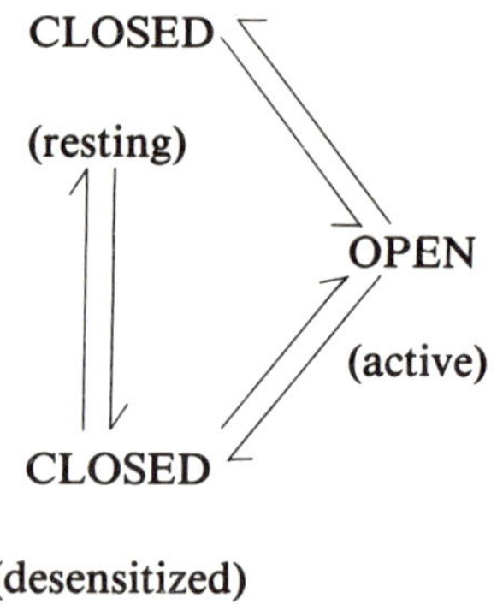

More complicated schemes are necessary to accommodate some of the subtleties of channel function (Hucho 1986).

Radiolabelled non-competitive antagonists such as chlorpromazine can be used as probes to study the channel (Heidmann *et al.* 1983); such compounds label a single, high-affinity (micromolar) binding site per pentameric receptor complex. Further information about structural aspects of the channel comes from the primary structure revealed by molecular genetics techniques.

7.2.4 *Molecular genetics*

7.2.4.1 *Protein sequencing* A milestone in the molecular characterization of the nicotinic receptor occurred in 1982 when the complete amino acid sequence of the precursor α subunit from *Torpedo* electroplax was published

(Sumikawa *et al.* 1982; Noda *et al.* 1982). This was achieved through the application of recombinant DNA technology. Thus problems of protein chemistry can be solved by resorting to the nucleic acids governing the protein in question: molecular genetic methodologies allow the large scale production of the nucleic acids and their protein products, and subtle manipulation to probe the functional and structural significance of the protein structure. The principles governing recombinant DNA techniques are outlined in Fig. 7.2. Essentially messenger RNA (mRNA) is used as a nucleic acid template for generating a complementary strand of DNA (hence cDNA) that can be multiplied in bacterial or other host cells. Cells expressing the desired cDNA can be selected for (cloned) and grown up to provide cDNA for sequencing. The nucleic acid sequence provides the amino acid sequence (primary structure) of the polypeptide for which it codes, which should be regarded as a precursor protein because of the post-translational modifications that commonly take place. For example, membrane proteins (including all the nicotinic receptor subunits) initially bear a so-called leader sequence of amino acids that is important for their insertion into the membrane, but which is subsequently cleaved. Sequencing of the nicotinic receptor subunits by recombinant DNA technology has been complemented by other powerful techniques in order to elucidate the structural and functional domains of the receptor.

7.2.4.2 *Peptide mapping* Monoclonal antibodies raised against whole subunits or synthetic peptides, based on the deduced amino-acid sequences of the subunits (see Lindstrom 1986), have been used to define the transmembrane orientation of discrete regions, and to determine the functional domains related to the acetylcholine and αbungarotoxin binding sites. Synthetic peptides, corresponding to short stretches of the protein, have been used to localize the αbungarotoxin binding site (Mulac-Jericevic and Atassi 1987; Wilson and Lentz 1988).

7.2.4.3 *Functional expression* Recombinant DNA technology allows the production of quantities of mRNA from the corresponding cDNA. The mRNA can subsequently be translated into protein. Because the nicotinic-acetylcholine receptor is a multi-subunit membrane protein, functionally competent receptor requires post-translational processing and insertion into a membrane. Thus the commonly used cell-free translation systems are inappropriate. Amphibian oocytes, particularly those of the clawed frog *Xenopus laevis*, provide a useful system for the expression and study of membrane proteins (Snutch 1988). These cells are large (up to 1.3 mm diameter) and therefore amenable to microdissection (to give single cells), microinjection (of mRNA) and electrophysiological techniques (for studying the expressed protein). Microinjection of isolated mRNA corresponding to the α, β, γ, and δ subunits into *Xenopus* oocytes results in the assembly and

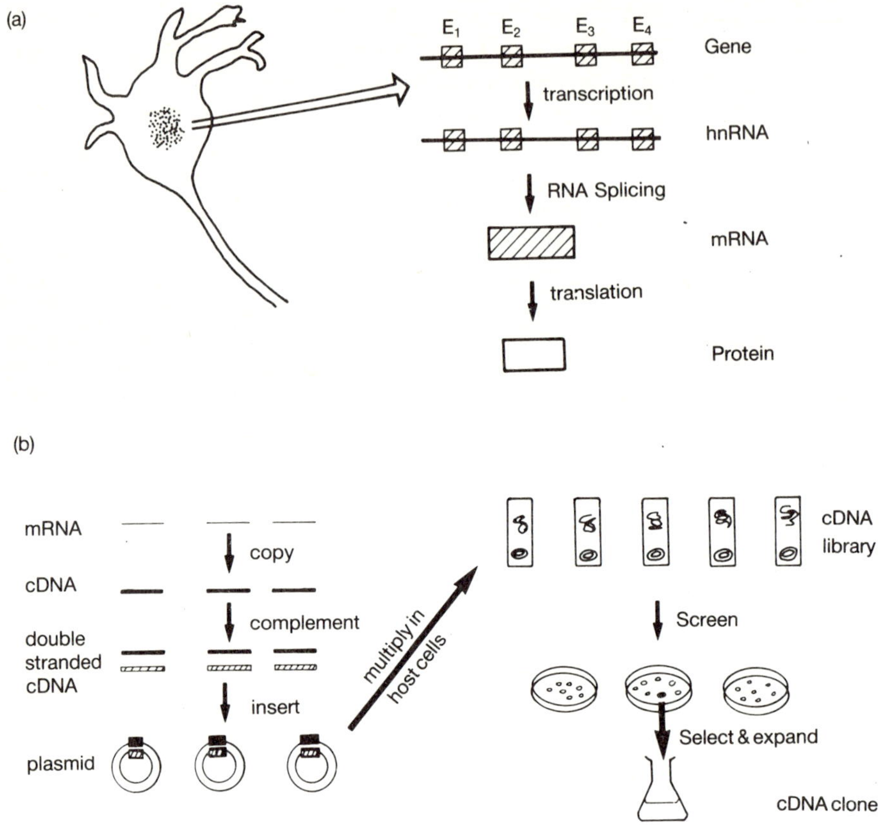

Fig. 7.2. Principles of recombinant D N A technology. (a) Schematic view of the relationship between nucleic acids and protein synthesis. (b) Manipulation of nucleic acids in gene cloning.

Genes are represented on double-stranded D N A that constitutes the chromosomes found in the nucleus. Each gene is composed of exons (coding regions indicated by the boxes E), separated by non-coding stretches of D N A, termed introns. The whole gene is first transcribed to produce single stranded heterogeneous nuclear R N A (hnR N A). Messenger R N A (mR N A) is produced from this by eliminating the non-coding segments and fusing the coding regions together (splicing). Each triplet of R N A bases in the mR N A codes for one amino acid; the mR N A is translated into protein in the cytoplasm.

In gene cloning mR N A is used a as a template for proteins produced by the tissue under investigation. Total mR N A (comprising all the separate mR N A's transcribed by the tissue) is copied into D N A (hence complementary or cD N A) which is then made double stranded so that it can be reproduced. Each double-stranded fragment (corresponding to a piece of mR N A) is inserted into a plasmid or vector that is then grown up in bacterial or other host cells. The immortalization of the protein repertoire of a given tissue or cell type constitutes a cD N A library. The cD N A library can be screened using a labelled piece of known cD N A (oligonucleotide probe) to identify plasmids containing the same or related nucleotide sequence, which are then isolated and multiplied (in the host cells) to produce a specific cD N A clone. In the screening procedure the extent of identity between the cD N A clone and the labelled obligonucleotide probe can be ascertained from the rigorousness (stringency) of the hybridization procedure: under low-stringency conditions (high salt and low temperature) complementary nucleotide sequences will anneal (hybridize) despite several mismatches, thus related but not identical sequences can be recognized. At high membrane incorporation of functional nicotinic receptor (Mishina *et al.* 1984). Thus the receptor can be investigated for ligand binding (using [^{125}I]αbungarotoxin; Mishina *et al.* 1984) and its channel properties can be studied electrophysiologically (Mishina *et al.* 1984; Sakmann *et al.* 1985). The requirement for all four subunits for receptor function (defined as cation conductance in response to acetylcholine) was confirmed by injecting into oocytes different combinations of mRNA's for three of the four subunits. In contrast, only the α subunit was indispensable for αbungarotoxin binding, although levels of binding were markedly lower in the absence of any one of the other subunits (Mishina *et al.* 1984).

7.2.4.4 *Site-directed mutagenesis* Another facet of recombinant DNA technology is the manipulation of DNA to produce point mutations (the substitution of one DNA base for another to introduce a different amino acid into the protein) or deletions of bases (resulting in the omission of one or more amino acids) by site-directed mutagenesis (White 1985). The effects of such changes on receptor assembly and function can be ascertained in the oocyte expression system (Mishina *et al.* 1985; Imoto *et al.* 1988). Thus important functional domains can be pin-pointed, although it must be remembered that small amino acid changes or omissions at one locus can have profound effects on the tertiary structure of other regions of the receptor.

7.2.4.5 *A molecular description of the receptor from Torpedo electroplax* Comparison of the cDNA sequences for all four precursor chains of the receptor from *Torpedo Californica* (Noda *et al.* 1983*a*) reveals the following features, summarized in Fig. 7.3)

1. The sequences are highly homologous indicating that the subunits have evolved from a single ancestral gene.

2. They all contain four stretches of about 26 hydrophobic amino acids designated M_1, M_2, M_3, M_4. These are proposed to be membrane-spanning domains: each segment could form an αhelix long enough to traverse the lipid bilayer, and the exposed hydrophobic groups would stabilize the protein in the membrane by interacting with the hydrophobic environment.

3. A fifth domain with alternating polar and hydrophobic amino acids has also been recognized in all the receptor subunits. This is postulated to form an

---

stringency (low salt and high temperatures) only perfectly matched sequences will remain hybridized. Sometimes a genomic DNA library is employed. In this case the initial template is not mRNA but the cell's DNA. This will contain information for all the genes in the organism, not only those transcribed by the cell in question. As genomic DNA is already double stranded, it is simply cut into fragments and inserted directly into vectors as described above. However, genomic DNA contains both coding and non-coding sequences [see (a) above] and cannot be directly related to its protein product. Instead it is most useful for gene mapping and enquiry into gene regulation.

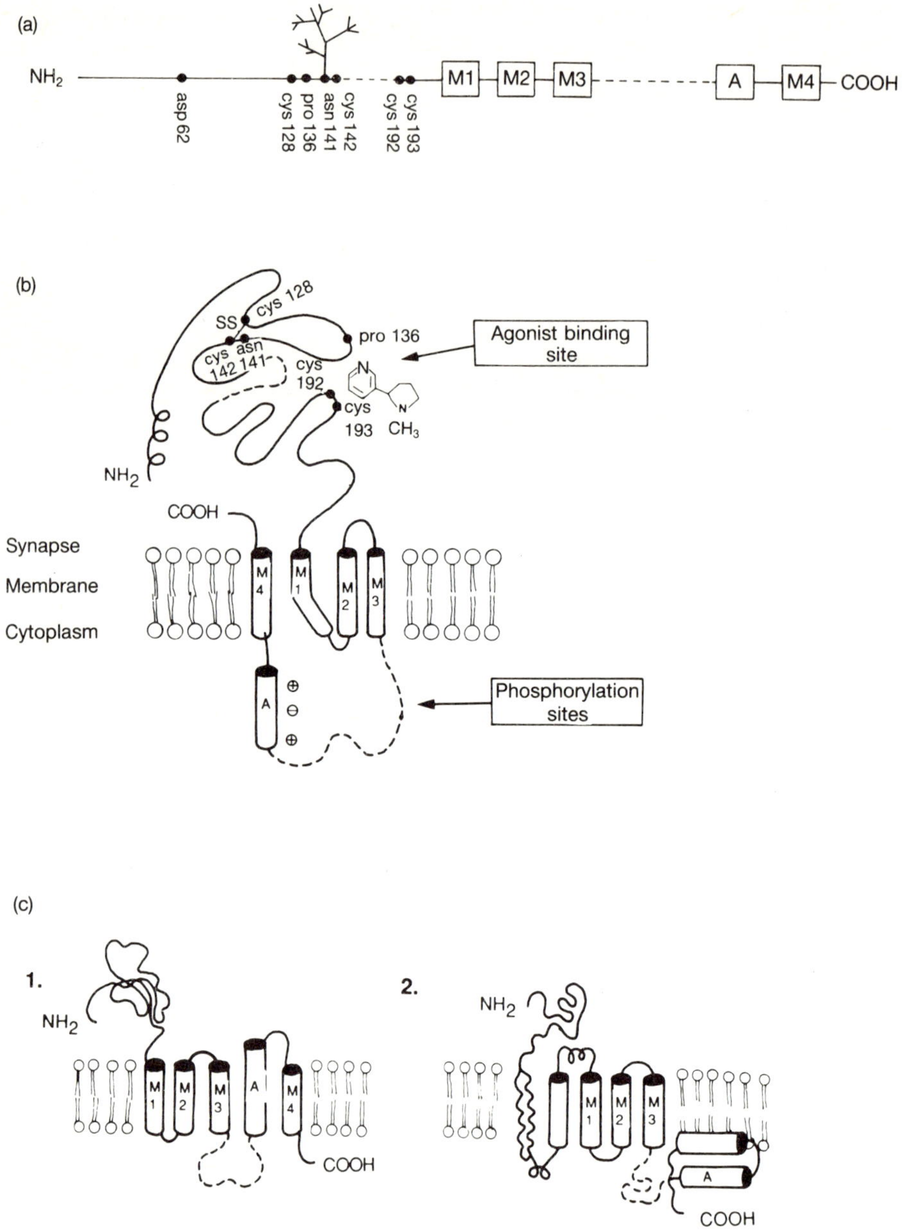

Fig. 7.3. Primary sequence and possible transmembrane configurations of nicotinic receptor subunits. (a) Primary sequence typical of nicotinic receptor subunits, showing the relative positions of conserved residues (note that cys 192 and 193 are only found in α subunits), hydrophobic spans capable of membrane insertion (designated M_1, M_2, M_3 and M_4) and the amphipathic helix (A). Dashed lines indicate regions of variable length

amphipathic helix (A): in αhelical conformation this would present one highly charged surface and one hydrophobic surface.

4. The amino terminus is extracellular, together with a long hydrophilic sequence prior to the first predicted crossing of the membrane (M_1; see Fig. 7.3b). This segment includes two invariant cysteine residues at positions 128 and 142, that form a disulphide bridge, and an invariant asparagine at position 141 that can be *N*-glycosylated. In the α subunit, the acetylcholine binding site has also been localized to this region, in the vicinity of cysteines 192 and 193 (Kao *et al.* 1984), which are unique to all α subunits.

5. There is a large cytoplasmic loop between M_3 and A (Fig. 7.3b). Phosphorylation sites (serine and tyrosine residues) are located in a short stretch (20 amino acids) of this loop and are present in homologous positions in all four subunits (Huganir and Greengard 1987). Three endogenous protein kinases phosphorylate discrete sites at this locus; phosphorylation may regulate the rate of desensitization (Huganir *et al.* 1986; Hopfield *et al.* 1988).

Although the primary sequences suggest a complex folding pattern with each subunit traversing the membrane several times, the precise tertiary structures of the subunits are controversial. At least three models, each having a different number of membrane crossings, have been proposed (Fig. 7.3b and c; see Guy and Hucho 1987). An odd or even number of crossings will be dictated by the intracellular or extracellular localization of the carboxy-termini of the subunits; this too is a contentious issue. The manner in which the subunits cross the membrane is critical in the formation of the ion channel. Early models (Guy 1984; Finer-Moore and Stroud 1984) predicted that the amphipathic helix (A) would span the membrane, stabilized by interaction with its hydrophobic face, and contribute to the channel, lining it with charged amino acid side chains (Fig. 7.3c). This has been challenged by the observation that the non-competitive channel blockers chlorpromazine (Giraudat *et al.* 1986) and $TPMP^+$ (triphenyl-methylphosphonium; Hucho *et al.* 1986)

---

and low-sequence conservation between subunits. Amino acid numbering is based on the *Torpedo* α subunit sequence. (b) Model of the transmembrane configuration of an α subunit, showing the relative positions of conserved amino acids and domains indicated in (a). This simplified model is derived (V. B. Cockcroft, unpublished) from available biochemical and physicochemical data; the cytoplasmic and extracellular protrusions are consistent with the dimensions of the nicotinic receptor derived from electron imaging (see Fig. 7.1). The agonist binding site would face into the mouth of the receptor channel which is formed from 5 subunits, creading a cylindrical structure. Four membrane crossings per subunit and an extracellular C-terminus have previously been proposed by Noda *et al.* (1982), Devillers-Thiery *et al.* (1983), and Claudio *et al.* (1983). (c) Alternative models proposed (1) by Guy (1984) and Finer-Moore and Stroud (1984), incorporating the amphipathic segment (A) as a membrane spanning region, and (2) by Ratnam *et al.* (1986). See Guy and Hucho (1987).

photoaffinity label serine 262 of the δ subunit and homologous amino acids of the α and β chains. These residues are all found in the M_2 segments of the respective subunits. M_2 is the most highly conserved domain in the entire sequence (Guy and Hucho 1987), which might be expected if it is responsible for the ubiquitous functioning of nicotinic receptors. The importance of M_2 in influencing the rate of ion flux was demonstrated by the construction of chimeric δ subunits (Imoto *et al.* 1986). In these elegant experiments, hybrid receptors made up of bovine and *Torpedo* subunits indicated that the δ subunit is important in determining channel conductance; the formation of chimeric δ subunits (where cDNA constructs were generated by replacing portions of the *Torpedo* δ subunit by homologous regions of the bovine δ subunit) localized the M_2 segment and the bend linking M_2 and M_3 as the domains responsible for differences in conductance. This has been confirmed by site-directed mutagenesis experiments (see section 7.2.4.4) that have identified clusters of negatively-charged amino acids in the M_2 segment of α, β, γ and δ subunits as major determinants of the rate of ion transport (Imoto *et al.* 1988). The involvement of M_2 in the ion channel does not necessarily exclude A from making a contribution to the channel (Guy and Hucho 1987), but the transmembrane orientation of A remains questionable (Ratnam *et al.* 1986; see Figs. 7.3b and c).

7.2.5 *Overview of the end-plate nicotinic acetylcholine receptor*

In summary, the nicotinic receptor is composed of five transmembrane subunits around a central channel. Two functional sites on the receptor can be defined:

(1) the agonist recognition site; competitive antagonists including αbungarotoxin also bind at or near this site, which is located on the αsubunit, present in two copies in the oligomer.

(2) the cation-selective channel; this is the site of action of non-competitive antagonists.

The nicotinic receptor mediates the transmission of electrical signals by responding to the binding of acetylcholine (or nicotine) by opening the channel via a fast (millisecond) allosteric transition.

Although much of this work has derived from studies on *Torpedo* electroplax, this proves to be an excellent model for the receptor at the vertebrate neuromuscular junction. All receptors isolated recently from vertebrate muscle have four subunits of relative molecular masses comparable to the α, β, γ and δ subunits of *Torpedo* receptor (see Dolly and Barnard 1984; Table 1). This resemblance is confirmed by the remarkable homology in the amino acid sequences deduced from cDNA clones of the α subunits from *Torpedo* electroplax and calf and human muscle: the sequences from mammalian sources are 98 per cent homologous with each other and 80 per cent homologous with the *Torpedo* α sequence (Noda *et al.* 1983*b*). The β,

γ, and δ chains have also been shown to be highly conserved (see Kubo *et al.* 1985).

It is against this backdrop of a well characterized and highly conserved receptor at the neuromuscular junction that its counterparts in the brain must be pursued.

7.3 αBungarotoxin binding sites in the brain

7.3.1 *Ligand binding studies*

The initial requirement for the identification and characterization of a receptor molecule is the availability of a specific, high-affinity ligand to label the sites in question. Therefore it is not surprising that αbungarotoxin should have been favoured as the ideal ligand in view of its contribution to the study of the muscle nicotinic receptor. Indeed this rationale appeared to bear fruit when it was shown that radiolabelled αbungarotoxin identified a small number of sites in the brain, and these displayed a non-uniform distribution and a

Table 7.2. Potency of nicotinic drugs for binding sites labelled by $[^3H](-)$nicotine or $[^{125}I]$αbungarotoxin in rat brain

Competing ligand	Binding site	
	$[^3H](-)$nicotine	$[^{125}I]$αbungarotoxin
agonists:	$K_I(\mu M)$	
(+)anatoxin[1]	0.0003	0.09
(−)anatoxin[1]	0.39	21.0
cytisine[2]	0.003	1.4
(−)nicotine[2]	0.014	0.82
(+)nicotine[2]	0.28	3.0
acetylcholine[2]	0.12	4.0
DMPP[2]	0.20	0.47
(−)lobeline[2]	0.24	11.0
carbamylcholine[2]	1.2	12.0
antagonists:		
neosurugatoxin[3]	0.61	>100
dihydroβerythroidine[1]	1.3	25
methyllycaconitine[1]	3.7	0.00014
αbungarotoxin[2]	>100	0.0002
d-tubocurarine[2]	33	2.8
decamethonium[2]	16	25
hexamethonium[2]	720	540
pempidine[1]	2600	2500
mecamylamine[1]	10000	1500

Data from: (1) MacAllan *et al.* (1988) (2) Marks *et al.* (1986) (3) Wonnacott *et al.* (1988).

cholinergic pharmacology (reviewed by Schmidt *et al.* 1980; Oswald and Freeman 1981). Thus the binding could be prevented by incubation with acetylcholine or nicotine (Table 7.2) which must therefore compete for the same site, a fundamental prerequisite of a nicotinic receptor. Other features of the binding site are also reminiscent of the muscle nicotinic receptor:

(1) it is a glycoprotein (Salvaterra *et al.* 1977; Dolly and Barnard 1984);

(2) inhibition of [^{125}I]αbungarotoxin binding by acetylcholine is enhanced by preincubation with non-competitive nicotinic antagonists (Lukas and Bennett 1979), consistent with a co-operative interaction between the agonist binding site and non-competitive antagonist site;

(3) [^{125}I]αbungarotoxin binding is reversibly diminished by sulphydral reduction (Lukas and Bennett 1980; Marks *et al.* 1986) implicating a critical disulphide bond in the vicinity of the binding site;

(4) binding can be blocked by the affinity reagents MBTA and bromacetylcholine (Lukas and Bennett 1980; see section 7.2.2).

However, the αbungarotoxin binding site in vertebrate brain also differs in certain respects from its presumed counterpart at the neuromuscular junction:

(1) the apparent dissociation constant for αbungarotoxin binding to nervous tissue is higher (i.e. the toxin binds with lower affinity). This is a reflection of the much more rapid rate of dissociation of αbungarotoxin from neuronal binding sites (see Oswald and Freeman 1981) so that its binding is reversible.

(2) hydrodynamic measurements indicate that the brain protein has a larger molecular mass (350 000 daltons) than the muscle nicotinic receptor (250 000 daltons; Oswald and Freeman 1981; Conti-Tronconi *et al.* 1985).

7.3.2 *Subunit composition and structural relationship with the muscle nicotinic receptor*

Affinity purification protocols have been successfully employed to isolate the brain αbungarotoxin binding protein, using αbungarotoxin as an affinity ligand. Initially a single subunit of apparent relative molecular mass M_r 50 000 was reported for mouse (Seto *et al.* 1981) and chick brain (Norman *et al.* 1982). More recently, by paying greater attention to the problem of proteolytic degradation during isolation, preparations of αbungarotoxin binding protein from brain have revealed at least two (Conti-Tronconi *et al.* 1985; Kemp *et al.* 1985) or three (Whiting and Lindstrom 1987*a*) additional subunits (Table 7.1; Fig. 7.4). Thus it may be considered that the brain protein resembles the muscle nicotinic receptor in having an $\alpha\ \beta\ \gamma\ \delta$ subunit composition. However, affinity-labelling experiments, using covalently bound agonists [^{3}H]bromacetylcholine or [^{3}H]MBTA (section 7.2.2) indicate that the acetylcholine binding site resides on the M_r 56 000 subunit (Norman *et al.* 1982;

Conti-Tronconi *et al.* 1985; Kemp *et al.* 1985) rather than the smallest (α) subunit.

Immunological studies corroborate the concept of limited identity with the muscle receptor. Antibodies raised against muscle nicotinic receptor recognize only weakly neuronal [^{125}I]αbungarotoxin binding sites (Norman *et al.* 1982; Wonnacott *et al.* 1982), and antibodies raised against electroplax nicotinic-acetylcholine receptors commonly fail to recognize αbungarotoxin-labelled sites in mammalian brain (Morley *et al.* 1983; Mills and Wonnacott 1984; see Lindstrom *et al.* 1987).

An insight into the relationship between the brain αbungarotoxin binding protein and the muscle nicotinic receptor has come from a comparison of their amino acid sequences. Taking advantage of the high concentration of αbungarotoxin binding sites in the optic lobe of the newly-hatched chick, Barnard and co-workers were able to purify sufficient protein from this brain area to undertake amino-terminal amino acid sequence analysis (Conti-Tronconi *et al.* 1985). Comparison of the 25 amino-acid sequence, derived from the smallest (M_r 48 000) subunit, with the known amino terminal sequences for electroplax and muscle nicotinic receptor subunits revealed that although different, there is considerable homology. For example 35 per cent of the amino acids were identically positioned in chick optic tectum and chick muscle αsubunit sequences. Thus the αbungarotoxin binding protein α subunit must be the product of a separate, albeit related, gene from that coding for the muscle α subunit, and genes for subunits of the αbungarotoxin binding protein from chick brain have recently been cloned (Schoepfer *et al.*, 1989; see section 7.5.5). However, the neuronal protein showed greater homology (43 and 48 per cent respectively) with the α subunits of *Torpedo* and *Electrophorus* electroplax (Conti-Tronconi *et al.* 1985), indicating that the divergence of muscle and neuronal nicotinic proteins (as a result of gene duplication) probably occurred very early during vertebrate evolution.

7.3.3 Functional significance

The structural differences between brain αbungarotoxin binding proteins and muscle nicotinic receptors, arising from their independent genetic control, raise questions about the functional significance of this divergence. This brings us to the major problem concerning this brain protein. In contrast to the formidable potency of αbungarotoxin as an antagonist at the neuromuscular junction, it is markedly ineffective in blocking nicotinic transmission in the brain and spinal cord, with only a few exceptions (see Chiappinelli 1985; Clarke, chapter 5, this volume). One system that is sensitive to αbungarotoxin is the visual pathway (optic tectum) of lower vertebrates, where a clear-cut blockade of transmission has been demonstrated (see Henley and Oswald 1987). In the mammalian brain, the superior colliculus is homologous to the optic tectum for processing visual input en route to the visual cortex, and this nucleus is also densely labelled by [^{125}I]αbungarotoxin (Clarke *et al.* 1985). As discussed above, (section 7.3.2), Barnard and co-workers have characterized the αbungarotoxin binding protein from the optic tectum of the chick, where it may be considered to be a nicotinic receptor. By all the criteria used, this protein is identical to the αbungarotoxin binding protein present in the rest of the chick brain, where its function is uncertain (Norman *et al.* 1982; Conti-Tronconi *et al.* 1985).

Another brain region that responds to visual input is the suprachiasmatic nucleus of the hypothalamus. This nucleus is involved in circadian pacemaking. αBungarotoxin binding to the suprachiasmatic nucleus has been investigated following the observations that a cholinergic-nicotinic mechanism attributed to this nucleus mediated the photic stimulation of the pineal gland (Zatz and Brownstein 1979), and that local injection of αbungarotoxin blocked the light-induced changes (Zatz and Brownstein 1981). More recently (Miller *et al.* 1987*a*) it has been shown that systemically administered nicotine ($20 \mu g\ kg^{-1}$) resulted in the excitation of suprachiasmatic nucleus neurons, and this effect could be blocked by mecamylamine; αbungarotoxin was not tested. However, the systemic route of administration does not permit a distinction to be made between direct nicotinic stimulation of neurons in the suprachiasmatic nucleus and indirect stimulation arising from an activation of the retinal ganglion cells. Binding studies have shown that [^{125}I]αbungarotoxin densely labels the hypothalamus, and binding sites are most concentrated in the suprachiasmatic nucleus (Block and Billiar 1981; Meeker *et al.* 1986); the numbers of binding sites are under hormonal control (Miller *et al.* 1982, see section 7.4.3.1.). Ultrastructural examination of the cellular localization of αbungarotoxin binding sites suggested that they are present at axo-dendritic synapses (Miller *et al.* 1987*b*) but the differential localization of binding sites within the suprachiasmatic nucleus, with densest labelling of the dorsolateral regions (Miller and Billiar 1986; Fuchs and Hoppens 1987) is not compatible with the termination of the retinal projections in the ventromedial areas (Fuchs and Hoppens 1987). Moreover, the lack of effect of enucleation on [^{125}I]αbungarotoxin binding levels prompted Fuchs and Hoppens (1987) to question the significance of [^{125}I]αbungarotoxin binding sites in the suprachiasmatic nucleus with respect to nicotinic transmission.

The relationship between αbungarotoxin binding proteins and functional nicotinic receptors has been easier to clarify in autonomic ganglia and related cell lines, and is worth a brief mention here because of its possible relevance to the CNS. Using PC12 cells (a cell line that resembles sympathetic neurones), Patrick and Stallcup (1977*a*, *b*) were able to convincingly distinguish αbungarotoxin binding sites from functional nicotinic receptors (the latter assessed by measuring Na^+ fluxes in response to nicotinic agonists). Firstly, αbungarotoxin failed to inhibit carbamylcholine-evoked ion flux (Patrick and Stallcup 1977*a*). Secondly, antibodies capable of inhibiting the ion flux failed to recognize the αbungarotoxin binding protein (Patrick and Stallcup 1977*b*). In intact avian and mammalian ganglia, αbungarotoxin is also ineffective as an antagonist (see Schmidt *et al.* 1980) and autoradiographic studies have indicated that the αtoxin binding sites are located extrasynaptically (Jacob and Berg 1983; Messing and Gonatas 1983).

Many of the behavioural effects of nicotine and nicotinic responses examined in the brain and spinal cord appear to be mediated by ganglionic (C6) rather than muscle (C10) type nicotinic receptors (see Clarke, chapter 5, and Stolerman, chapter 8, this volume), but the distinctions between functional receptors and extrasynaptic αbungarotoxin binding proteins in autonomic ganglia may not necessarily apply in the functionally and structurally more complex central nervous system. Despite the many incompatibilities between [^{125}I]αbungarotoxin labelling and nicotinic transmission in the brain (see Schmidt *et al.* 1980; Clarke *et al.* 1985), αbungarotoxin binding sites are localized at synapses in some brain regions (Schmidt *et al.* 1980) and may indeed mediate nicotinic transmission in some cases. The purpose of other αbungarotoxin binding sites in the brain is unknown, although it has been speculated that they may subserve a trophic (Oswald and Freeman 1981) or slow, long-term function (Jacob and Berg 1983). However, in the context of our present concern with the molecular sites of action of nicotine in the brain, αbungarotoxin binding proteins should not be dismissed: they are both structurally and pharmacologically nicotinic in character, and clearly sensitive to nicotine. Moreover, numbers of [^{125}I]αbungarotoxin binding sites in rat brain regions are increased in response to chronic nicotine treatment (Marks *et al.* 1985; Fig. 7.5; see section 7.4.3.2). Therefore, whatever their role(s) in the brain, they may be relevant targets for nicotine in both smoking and therapy.

7.4 [^{3}H]nicotine binding sites

In considering the molecular locus of nicotine's action in the CNS, radiolabelled nicotine is an obvious probe. The first description of tritiated nicotine binding to brain tissue appeared in 1974 (Schleifer and Eldefrawi 1974) and this account gave a binding affinity and site density in remarkably good agreement with recent observations. Although a few other reports followed (Abood *et al.* 1979; Yoshida and Imura 1979; Nordberg and Larsson 1980; Vincek *et al.* 1980) the low specific activity of the radiolabelled nicotine (0.25–5 Ci mmol^{-1}) precluded detailed studies. However, the increasing concern about the significance of [^{125}I]αbungarotoxin binding sites in the brain provided the impetus for seeking an alternative ligand for neuronal nicotinic receptors, and in 1980 Romano and Goldstein published their classic paper entitled 'Stereospecific nicotine receptors on rat brain membranes'. This study used [^{3}H](±)nicotine of markedly higher specific activity (24 Ci mmol^{-1}) that subsequently became commercially available, and an abundance of binding studies has ensued (see Wonnacott 1987*b*).

7.4.1 *Radioligand binding studies*

Romano and Goldstein (1980) reported that [^{3}H]($\pm$)nicotine labelled two sites in rat brain membranes, one with high affinity (K_d = 43 nM) and low density (4.4 fmol mg^{-1} tissue), and a second, low-affinity site (K_d = 46 μM) of higher density that was difficult to quantitate reproducibly. The binding was stereo-selective in favour of the naturally occurring (–) form of nicotine and could be displaced by a number of nicotinic agonists including acetylcholine. Nicotinic antagonists, however, were largely ineffective in competition assays. The latter fact probably accounts for contemporary reports of [^{3}H]nicotine binding to brain membranes described as non-cholinergic (Abood *et al.* 1980; Sershen *et al.* 1981).

The description of [^{3}H]($\pm$)nicotine binding by Romano and Goldstein (1980) has been reproduced in many subsequent studies (reviewed by Wonnacott 1987*b*). With the commercial availability of [^{3}H](–)nicotine instead of the racemic mixture, and attention to other factors including radio-ligand purity, extent of washing of the tissue preparation to remove endogenous ligands (see section 7.4.3.3), and speed of assay method, affinity constants (K_d) for high-affinity [^{3}H](–)nicotine binding to mammalian brain are generally in the range 1–10 nM. The density of high-affinity [^{3}H]nicotine binding sites in the brain is about a hundred times lower than the density of muscarinic-acetylcholine receptors: typically 50–80 fmol [^{3}H]nicotine binding sites per mg protein in rodent brain (for example, Marks *et al.* 1986; Wonnacott 1986; Martino-Barrows and Kellar 1987), but levels in human brain are five times lower (Flynn and Mash 1986; Whitehouse *et al.* 1986; Perry *et al.* 1987). [^{3}H]nicotine binding sites are not uniformly distributed throughout the brain but are concentrated in some regions (Clarke *et al.* 1985; London *et al.* 1985*a*), notably the interpeduncular nucleus, thalamic nuclei, superior colliculus, medial habenula, cerebral cortex, and substantia nigra (see Clarke, chapter 5 and Table 5.1, this volume).

7.4.1.1 *Multiple binding sites* The number of classes of binding site for [^{3}H]nicotine in the brain is controversial. Some researchers (for example, Abood *et al.* 1985; Benwell and Balfour 1985; Larsson and Nordberg 1985; Perry *et al.* 1987) report binding data best interpreted by two binding sites, others (for example, Marks *et al.* 1986; Lippiello and Fernandes 1986; Martino-Barrows and Kellar 1987; Reavill *et al.* 1988) find a single, high-affinity site. Marks and Collins (1982) observed a low-affinity site ($IC_{50} = 5 \times 10^{-5}$M) only when binding was carried out at low temperature. This site accounted for 28 per cent of the total number of binding sites and, in contrast to the high-affinity site, was equally sensitive to (+)– and (–)nicotine. The low-affinity for nicotine, site density, and lack of stereo-specificity suggests that this may be the same site that is labelled with [^{125}I]αbungarotoxin (Wonnacott 1986). However, the pharmacological significance of low-affinity binding sites for [^{3}H]nicotine has been questioned (Clarke 1987) because they display no clear-cut anatomical distribution (Benwell and Balfour 1985). Moreover, they are not affected by treatments that modulate numbers of high-affinity binding sites (Costa and Murphy 1983; Fuxe *et al.* 1988). Lippiello and Fernandes (1986) observed that the inclusion of protease inhibitors during tissue preparation resulted in the identification of a single class of binding site, compared with the appearance of a second site in the absence of anti-proteases. Thus proteolytic damage to the receptor, and other factors such as radioligand impurities, may account for the second, lower-affinity binding site, although under favourable conditions binding to the low-affinity site identified by [^{125}I]αbungarotoxin may be demonstrable. Sloan *et al.* (1985) have described up to 5 classes of binding site for [^{3}H]nicotine on the basis of their affinity, stereoselectivity, and co-operativity. It is not clear if these distinctions are physiologically meaningful. However, the multiplicity of binding sites observed under some conditions may merit re-evaluation in view of the molecular biology evidence (section 7.5.1) for heterogeneity of neuronal nicotinic receptors. The remainder of this section will focus on the high affinity binding site for [^{3}H]nicotine.

7.4.1.2 *Pharmacological specificity* The pharmacological specificity of [^{3}H]nicotine binding sites is illustrated in Table 7.2. Thus the agonists cytisine, (−)nicotine, acetylcholine, and lobeline are particularly potent at inhibiting [^{3}H]nicotine binding; in addition, the novel neurotoxin (+)anatoxin-a is also a potent competitor at this site (Zhang *et al.* 1987; MacAllan *et al.* 1988). All of these agonists discriminate between the [^{3}H]nicotine and [^{125}I]αbungarotoxin binding sites, showing markedly higher affinity for the former. In contrast, DMPP, (+)nicotine, anabasine, and carbamylcholine are approximately equipotent at both sites. Thus the [^{3}H]nicotine binding site shows considerably greater stereoselectivity for (−)nicotine (Wonnacott 1986). The nanomolar affinity for some agonists (notably nicotine and acetylcholine) has been reconciled with the micromolar concentrations that are physiologically active by proposing that the prolonged incubation with agonist during the binding assay converts the binding site into a high-affinity, desensitized configuration (Schwartz *et al.* 1982; see section 7.2.3). Some support for such a conversion comes from a kinetic analysis of the binding of [^{3}H]nicotine (Lippiello *et al.* 1987), and from a comparison of [^{3}H]nicotine binding and desensitization of nicotinic function in adrenal chromaffin cells (Higgins and Berg 1988). Alternatively, agonists may bind with high affinity to a desensitization site that is distinct from the recognition site that mediates receptor activation (Conti-Tronconi *et al.* 1982).

Although most nicotinic antagonists are ineffective competitors of ligand binding (Table 7.2), as originally reported by Romano and Goldstein (1980), there are a few notable exceptions, namely dihydroβerythroidine (Clarke *et al.* 1984; Martino-Barrows and Kellar 1987; MacAllan *et al* 1988; Reavill *et al.* 1988), neosurugatoxin (Hayashi *et al.* 1984; Rapier *et al.* 1985), and methyllycaconitine (MacAllan *et al.* 1988). Dihydroβerythroidine is effective at both end-plate and ganglionic synapses, whereas neosurugatoxin appears to be highly selective for ganglionic nicotinic receptors (Hayashi *et al.* 1984; see Wonnacott 1987*a*). Methyllycaconitine has been little studied to date but is unusual in showing a preference for the [^{125}I]αbungarotoxin binding site (Table 7.2). The ineffectiveness of classical ganglionic antagonists such as hexamethonium and mecamylamine in inhibiting [^{3}H]nicotine binding (Table 7.2) was originally attributed to the conversion of the binding site by nicotine into an 'agonist selective' conformation (Romano and Goldstein 1980); this hypothesis must be rejected in view of the inhibition by dihydro-βerythroidine and neosurugatoxin. An alternative explanation is that those antagonists that are poor competitors of [^{3}H]nicotine binding may be non-competitive antagonists acting at a separate locus on the receptor complex. Although there is evidence that some non-competitive antagonists allosterically modify agonist binding to the αbungarotoxin binding protein (Lukas and Bennett 1979; see section 7.3.1), comparable experiments where the radioligand is an agonist (as in the case of [^{3}H]nicotine binding) are not feasible. Nevertheless, it is increasingly recognized that many of these drugs block nicotinic transmission by interacting with the ion channel of the nicotinic receptor (Adams and Sakman 1978; Ascher *et al.* 1979), and recent electrophysiological (Aracava *et al.* 1987), pharmacological (Rapier *et al.* 1987), and molecular biology evidence (Wada *et al.* 1988; see section 7.5.3) points to the presence of a similar ion channel associated with brain nicotinic receptors. Thus the ability of these drugs to antagonize nicotinic function (see Clarke, chapter 5, and Stolerman, chapter 8, this volume) can be reconciled with their failure to inhibit agonist binding.

However, the lack of inhibition of [^{3}H]nicotine binding by αbungarotoxin (Table 7.2) is because it binds to a distinct site in the brain. The evidence for this comes from the anatomical distributions of binding sites for [^{3}H]nicotine and [^{125}I]αbungarotoxin, which are clearly different with few regions of overlap (Clarke *et al.* 1985; see Clarke, Table 5.1, this volume), while [^{3}H]nicotine and [^{3}H]acetylcholine distributions are coincident (see section 7.4.2); and the separation of [^{3}H]nicotine and [^{125}I]αbungarotoxin binding proteins by affinity purification (Whiting and Lindstrom 1986; Wonnacott 1986). This distinction has enhanced the credibility of [^{3}H]nicotine binding sites as nicotinic receptors in the CNS, and accelerated the demise of interest in [^{125}I]αbungarotoxin binding sites.

7.4.2 *Purification and subunit composition*

Early attempts (Abood *et al.* 1983) to isolate a nicotine binding protein from rat brain employed an affinity chromatography approach using a nicotine analogue. A major polypeptide of M_r 56 000 was observed, but the extent of purification and yield were low. Abood and coworkers have repeated and

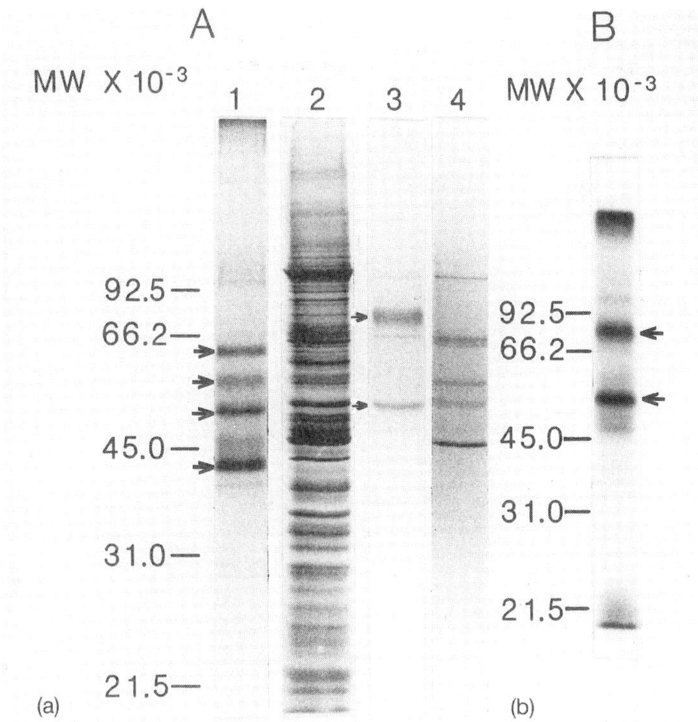

Fig. 7.4. Subunit composition of putative neuronal nicotinic receptors from rat brain. (a) Polyacrylamide gel electrophoresis analysis, under denaturing conditions. The protein bands were silver stained. Positions of molecular weight (M W) standards are shown on the left. (1) nicotinic receptor purified from *Torpedo Californica* electroplax showing 4 major bands (arrows). (2) detergent extract of rat brain membranes. (3) immunoaffinity purified [^{3}H]nicotine binding protein from rat brain showing two bands of M_r 52 000 and 72 000 (arrows); the faint bands of M_r 66 000 and 95 000 are trace contaminants. (4) affinity purified [^{125}I]αbungarotoxin binding protein from rat brain, showing 4 bands M_r 45 000, 52 000, 56 000, 65 000, and the same M_r 95 000 contaminant as in lane 3. (b) Polyacrylamide gel electrophoresis of [^{125}I]-labelled [^{3}H]nicotine binding protein after immunoaffinity isolation from rat brain. Autoradiography identifies 2 labelled subunits corresponding to the 2 subunits stained in (a) lane 3; labelling at the top of the gel represents aggregated protein. (From Whiting and Lindstrom 1987*a* with permission.)

extended this approach (Abood *et al.* 1987), and with the addition of protease inhibitors during the extraction they now report two subunits of M_r 62 000 and 57 000. The results were substantiated using anti-idiotypic antibodies.

In a more detailed series of studies, Whiting and Lindstrom have used immunoaffinity purification techniques to isolate from brain tissue a protein that binds [^{3}H]nicotine with high affinity, and displays the pharmacological specificity characteristic of [^{3}H]nicotine binding sites (Table 7.2; see

Lindstrom *et al.* 1987). Initially, an antibody raised against electroplax nicotinic receptor, and shown to cross-react with avian (but not mammalian) brain, was used as a high-affinity ligand to isolate a [^{3}H]nicotine binding protein from chick brain (Whiting and Lindstrom 1986). An antibody (mAb270) against this brain protein was subsequently utilized to isolate a second [^{3}H]nicotine binding protein from chick brain (Whiting *et al.* 1987*a*), and the corresponding protein from rat brain (Whiting and Lindstrom 1987*a*; Fig. 7.4), which in turn has generated antibodies that recognize similar receptors in bovine and human brain (Whiting and Lindstrom 1988). In each instance, the isolated protein consists of only two types of subunit (Table 7.1), of molecular weights ~ 50 000, and ~ 75 000 (except the first chick brain protein to be isolated, Whiting and Lindstrom 1986, which had subunits of M_r 48 000 and 59 000). Thus the [^{3}H]nicotine binding proteins (which are good candidates for neuronal nicotinic receptors) differ markedly from the prototype nicotinic acetylcholine receptor from muscle. This does not appear to be an artefact of proteolytic degradation because, under identical conditions, the αbungarotoxin binding protein from rat brain consists of four polypeptides (Whiting and Lindstrom 1987*a*; Table 7.1 and Fig. 7.4). Evidence that two types of subunit are sufficient for receptor function comes from molecular biology experiments (see section 7.5.3). The stoichiometry of the two types of subunit in the [^{3}H]nicotine binding protein is not established, but by analogy with the pentameric endplate receptor, a ratio of 2:3 or 3:2 is probable (Whiting *et al.* 1987*b*), although a tetrameric arrangement of alternating subunits is plausible (see Lindstrom *et al.* 1987).

Despite the differences in subunit size and pattern (Fig. 7.4), immunological cross-reactivity between the M_r 48 000 subunit of the chick brain [^{3}H]nicotine binding protein and the αsubunit of *Torpedo* nicotinic acetylcholine receptor (Whiting and Lindstrom 1986) testifies to at least some structural similarities between the two proteins. This is also implicit in the susceptibility of the brain [^{3}H]nicotine binding proteins to affinity alkylation with the reagents MBTA and bromacetylcholine (see section 7.2.2), consistent with the presence of cysteine residues in the vicinity of the agonist binding site. Whereas this site resides on the smallest (α) subunit of the nicotinic receptor from *Torpedo* and muscle, it is the larger subunit of the brain [^{3}H]nicotine binding proteins that is labelled (Whiting and Lindstrom 1987*b*). This is reminiscent of the affinity labelling of the second subunit of the brain αbungarotoxin binding protein (see Table 7.1, section 7.3.2). To overcome the confusion of nomenclature based on α and β designations for the [^{3}H]nicotine binding protein subunits, it is proposed (Whiting *et al.* 1987*b*) that the larger, affinity alkylated subunit is referred to as the 'agonist-binding subunit', whereas the smaller subunit may be called the 'structural subunit'. However, the convention in molecular biology is to designate putative agonist binding subunits 'α', regardless of their size (see section 7.5.2, and Steinbach and Ifune 1989).

A further complexity is evident in chick brain where two distinct

[^{3}H]nicotine binding proteins have been recognized (Whiting *et al.* 1987*a*). These share the same structural subunit (M_r 48 000) but have different agonist binding subunits (M_r 59 000 and 75 000; see Table 7.1), although they do not differ in their pharmacological specificity for the nicotinic drugs tested to date (Whiting *et al.* 1987*b*). Such heterogeneity has not been found in mammalian brain (Whiting and Lindstrom 1988), but nicotine receptors in brain and autonomic ganglia appear to be distinct. Chick ciliary ganglia in culture and PC12 cells do not show high-affinity [^{3}H]agonist binding (Halvorsen and Berg 1987; Kemp and Morley 1986; Whiting *et al.* 1987*c*). This does not seem to be an artefact of cell culture conditions because primary cultures of rat cortical neurons bind [^{3}H]nicotine with high affinity (Lippiello and Fernandes 1988). However, ganglion cells are recognized by antibodies against the structural subunit of the brain [^{3}H]nicotine binding protein (Whiting *et al.* 1987*c*; Halvorsen and Berg 1987), suggesting that the structural subunit may be shared by subtypes of neuronal nicotinic receptor. Parallels of this heterogeneity and commonality are emerging from molecular biology studies (see section 7.5).

7.4.3 *Modulation of [^{3}H]nicotine binding sites*

7.4.3.1 *Down-regulation* *In vivo* regulation of [^{3}H]nicotine binding sites in response to pharmacological or physiological manipulations supports a functional receptor role for these binding proteins. The chronic administration of an anti-cholinesterase to rats resulted in decreased numbers of [^{3}H]nicotine binding sites in the brain (Costa and Murphy 1983; Martino-Barrows and Kellar 1987), and this was accompanied by a diminution in the anti-nociceptive effect of nicotine (Costa and Murphy 1983). This down-regulation of [^{3}H]nicotine binding sites is presumed to arise from the increased synaptic availability of acetylcholine following cholinesterase inhibition, and is consistent with the localization of nicotine recognition sites at synapses receiving a cholinergic input (Martino-Barrows and Kellar 1987). Indeed, the non-hydrolyzable agonist carbamylcholine decreases the numbers of functional receptors in PC12 cells (Robinson and McGee 1985) and αbungarotoxin binding sites in chick ciliary ganglia (Messing 1982). αBungarotoxin binding sites in the suprachiasmatic nucleus of the hypothalamus are also modulated by steroid hormones: ovariectomy resulted in a loss of binding sites (Miller *et al.* 1982), which could be restored by oestradiol (Miller *et al.* 1984).

7.4.3.2 *Up-regulation* If treatments that increase the availability of agonist for its receptor (such as anti-cholinesterase administration) promote down-regulation of the binding sites, procedures that diminish the availability or access of agonist might be expected to have the opposite effect. However, *in vivo* treatment for 10 days with mecamylamine or dihydroβerythroidine (both at 1 mg kg^{-1}) failed to alter numbers of nicotinic binding sites in rat brain (Schwartz and Kellar 1985). But up-regulation of brain [^{3}H]nicotine binding sites in response to nicotine itself is well documented (Fig. 7.5; Marks *et al.* 1985; Martino-Barrows and Kellar 1987; see Wonnacott 1987*b*). The changes, which reflect an increase in numbers of sites with no change in their affinity for nicotine, are more pronounced in certain brain regions (notably the cortex, hippocampus, and hypothalamus, Marks *et al.* 1985, 1986), and this, together with methodological considerations of nicotine dose and route of administration (Murrin *et al.* 1987) may explain the negative reports in the literature (see Larsson *et al.* 1986). Increased numbers of [^{3}H]nicotine binding sites have also been identified in the brains of smokers versus non-smokers (Benwell *et al.* 1988).

Up-regulation by nicotine of its recognition sites appears to contradict the hypothesis, outlined in the previous section, that agonists promote down-regulation. This effect of nicotine is generally ascribed to its propensity to desensitize its receptors, resulting in blockade (see Schwartz and Kellar 1985). Thus nicotine may be perceived as an antagonist, and this may explain the failure of mecamylamine and dihydroβerythroidine to counteract the stimulation by nicotine of numbers of binding sites (Schwartz and Kellar 1985). However, this rationale is not consistent with the inability of recognized antagonists (particularly the competitive antagonist dihydroβerythroidine) to elicit an increase in binding site density themselves (Schwartz and Kellar 1985; see above). The well-established phenomenon of up-regulation of specific binding sites by chronic nicotine administration (Fig. 7.5) cannot be readily explained; a unique action of the drug in this respect could be fundamental to its ability to produce dependence. Chronic treatment with other nicotinic agonists needs to be undertaken, and *in vitro* systems amenable to long-term manipulation could prove useful for distinguishing the direct and indirect effects of drug action. Adrenal chromaffin cells, which possess high-affinity [^{3}H]nicotine binding sites and functional nicotinic receptors that can be shown to desensitize (Higgins and Berg 1988), constitute a useful model system for such studies.

Up-regulation of [^{3}H]nicotine binding sites has some physiological credibility from its correlation with tolerance to nicotine in behavioural tests (Fig. 7.5; Marks *et al.* 1985; see Swedberg *et al.*, chapter 2, this volume); the apparent contradiction of increased receptor number and diminished response can be accommodated if many of those receptor sites are inactive (desensitized). A correlation between enhanced locomotory activity in response to nicotine and increased numbers of brain nicotinic binding sites in rats following chronic exposure to low doses (0.01–0.3 $mg\,kg^{-1}$) of nicotine has also been noted (Ksir *et al.* 1987). However, this relationship was not evident after a high-dose regime (1.6 $mg\,kg^{-1}$ twice daily): perhaps the stimulatory action of nicotine was counteracted by a tolerance phenomenon as described by Marks *et al.* (1985). Alternatively, heterogeneity of functional receptors with different sensitivities to nicotine may account for this disparity (see Ksir *et al.* 1987).

The mechanisms underlying modulation of binding site numbers are not

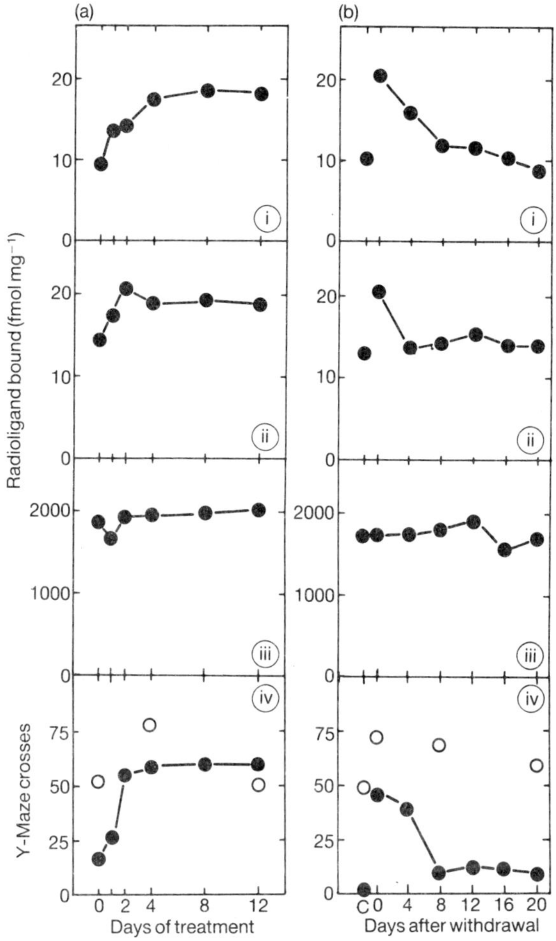

Fig. 7.5. Timecourse of (a) chronic nicotine treatment and (b) withdrawal on receptor numbers and behavioural response. D B A mice were infused with nicotine (4 mg kg^{-1} h^{-1}) for the days indicated in (a). Measurcments were continued for up to 20 days after cessation of nicotine infusion, as shown in (b). Numbers of cholinergic binding sites in the cortex were measured using (i) [^{3}H](−)nicotine (4 nM); (ii) [^{125}I]αbungarotoxin (0.67 nM): (iii) [^{3}H]quinuclidinyl benzilate (0.15 nM). Behavioural responses, Y-maze crosses illustrated here (iv), were measured following a challenge (intraperitoneal injection) of saline (○) or nicotine (●; see Marks *et al.* 1985).

known, but changes in binding-site density presumably reflect altered turnover rates: in the case of up-regulation either increased synthesis and/or membrane incorporation, or decreased degradation. Changes in [^{3}H]nicotine binding sites are slow to appear: single, acute doses of anticholinesterase (Costa and Murphy 1983) or nicotine (Marks *et al.* 1985) do not alter receptor numbers. A time course of the effects of chronic nicotine infusion ($4\,mg\,kg^{-1}\,h^{-1}$) on [^{3}H]nicotine binding sites in mouse brain is shown in Fig. 7.5; about two days were required for 50 per cent of the maximal change to occur; the maximal increase was seen after four days (Marks *et al.* 1985). Elevated numbers of binding sites persisted for at least four days after withdrawal of nicotine. A similar time course (but with slightly slower onset and offset) was observed in rats receiving twice daily injections of nicotine ($1\,mg\,kg^{-1}$; Schwartz and Kellar 1985).

In addition to the up-regulation of [^{3}H]nicotine (and nicotinic-[^{3}H]acetylcholine) binding sites, chronic administration of high doses of nicotine also increased the number of [^{125}I]αbungarotoxin binding sites in some brain regions (Marks *et al.* 1985; Fig. 7.5). Other cholinergic parameters (acetylcholinesterase and cholineacetyltransferase activities; muscarinic binding sites) were unaffected (Marks *et al.* 1986; Fig. 7.5), although a decreased affinity of muscarinic sites in the cerebral cortex has been noted (Yamanaka *et al.* 1987). Moreover, chronic administration of the muscarinic antagonist scopolamine ($10\,mg\,kg^{-1}$ per day) upregulates both muscarinic and [^{3}H]nicotine binding sites in rat cortex (Vige and Briley 1988). These authors propose that the smaller increase in numbers of nicotinic sites (12 per cent, versus 27 per cent increase in muscarinic sites) is a consequence of muscarinic blockade, which may influence levels of acetylcholine at the synapse, rather than a direct action of scopolamine on [^{3}H]nicotine sites. An increase in the density of α_1 and α_2 adrenoreceptors in response to chronic nicotine administration has been reported (Yamanaka *et al.* 1985), as well as metabolic changes (Shallom and Katyare 1985; Katyare and Shallom 1988) and genomic activation (Greenberg *et al.* 1986).

Other treatments may also up-regulate neuronal nicotinic binding sites. Oral choline elicited a dose-dependent increase in the numbers of [^{125}I]αbungarotoxin binding sites in rat brain (Morley and Fleck 1987); [^{3}H]nicotine binding sites were not measured. Chronic ethanol administration increased numbers of [^{3}H]nicotine binding sites in some brain areas in one study (Yoshida *et al.* 1982), but no changes were observed in a long-term study by Nordberg *et al.* (1985). β Nerve growth factor up-regulates nicotine receptors, but not αbungarotoxin binding sites, on PC12 cells (Whiting *et al.* 1987*c*).

7.4.3.3 *Endogenous ligands* There is some speculation that endogenous factors other than acetylcholine may interact with nicotinic receptors. Such factors could modify the effects of nicotine or contribute to the functional consequences of nicotine's action. This concept arises from the observation that brain supernatant fractions inhibit [^{3}H]nicotine (Sershen *et al.* 1984; Perry *et al.* 1987) and [^{125}I]αbungarotoxin (Quik 1982; Polz-Tejera and Schmidt 1983) binding to brain membranes. The endogenous substance is of low molecular weight, heat stable, and does not affect the binding of ligands for other neurotransmitter receptors (Sershen *et al.* 1984; Quik 1982). Polz-Tejera and Schmidt (1983) identified their inhibitory factor as choline, which is a weak inhibitor of [^{3}H]nicotine binding to brain membranes (Costa and Murphy 1984), and has both agonist and antagonist actions at nicotinic receptors (Holz and Senter 1981; Sterz *et al.* 1986). Such an interaction might account for the up-regulation of sites induced by oral choline (Morley and Fleck 1987), mentioned in the previous section. However, choline has been discounted by Quik *et al.* (1986). Other contenders include histamine (Ohta *et al.* 1984), opiates (Kumakura *et al.* 1980), substance P (see Molinaro and Del Rio 1987), and piperidine (Yamada *et al.* 1986), considered to be an endogenous nicotine-like substance. Of these, substance P has received most attention (see Clarke, chapter 5, this volume), and the recent demonstration (Molinaro and Del Rio 1987) of its interaction with brain nicotinic receptors in the mediation of anti-nociception, produced by both substance P itself and nicotine, suggests that this may be an interaction of physiological significance. As views of neurocommunication change, with recognition of the prevalence of co-transmission, for example, so the concepts of receptor specificity must be modified.

7.4.4 *Comparison of [^{3}H]nicotine with other radioligands*

In the light of the heterogeneity of nicotinic receptor candidates in the brain (see section 7.5) and the likelihood that some of these will not display high-affinity agonist binding (cf., nicotinic receptors in autonomic ganglia, Kemp and Morley 1986), alternative radioligands will be necessary for their identification.

7.4.4.1 *Antagonists* [^{125}I]αBungarotoxin and [^{3}H]nicotine have already been compared (section 7.4.1), and it is clear that they label distinct proteins (see Figs. 7.4 and 7.6). Other components of snake venoms have been examined for nicotinic potency. While the most abundant activity is neuromuscular blockade, minor fractions that show functional antagonism of nicotinic responses in the central and autonomic nervous systems have been described (see Chiappinelli 1985; Wonnacott 1987*a*). The best characterized of these is neuronal bungarotoxin (formerly referred to as BgT 3.1, Ravdin and Berg 1979; kappa toxin, Chiappinelli 1985; and toxin F, Loring and Zigmond 1988; see Lindstrom *et al.* 1987). Neuronal bungarotoxin has been most widely studied in autonomic ganglia, where it potently inhibits nicotinic transmission, in contrast to αbungarotoxin. However, it binds to both αbungarotoxin binding sites (which are extrasynaptic in ganglia, Jacob and Berg 1983) and synaptic nicotinic receptors. Binding to the synaptic sites is inhibited by nicotinic ligands (Halvorsen and Berg 1986), but agonist affinities are micromolar in contrast to their nanomolar affinities at brain [^{3}H]nicotine binding sites (see Table 7.2). This is consistent with the absence of high-affinity [^{3}H]agonist binding to ganglionic neurones (Kemp and Morley 1986). Conversely, in the CNS neuronal bungarotoxin does not inhibit [^{3}H]nicotine binding (Wolf *et al.* 1988), although it does inhibit the nicotine-induced release of dopamine from striatal slices (Schulz and Zigmond 1989). Neuronal bungarotoxin potently blocks the functional expression of only one of the three brain nicotinic receptor subtypes that have been examined, the α3 subtype that predominates in ganglia and is only weakly represented in the brain (see section 7.5.3). Neuronal bungarotoxin may be essentially a ganglion-specific antagonist.

Antagonists as labelled probes have the advantage that the seemingly unphysiological high-affinity binding encountered with agonists (see section 7.4.1.2 and below) is avoided. However, few centrally-active antagonists of sufficient potency have been recognized. Tritiated *d*-tubocurarine (Larsson and Nordberg 1985) and dihydroβerythroidine (Williams and Robinson 1984) have been examined in brain preparations. Both antagonists label two sites, but their broad specificity for both end-plate and ganglionic nicotinic receptors limits their usefulness as probes for receptor subtypes (see Clarke 1987; Wonnacott 1987*a*). An ideal candidate would be neosurugatoxin, an antagonist that is both highly specific for [^{3}H]nicotine binding sites and very potent in ganglion and CNS blockade (section 7.4.1.2; Hayashi *et al.* 1984; see Wonnacott 1987*a*), but its complex chemical structure presently precludes the synthesis of radiolabelled derivatives.

7.4.4.2 *Agonists* Although [^{3}H]nicotine has been the most widely used agonist for identifying nicotinic sites in the brain, [^{3}H]acetylcholine ([^{3}H]ACh) (in the presence of a muscarinic antagonist and anti-cholinesterase) has been employed as a nicotinic ligand by several laboratories (Schwartz *et al.* 1982; Larsson *et al.* 1987). Both agonists label a high-affinity site with nanomolar affinity, and comparison of their binding characteristics (Clarke *et al.* 1985; Marks *et al.* 1986; Martino-Barrows and Kellar 1987) indicates that they label the same binding protein (see Wonnacott 1987*b*). This is exemplified by the highly correlated pharmacological specificity and regional distribution of the two sites, shown in Fig. 7.6, which contrasts with the poor correlation with [^{125}I]αbungarotoxin binding sites. However, some differences in the responses of [^{3}H]nicotine and [^{3}H]ACh-labelled sites to chronic nicotine treatment have been noted (Larsson *et al.* 1986), although parallel up-regulation was observed by other groups (Marks *et al.* 1986; Martino-Barrows and Kellar 1987).

As a non-hydrolyzable alternative to acetylcholine, a tritiated derivative of carbamylcholine, has been developed. Surprisingly, [^{3}H]-N-methylcarbamylcholine is devoid of muscarinic activity, and displays nanomolar affinity for nicotinic sites (Abood and Grassi 1986; Boksa and Quirion 1987). The pharmacological specificity and regional distribution (Yamada *et al.* 1987; Araujo *et al.* 1988) of binding sites of this ligand and [^{3}H]nicotine are very similar, indicative that [^{3}H]methylcarbamylcholine labels the same protein. It has not been tested in autonomic ganglia, but one would predict that like [^{3}H]nicotine (Whiting *et al.* 1987*c*) and [^{3}H]ACh (Kemp and Morley

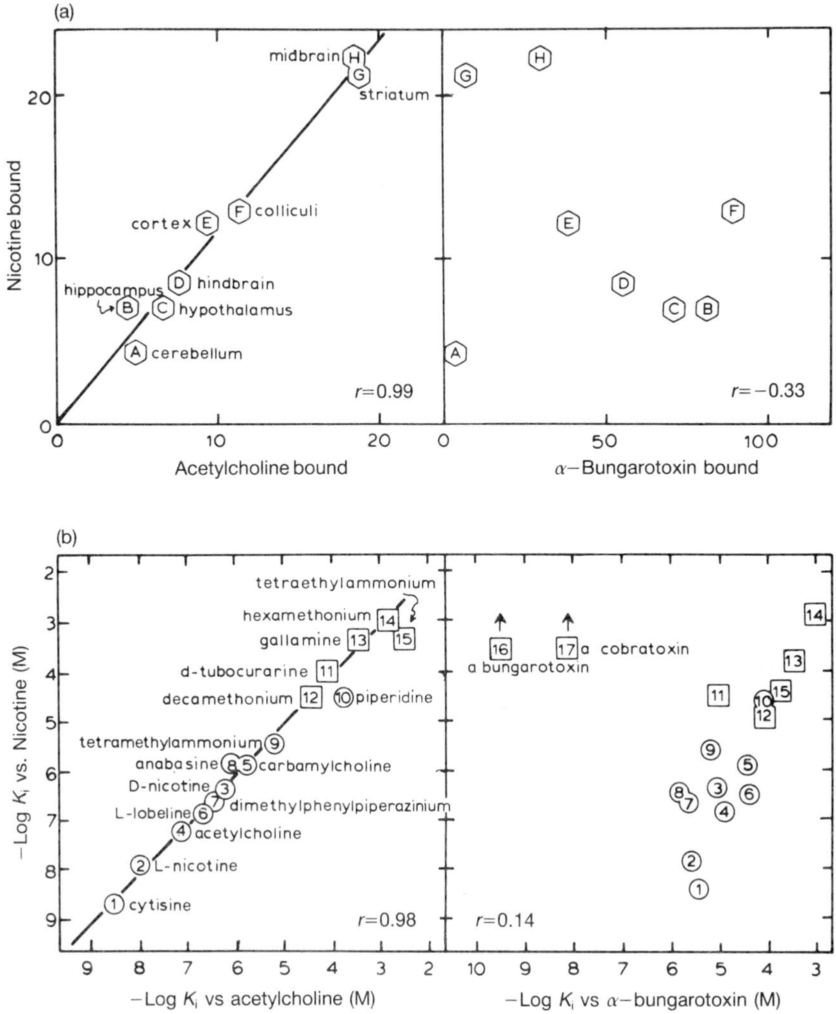

Fig. 7.6. Comparison of (a) regional distribution and (b) the pharmacological specificity of nicotinic ligand binding sites in rat brain. (a) Numbers of binding sites (fmol mg^{-1} protein) in dissected rat brain regions were measured using [^{3}H](−)nicotine (5 nM), [^{3}H]acetylcholine (10 nM), or [^{125}I]αbungarotoxin (1.7 nM) (Marks *et al.* 1986). The left-hand panel shows that the regional distributions of [^{3}H]nicotine and [^{3}H]acetylcholine binding sites are highly correlated ($r = 0.99$), whereas there is little correlation with [^{125}I]αbungarotoxin binding sites ($r = 0.33$; right-hand panel). (b) Inhibition constants (K_i) were determined for the 17 drugs indicated, in competition assays with [^{3}H](−)nicotine (4.5 nM); [^{3}H]acetylcholine (11 nM), or [^{125}I]αbungarotoxin (1.9 nM) (Marks *et al.* 1986; Table 7.2). The left-hand panel shows excellent correlation between [^{3}H]nicotine and [^{3}H]acethylcholine binding sites with respect to drug sensitivity ($r = 0.98$), whereas there is no correlation with [^{125}I]αbungarotoxin binding sites ($r = 0.14$; right-hand panel). Omitting the αtoxins (16 and 17) from the comparison does improve the correlation between [^{125}I]αbungarotoxin and [^{3}H]nicotine binding sites ($r = 0.79$). Note that agonists (circles) are more potent than antagonists (squares).

1986), it will not label ganglionic receptors. A more potent agonist that may have sufficient affinity to identify such receptors in binding assays is (+)anatoxin-a (Zhang *et al.* 1987; MacAllan *et al.* 1988); radiolabelled anatoxin-a is eagerly awaited.

7.4.4.3 *Antibodies* Antibodies provide an alternative to pharmacological probes. Although polyclonal antisera raised against the peripheral nicotinic receptor show little (Norman *et al.* 1982; Wonnacott *et al.* 1982) or no cross-reactivity (Morley *et al.* 1983) with related brain proteins, the generation of monoclonal antibodies allows the selection of very specific, high-affinity probes (see Lindstrom 1986). In addition to their use as ligands for affinity chromatography purification (see section 7.4.2), antibodies can be iodinated to high specific activity and used as radioligands for assaying and mapping their antigen, or immunohistochemical techniques can be employed. For example, [^{125}I]mAb 270, an antibody that recognizes the [^{3}H]nicotine binding protein in rat brain (section 7.4.2; see Lindstrom *et al.* 1987), has been used in autoradiographic studies of the rat CNS (Swanson *et al.* 1987). The pattern of immunolabelling corresponds very closely to that of [^{3}H]nicotine (Clarke *et al.* 1985; London *et al.* 1985*a*). Immunohistochemical localization of nicotinic sites in rat brain using another monoclonal antibody (Deutch *et al.* 1987) gave the same pattern of labelling. In addition, cytoplasmic labelling was observed (Deutch *et al.* 1987; Swanson *et al.* 1987); this is assumed to represent precursor receptor protein in transit to sites of membrane incorporation, or internalized receptor prior to degradation. It is unlikely that such intracellular receptors would be detected by [^{3}H]agonist binding, because this is conformationally dependent upon insertion of the protein in the plasma membrane.

Monoclonal antibodies specific for each of the constituent subunits in the [^{3}H]nicotine binding protein have been characterized (Whiting *et al.* 1987*b*). Thus mAb 270 recognizes an epitope on the smaller, structural subunit, and the binding of this antibody to PC12 cells (Whiting *et al.* 1987*c*) is evidence that the ganglionic nicotinic receptor shares this subunit with the brain protein, although they differ in their agonist-binding subunits (see section 7.4.2). Now that the deduced amino acid sequences of neuronal nicotinic subunits are known (see section 7.5), monoclonal antibodies can be raised against synthetic peptides corresponding to putative structural or functional domains in the protein, in order to pursue its molecular characterization (cf., section 7.2.4.2).

7.4.5 *[^{3}H]nicotine binding sites and nicotinic receptor function in the brain*

Caution should be exercised in elevating ligand-binding sites to functional receptor status, especially in view of the fall from grace of αbungarotoxin sites. Nevertheless, there are several lines of evidence that favour a synaptic receptor role for [^{3}H]nicotine binding sites. These have largely been dealt with elsewhere in this volume and will only be briefly mentioned here.

7.4.5.1 *Electrophysiological studies* Electrophysiological data have been reviewed by Clarke (chapter 5); correlation with binding data is hampered by the reliance on antagonists that are either non-selective (dihydroβerythroidine, *d*-tubocurarine), or presumed to be non-competitive (mecamylamine, hexamethonium). However, there is, in general, a preponderance of C6 type nicotinic responses in brain regions displaying the densest labelling by [^{3}H]nicotine (see Clarke, Table 5.1, this volume).

7.4.5.2 *Behavioural studies* Behavioural effects of nicotine in relation to central nicotinic receptors have been discussed in detail by Stolerman (chapter 8). Although behavioural studies fail to satisfy all the criteria for their mediation by [^{3}H]nicotine binding sites (because the studies are often incomplete, antagonists are non-competitive, and other factors are unknown, such as the synaptic concentration of systemically administered nicotine), careful comparison of nicotinic agonists can generate a highly significant correlation between behavioural potency and binding affinity at the [^{3}H]nicotine site (see Stolerman, Fig. 8.6, this volume).

7.4.5.3 *Biochemical studies* Similarities between brain ligand binding sites and the well-characterized nicotinic receptor from *Torpedo* electroplax and muscle include affinity alkylation with MBTA and bromacetylcholine (see sections 7.2.2 and 7.4.2); immunological cross-reactivity (section 7.4.2; see Lindstrom *et al.* 1987), and sequence homology (sections 7.3.2, 7.5.2). Together with the up- and down-regulation of binding sites (section 7.4.3), these features provide circumstantial evidence that [^{3}H]nicotine binding sites are nicotinic receptors. However, the same properties are also shared by αbungarotoxin binding sites. Clearly, measurements of receptor function are crucial before such interpretations can be made. An indirect assessment of nicotinic receptor activation in the brain has used the 2-deoxy-*D*-[I-^{14}C]glucose technique to monitor the *in vivo* effects of nicotine on local cerebral glucose utilization (London *et al.* 1985*b*; Grunwald *et al.* 1987). The regional distribution of nicotine-stimulated cerebral glucose utilization correlates closely with anatomical maps of [^{3}H]nicotine binding sites. More direct measurements of receptor activation can be obtained from an *in vitro* functional assay, applicable to brain regions, which makes use of synaptosome preparations to study nicotine-evoked transmitter release, regulated through presynaptic nicotinic receptors (Rapier *et al.* 1988; see below).

7.4.5.4 *Presynaptic nicotinic receptors* It is increasingly recognized that nicotine acts presynaptically to promote the release of neurotransmitters in many brain regions (Balfour 1982; Rowell 1987), and that a significant proportion of [^{3}H]nicotine binding sites are located presynaptically. Evidence for the latter point comes from lesion studies, which demonstrate marked decrements in [^{3}H]agonist binding sites on degeneration of axons (Schwartz *et al.* 1984; Clarke and Pert 1985; Clarke *et al.* 1986), and from immunohistochemical evidence for axonal transport of the receptor protein (Deutch *et al.* 1987; Swanson *et al.* 1987). A presynaptic localization of nicotinic receptors is also consistent with the reduction in the numbers of high-affinity binding sites for nicotinic agonists in the brains of Alzheimer patients, a loss which parallels the degeneration of cholinergic projections to the cortex and hippocampus (see Kellar and Wonnacott, Fig. 10.2, this volume).

The presynaptic nicotinic receptor that has been most studied is that present on dopaminergic nerve terminals in the striatum (Chesselet 1984; see Reavill, chapter 9, this volume). In brief, micromolar concentrations of nicotine stimulate dopamine release from striatal nerve terminals (Giorgiueff-Chesselet *et al.* 1979; Rapier *et al.* 1988), and the effect is concentration-dependent with an EC_{50} of 4 μM. This action of nicotine is blocked by ganglionic nicotinic antagonists (Chesselet 1984), but is not inhibited by αbungarotoxin (Rapier *et al.* 1985; see Wonnacott *et al.* 1988). Nicotine-evoked dopamine release is also inhibited by the nicotinic channel blocker histrionicotoxin (Rapier *et al.* 1987) and by neuronal bungarotoxin (Schulz and Zigmond 1989). The presynaptic receptor mediating this action correlates with [^{3}H]nicotine binding sites in the brain as judged by its stereospecificity for (−)nicotine (Rapier *et al.* 1988), and its sensitivity to neosurugatoxin (Rapier *et al.* 1985; see Wonnacott *et al.* 1988). 6-hydroxydopamine lesions of the nigrostriatal tract in rats decreased [^{3}H]ACh or [^{3}H]nicotine binding in the striatum by 30 per cent (Schwartz *et al.* 1984; Clarke and Pert 1985).

Another dopaminergic pathway important in the physiological effects of nicotine is the mesolimbic system (see Fuxe *et al.*, chapter 6, and Reavill, chapter 9, this volume). Micromolar concentrations of nicotine stimulated the release of dopamine from the nucleus accumbens (Rowell *et al.* 1987; see Reavill, Fig. 9.2, this volume), and 6-hydroxydopamine lesions of the mesolimbic tract markedly reduced the numbers of [^{3}H]nicotine binding sites in this brain region (Clarke and Pert 1985). Nicotine also influences the release of other transmitters in the brain, including noradrenaline (Yoshida *et al.* 1980), 5-HT (Balfour 1973), GABA (Wonnacott *et al.* 1988), and acetylcholine (Rapier *et al.* 1987) release from hippocampal nerve terminals, acetylcholine release in the cortex (Beani *et al.* 1985; Rowell and Winkler 1986), and 5-HT release in the hypothalamus (Hery *et al.* 1977). In some of these instances, high nicotine concentrations were used and nicotinic pharmacology was not fully established. However, lesion studies corroborate the presence of high-affinity nicotinic binding sites on hypothalamic 5-HT terminals (Schwartz *et al.* 1984; see Fuxe *et al*, chapter 6, this volume). In contrast, degeneration of brain dopaminergic and serotonergic pathways did not alter the numbers of nicotinic binding sites in the cortex or thalamus (Schwartz *et al.* 1984), suggesting that nicotinic receptors are not ubiquitously associated with these nerve terminals. This also appears to be the case with respect to cholinergic nerve terminals.

The nicotinic agonist *N*-methylcarbamylcholine (see section 7.4.4.2) increased acetylcholine release from hippocampal and frontal cortical brain slices, but had no effect on acetylcholine release from striatal slices (Araujo *et al.* 1988). This must confer some regional specificity to the actions of nicotine.

The prevalence of nicotinic receptors at presynaptic locations may explain the scarcity of nicotinic responses demonstrable by electrophysiological techniques (such techniques are not applicable to axon terminals). Except for nicotinic autoreceptors on cholinergic nerve terminals, the source of endogenous acetylcholine presumed to stimulate these receptors *in vivo* is not well established. In the case of striatal dopaminergic terminals, the presynaptic nicotinic receptors may be post-synaptic to cholinergic interneurons (McGeer *et al.* 1975). In other locations it is possible that the receptors respond to acetylcholine released at a distance (see Lindstrom *et al.* 1987), or they may be stimulated by other endogenous ligands (see section 7.4.3.3). The variety of nicotinic functions implicit in these considerations (presynaptic, both autoreceptors and heteroreceptors; postsynaptic and possibly extra-synaptic) is matched by the structural heterogeneity revealed by molecular genetics.

7.5 Molecular biology approaches to neuronal nicotinic receptors

7.5.1 *Isolation of multiple genes for neuronal nicotinic receptor subunits*

One of the remarkable findings to emerge from the application of recombinant DNA techniques to the study of neuronal nicotinic receptors is the evidence for heterogeneity of receptors at both the gene and protein level (see Steinbach and Ifune 1989). The strategy adopted assumed that there must be some sequence homology between muscle and neuronal nicotinic receptors, so that cDNA probes derived from muscle receptor subunit genes could be used to recognize related sequences in neuronal cDNA or genomic DNA libraries (see Fig. 7.2).

Thus the group of Patrick and Heinemann successfully isolated a cDNA clone coding for a putative nicotinic receptor α subunit from PC12 cells (derived from sympathetic neurones) by low-stringency hybridization with a mouse muscle α subunit cDNA (Boulter *et al.* 1986). The neuronal clone coded for a protein that is 47 per cent homologous with the mouse muscle receptor α subunit (contrast the high homology between muscle receptor subunits from different species, see section 7.2.5). This clone hybridized strongly with mRNA from dissected rat brain regions, indicating that the same gene (designated $\alpha 3$) is expressed in the central nervous system, but it did not hybridize with mRNA from skeletal muscle. The $\alpha 3$ clone was subsequently used to probe cDNA libraries prepared from rat brain hypothalamus and hippocampus, and two further gene products were identified and designated

Table 7.3. Comparison of deduced amino acid sequences of cloned neuronal nicotinic receptor subunits and muscle receptor αsubunit

Tissue	Subunit	per cent homology versus α1	per cent homology versus α2	total number of amino acids	number of amino acids inserted in cytoplasmic domain
chick muscle	α1	[100]	50	437	—
chick brain	α2	50	[100]	505	67
"	α3	53	65	473	42
"	α4	51	80	599	165
"	non-α	44	55	473	28

Data from Nef *et al.* (1988)

α4 (Goldman *et al.* 1987) and β2 (Deneris *et al.* 1988), on the basis of the presence and absence respectively of cysteine residues at positions 192 and 193 ('vicinal' cysteines) that are characteristic of α subunits (see sections 7.2.4.5 and 7.5.2).

A corresponding family of nicotinic receptor subunit genes has been identified in avian brain (Nef *et al.* 1988), including an additional clone coding for another α-type subunit, designated α2 (as this was the first neuronal subunit to be identified). The avian α2 gene was subsequently employed to screen a rat genomic DNA library, and the α2 gene from rat brain was cloned and analyzed (Wada *et al.* 1988).

Thus genes for 3 putative agonist binding (α) subunits and one non-alpha (β) subunit have been recognized in mammalian and avian brain (Table 7.3, see Steinbach and Ifune 1989); other neuronal nicotinic receptor-related genes are almost certainly present (McKinnon *et al.* 1988; Deneris *et al.* 1989). The heterogeneity of nicotinic receptor subunits in the brain is further increased by the identification of two transcripts of the α4 gene (clones 4-1 and 4-2; Goldman *et al.* 1987), proposed to arise by alternative splicing (i.e. alternative reading of the DNA, see Fig. 7.2a, to give, in this case, two mRNA species that differ at their 3′ ends, which correspond to the carboxy termini of the translated proteins). This is another mechanism for generating receptor diversity.

7.5.2 *Structural features of cloned nicotinic receptor subunits*

What is the evidence that the nervous system-derived clones encode nicotinic receptor subunits? The initial evidence comes from the deduced amino acid sequences of the proteins which show the conservation of structural domains characteristic of muscle and electroplax receptor subunits (Goldman *et al.* 1986, 1987; Nef *et al.* 1988; see section 7.2.4.5, Fig. 7.3, Table 7.4). These are:

1. The four putative membrane-spanning sequences (M_1, M_2, M_3, M_4), which exhibit between 50 and 100 per cent homology with corresponding sequences in either muscle or other neuronal receptor subunit candidates.

2. The amphipathic helix, originally proposed to contribute to the ion channel of the receptor (see section 7.2.4.5). The homology of this region between neuronal and muscle sequences is not as high as for the hydrophobic sequences M_1 to M_4, but the distribution of charged and uncharged residues about the helix is maintained.

3. The extracellular domain containing two invariant cysteine residues at positions 128 and 142 that are covalently linked through a disulphide bridge to create a loop. This structural element is shared by all known nicotinic receptor subunits, and is also present in the glycine (Grenningloh *et al.* 1987) and $GABA_A$ receptors (Schofield *et al.* 1987; Table 7.4). Additionally, the vicinal cysteines at positions 192 and 193 are characteristic of nicotinic receptor α subunits, where they are believed to contribute to the agonist binding site (see section 7.2.4.5). The presence of these cysteines clearly distinguishes the $\alpha2$, $\alpha3$ and $\alpha4$ clones from the $\beta2$ clone, and accounts for their classification as α or non-alpha subunits respectively.

4. All known subunits from end-plate nicotinic receptors have a potential *N*-linked glycosylation site at asparagine 141 (α subunit numbering). All the neuronal clones also have an asparagine residue in this position, although it is not part of a glycosylation sequence in $\alpha2$ and $\alpha4$. Instead, these proteins have a glycosylation site at asparagine 81 and 79 respectively.

Table 7.4. Structural features conserved throughout the nicotinic gene family

*(1)	Four hydrophobic, putative membrane-spanning sequences (M_1, M_2, M_3, M_4)
(2)	Amphipathic helix (M_A) between M_3 and M_4
(3)	large extracellular domain at the *N*-terminus, including:
	(i) invariant cysteines at positions *128, *142, (192, 193)+
	(ii) invariant asparagine at position *141 (potential glycosylation site, except in $\alpha2$ and $\alpha4$).
	(iii) invariant proline at position *136
*(4)	Intracellular phosphorylation site(s) present in cytoplasmic domain between M_3 and M_A

* Also common to $GABA_A$ and glycine receptor subunits
\+ Unique to all nicotinic receptor α subunits

Although these features are conserved in the neuronal sequences, there are also some consistent differences between the neuronal and muscle subunits:

1. All the neuronal sequences analysed to date have a second glycosylation site at about position 30.

2. The four neuronal proteins α2, α3, α4 and β2 share extensive sequence homology in their extracellular domains, but apart from the conserved cysteines mentioned above, this region is poorly conserved with respect to the muscle receptor subunits. The binding site for αbungarotoxin has been localized to the region 173–204 of the end-plate α subunit (Wilson and Lentz 1988). The lack of conservation of this region in the neuronal α sequences most probably accounts for the insensitivity of these proteins to αbungarotoxin in a functional expression system (see section 7.5.3), and indicates that none of these clones code for the αbungarotoxin binding protein in brain. This is supported by the lack of identity between the amino-terminal amino-acid sequence of the smallest subunit of the αbungarotoxin binding protein from chick optic lobe (Conti-Tronconi *et al.* 1985; see section 7.3.2) and any of sequences deduced for the neuronal clones.

3. All of the neuronal clones include insertions in the cytoplasmic domain, between the third hydrophobic segment (M_3) and the amphipathic helix. These vary in length (Table 7.3) and are little conserved between subunits, although phosphorylation sites are retained (Goldman *et al.* 1987).

The cytoplasmic region is likely to be important in directing the localization of the receptor in the cell, for example, by interaction with the cytoskeletal elements responsible for axonal transport to the nerve terminals. The ensuing additional length of the neuronal α subunits, compared with the invariant 437 amino acids in *all* muscle nicotinic receptor α subunits (Table 7.3) is consistent with the greater molecular mass of the agonist-binding subunit of the brain [^{3}H]nicotine binding protein (Whiting and Lindstrom 1986, 1987*a*; see section 7.4.2) compared with the muscle receptor α subunit. Indeed, amino-terminal amino acid sequencing of this subunit of the rat brain [^{3}H]nicotine receptor (Whiting *et al.* 1987*b*) shows it to be identical (at least over the stretch of sequence determined) to the protein coded by gene α4. Moreover, the structural subunit identified by Whiting and Lindstrom (1986, 1987*a*) is of lower molecular weight than the agonist binding subunit, but larger that the α subunit from *Torpedo* or muscle (Table 7.1, Fig. 7.4), consistent with the intermediate size of the product of the β2 gene (Table 7.3), and their equivalence has been demonstrated (Schoepfer *et al.* 1988). Evidence that the two types of subunit described for the purified protein are sufficient for a functional neuronal nicotinic receptor comes from the functional expression of the cloned subunits.

7.5.3 *Functional expression of cloned neuronal nicotinic receptors*

The ultimate test of the nicotinic nature of the cDNA clones described above (Table 7.3) is that they should give rise to nicotinic responses when expressed in *Xenopus* oocytes (see section 7.2.4.3). Oocytes injected with α2, α3 or α4 mRNA in combination with β2 mRNA produced strong depolarizing responses to micromolar concentrations of acetylcholine or nicotine (Boulter *et al.* 1987; Wada *et al.* 1988; Fig. 7.7). Injection of mRNA for a single subunit failed to produce nicotinic channels (Boulter *et al.* 1987; Deneris *et al.* 1988), with the exception of $\alpha4$ which gave a weak response to $100\,\mu M$ acetylcholine (Boulter *et al.* 1987). This supports the notion that two types of subunit are necessary for full nicotinic receptor activity, although the $\alpha4$ subunit may be able to substitute, rather inefficiently, for $\beta2$, a reflection of the high degree of conservation of the membrane-spanning sequences that make up the ion channel. From the preliminary agonist-response data (Boulter *et al.* 1987), the $\alpha4$, + $\beta2$ combination gave good responses to lower concentrations of acetylcholine and nicotine than $\alpha3 + \beta2$ (Fig. 7.7). This may reflect a greater agonist sensitivity of the $\alpha4 + \beta2$ receptor, but could equally result from a higher density of these receptors expressed in the oocyte. However, it is noteworthy that $\alpha4$ corresponds to the high-affinity [^{3}H]agonist binding protein in rat brain from amino acid sequencing (Whiting *et al.* 1987*b*; section 7.5.4.2), whereas $\alpha3$ may encode the agonist binding subunit of the nicotinic receptor in autonomic ganglia (Boulter *et al.* 1986).

The nicotinic character of the expressed neuronal receptors had been further explored by examining the effect of antagonists on acetylcholine-evoked responses (Fig. 7.7b). In no case did αbungarotoxin ($0.1\,\mu M$) have any effect, whereas neuronal bungarotoxin ($0.1\,\mu M$) completely blocked the response of $\alpha3 + \beta2$ receptors (Boulter *et al.* 1987). The $\alpha4 + \beta2$ receptor is less sensitive to neuronal bungarotoxin (Luetje and Boulter, personal communication); this may accord with the failure of this toxin to inhibit [^{3}H]nicotine binding in chick optic lobe (Wolf *et al.* 1988; see section 7.4.4.1). However, $\alpha4 + \beta2$ receptors expressed in *Xenopus* oocytes could be blocked by hexamethonium ($50\,\mu M$; Ballivet *et al.* 1988). The $\alpha2 + \beta2$ receptor was totally resistant to neuronal bungarotoxin (Wada *et al.* 1988) but its sensitivity to hexamethonium ($100\,\mu M$) and *d*-tubocurarine ($100\,\mu M$, Wada *et al.* 1988), support its nicotinic character. More quantitative pharmacological profiles are now required to pursue the comparisons with nicotinic receptors characterized by other methods (see section 7.4.3.5).

Thus the threc neuronal α subunits described to date by molecular biology techniques are capable of forming nicotinic receptor channels when combined with a common non-alpha ($\beta2$) subunit. It remains to be established that $\beta2$ is the partner of all of them *in situ*, although the evidence from protein chemistry of a common structural subunit shared by two distinct [^{3}H]nicotine binding subunits in chick brain (Whiting *et al.* 1987*a*) is consistent with such a view. However, additional α and β subunits are present in the nervous system (McKinnon *et al.* 1988; Deneris *et al.* 1989).

7.5.4 *Regional distribution of nicotinic receptor subunit expression in the brain*

Transcripts of $\alpha4$ and $\beta2$ genes are much more abundant in chick brain (Nef *et al.* 1988) and rat brain (Wada *et al.* 1988) than transcripts of the $\alpha2$ and $\alpha3$

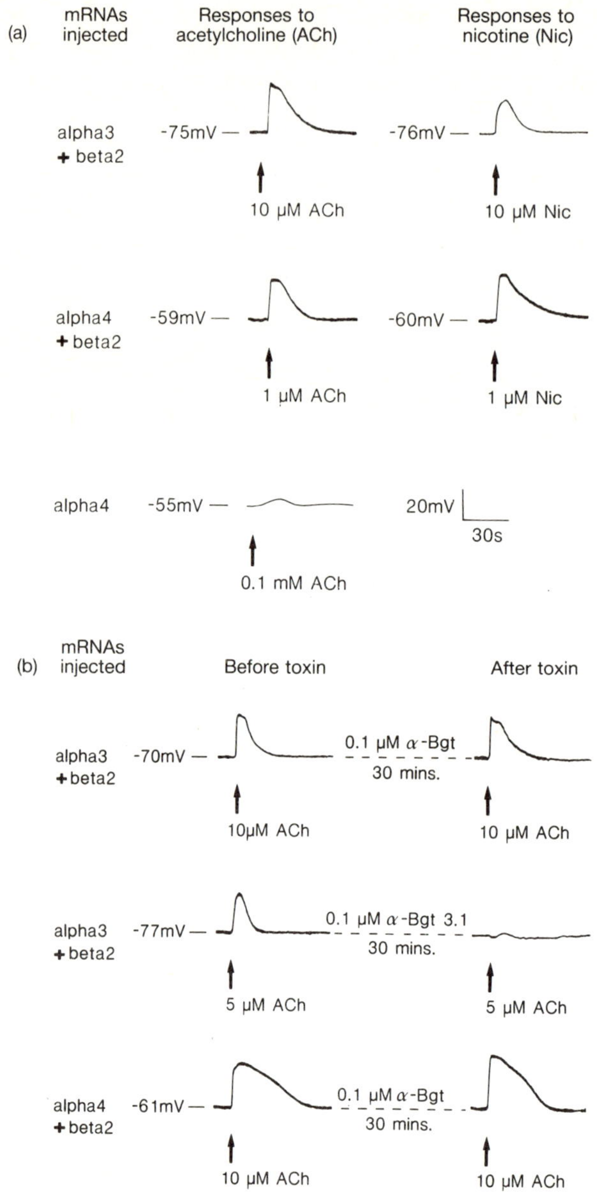

Fig. 7.7. Functional expression of neuronal nicotinic receptors in *Xenopus* oocytes. (a) Voltage traces obtained from *Xenopus* oocytes injected with mR N A derived from the neuronal alpha and beta genes. Representative responses to applied acetylcholine and (–)nicotine, recorded from oocytes injected with the mR N A combinations shown on the left, are illustrated. (b) Effect of two different neurotoxins on the activation by acetylcholine (ACh) of two neuronal nicotinic-acetylcholine receptor subtypes, α3 +

genes. The α3 gene was originally cloned from PC12 cells (Boulter *et al.* 1986), and is also expressed in adrenal medulla. The α2 and α4 genes are not expressed in PC12 cells. Therefore the agonist binding subunit encoded by the α3 gene may be considered to be part of the definitive ganglionic nicotinic receptor, although this correlation has been challenged (Whiting *et al.* 1987*c*).

In the CNS, the technique of *in situ* hybridization histochemistry (see legend to Fig. 7.8) has been employed to produce maps of regions expressing the various subunits. As already indicated, the α3 gene is expressed in the CNS, as well as in the autonomic nervous system, and the hybridization signal was strongest in the medial habenula (Boulter *et al.* 1986; Goldman *et al.* 1986; see Fig. 7.8). Caution is necessary in interpreting weak hybridization signals becuase of the possibility of cross-hybridization between related but non-identical mRNA's (Goldman *et al.* 1986); this is analogous to the cross-reactivity of an antibody with two or more related antigens. However, it appears that the α3 gene is expressed at lower levels in the ventral tegmental area, substantia nigra pars compacta, neocortex, certain thalamic nuclei, and hypothalamus. In contrast (Fig. 7.8) the most intense signal for α2 transcription was detected in the interpeduncular nucleus (Wada *et al.* 1988). The α4 gene product is much more widely distributed in the brain (Goldman *et al.* 1987).

The differential localization of α2, α3, and α4 gene expression in the CNS is evidence for three nicotinic-receptor subtypes in the brain, rather than a single receptor having more than one type of α-subunit. The β2 gene is widely expressed throughout the brain as well as in PC12 cells (Deneris *et al.* 1988) indicating co-expression with all three α genes (Wada *et al.* 1988), and validating the combinations of subunits in the functional expression studies (section 7.5.3). However, in some instances there is imperfect matching of α and β co-expression, which favours the presence of additional nicotinic receptor subunit genes in the brain (Deneris *et al.* 1989).

Interpretation of the anatomical maps of gene expression in terms of the physiological role of the receptor product is handicapped by the fact that *in situ* hybridization identifies the site of receptor subunit synthesis (i.e. the neuronal cell body) which may be far removed from the sites where the receptor functions (dendrites, axon terminals). Thus it is difficult to directly compare the distribution of ligand binding sites (Clarke *et al.* 1985) with *in situ* hybridization maps, although in general regions expressing the neuronal α4 and β2 genes also exhibit high [^{3}H]agonist binding and low [^{125}I]αbungarotoxin binding (cf. medial habenula). Indeed, α4 gene expression correlates quite well with [^{3}H]-nicotine binding sites (Goldman *et al.* 1987). However, β2 gene

---

β2 and α4 + β2, indicated on the left. The left-hand voltage tracing shows the response before application of the toxin, and the voltage tracing on the right shows the response following 30-min incubation with αbungarotoxin (α-Bgt) or neuronal bungarotoxin (Bgt 3.1). (From Boulter *et al.* 1987, with permission.)

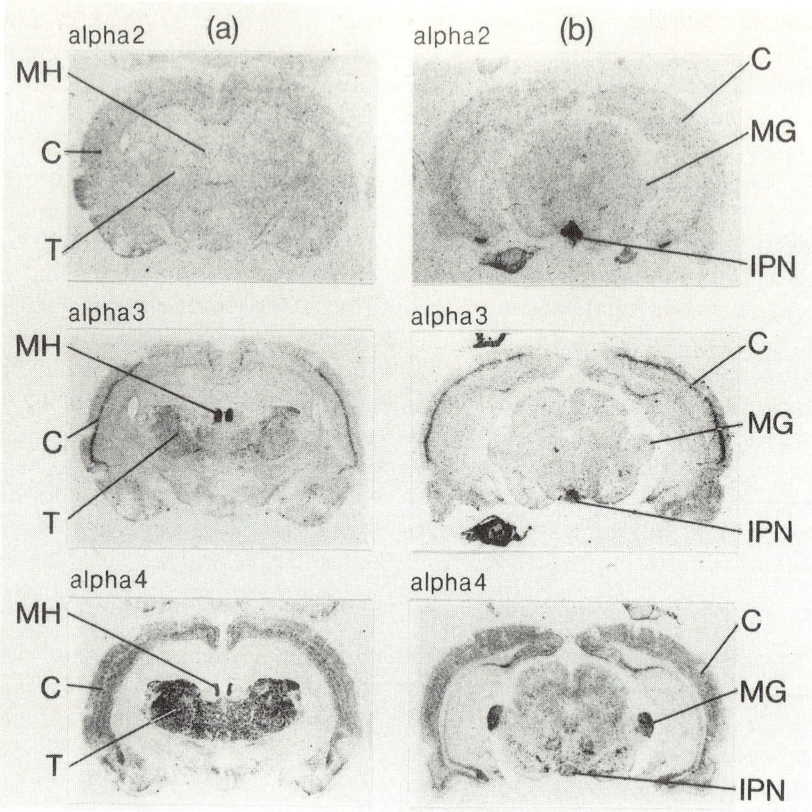

Fig. 7.8. Comparison of *in situ* hybridization maps for neuronal nicotinic receptor α subunit gene products. Serial coronal sections of rat brain taken through the medial habenula (a) and the interpeduncular nucleus (b) were hybridized with probes for $\alpha2$, $\alpha3$, and $\alpha4$ mR N A's. Fixed sections were incubated with [^{35}S]-labelled probes complementary to the target mR N A's, washed, dehydrated, and exposed to X-ray film to visualize the sites of hybridization. Abbreviations: C, cortex; I P N, interpeduncular nucleus; M H, medial habenula; M G, medial geniculate nucleus; T, thalamus. (From Wada *et al.* 1988, with permission.)

expression was also observed in the absence of $\alpha2$, $\alpha3$, or $\alpha4$ gene products in regions not associated with [^{3}H]nicotine binding (Deneris *et al.* 1988). One example is the supraoptic nucleus that is densely labelled by [^{125}I]αbungarotoxin (Clarke *et al.* 1985), so it is possible that the $\beta2$ gene product is also a subunit of the neuronal αbungarotoxin binding protein, or contributes to a non-cholinergic receptor in the super-gene family (see section 7.5.5; Table 7.4).

The disparity between the two mapping techniques (hybridization versus ligand binding) is useful in tracing putative nicotinic-cholinergic pathways expressing presynaptic receptors. Thus three tracts have been identified (Goldman *et al.* 1987):

(1) nigrostriatal pathway projecting from the substantia nigra to the striatum;
(2) mesolimbic pathway projecting from the ventral tegmental area to the nucleus accumbens;
(3) the projection from the medial habenula to the interpeduncular nucleus.

The substantia nigra, ventral tegmental area, and medial habenula express both $\alpha 3$ and $\alpha 4$ genes, leading to the proposition (Goldman *et al.* 1987) that the $\alpha 3$ gene may encode the α subunit of a presynaptic nicotinic receptor synthesized in these brain regions, and subsequently transported to axon terminals in the striatum, nucleus accumbens, and interpeduncular nucleus respectively. The $\alpha 4$ gene would encode an α subunit of a post-synaptic receptor expressed on the cell bodies. Evidence in favour of pre-synaptic nicotinic receptors associated with these projections has already been presented (section 7.4.5.4), and is supported by immunohistochemical evidence for axonal transport of nicotinic receptors along these pathways (Deutch *et al.* 1987). The functions of these nuclei have been discussed elsewhere (see Clarke, chapter 5, this volume; Reavill, chapter 9, this volume).

7.5.5 *Nicotinic receptor heterogeneity and the super-gene family*

The experiments reviewed in the preceding sections point to the presence of at least three nicotinic receptor subtypes in the brain. The occurrence of an $\alpha 5$ subunit (McKinnon *et al.* 1988) and a second non-alpha ($\beta 3$) subunit (Deneris *et al.* 1989) increases the number of subunit permutations and potential nicotinic receptors. The structural heterogeneity evident at the gene and protein level should be paralleled by functional diversity. This diversity need not necessarily be manifested in pharmacologically distinct agonist binding characteristics; structural differences could dictate the channel properties of the receptors, their cellular localization, or their propensity to desensitize, for example. There is evidence that the $\alpha 3$ and $\alpha 4$ gene products do display different agonist sensitivities (see Fig. 7.7) which may correlate with the lack of [^{3}H]agonist binding to autonomic ganglia, which expresses the $\alpha 3$ gene, and the identity of the $\alpha 4$ gene product with the agonist binding subunit of the [^{3}H]nicotine binding protein from brain. There is also evidence in favour of both pre- and postsynaptic locations for nicotinic receptors (sections 7.4.5.4 and 7.5.4), and different α gene products have been tentatively assigned to these roles (Goldman *et al.* 1987). There is insufficient data available to compare the pharmacological specificities and sensitivities of these two receptor classes, but the structural differences in the constituent subunits may be chiefly concerned with the disposition of the receptors in the cell. In this respect the highly variable cytoplasmic domain (section 7.5.2) is likely to be important in determining transport and membrane anchoring phenomena.

A further potential candidate for mediating some of nicotine's actions is the αbungarotoxin binding protein described in section 7.3. In view of the nicotinic character of this protein, with respect to its pharmacology, amino-terminal amino-acid sequence and subunit composition, it was anticipated that it is coded by a set of related genes, and this has recently been confirmed (Schoepfer *et al.* 1989), using the known anino acid sequence (see section 7.3.2) as a probe. However, the failure to isolate the αbungarotoxin-binding subunit gene (presumably an α-type subunit) by cross-hybridization with any of the neuronal or muscle-derived clones is surprising, particularly as this protein is reasonably abundant in brain and ganglia. Evidently the αbungarotoxin binding protein is less closely related to the proven nicotinic receptors and may have diverged both structurally and functionally from the classical nicotinic receptor.

The breadth of divergence within this gene family is evidenced by the inclusion of glycine (Grenningloh *et al.* 1987) and $GABA_A$ (Schofield *et al.* 1987) receptors as distant relatives of the nicotinic receptor, based on considerable structural and amino-acid sequence homology (see Table 7.4). Thus receptors that operate as neurotransmitter-gated ion channels may have evolved from a single ancestral gene, providing a unifying concept of common origin together with scope for hitherto unimagined heterogeneity. The complexity of nicotine's subtle yet profound effects will ultimately be explained by the interaction of this drug with a spectrum of nicotinic receptor subtypes.

Acknowledgements

I am indebted to Vic Cockroft for helpful discussions throughout the preparation of this manuscript, and for developing the receptor model in Fig. 7.3b. I am also grateful to Michael Marks for preparing Figs. 7.5 and 7.6

References

Abood L.G. and Grassi S. (1986). [^{3}H]methylcarbamylcholine, a new radioligand for studying brain nicotinic receptors. *Biochemical Pharmacology* **35**, 4199–202.

Abood L.G., Lowy K., Tometsko A., and MacNeil M. (1979). Evidence for noncholinergic site for nicotine's action in the brain: pharmacological, electrophysiological and receptor binding studies. *Archives of International Pharmacodynamics* **237**, 213–29.

Abood L.G., Reynolds D.T., and Bidlack J.M. (1980). Stereospecific ^{3}H-nicotine binding to intact and solubilised rat brain membranes and evidence for its noncholinergic nature. *Life Sciences* **27**, 1307–14.

Abood L.G., Grassi S., and Costanza M. (1983). Binding of optically pure (–) [^{3}H]nicotine to rat brain membranes. *FEBS Letters* **157**, 147–9.

Abood L.G., Grassi S., and Noggle H.D. (1985). Comparison of the binding of optically pure (–) and (+)[^{3}H]nicotine to rat brain membranes. *Neurochemical Research* **10**, 259–67.

Abood L.G., Langone J.J., Bjercke R., Lu X., and Banerjee S. (1987). Characterization of a purified nicotinic receptor from rat brain by using idiotypic and anti-idiotypic antibodies. *Proceedings of the National Academy of Sciences USA* **84**, 6587–90.

Adams P.R., and Sakman B. (1978). Decamethonium both opens and blocks end-plate channels. *Proceedings of the National Academy of Sciences USA* **75**, 2994–8.

Aracava Y., Desphande S.S., Swanson K.L., Rapoport H., Wonnacott S., Lunt G.G., and Albuquerque E.X. (1987). Nicotinic acetylcholine receptors in cultured neurons from the hippocampus and brain stem of the rat characterized by single channel recording. *FEBS Letters* **222**, 63–70.

Araujo D.M., Lapchak P.A., Collier B., and Quirion R. (1988). Characterization of $[^3H]$*N*-methylcarbamylcholine binding sites and effect of *N*-methylcarbamylcholine on acetylcholine release in rat brain. *Journal of Neurochemistry* **51**, 292–9.

Ascher P., Large W.A., and Rang H.P. (1979). Studies on the mechanism of action of acetylcholine on rat para-sympathetic ganglion cells. *Journal of Physiology* **295**, 139–70.

Balfour D.J.K. (1973). Effects of nicotine on the uptake and retention of ^{14}C-noradrenaline and ^{14}C-5hydroxytryptamine by rat brain homogenates. *European Journal of Pharmacology* **23**, 19–26.

Balfour D.J.K. (1982). The effects of nicotine on brain neurotransmitter systems. *Pharmacology and Therapeutics* **16**, 269–82.

Ballivet M., Nef P., Couturier S., Rungger D., Bader C.R., Bertrand D. and Cooper E. (1988). Electrophysiology of a chick neuronal nicotinic acetylcholine receptor expressed in *Xenopus* oocytes after cDNA injection. *Neuron* **1**, 847–52.

Barnard E.A., Beeson D.M., Cockcroft V.B., Darlison M.G., Hicks A.A., Lai F.A., Moss S.J., and Squire M.D. (1986). Molecular biology of nicotinic acetylcholine receptors from chicken muscle and brain. In *Nicotinic acetylcholine receptor*, (A. Maelicke, ed.), pp. 389–415. Springer-Verlag, Berlin.

Beani L., Bianchi C., Nilsson L., Nordberg A., Romanelli L., and Sivilotti L. (1985). The effects of nicotine and cytisine on ^3H-acetylcholine release from cortical slices of guinea pig. *Naunyn-Schmiedeberg's Archives of Pharmacology* **331**, 323–33.

Benwell M.E.M. and Balfour D.J.K. (1985). Nicotine binding to brain tissue from drug-naive and nicotine-treated rats. *Journal of Pharmacy and Pharmacology* **37**, 405–9.

Benwell M.E.M., Balfour D.J.K., and Anderson J.M. (1988). Evidence that tobacco smoking increases the density of $(-)[^3H]$-nicotine binding sites in human brain. *Journal of Neurochemistry* **50**, 1243–7.

Block G.A. and Billiar R.B. (1981). Properties and regional distribution of nicotinic cholinergic receptors in the rat hypothalamus. *Brain Research* **212**, 152–8.

Boksa P. and Quirion R. (1987). $[^3H]$*N*-methylcarbamylcholine, a new radioligand specific for nicotinic acetylcholine receptors in brain. *European Journal of Pharmacology* **139**, 323–33.

Boulter J., Evans K., Goldman D., Martin G., Treco D., Heinemann S., and Patrick J. (1986). Isolation of a cDNA clone coding for a possible neural nicotinic acetylcholine receptor α-subunit. *Nature* **319**, 368–74.

Boulter J., Connolly J., Deneris E., Goldman D., Heinemann S., and Patrick J. (1987). Functional expression of two neuronal nicotinic acetylcholine receptors from cDNA clones identifies a gene family. *Proceedings of the National Academy of Sciences USA* **84**, 7763–7.

Brisson A. and Unwin P.N.T. (1985). Quaternary structure of the acetylcholine receptor. *Nature* **315**, 474–7.

Changeux J.P., Devillers-Thiery A., and Chemouilli P. (1984). Acetylcholine receptor: an allosteric protein. *Science* **225**, 1335–45.

Chesselet M.F. (1984). Presynaptic regulation of neurotransmitter release in the brain. *Neuroscience* **12**, 347–75.

Chiappinelli V.A. (1985). Actions of snake venom toxins on neuronal nicotinic receptors and other neuronal receptors. *Pharmacology and Therapeutics* **31**, 1–32.

Clarke P.B.S. (1987). Recent progress in identifying nicotinic cholinoreceptors in mammalian brain. *Trends in Pharmacological Sciences* **8**, 32–5.

Clarke P.B.S. and Pert A. (1985). Autoradiographic evidence for nicotine receptors on nigrostriatal and mesolimbic dopaminergic neurons. *Brains Research* **348**, 355–8.

Clarke P.B.S., Pert C.P., and Pert A. (1984). Autoradiographic distribution of nicotine receptors in rat brain. *Brain Research* **323**, 390–5.

Clarke P.B.S., Schwartz R.D., Paul S.M., Pert C.B., and Pert A. (1985). Nicotinic binding in rat brain: autoradiographic comparison of [^{3}H]acetylcholine, [^{3}H]nicotine and [^{125}I]α-bungarotoxin. *Journal of Neuroscience* **5**, 1307–15.

Clarke P.B.S., Hamill G.S., Nadi N.S., Jacobowitz D.M., and Pert A. (1986). [^{3}H]-nicotine and [^{125}I]-αbungarotoxin-labelled nicotinic receptors in the interpeduncular nucleus of rats II. Effects of habenular deafferentation. *Journal of Comparative Neurology* **251**, 407–13.

Claudio, T., Ballivet, M., Patrick, J., and Heinemann, S. (1983). Nucleotide and deduced amino acid sequences of *Torpedo californica* acetylcholine receptor α subunit. *Proceedings of the National Academy of Sciences, U.S.A.* **80**, 1111–5.

Conti-Tronconi B.M., Dunn S.M.J., and Raftery M.A. (1982). Independent sites of low and high-affinity for agonists on *Torpedo Californica* acetylcholine receptor. *Biochemistry and Biophysics Research Communications* **107**, 23–9.

Conti-Tronconi B.M., Dunn S.M.J., Barnard, E.A., Dolly J.O., Lai F.A., Ray N., and Raftery M.A. (1985). Brain and muscle acetylcholine receptors are different but homologous proteins. *Proceedings of the National Academy of Sciences USA* **82**, 5208–12.

Costa L.G. and Murphy S.D. (1983). [^{3}H]-nicotine binding in rat brain: alteration after chronic acetylcholinesterase inhibition. *Journal of Pharmacology and Experimental Therapeutics* **226**, 392–7.

Costa L.G. and Murphy S.D. (1984). Interaction of choline with nicotinic and muscarinic cholinergic receptors in the rat brain *in vivo. Clinical and Experimental Pharmacology and Physiology* **11**, 649–54.

Dale H.H. (1914). The action of certain esters and ethers of choline and their relation to muscarine. *Journal of Physiology* **6**, 147–90.

Deneris E.S., Connolly J., Boulter J., Wada E., Wada K., Swanson L.W., Patrick J., and Heinemann S. (1988). Primary structure and expression of β2: a novel subunit of neuronal nicotinic receptors. *Neuron* **1**, 45–54.

Deneris E.S., Boulter J., Patrick J., Swanson, L.W., and Heinemann S. (1989). β_3: A new member of nicotinic acetylcholine receptor gene family is expressed in brain. *Journal of Biological Chemistry*, **264**, 6268–72.

Deutch A.Y., Holliday J., Roth R.H., Chun L.L.Y., and Hawrot E. (1987). Immunohistochemical localization of a neuronal nicotinic acetylcholine receptor in mammalian brain. *Proceedings of the National Academy of Sciences USA* **84**, 8697–701.

Devillers-Thiery, A., Giraudat, J., Bentabonlet, M., and Changeux, J-P. (1983). Complete mRNA coding sequence for the acetylcholine binding subunit of Torpedo marmorata acetylcholine receptor: A model for the transmembrane orientation of the polypeptide chain. *Proceedings of the National Academy of Sciences, U.S.A.* **80**, 2067–71.

Dolly J.O. and Barnard E.A. (1984). Nicotinic acetylcholine receptors: an overview. *Biochemical Pharmacology* **33**, 841–58.

Finer-Moore J. and Stroud R. (1984). Amphipathic analysis and possible formation of the ion channel in an acetylcholine receptor. *Proceedings of the National Academy of Sciences USA* **81**, 155–9.

Flynn D.D. and Mash D.C. (1986). Characterization of *L*-[^{3}H] nicotine binding in human cerebral cortex: comparison between Alzheimer's disease and the normal *Journal of Neurochemistry* **47**, 1948–54.

Fuchs J.L. and Hoppens K.S. (1987). α-Bungarotoxin binding in relation to functional organisation of the rat suprachiasmatic nucleus. *Brain Research* **407**, 9–16.

Fuxe K., Von Euler G., and Agnati L.F. (1988). Reduction of [^{3}H]-nicotine binding by clonidine in membrane preparations of the rat cerebral cortex. *Acta Physiologica Scandinavica* **132**, 265–6.

Giorguieff-Chesselet M.F., Kemel M.L., Wandscheer D., and Glowinski J. (1979). Regulation of dopamine release by presynaptic nicotinic receptors in rat striatal slices: effect of nicotine in a low concentration. *Life Sciences* **25**, 1257–62.

Giraudat J., Dennis M., Heidmann T., Haumont P.T., Lederer F., and Changeux J.P. (1986). Structure of the high-affinity binding site for noncompetitive blockers of the acetylcholine receptor: serine-262 of the delta subunit is labelled by [^{3}H]chlorpromazine. *Proceedings of the National Academy of Sciences USA* **83**, 2719–23.

Goldman D., Simmons D., Swanson L.W., Patrick J., and Heinemann S. (1986). Mapping of brain areas expressing RNA homologous to two different acetylcholine receptor α-subunit cDNA's. *Proceedings of the National Academy of Sciences USA* **83**, 4076–80.

Goldman D., Deneris E., Luyten W., Kochhar A., Patrick J., and Heinemann S. (1987). Members of a nicotinic acetylcholine receptor gene family are expressed in different regions of the mammalian central nervous system. *Cell* **48**, 965–73.

Greenberg M.E., Ziff E.B., and Greene L.A. (1986). Stimulation of neuronal acetylcholine receptors induces rapid gene transcription. *Science* **234**, 80–3.

Grenningloh G., Rientiz A., Schmitt B., Methfessel C., Zenzen M., Beyreuther K., Gundelfinger E.D., and Betz H. (1987). The strychnine-binding subunit of the glycine receptor shows homology with nicotinic acetylcholine receptors. *Nature* **328**, 215–20.

Grunwald F., Schrock H., and Kuschinsky W. (1987). The effect of an acute nicotine infusion on the local glucose utilisation of the awake rat. *Brain Research* **400**, 232–8.

Guy H.R. (1984). A structural model of the acetylcholine receptor channel based on partition energy and helix packing calculations. *Biophysics Journal* **45**, 249–61.

Guy H.R. and Hucho F. (1987). The ion channel of the nicotinic acetylcholine receptor. *Trends in Neuroscience* **10**, 318–21.

Halvorsen S.W. and Berg D.K. (1986). Identification of a nicotinic acetylcholine receptor on neurons using an α-neurotoxin that blocks receptor function. *Journal of Neuroscience* **6**, 3405–12.

Hayashi E., Isogai M., Kagawa Y., Takayanagi N., and Yamada S. (1984). Neosurugatoxin, a specific antagonist of nicotinic acetylcholine receptors. *Journal of Neurochemistry* **42**, 1491–4.

Heidmann T., Oswald R.E., and Changeux J.P. (1983). Multiple sites of action for noncompetitive blockers on acetylcholine receptor-rich membrane fragments from *Torpedo marmorata. Biochemistry* **22**, 3112–27.

Henley, J. and Oswald, R. (1987). Two distinct (−)nicotine binding sites in goldfish brain. Identification and characterisation of putative neuronal nicotinic receptor subtypes. *Journal of Biological Chemistry* **262**, 6691–8.

Hery F., Bourgoin S., Hamon M., Ternaux J.P., and Glowinski J. (1977). Control of the release of newly synthesized ^{3}H-5-hydroxytryptamine by nicotinic and muscarinic receptors in rat hypothalamic slices. *Archives of Pharmacology* **296**, 91–7.

Higgins L.S. and Berg D.K. (1988). A desensitized form of neuronal acetylcholine receptor detected by [^{3}H]-nicotine binding on bovine adrenal chromaffin cells. *Journal of Neuroscience* **8**, 1436–46.

Holz W.R. and Senter R.A. (1981). Choline stimulates nicotinic receptors on adrenal medullary chromaffin cells to induce catecholamine secretion. *Science* **214**, 466–8.

Hopfield J.F., Tank D.W., Greengard P., and Huganir R.L. (1988). Functional modulation of the nicotinic acetylcholine receptor by tyrosine phosphorylation. *Nature* **336**, 677–80.

Hucho F. (1986). The nicotinic acetylcholine receptor and its ion channel. *European Journal of Biochemistry* **158**, 211–26.

Hucho F., Oberthur W., and Lottspeich F. (1986). The ion channel of the nicotinic acetylcholine receptor is formed by the homologous helices MII of the receptor subunits *FEBS Letters* **205**, 137–42.

Huganir R.L. and Greengard P. (1987). Regulation of receptor function by protein phosphorylation. *Trends in Pharmacological Sciences* **8**, 472–5.

Huganir R.L., Delcour A.H., Greengard P., and Hess G.P. (1986). Phosphorylation of the nicotinic acetylcholine receptor regulates its rate of desensitization. *Nature* **321**, 774–6.

Imoto K., Methfessel C., Sakmann B., Mishina M., Konno T., Fukuda K., Kurasaki M., Bujo H., Fujita Y., and Numa S. (1986). Location of a β-subunit region determining ion transport through the acetylcholine receptor. *Nature* **324**, 670–4.

Imoto K., Busch C., Sakmann B., Mishina M., Konno T., Nakai J., Bujo H., Mori Y., Fukuda K., and Numa S. (1988). Rings of negatively charged amino acids determine the acetylcholine receptor channel conductance. *Nature* **335**, 645–8.

Jacob M.H. and Berg D.K. (1983). The ultrastructural localization of alpha-bungarotoxin binding sites in relation to synapses on chick ciliary ganglion neurons. *Journal of Neuroscience* **3**, 260–71.

Kao P., Dwork A.J., Kaldany R.R.J., Silver M.L., Wideman J., Stein S., and Karlin A. (1984). Identification of the α-subunit half-cystine specifically labelled by an affinity reagent for the acetylcholine receptor binding site. *Journal of Biological Chemistry* **259**, 11662–5.

Karlin A. (1980). Molecular properties of nicotinic acetylcholine receptors. In *The cell surface and neuronal function* (C.W. Cotman, G. Poste, and G.L. Nicolson, ed.), pp. 191–260 Elsevier/North-Holland Biomedical Press, New York.

Katyare S.S. and Shallom J.M. (1988). Altered cerebral protein turnover in rats following prolonged *in vivo* treatment with nicotine. *Journal of Neurochemistry* **50**, 1356–63.

Kemp G. and Morley B. (1986). Ganglionic AChR's and high-affinity nicotinic binding sites are not equivalent. *FEBS Letters* **205**, 265–8.

Kemp G., Bentley L., McNamee M.G., and Morley B.J. (1985). Purification and characterization of the α-bungarotoxin binding protein from rat brain. *Brain Research* **347**, 274–83.

Ksir C., Hakan R.L., and Kellar K.J. (1987). Chronic nicotine and locomotor activity:

influences of exposure dose and test dose. *Psychopharmacology* **92**, 25–9.

Kubo T., Noda M., Takai T., Kayano T., Shimizu S., Tanaka K., Takahashi H., Hirose T., Inayama S., Kikuno R., Miyata T., and Numa S. (1985). Primary structure of delta subunit precursor of calf muscle acetylcholine receptor deduced from cDNA sequence. *European Journal of Biochemistry* **149**, 5–13.

Kumakura K., Karoum F., Guidotti A., and Costa E. (1980). Modulation of nicotinic receptors by opiate receptor agonists in cultured adrenal chromaffin cells. *Nature* **283**, 489–92.

Langley J.N. (1907). On the contraction of muscle, chiefly in relation to the presence of 'receptive' substances. *Journal of Physiology* **36**, 347–84.

Larsson C. and Nordberg A. (1985). Comparative analysis of nicotine-like receptor–ligand interactions in rodent brain homogenate. *Journal of Neurochemistry* **45**, 24–31.

Larsson C., Nilsson L., Halen A., and Nordberg A. (1986). Subchronic treatment of rats with nicotine: effects on tolerance and on [^{3}H]acetylcholine and [^{3}H] nicotine binding in the brain. *Drug and Alcohol Dependence* **17**, 37–45.

Larsson C., Lundberg P.-A., Halen A., Adem A., and Nordberg A. (1987). *In vitro* binding of [^{3}H]acetylcholine to nicotinic receptors in rodent and human brain. *Journal of Neuronal Transmission* **69**, 3–18.

Lindstrom J. (1986). Probing nicotinic acetylcholine receptors with monoclonal antibodies. *Trends in Neuroscience* **9**, 401–7.

Lindstrom J., Schoepfer R., and Whiting P. (1987). Molecular studies of the neuronal nicotinic acetylcholine receptor family. *Molecular Neurobiology* **1**, 281–337.

Lippiello P.M. and Fernandes K.G. (1986). The binding of *L*-[^{3}H] nicotine to a single class of high-affinity sites in rat brain membranes. *Molecular Pharmacology* **29**, 448–54.

Lippiello P.M., Sears S.B., and Fernandes K.G. (1987). Kinetics and mechanism of *L*-[^{3}H] nicotine binding to putative high-affinity receptor sites in the brain. *Molecular Pharmacology* **31**, 392–400.

Lippiello P.M. and Fernandes K.G. (1988). Identification of putative high-affinity nicotinic receptors on cultured cortical neurons. *Journal of Pharmacology and Experimental Therapeutics* **246**, 409–15.

London E.D., Waller S.B., and Wamsley J.K. (1985*a*). Autoradiographic localization of [^{3}H] nicotine binding sites in the rat brain. *Neuroscience Letters* **53**, 179–84.

London E.D., Connolly R.J., Szikszay M., and Wamsley J.K. (1985*b*). Distribution of cerebral metabolic effects of nicotine in the rat. *European Journal of Pharmacology* **110**, 391–2.

Loring R.H. and Zigmond R.E. (1988). Characterization of neuronal nicotinic receptors by snake venom neurotoxins. *Trends in Neuroscience* **11**, 73–8.

Lukas R.J. and Bennett E.L. (1979). Agonist-induced affinity alterations of a central nervous system α bungarotoxin receptor. *Journal of Neurochemistry* **33**, 1151–7.

Lukas R.J. and Bennett E.L. (1980). Chemical modification and reactivity of sulphydrals and disulphides of rat brain nicotinic-like acetylcholine receptors. *Journal of Biological Chemistry* **255**, 5573–7.

MacAllan D.R.E., Lunt G.G., Wonnacott S., Swanson K.L., Rapoport H., and Albuquerque E.X. (1988). Methyllycaconitine and anatoxin-a differentiate between nicotinic receptors in vertebrate and invertebrate nervous systems. *FEBS Letters* **226**, 357–63.

Marks M.J. and Collins A.C. (1982). Characterization of nicotine binding in mouse brain and comparison with the binding of α-bungarotoxin and quinuclidinyl benzylate. *Molecular Pharmacology* **22**, 554–65.

Marks M.J., Stitzel J.A., and Collins A.C. (1985). Time course study of the effects of chronic nicotine infusion on drug response and brain receptors. *Journal of Pharmacology and Experimental Therapeutics* **235**, 619–28.

Marks M.J., Stitzel J.A., Romm E., Wehner J.M., and Collins A.C. (1986). Nicotinic binding sites in rat and mouse brain: comparison of acetylcholine, nicotine and α-bungarotoxin. *Molecular Pharmacology* **30**, 427–36.

Martino-Barrows A.M. and Kellar K.J. (1987). [^{3}H]acetylcholine and [^{3}H]nicotine label the same recognition site in rat brain. *Molecular Pharmacology* **31**, 169–74.

McKinnon D., Boulter J., Wada E., Heinemann S., and Patrick J. (1988). Isolation of a cDNA clone for a new neuronal nicotinic acetylcholine receptor alpha-subunit. *Society for Neuroscience Abstracts* **14**, 95.10.

Meeker R.B., Michels K.M., Libber M.T., and Hayward J.N. (1986). Characteristics and distribution of high- and low-affinity alpha bungarotoxin binding sites in the rat hypothalamus. *Journal of Neuroscience* **6**, 1866–75.

McGeer E.G., McGeer P.L., Grewaal D.S., and Singh V.K. (1975). Striatal cholinergic interneurons and their relation to dopaminergic nerve endings. *Journal of Pharmacology (Paris)* **6**, 143–52.

Messing A. (1982). Cholinergic agonist-induced down regulation of neuronal α-bungarotoxin receptors. *Brain Research* **232**, 479–84.

Messing A. and Gonatas N.K. (1983). Extra-synaptic localization of α-bungarotoxin receptors in cultured chick ciliary ganglion neurons. *Brain Research* **269**, 172–6.

Miller M.M. and Billiar R.B. (1986). A quantitative and morphometric evaluation of [^{125}I]-alphabungarotoxin binding in the rat hypothalamus. *Brain Research Bulletin* **16**, 681–8.

Miller M.M., Silver J., and Billiar R.B. (1982). Effects of ovariectomy on the binding of [^{125}I]α-bungarotoxin to the suprachiasmatic nucleus of the hypothalamus: an *in vivo* autoradiographic analysis. *Brain Research* **247**, 355–64.

Miller M.M., Silver J., and Billiar R.B. (1984). Effects of gonadal steroids on the *in vivo* binding of [^{125}I]α-bungarotoxin to the suprachiasmatic nucleus. *Brain Research* **290**, 67–75.

Miller J.D., Murakami D.M., and Fuller C.A. (1987*a*). The response of suprachiasmatic neurons of the rat hypothalamus to photic and nicotinic stimuli. *Journal of Neuroscience* **7**, 978–86.

Miller M.M., Billiar R.B., and Beaudut A. (1987*b*). Ultrastructural distribution of alpha-bungarotoxin binding sites in the suprachiasmatic nucleus of the rat hypothalamus. *Cell and Tissue Research* **250**, 13–20.

Mills A. and Wonnacott S. (1984). Antibodies to nicotinic acetylcholine receptors used to probe the structural and functional relationship between brain αbungarotoxin binding sites and nicotinic receptors. *Neurochemistry International* **6**, 249–57.

Mishina M., Kurosaki T., Tobimatsu T., Mortimoto Y., Noda M., Yamamoto T., Terao M., Lindstrom J., Takahashi T., Kuno M., and Numa S. (1984). Expression of functional acetylcholine receptor from cloned cDNAs. *Nature* **307**, 604–8.

Mishina M., Tobimatsu T., Imoto K., Tanaka K., Fujita Y., Fukuda K., Kurasaki M., Takahashi H., Moritomo Y., Hirose T., Inayama S., Takahashi T., Kuno M., and Numa S. (1985). Location of functional regions of acetylcholine receptor α-subunit by site-directed mutageneis. *Nature* **313**, 364–9.

Molinaro, M.T. and Del-Rio, J. (1987). Substance P, nicotinic acetylcholine receptors and antinociception in the rat. *Neuropharmacology* **26**, 1715–20.

Morley B.J. and Fleck D.L. (1987). A time course and dose–response study of the regulation of brain nicotinic receptors by dietary choline. *Brain Research* **421**, 21–9.

Morley, B.J., Dwyer, D.S., Strang-Brown, P.F., Bradley, R.J., and Kemp, G.E. (1983). Evidence that certain peripheral anti-acetylcholine receptor antibodies do not interact with brain BuTX binding sites. *Brain Research* **262**, 109–16.

Mulac-Jericevic, B. and Atassi M.Z. (1987). α-Neurotoxin binding to acetylcholine receptor: localization of the full profile of the cobratoxin-binding regions on the α-chain of *Torpedo Californica* acetylcholine receptor by a comprehensive synthetic strategy. *Journal of Protein Chemistry* **6**, 365–73.

Murrin, L.C., Ferrer, J.R., Wanyun, Z., and Haley N.J. (1987). Nicotine administration to rats: methodological considerations. *Life Sciences* **40**, 1699–1708.

Nef, P., Oneyser, C., Alliod, C., Couturier, S., and Ballivet M. (1988). Genes expressed in the brain define three distinct neuronal nicotinic acetylcholine receptors. *EMBO Journal* **7**, 595–601.

Noda, M., Takahashi, H., Tanabe, T., Toyosata, M., Furutani, Y., Hirose, T., Asai, M., Inayama, S., Miyata, T., and Numa, S. (1982). Primary structure of α-subunit precursor of *Torpedo Californica* acetylcholine receptor deduced from a cDNA sequence. *Nature* **299**, 793–7.

Noda, M., Furutani, Y., Takahashi, H., Toyosato, M., Tanabe, T., Shimizu, S., Kikyotani, S., Kayano, T., Hirose, T., Inayama, S., and Numa, S. (1983*a*). Cloning and sequence analysis of calf cDNA and human genomic DNA encoding α-subunit precursor of muscle acetylcholine receptor. *Nature* **305**, 818–23.

Noda, M., Takahashi, H., Tanabe, T., Toyosata, M., Kikyotani, S., Furutani, Y., Hirose, T., Takashima, H., Inayama, S., Miyata, T., and Numa S. (1983*b*). Structural homology of *Torpedo Californica* acetylcholine receptor subunits. *Nature* **302**, 528–32.

Norman, R.I., Mehraban, F., Barnard, E.A., and Dolly, J.O. (1982). Nicotinic acetylcholine receptor from chick optic lobe. *Proceedings of the National Academy of Sciences USA* **79**, 1321–5.

Nordberg, A., and Larsson, C. (1980). Studies of muscarinic and nicotinic binding sites in brain. *Acta Physiological Scandinavia* Suppl. **479**, 19–23.

Nordberg, A., Wahlstrom, G., and Eriksson, B. (1985). Relations between muscimol, quinuclidinyl benzilate and nicotine binding sites in brain after very long treatment with ethanol in rats. *European Journal of Pharmacology* **115**, 301–4.

Ohta, Y., Ariyoshi, M., and Koketsu, K. (1984). Histamine as an endogenous antagonist of nicotinic ACh-receptor. *Brain Research* **306**, 370–3.

Oswald, R.E., and Freeman, J.A. (1981). α-Bungarotoxin binding and central nervous system nicotinic acetylcholine receptors. *Neuroscience* **6**, 1–14.

Paton, W.D.M., and Zaimis E.J. (1949). The pharmacological actions of polymethylene bistrimethylammonium salts. *British Journal of Pharmacology* **4**, 381–400.

Patrick, J., and Stallcup, B. (1977*a*). α-Bungarotoxin binding and cholinergic receptor function in a rat sympathetic nerve line. *Journal of Biological Chemistry* **252**, 8629–33.

Patrick, J., and Stallcup, W.B. (1977*b*). Immunological distinction between acetylcholine receptor and the α-bungarotoxin binding component on sympathetic neurons. *Proceedings of the National Academy of Sciences USA* **76**, 4689–92.

Perry, E.K., Perry, R.H., Smith, C.J., Dick, D.J., Candy, J.M., Edwardson, J.A., Fairbairn, A., and Blessed, J.G. (1987). Nicotinic receptor abnormalities in Alzheimer's and Parkinson's diseases. *Journal Neurology Neurosurgery and Psychiatry* **50**, 806–9.

Polz-Tejera, G., and Schmidt, J. (1983). Search for ligands of neuronal α-bungarotoxin receptors. *Biochemical and Biophysical Research Communications* **111**, 82–8.

Quik, M. (1982). Presence of an endogenous factor which inhibits binding of α-bungarotoxin 2.2 to its receptor. *Brain Research* **245**, 57–67.

Quik, M., Fournier, S., and Trifaro J.M. (1986). Modulation of the nicotinic α-bungarotoxin site in chromaffin cells in culture by a factor(s) endogenous to neuronal tissue. *Brain Research* **373**, 11–20.

Rapier, C., Harrison, R., Lunt, G.G., and Wonnacott, S. (1985). Neosurugatoxin blocks nicotinic acetylcholine receptors in the brain. *Neurochemistry International* **7**, 389–96.

Rapier, C., Wonnacott, S., Lunt, G.G., and Alburquerque, E.X. (1987). The neurotoxin histrionicotoxin interacts with the putative ion channel of the nicotinic acetylcholine receptors in the central nervous system. *FEBS Letters* **212**, 292–6.

Rapier, C., Lunt, G.G., and Wonnacott, S. (1988). Stereoselective nicotine-induced release of dopamine from striatal synaptosomes: concentration dependence and repetitive stimulation. *Journal of Neurochemistry* **50**, 1123–30.

Ratnam, M., Le Nguyen, D., Rivier, J., Sargent, P.B., and Lindstrom, J. (1986). Transmembrane topography of nicotinic acetylcholine receptor: immunochemical tests contradict theoretical predictions based on hydrophobicity profiles. *Biochemistry* **25**, 2633–43.

Ravdin, P.M., and Berg, D.K. (1979). Inhibition of neuronal acetylcholine sensitivity by α-toxins from *Bungarus multicintus* venom. *Proceedings of the National Academy of Sciences USA* **76**, 2072–6.

Reavill, C., Jenner, P., Kumar, R., and Stolerman, I.P. (1988). High-affinity binding of [^{3}H](–)nicotine to rat brain membranes and its inhibition by analogues of nicotine. *Neuropharmacology* **27**, 235–41.

Robinson, D., and McGee, R. (1985). Agonist-induced regulation of the neuronal nicotinic acetylcholine receptor of PC12 cells. *Molecular Pharmacology* **27**, 409–17.

Romano, C., and Goldstein, A.M. (1980). Stereospecific nicotine receptors in rat brain membranes. *Science* **210**, 647–9.

Rowell, P.P. (1987). Current concepts on the effects of nicotine in the central nervous system. In *Tobacco smoking and health* (W.R. Martin, G.R. Van Loon, E.T. Iwamoto, and L. Davis, ed.), pp. 191–207 Plenum, New York.

Rowell, P.P. and Winkler, D.L. (1986). Nicotinic stimulation of [^{3}H]acetylcholine release from mouse cerebral cortical synaptosomes. *Journal of Neurochemistry* **43**, 1593–8.

Rowell, P.P., Carr, L.A., and Garner, A.C. (1987). Stimulation of [^{3}H]dopamine release by nicotine in rat nucleus accumbens. *Journal of Neurochemistry* **49**, 1449–54.

Sakmann, B., Methfessel, C., Mishina, M., Takahashi, T., Takai, T., Kurasaki, M., Fukuda, K., and Numa, S. (1985). Role of acetylcholine receptor subunits in gating of the channel. *Nature* **318**, 538–43.

Salvaterra, P.M., Gurd, J.M., and Mahler, H.R. (1977). Interaction of the nicotinic acetylcholine receptor from rat brain with lectins. *Journal of Neurochemistry* **29**, 345–8.

Schleifer, L.S., and Eldefrawi, M.M.E. (1974). Identification of the nicotinic and muscarinic acetylcholine receptors in subcellular fractions of mouse brain. *Neuropharmacology* **13**, 53–63.

Schmidt, J., Hunt, S., and Polz-Tejera, G. (1980). Nicotinic receptors of the central and autonomic nervous system. In *Neurotransmitters, receptors and drug action*, (W.B. Essman, ed.), pp. 1–45 Spectrum, New York.

Schoepfer, R., Whiting, P., Esch, F., Blacher, R., Shimasaki, S., and Lindstrom, J. (1988). cDNA clones coding for the structural subunit of a chicken brain nicotinic acetylcholine receptor. *Neuron* **1**, 241–8.

Schoepfer, R., Whiting, P., Conroy, W.G., and Lindstrom, J. (1989). cDNA clones coding for subunits of the neuronal αbungarotoxin-binding protein. *Society for Neuroscience Abstracts* **15**.

Schofield, P.R., Darlison, M.G., Fujita, N., Burt, D.R., Stephenson, F.A., Rodriquez, H., Rhee, L.M., Ramachandran, J., Reale, V., Glencourse, T.A., Seeburg, P.H., and Barnard, E.A. (1987). Sequence and functional expression of the GABA-A receptor shows a ligand-gated receptor super-family. *Nature* **328**, 221–7.

Schulz, D.W., and Zigmond, R.E. (1989). Neuronal bungarotoxin antagonises nicotinic function in rat caudate-putamen. *Neuroscience Letters*, **98**, 310–6.

Schwartz, R.D., and Kellar, J. (1985). *In vivo* regulation of [^{3}H]acetylcholine recognition sites in brain by nicotinic cholinergic drugs. *Journal of Neurochemistry* **45**, 427–33.

Schwartz, R.D., McGee, R., and Kellar, K.J. (1982). Nicotinic cholinergic receptors labelled by [^{3}H]acetylcholine in rat brain. *Molecular Pharmacology* **22**, 56–62.

Schwartz, R.D., Lehman, J., and Kellar, K.J. (1984). Presynaptic nicotinic cholinergic receptors labelled by [^{3}H]acetylcholine on catecholamine and serotonin axons in brain. *Journal of Neurochemistry* **42**, 1495–8.

Sershen, H., Reith, M.E.A., Lajtha, A., and Gennaro, J. (1981). Noncholinergic, saturable binding of (+ −)-[^{3}H] nicotine to mouse brain. *Journal of Receptor Research* **2**, 1–15.

Sershen, H., Reith, M.E.A., Hashim, A., and Lajtha, A. (1984). Endogenous material in brain inhibiting [^{3}H] nicotine and [^{3}H]acetylcholine binding. *Journal of Neuroscience Research* **12**, 563–9.

Seto, A., Arimatsu, Y., and Amano, T. (1981). Subunit structure of α-Bungarotoxin binding component in mouse brain. *Journal of Neurochemistry* **37**, 210–6.

Shallom, J.M., and Katyare, S.S. (1985). Altered synaptosomal ATPase activity in rat brain following prolonged *in vivo* treatment with nicotine. *Biochemical Pharmacology* **34**, 3445–9.

Sloan, J.W., Martin, W.R., Hernandez, J., and Hook, R. (1985). Binding characteristics of (−)- and (+)-nicotine to the rat brain P2 fraction. *Biochemistry, Pharmacology and Behavior* **23**, 987–93.

Snutch, T.P. (1988). The use of *Xenopus* oocytes to probe synaptic communications. *Trends in Neuroscience* **11**, 250–6.

Steinbach, J.H., and Ifune, C. (1989). How many kinds of nicotinic acetylcholine receptor are there? *Trends in Neuroscience* **12**, 3–6.

Sterz, R., Peper, K., Simon, J., Ebert, J.P., Edge, M., Pagala, M., and Bradley, R.J. (1986). Agonist and blocking effects of choline at the neuromuscular junction. *Brain Research* **385**, 99–114.

Sumikawa, K., Houghton, M., Smith, J.C., Bell, L., Richards, B.H., and Barnard, E.A. (1982). The molecular cloning and characterization of cDNA coding for the α-subunit of the acetylcholine receptor. *Nucleic Acids Research* **10**, 5809–22.

Swanson L., Simmons D., Whiting P., and Lindstrom J. (1987). Immunohistochemical localization of neuronal nicotinic receptors in the rodent central nervous system. *Journal of Neuroscience* **7**, 3334–42.

Vige, X., and Briley, M. (1988). Scopolamine induces up-regulation of nicotinic receptors in intact brain but not in nucleus basalis lesioned rats. *Neuroscience Letters* **88**, 319–24.

Vincek, W.C., Martin, B.R., Aceto, M.D., and Bowman, E.R. (1980). Synthesis and preliminary binding studies of 4.4-ditritio-(–)-nicotine of high specific activity. *Journal of Medicinal Chemistry* **23**, 962–4.

Wada, K., Ballivet, M., Boulter, J., Connolly, J., Wada, E., Deneris, E.S., Swanson, L.W., Heinemann, S., and Patrick, J. (1988). Functional expression of a new pharmacological subtype of brain nicotinic acetylcholine receptor. *Science* **240**, 330–4.

White, M.M. (1985). Designer channels: site-directed mutagenesis as a probe for structural features of channels and receptors. *Trends in Neuroscience* **8**, 364–8.

Whitehouse, P.J., Martino, A.M., Autuono, P.G., Lowensstein, P.R., Coyle, J.T., Price, D.L., and Kellar, K.J. (1986). Nicotinic acetylcholine binding in Alzheimer's disease. *Brain Research* **371**, 146–51.

Whiting, P.J., and Lindstrom, J.M. (1986). Purification and characterization of a nicotinic acetylcholine receptor from chick brain. *Biochemistry* **25**, 2082–93.

Whiting, P., and Lindstrom, J. (1987*a*). Purification and characterization of a nicotinic acetylcholine receptor from rat brain. *Proceedings of the National Academy of Sciences USA* **84**, 595–9.

Whiting, P., and Lindstrom, J. (1987*b*). Affinity labelling of neuronal acetylcholine receptors localizes acetylcholine-binding site to their β-subunits. *FEBS Letters* **213**, 55–60.

Whiting, P., and Lindstrom, J. (1988). Characterization of bovine and human neuronal nicotinic acetylcholine receptors using monoclonal antibodies. *Journal of Neuroscience* **8**, 3395–404.

Whiting, P.J., Liu, R., Morley, B.J., and Lindstrom, J.M. (1987*a*). Structurally different neuronal nicotinic acetylcholine receptor subtypes purified and characterized using monoclonal antibodies. *Journal of Neuroscience* **7**, 4005–16.

Whiting, P., Esch, F., Shimasaki, S., and Lindstrom J. (1987*b*). Neuronal nicotinic acetylcholine receptor β-subunit is coded for by the cDNA clone α4. *FEBS Letters* **219**, 459–63.

Whiting, P.J., Schoepfer, R., Swanson, L.W., Simmons, D.M., and Lindstrom, J.M. (1987*c*). Functional acetylcholine receptor in PC12 cells reacts with a monoclonal antibody to brain nicotinic receptors. *Nature* **327**, 515–7.

Williams, M., and Robinson, J. (1984). Binding of the nicotinic cholinergic antagonist dihydro-β-erythroidine to rat brain tissue. *Journal of Neuroscience* **4**, 2906–11.

Wilson, P.T., and Lentz, T.L. (1988). Binding of α-bungarotoxin to synthetic peptides corresponding to residues 173–204 of the α-subunit of *Torpedo*, calf and human acetylcholine receptor and restoration of high-affinity binding by sodium dodecyl sulfate. *Biochemistry* **27**, 6667–74.

Wolf, K.M., Ciarlegio, A., and Chiappinelli, V.A. (1988). K-Bungarotoxin: binding of a neuronal nicotinic receptor antagonist to chick optic lobe and skeletal muscle. *Brain Research* **439**, 249–58.

Wonnacott, S. (1986). α-Bungarotoxin binds to low-affinity nicotine binding sites in rat brain. *Journal of Neurochemistry* **47**, 1706–12.

Wonnacott, S. (1987*a*). Neurotoxin probes for neuronal nicotinic receptors. In

Neurotoxins and their pharmacological implications. (P. Jenner, ed.), pp. 209–231 Raven Press, New York.

Wonnacott, S. (1987*b*). Brain nicotine binding sites. *Human Toxicology* **6**, 343–53.

Wonnacott, S., Harrison, R., and Lunt, G.G. (1982). Immunological cross-reactivity between the α-bungarotoxin binding component from rat brain and nicotinic acetylcholine receptor. *Journal of Neuroimmunology* **3**, 1–13.

Wonnacott, S., Irons, J., Lunt, G.G., Rapier, C.M., and Albuquerque, E.X. (1988). α-Bungarotoxin and presynaptic nicotinic receptors: functional studies. In *Nicotinic acetylcholine receptors in the nervous system* (F. Clementi, C. Gotti, and E. Sher, ed.) pp. 41–60 Springer-Verlag, Heidelberg.

Yamada, S., Kagawa, Y., Ushijima, H., Takayanagi, N., and Hayashi, E. (1986). Effect of piperidine and related alicyclic amines on nicotinic and muscarinic agonist binding sites in the mammalian brain. *Journal of Pharmacobio-Dynamics* **9**, 620–5.

Yamada, S., Gehlert, D.R., Hawkins, K.M., Nakayama, K., Roeske, W.R., and Yamamura, H.I. (1987). Autoradiographic localization of nicotinic receptor binding in rat brain using [^{3}H]methylcarbamylcholine, a novel radioligand. *Life Sciences* **41**, 2851–61.

Yamanaka, K., Muramatsu, I., and Kigoshi S. (1987). Muscarinic agonist binding in rat brain following chronic nicotine treatment. *Brain Research* **409**, 395–7.

Yamanaka, K., Oshita, M., and Muramatsu, I. (1985). Alteration of alpha and muscarinic receptors in rat brain and heart following chronic nicotine treatment. *Brain Research* **348**, 241–8.

Yoshida, K., and Imura, H. (1979). Nicotinic cholinergic receptors in brain synaptosomes. *Brain Research* **172**, 453–9.

Yoshida, K., Kato, Y., and Imura, H. (1980). Nicotine-induced release of noradrenaline from hypothalamic synaptosomes. *Brain Research* **182**, 361–8.

Yoshida, K., Engel, J. and Lilequist, S. (1982). The effect of chronic ethanol administration on high affinity ^{3}H-nicotine binding brain. *Naunyn Schmiedeberg's Archives of Pharmacology* **321**, 74–6.

Zatz, M., and Brownstein, M.J. (1979). Intraventricular carbachol mimics the effects of light on the circadian rhythm in the rat pineal gland. *Science* **203**, 358–61.

Zatz, M., and Brownstein, M.J. (1981). Injection of α-bungarotoxin near the suprachiasmatic nucleus blocks the effects of light on nocturnal pineal enzyme activity. *Brain Research* **213**, 438–42.

Zhang, X., Stjernlof, P., Adem, A., and Nordberg, A. (1987). Anatoxin-a: a potent ligand for nicotinic cholinergic receptors in rat brain. *European Journal of Pharmacology* **135**, 457–8.

8 Behavioural pharmacology of nicotine in animals

I.P. Stolerman

8.1 Introduction

Nicotine resembles many other psychoactive drugs in the sense that it has a very wide range of psychological effects. In this chapter, some of the more important of these effects are discussed in order, firstly, to assess what they can tell us that is relevant to the uses of nicotine and, secondly, to see how far the psychological effects can be related to the neurophysiological and biochemical actions of nicotine that have been surveyed elsewhere in the volume.

The major human use of nicotine has, of course, been in the form of tobacco, and many attempts have been made to determine a psychopharmacological basis for this use. Interest in the nicotinic mechanisms of the CNS has recently broadened to include other psychiatric and neurological disorders that are associated with disturbances in functions of the neurotransmitter acetylcholine. In the peripheral nervous system, most of the effects of nicotine are mediated through a subpopulation of acetylcholine receptors that are, for that reason, defined as nicotinic-cholinergic sites (in contrast to muscarinic-cholinergic sites). Knowledge of the structure and function of peripheral nicotinic receptors is more advanced than that of the receptor for any other neurotransmitter; the structure is largely established at the molecular level and its three-dimensional form in relation to the cell membrane has been summarized by Wonnacott (chapter 7, this volume). Whether similar nicotinic-cholinergic receptors exist in the central nervous system (CNS), and whether such sites constitute the main or only receptors for nicotine, has been the focus of much attention. A related question is whether the CNS sites upon which nicotine acts are heterogeneous; a variety of subtypes of receptor would provide a sound basis for the development of drugs with effects much more selective than nicotine and which might, therefore, have novel therapeutic applications. The effects of nicotine may also be mediated indirectly, through the release of other neurotransmitters.

Much knowledge of the psychological effects of nicotine has come from quantifying its effects on the behaviour of either trained or untrained animal subjects. One approach has involved studies of ways in which the effects of nicotine can function as stimuli in the psychological sense; by this means, it has been possible to develop experimental models for the rewarding, aversive, and other effects of nicotine that may be very relevant to the problem of tobacco dependence. Thus the effects of drugs such as nicotine can serve as internal, positively-reinforcing (rewarding) stimuli in much the same way that food can serve as a reward, and this effect can lead to behavioural dependence as shown by self-administration (Schuster and Thompson 1969). The general principles underlying applications of these methods to the behavioural analysis of drug dependence have been surveyed elsewhere (Goldberg and Stolerman 1986). Stolerman (1987) and Henningfield and Goldberg (1984) have reviewed the work on nicotine in greater detail than is possible in the space available now. Some of the procedures can serve as assays for studying correlations between biochemical and behavioural effects.

Although it is of fundamental significance to know that nicotine can be self-administered by virtue of its positively reinforcing effect, the nature and mechanism of reinforcement needs further elucidation. Studies of nicotine on a wider range of behaviours provide valuable guides to the nature of its effects, and thus are complementary to knowledge of its properties as a stimulus. Effects on both conditioned and unconditioned behaviours may both provide vital clues. It is important to distinguish between behavioural effects *per se*, such as changes in response rates, and inferred changes in psychological processes, such as attention, learning, and memory. The latter are much more difficult to establish convincingly. Very frequently, the practical, human problem is expressed in terms of these processes, in which case great care must be exercised when extrapolating from the results of specific experiments. General reviews of this work have been published (Clarke 1987; Stolerman 1987).

8.2 The stimulus properties of nicotine

8.2.1 *Positive Reinforcement*

The finding that pure nicotine can serve as a positive reinforcer is at the core of the present understanding of tobacco use as nicotine dependence. It shows an effect common to nicotine and classical misused drugs such as opiates, psychomotor stimulants, and depressants (Jaffe, chapter 1, this volume). It provides the best available model for studying the primary addiction potential of drugs in animal experiments and it is reviewed fully by Swedberg *et al.* (chapter 2, this volume). Remarkably, it has taken some 10–15 years of research to identify the conditions under which animal subjects respond reliably, at high rates, to obtain nicotine. The main differences between nicotine and the classical misused drugs demonstrated by this work have been the surprisingly narrow range of conditions in which nicotine is reinforcing, and the slow rate of acquisition. There is very limited evidence that rats show a preference for distinctive environments that have been associated with nicotine (Fudala and Iwamoto 1986), and other carefully controlled studies with this technique have yielded negative results (Clarke and Fibiger 1987). It may be that some key factor which induces or predisposes people to smoke has not yet been adequately modelled in the experiments. Warburton (chapter 3, this volume) argues that people smoke to obtain improvements produced by nicotine in the performance of certain tasks. This idea is compatible with the belief that the positive reinforcing effect of nicotine underlies smoking, and indicates that it may have a psychological basis.

As with other behavioural findings, equivalence of doses with those received by tobacco users is an important consideration. The smallest dose of nicotine self-administered (intravenously) by monkeys is 0.03 mg kg^{-1} (Swedberg *et al.* chapter 2, this volume); this dose seems about 30 times larger than the 0.001 mg kg^{-1} doses that tobacco users may typically obtain by inhaling the smoke from a single puff, although not much larger than the total dose from a cigarette. However, a better criterion for equivalence would be the comparative plasma concentrations of nicotine in monkey and man, and the animal data are not available to enable this comparison to be made.

8.2.2 *Aversive effects*

Nicotine has an aversive as well as a positive reinforcing effect and this can be shown in four paradigms for assessing aversive effects of drugs. Infusions of nicotine can be used to punish responding for food rewards in squirrel monkeys, in much the same way that brief electric shocks can serve as punishing stimuli (Goldberg and Spealman 1983). Monkeys will also work to prevent the occurrence of nicotine infusions or to terminate infusions that are in progress; thus nicotine can serve as a negative reinforcer, maintaining

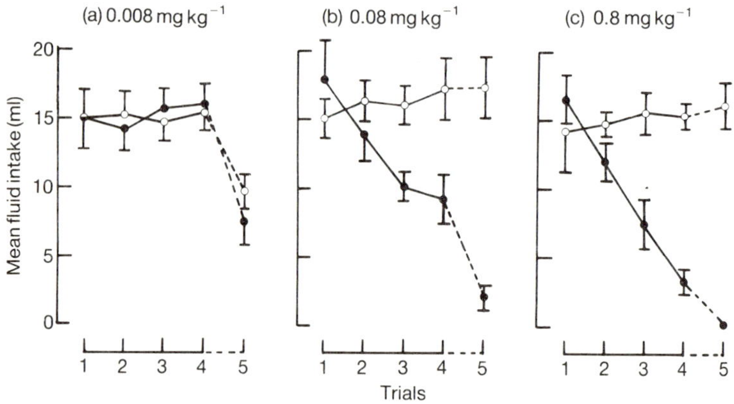

Fig. 8.1. Conditioned taste aversions to nicotine-paired flavoured solutions in three groups of rats (●, n = 8). In the same rats, intakes of saline-paired flavoured solutions were not suppressed (○). Trials 1–4 were conditioning sessions. Trial 5 was a different and more sensitive type of test in which both flavoured solutions were presented simultaneously, hence the apparent fall in consumption of each solution in (a). Vertical bars show ± standard error of mean. Overlapping bars and those smaller than diameters of symbols are omitted for clarity (from Kumar *et al.* 1983).

escape and avoidance behaviour in a manner similar to termination of electric shock or other noxious events (Spealman 1983). Thirsty rodents will not consume normally palatable fluids when their consumption has been associated with administrations of nicotine, thus showing the so-called 'conditioned taste aversion' that has been used as an index of aversive effects (Kumar *et al.* 1983; Iwamoto and Williamson 1984). However, conditioned taste-aversion experiments can be interpreted without assuming that the drug used is aversive (Stolerman and D'Mello 1981). Finally, there is some recent evidence that rodents will stay away from distinctive environments that have been associated with nicotine (Fudala and Iwamoto 1987).

The doses of nicotine that have aversive effects are in the range of 0.03–0.10 mg kg^{-1} in monkeys, and these are the same doses that also have positive reinforcing effects (see above). The equivalence of dose with that in the human situation has been demonstrated more clearly in rats; Fig. 8.1 shows that small doses of nicotine can produce conditioned taste aversions; Fig. 8.2(c) shows that these doses are associated with plasma concentrations of nicotine similar to those found in cigarette smokers who inhale (Kumar *et al.* 1983).

The ability of nicotine to act as either a positive reinforcer or as an aversive stimulus depending on the circumstances surrounding its administration is not unique; it is a characteristic shared with many other psychoactive addictive agents, including amphetamine, cocaine, and morphine (reviewed by Stolerman and D'Mello 1981). However, the contrast is perhaps more striking with

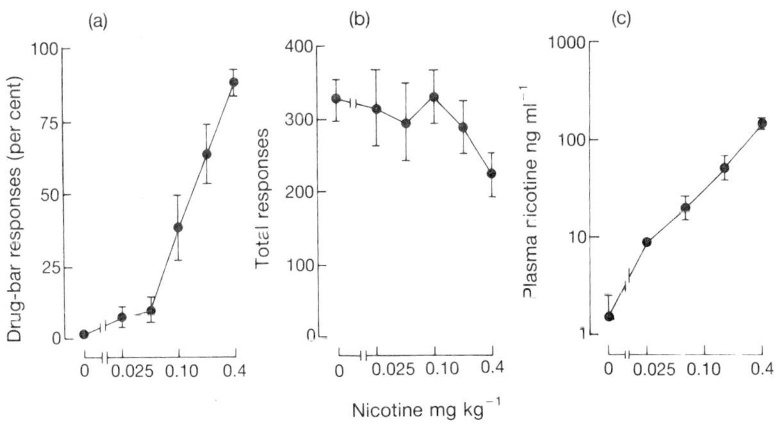

Fig. 8.2. Dose–response curves in rats trained to discriminate 0.4 mg kg^{-1} of nicotine from saline (n = 6–8). The discriminative stimulus (cue) effect of nicotine is shown in (a) by dose-related increases in percentages of responding on the drug-appropriate bar. The weak reduction in overall response rate is shown in (b) by total numbers of responses on both bars. Corresponding mean plasma nicotine concentrations are shown in (c). All injections (subcutaneous) 15 min before sampling behaviour or plasma. Vertical bars show ± standard error of mean. Overlapping bars and those smaller than diameters of symbols are omitted for clarity (from Pratt *et al.* 1983).

nicotine than with most other drugs because its aversive effect is so clearly established by several methods. The aversive effect of nicotine may set an upper limit to smoke exposure in humans, as discussed by Russell (chapter 11, this volume). Little is known about the occurrence of tolerance to aversive effects of nicotine although such a phenomenon could usefully be explored as a potential model for tolerance to subjectively unpleasant effects in humans, such as nausea and vomiting.

8.2.3 *Discriminative stimulus properties*

Nicotine can serve as a discriminative stimulus (cue) in conditioning experiments. The subjects of these experiments are trained to report, by making one behavioural response rather than another, whether or not they have received a dose of nicotine. Nicotine can be reliably distinguished from saline in animal experiments. This effect of nicotine does not model its primary addictive potential, but merely indicates that the subjects can identify the presence of the drug in the body. However, the nicotine cue may have a role in relapse to tobacco use after periods of cessation. Any exposure to nicotine in the former tobacco user will create part of the stimulus situation previously associated with smoking, and thus might serve to facilitate resumption of the habit. Such effects have been demonstrated with other addictive agents (de Wit and Stewart 1983) but work with nicotine is lacking. It has also been suggested that the performance of other learned tasks may be cued to and thus facilitated by nicotine, an effect called state-dependent learning (Rosecrans *et al.* 1978; Lowe 1986; Peters and McGee 1982). However, any state-dependent effects seem to contribute only a minor part of the total improvement in performance induced by nicotine (Warburton, chapter 3, this volume). The main value of the nicotine discriminative stimulus has been as a behavioural assay.

Early work by Morrison and Stephenson (1969) and Rosecrans and his colleagues (Rosecrans *et al.* 1978) established the nicotine cue as an exceptionally reliable method for assessing the sensitivity of animal subjects to central effects of nicotine. A major achievement was the clear demonstration that central muscarinic and nicotinic receptors were totally distinguishable; drugs acting upon each of these subtypes of acetylcholine receptor had completely different discriminative stimulus effects, and muscarinic and nicotinic antagonists had the selective blocking actions expected from their classical pharmacology at sites in the autonomic nervous system.

Figure 8.2 illustrates discriminative effects of nicotine in rats in relation to changes in overall rates of bar-pressing (a non-discriminative index) and to plasma concentrations of nicotine. These animals were trained to press bars to obtain food rewards. After administration of nicotine ($0.4\,mg\,kg^{-1}$ subcutaneously), responses on one of two bars were rewarded, whereas presses on the other bar had no consequences (Pratt *et al.* 1983). After administration of saline, these arrangements were reversed. Training sessions with nicotine

Table 8.1. Characteristics of the nicotine cue as determined from tests of responses to nicotinic and non-nicotinic drugs in rats trained to discriminate nicotine from saline[a]

Test drug	Comments
Drugs that produce the nicotine cue	
Anabasine	Minor tobacco alkaloid
Cytisine	Laburnum alklaloid
Isoarecolone H C l	Synthetic nicotine congener
(–)-Nicotine	The tobacco alkaloid
(+)-Nicotine	Synthetic stereoisomer
(–)-Nornicotine	Synthetic nicotine congener
(+)-Nornicotine	" " "
N-(3-pyridylmethyl)pyrrolidine	" " "
Amphetamine, apomorphine, SKF 38393	Partial responses only[c]
Drugs that do not produce the nicotine cue	
Cotinine	Major metabolite of nicotine
Nicotine methiodide	Does not penetrate to CNS
Piperidine	Proposed endogenous ligand
Mecamylamine, hexamethonium[b],	Ganglion-blocking drugs
Chlorisondamine[b], pentolinium[b]	" " "
Decamethonium[b]	Neuromuscular blocker
Dihydro-β-erythroidine[b]	Non-selective nicotinic blocker
Lobeline[b], 2- and 4-nicotine	Inactive nicotine analogues
Oxotremorine, arecoline	Muscarinic agonists
Atropine	Muscarinic antagonist
Cocaine	Psychomotor stimulant
Droperidol, haloperidol,	Dopamine antagonists
pimozide, Sch 23390	" "
Clenbuterol	β-adrenoceptor agonist
Clonidine	α-adrenoceptor agonist
Fenfluramine, quipazine	5-HT agonists
Methergoline	5-HT antagonist
Pentobarbitone	Sedative-hynotic agent
Chlordiazepoxide, midazolam	Benzodiazepine tranquillizers
FG-7142	Benzodiazepine inverse agonist
Morphine	Opiate analgesic
Phencyclidine	Dissociative anaesthetic
Physostigmine	Anticholinesterase
1-Phenylisopropyladenosine	Adenosine agonist
Dimenhydrinate, diphenhydramine	Antihistamines

[a] Modified from Stolerman (1987); Reavill *et al.* (1987, 1988); Goldberg *et al.* (1989), and unpublished data

[b] Includes direct, intracerobroventricular administration

[c] Discussed fully by Reavill (chapter 9, this volume).

and saline took place in random order until a criterion of discriminative performance was attained. In trained animals, the numbers of responses on the bar appropriate for nicotine were strongly related to the dose of nicotine, whereas the non-discriminative index was relatively insensitive. Nicotine in a dose of 0.1 mg kg^{-1} is associated with a mean plasma nicotine concentration of 35 ng ml^{-1}, which is in the middle of the range found for cigarette smokers who inhale (Russell, chapter 11, this volume). This very small dose of nicotine has been used as the standard in subsequent work. In tests with a wide range of different drugs, a number of nicotinic-cholinergic agonists and compounds structurally related to nicotine have been identified by rats as producing full nicotine-like effects (Table 8.1). In contrast, many different drugs from other pharmacological classes, such as opiates and tranquillizers, have not been recognized as like nicotine. Some dopamine agonists produce partial nicotine-like discriminative effects (Table 8.1), and the significance of these findings is discussed by Reavill (chapter 9, this volume). The effect of nicotine in the rat is not mediated through its major metabolite, cotinine (Goldberg *et al.* 1989).

8.3 Rates of conditioned behaviour

Nicotine has both facilitating and inhibitory effects on a wide range of different behaviours, in addition to its roles as a stimulus in the situations outlined above. These effects may provide clues as to how nicotine serves as a positive reinforcer and to its mechanism of action. For convenience, these behaviours are classified here as conditioned and unconditioned, but this is not intended to imply that there is a fundamental difference in the effects of the drug across these classes, or even that there is a sharp dividing line between them. There is evidence that metabolites of nicotine (cotinine and nornicotine) may make minor contributions to changes in response rates produced by the parent compound, but there are marked differences in the effects and potencies of these agents in different species of animal (Risner *et al.* 1985; Goldberg *et al.* 1989). Benowitz *et al.* (chapter 4, this volume) have reviewed studies on metabolites in humans in detail.

8.3.1 *Dose-response relations*

Nicotine has a complex range of effects that are dependent on factors that have been only partly identified. These findings may be exemplified by the changes in rates of responding to obtain rewards such as food pellets. Small doses of nicotine may increase response rates, up to maximum value which is rather less than the maximum effect of the classical psychomotor stimulants such as amphetamine (Fig. 8.3). Further increases in doses are then associated with marked reductions in the rate of responding, which may fall well below control rates. Sufficient doses of nicotine may completely suppress all responding; this effect of nicotine on responding is not pharmacologically specific, in the sense that large doses of almost any drug (even including amphetamine),

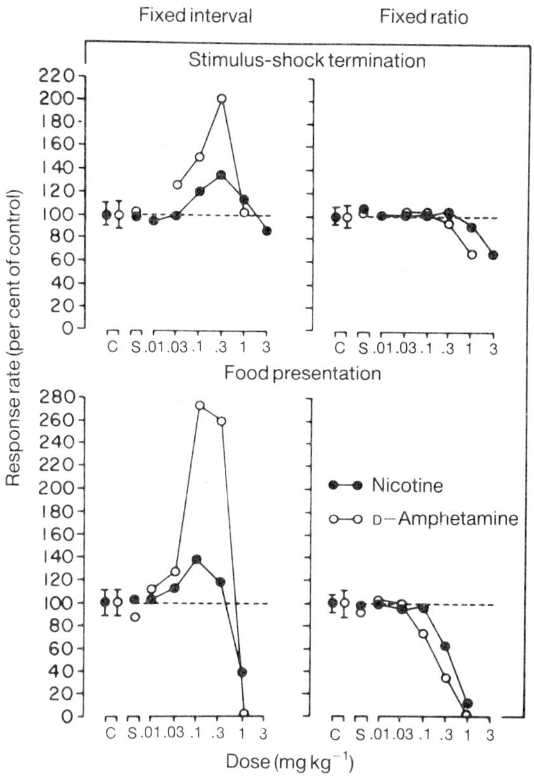

Fig. 8.3. Dose–response curves in monkeys (n = 3–4) trained to respond under multiple fixed-interval fixed-ratio schedules of stimulus-shock avoidance (upper panels) or food reinforcement (lower panels). Points above C and the horizontal dashed lines show the mean non-injection control rates of responding (taken as 100 per cent) and vertical bars show ± standard deviation. Points above S show effects of saline injections (from Spealman *et al.* 1981).

can disrupt patterns of behaviour that proceed with exquisite efficiency in undrugged subjects (Morrison 1967; Spealman *et al.* 1981). However, it is interesting to note that in cigarette smokers, the dose–response curve for the effects of nicotine on an electroencephalographic index related to arousal, the contingent negative variation, has a bimodal shape (Ashton *et al.* 1980). This effect is discussed further by Clarke (chapter 5, this volume).

The complex nature of nicotine's effects is a feature shared with many psychoactive drugs when tested under similar conditions. The decreases in rates of responding cannot be used as evidence for a tranquillizing action of nicotine that has been inferred mainly from smokers' self-reports (Gilbert 1979). In other experiments, nicotine has been examined in procedures specific for standard drugs widely used to treat anxiety neuroses. For example, anxiolytic drugs can reverse the suppression of behaviour produced by punishing stimuli. Nicotine has generally been inactive in such 'animal models' for anxiety (Balfour *et al.* 1986; Morrison and Stephenson 1970), although an atypical anxiolytic effect cannot be precluded. The activity of nicotine in certain procedures involving stress-inducing stimuli may be relevant here, but these procedures have not been validated as models for a human condition. Clarke (1987) has reviewed these studies.

Some degree of tolerance can develop to the depressant effects on rates of responding, but increases in response rates appear more resistant to tolerance; this matter is covered more fully by Swedberg *et al.* (chapter 2, this volume).

8.3.2 *Biphasic time-course*

Dose is not the only factor that determines the nature of the drug effect. Decreases in response rates are more marked in the first 30 min after administration of nicotine (by systemic injection) than in the remaining 60 min or so in which it acts. Thus certain doses of nicotine decrease rates of behaviour for 30 min, these decreases being followed by recovery and even increases in rate. Generally, the larger the dose of nicotine, the greater and more prolonged is the decreased responding. The nature of the behavioural baseline also determines the magnitude and even the direction of the drug effect. Thus responding at high rates (maintained by rewarding high rates on a 'fixed-ratio' schedule), is mainly inhibited by nicotine. In contrast, responding at lower rates (maintained by removing or reversing the relationship between response rates and reward in 'fixed-interval' and 'differential reinforcement of low rate' schedules), can be markedly facilitated by nicotine, but even this facilitation is lost at large doses of the drug (Spealman *et al.* 1981). There is no evidence that the effects of nicotine at different times after its administration are mediated through different types of receptor.

8.3.3 *Brain mechanisms of reward*

There has been much interest in the effects of addictive drugs, nicotine among them, on brain mechanisms underlying reinforcement and reward processes. One approach to these problems has examined the effects of nicotine on behaviour reinforced by intracranial electrical stimulation, where subjects press bars to obtain brief electrical stimuli delivered to areas of the brain thought to be involved in reward. Although nicotine alters the rate of responding for brain stimulation reward (Pradhan and Bowling 1971), it has not been established that this is a specific drug effect on the reward process, since nicotine has marked effects on response rates regardless of the type of reinforcer. Some ingenious studies have attempted to use 'rate-free' measures of reward, but nicotine has not been consistently effective in such paradigms (Clarke and Kumar 1983*b*, 1984*b*; Schaeffer and Michael 1986). Generally, the effects of nicotine seem not to be greatly dependent on the nature of the reward maintaining behaviour; results for food-reinforced behaviour have been similar to those for avoidance of, or escape from, shock, precluding, interpretations in motivational terms (for example, appetite) specific to particular rewards (Fig. 8.3).

8.3.4 *Learning and memory*

Studies have examined the effects of nicotine on the processes of learning and memory. Hunter *et al.* (1977) and Clarke (1987) have reviewed much of this work. The early studies showed that nicotine administered before behavioural tests could facilitate acquisition and performance of active avoidance and maze-learning tasks (Domino 1965: Bovet *et al.* 1966). Experiments of this type probably reflect general increases in locomotor activity or response rates produced by nicotine, effects that have nothing to do with learning or memory. Other interpretations of these findings could emphasize possible effects on arousal and attention to relevant stimuli, rather then associative factors.

Techniques to circumvent these problems have included measuring the accuracy of performance rather than simple rates of responding, and administering the drug *after* training, with the aim of facilitating the consolidation of newly-acquired information into long-term memory. Testing then takes place some days later, after nicotine is cleared from the system, so that direct effects on activity or arousal are excluded. Some experiments have yielded evidence that nicotine can enhance animal learning or memory (Garg and Holland 1968; Garg 1969), although this work employed only a single dose of nicotine that was slightly above the 'smoking' range. Other studies have used small 'smoking' doses of nicotine, and improvements in learning and memory have been found with some consistency (Battig 1970; Evangelista *et al.* 1970; Erickson 1971). However, studies have not demonstrated a clear dose–response relationship except in the work of Flood *et al.* (1981), which utilised 1,1,-dimethyl-4-phenylpiperazinium (DMPP), a nicotine analogue that has to be administered by intracerebroventricular injection since it does not penetrate into the brain after systemic administration. Figure 8.4 shows that small doses of DMPP administered three minutes *after* training mice in a shock-avoidance task produced a dose-related improvement of retention seven days later. Larger doses of DMPP either had no effect or impaired retention (Flood *et al.* 1981). However, equally clear dose-related effects of nicotine itself, administered systemically, remain to be demonstrated.

The preceding studies used various forms of avoidance task, whereas in one recent study, the effects of nicotine were determined on the acquisition of bar-pressing for food reinforcement (Mundy and Iwamoto 1988). There was no evidence that even small doses of nicotine enhanced acquisition when given either pre-session or post-session. Generally, larger doses of nicotine appear to lack facilitatory effects and may even be disruptive (Flood *et al.* 1981; Mundy and Iwamoto 1988). Nicotine also has important effects that are not detected

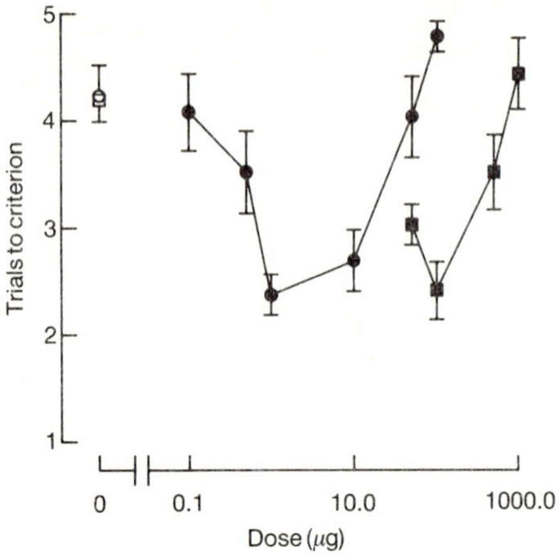

Fig. 8.4. Dose–response curves in mice (n = 19–25) trained in a T-maze shock-avoidance task 3 min before direct intracerebroventricular injections of D M P P (●), or arecoline (■). Data shown are mean numbers (± S E M) of retraining trials needed to reach an avoidance criterion in retention tests 7 days after training. Enhanced retention seemed to be a feature of both the nicotinic-cholinergic agonist D M P P and the muscarinic-cholinergic agonist arecoline. Points above 0 show retention after saline injections (drawn from data of Flood *et al.* 1981).

in studies where it is administered post-trial; after chronic administration it can improve performance of tasks requiring sustained attention, and this effect has been interpreted in terms of changes in arousal rather than of effects on associative factors or memory (Nelsen and Goldstein 1972). The effect was seen after the chronic administration of nicotine in small, $0.1\ \mathrm{mg\ kg^{-1}}$ 'smoking' doses.

The studies provide an indication that nicotine may have some beneficial effects in certain tasks, but their nature awaits confirmation. Nicotine does not have a general facilitatory action. However, some experiments may provide animal analogues for the improvements in human performance that nicotine can produce (Warburton, chapter 3, this volume). Some newer techniques, such as those involving the repeated acquisition of a sequence of operant responses, or radial and water mazes, promise to separate effects on learning and memory from those on other factors, but the effects of nicotine seem not to have been reported. The effect of nicotine on learning and memory is becoming an important issue, now that there is growing interest in the possible use of nicotinic agonists to ameliorate symptoms of Alzheimer's disease, including learning deficits (Kellar and Wonnacott, chapter 10, this volume).

8.4 Unconditioned behaviour

8.4.1 *Locomotor activity*

Investigations have focussed on two main areas, locomotor activity and feeding. Small doses of nicotine increase locomotor activity in rodents whereas larger doses are inhibitory. However, even small doses have an inhibitory

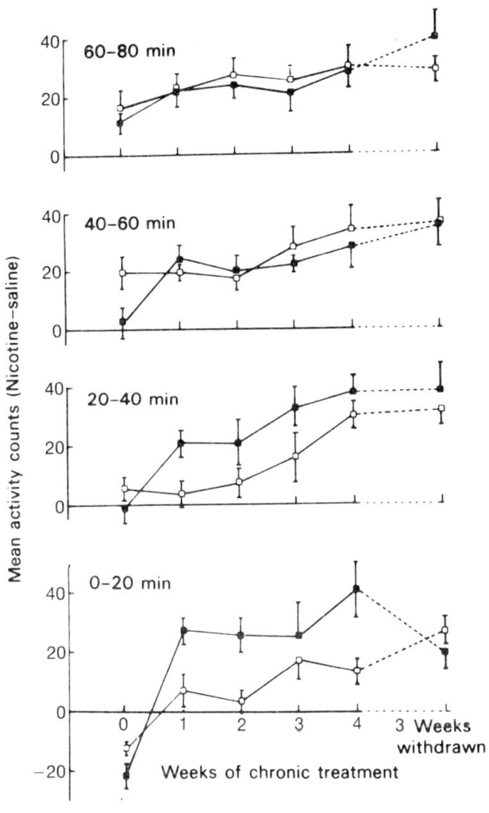

Fig. 8.5. Locomotor activity before, during, and after chronic treatment with nicotine. Every week, each rat was tested with 0.4 mg kg^{-1} of nicotine and with saline. Results from the 80-min tests in photocell activity cages are shown in 20-min periods. Rats in the chronic nicotine group (■) received the same dose of nicotine daily, whereas rats in the other group received nicotine only once each week, when they were tested (□). Data are shown as mean differences between nicotine and saline tests. With repeated testing, tolerance developed to the depressant effect of nicotine at 0–20 min, and concurrently, a stimulant effect appeared at all time periods, especially in the chronic nicotine group. Tolerance to the depressant effect can be seen 3 weeks after chronic administration of the drug was terminated. Vertical bars show ± SEM. Overlapping bars are omitted for clarity (from Clarke and Kumar 1983*a*).

effect during the first 30 min after injection, which is then followed by a longer period of facilitation (Clarke and Kumar 1983*a*). Figure 8.5 illustrates these effects. This biphasic time course of action is an unusual feature of nicotine's action and it is shared with few other drugs. The nature of the drug effect is also strikingly dependent on previous exposure to nicotine. Tolerance develops very rapidly to the initial depressant effect, so that after several administrations of nicotine, even quite large doses produce only increases in locomotor activity (Stolerman *et al.* 1973; Morrison and Stephenson 1972; Clarke and Kumar 1983*a*). The increases in locomotion are much more marked after chronic exposure to nicotine, perhaps because this effect of the drug is initially masked by the depressant action. These effects of nicotine occur at doses that are close to or only a little larger than estimated 'smoking' doses.

8.4.2 *Feeding and drinking*

In recent years there have been several studies on the effects of nicotine on food consumption and on associated changes in body weight. This work has been prompted by evidence that smokers weigh less than non-smokers of similar age, and that people who stop smoking put on weight. Three explanations have been proposed, that smokers eat less than non-smokers, that smoking alters metabolic rate, and that ex-smokers eat too much food that is sweet and high in calories. Grunberg (1982) has found that in humans and in male rats, nicotine decreases the intake of sweet, high-calorie foods whereas it has much less effect on the intake of bland foods (Fig. 8.6). Nicotine also decreases the normal weight gain of male rats, and this effect cannot be attributed entirely to the decreased caloric intake since it occurred in rats receiving either bland or sweet diets. Nicotine seems to preferentially reduce intake of foods that are either high in calories or very sweet (Grunberg *et al.* 1984).

After chronic administration of nicotine is terminated there is a tendency for increased consumption of sweet food and for weight to be gained rapidly. The consistency of the results in animal and human subjects suggest that changes in weight in human smokers may be due in part to the effects of nicotine on preferences for, and consumption of, specific types of food. The changes in weight of male rats, without concomitant changes in food or water intake, are consistent with evidence that nicotine may affect caloric expenditure by increasing thermogenesis in brown adipose tissue (Lupien and Bray 1988).

In subsequent studies, the effects were analyzed in female rats. Here, nicotine markedly decreased the intake of bland food (Grunberg *et al.* 1986; 1987). This was unexpected since other behavioural effects of nicotine have not been thought to differ greatly between the sexes, and also because no sex difference was reported in the human study (Grunberg 1982). Two other aspects of this series of experiments also deserve close attention. The nicotine was always administered by continuous, very slow, infusion from a miniature osmotic pump implanted subcutaneously. The amounts of nicotine released by

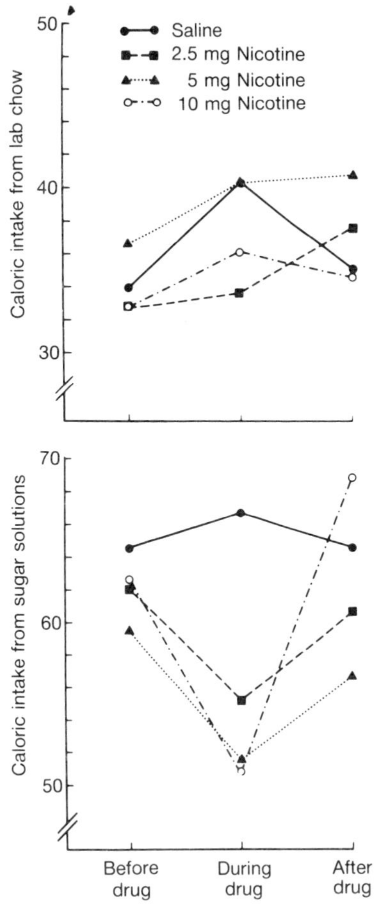

Fig. 8.6. Average caloric intake in rats (n = 5) before, during, and after administration of nicotine for 7 days via miniature pumps implanted subcutaneously. Results are based on intake of standard laboratory diet (upper panel) and on intake of sugar solutions (lower panel). (From Grunberg 1982.)

the pumps were calculated to be 6–12 mg kg^{-1} per day. Such doses of nicotine, when administered via minipumps in other studies, have produced plasma concentrations of nicotine of 80–160 ng ml^{-1}, which are rather larger than the plasma concentrations in smokers (Murrin *et al.* 1987). It is also debateable whether continuous, slow infusion is the best way to model the nicotine intake of humans. Other workers using injected doses of nicotine have found effects on the intake of bland foods in male rats (McNair and Bryson 1983; Clarke and Kumar 1984*a*), but the effects of injected nicotine on the intake of sweet, high-caloric foods seem not to have been determined.

8.4.3 *Effects in tests for analgesia*

Nicotine is active in some of the tests of responses to noxious stimuli that are widely used to evaluate analgesic drugs, although consistent effects were not reported in doses below 0.5 mg kg^{-1} (Sahley and Berntson 1979). Tripathi *et al.* (1982) reported that the ED_{50} for nicotine in a standard 'tail-flick' test was 0.7 mg kg^{-1} (subcutaneously), and that a 1 mg kg^{-1} dose was associated with plasma concentrations of nicotine about 300 ng ml^{-1}. It remains to be determined whether the effect is a real analgesic response since the doses of nicotine used produce marked effects on motor responses that have nothing to do with noxious stimuli (see above) and they are 5–10 times larger than 'smoking' doses of nicotine.

8.5 Receptors mediating behavioural effects

It must be apparent from the preceding sections of this chapter that nicotine has a wide spectrum of behavioural effects. Are these mediated through a single type of receptor, and to what extent are various neurotransmitter systems involved? Evidence from diverse approaches has implicated, with varying strength, acetylcholine, dopamine, 5-hydroxytryptamine, and opioid systems.

Some early studies on the binding of nicotine, taken together with other observations, have been interpreted as evidence for actions of nicotine at non-cholinergic sites (Abood *et al.* 1978), whereas others have interpreted binding data as evidence for nicotine acting at central cholinergic sites. There have also been difficulties in accepting the validity, as putative nicotine receptors, of the apparently cholinergic sites marked by αbungarotoxin. Wonnacott (chapter 7, this volume) has reviewed the main findings from these lines of work, from which the conclusion emerges that the high-affinity sites labelled with nicotine and with αbungarotoxin both have characteristics suggesting that they are functional receptors, and that acetylcholine is likely to be an important endogenous ligand in each case. The suggestion has been made that nicotine labels mainly those central nicotinic-cholinergic sites resembling receptors at autonomic ganglia, whereas αbungarotoxin labels a second type of central nicotinic-cholinergic site resembling that at the neuromuscular junction (Clarke *et al.* 1985). Additional sites binding nicotine with both lower and higher affinities have been reported as putative receptors by some workers, but cannot be identified by others. Low affinity sites also fail to saturate.

The question arises as to which of these sites constitute receptors through which the main behavioural effects of nicotine are mediated. This matter is now considered using five main criteria, as follows:

(1) the potency with which different nicotinic agonists inhibit binding must correlate with their potency in producing behavioural effects

(2) the effects of nicotinic antagonists should be consistent across the binding and behavioural procedures

(3) the brain concentrations of nicotine associated with the behavioural effects should be of the same order of magnitude as those needed to occupy the binding sites

(4) the brain sites associated with the behavioural effects should reflect the distribution of binding sites

(5) there should be behavioural correlates of the increased numbers of binding sites that occur as a function of chronic treatment with nicotine. At the present time, the information available is incomplete and only tentative conclusions can be reached.

8.5.1 *Nicotinic agonists*

In a number of behavioural models, the (−)-and (+)-stereoisomers of nicotine show a similar, small degree of stereoselectivity. The naturally occurring (−)-nicotine is consistently more potent than synthetic (+)-nicotine, although the ratio of potencies varies widely, from 4:1 to 50:1 (Martin 1987). For example, this applies to effects on rates of conditioned responding and locomotor activity, and to the conditioned taste aversion and discriminative stimulus procedures. Broadly similar results are found for binding of the two isomers to the high-affinity site labelled with [^{3}H]nicotine. In contrast, binding of the stereoisomers of nicotine to low-affinity sites

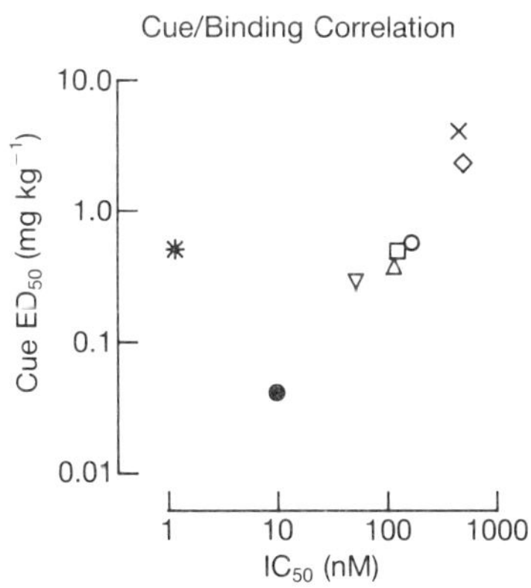

Fig. 8.7. Correlation of I C$_{50}$ values for inhibition of [^{3}H]nicotine binding with E D$_{50}$ values for producing the nicotine discriminative stimulus (cue) in rats. Cytisine *; (−)-nicotine ●; (+)nicotine □; (−)-nornicotine ○; (+)-nornicotine △; anabasine ◇; isoarecolone ×; *N*-(3-pyridylmethyl)pyrrolidine ▽. With data for cytisine discarded because this drug is thought to penetrate the brain poorly, the correlation coefficient $r_s = 0.96$ ($p<0.01$). (Modified from Reavill *et al.* 1988.)

labelled by [^{3}H]nicotine, and to sites labelled with αbungarotoxin, shows less stereoselectivity.

Several additional structural analogues of nicotine have been found to produce nicotine-like discriminative stimulus effects (see above, section 8.2.3). It is important to note the very high degree of pharmacological specificity shown by the nicotine cue; the great majority of non-nicotinic compounds tested to date are totally without effect (Table 8.1). Thus, only drugs that inhibit the binding of nicotine to a high-affinity site on rat brain membranes produce nicotine-like discriminative effects. Furthermore, Fig. 8.7 shows that the potencies of these drugs in inhibiting binding is correlated with their potencies in producing the behavioural effect. The most notable exception is lobeline, that inhibits binding but is inactive behaviourally. With this exception, an impressive correlation is emerging between the biochemical and behavioural results with nicotinic-cholinergic agonists. The reason for the discrepancy with lobeline is not known and it merits further investigation. A related series of compounds shows a correlation between ability to inhibit the binding of [^{3}H]nicotine and to produce severe motor impairment ('prostration') when injected directly into the ventricles of rat brains (Abood *et al.* 1987). To date, no correlation has been established between behavioural effects and the ability of drugs to inhibit low-affinity binding of [^{3}H]nicotine or the binding of αbungarotoxin.

8.5.2 *Nicotinic antagonists*

It has been known for many years that the ganglion-blocking drugs like mecamylamine block most if not all of the behavioural effects of nicotine. This is important evidence suggesting that the effects are mediated through central receptors that have at least some similarity to peripheral cholinoceptors. The validity of the evidence is entirely dependent on the assumption that the ganglion-blockers do not block the effects of drugs at other receptors. Mecamylamine and chlorisondamine have been found not to block responses to amphetamine, apomorphine, arecoline, cocaine, methysergide, midazolam, morphine, pentobarbitone, phencyclidine, quipazine, or scopolamine (reviewed by Stolerman 1987; Kumar *et al.* 1987; Reavill and Stolerman 1987). All these negative results have been obtained with the antagonists administered by doses and routes that block responses to nicotine, and there is currently no evidence that the antagonists act non-specifically at such doses. However, their mode of action is unclear and, therefore, conclusions about receptors that can be reached from studies with these antagonists is not completely convincing when considered in isolation from other data. As explained above (Wonnacott, chapter 7, this volume), ganglion-blocking drugs do not inhibit the binding of radiolabelled nicotine or αbungarotoxin *in vitro*. At least two explanations have been put forward to explain this puzzling finding, as discussed below.

The positive reinforcing effect of nicotine is apparently blocked by mecamylamine in both animals and humans, whereas mecamylamine does not attenuate the positive reinforcing effect of the non-nicotinic drug, cocaine (Stolerman *et al.* 1973; Spealman and Goldberg 1982). Mecamylamine blocks the actions of acetylcholine and nicotine at cholinergic receptors in the ganglia of the peripheral autonomic nervous system and it can also penetrate well into the brain. Drugs such as pentolinium and hexamethonium produce peripheral ganglionic blockade, but these drugs neither penetrate well into the CNS nor do they attenuate the positive reinforcing effect of nicotine. Thus the findings with antagonists provide some support for the view that the positive reinforcing effects of nicotine are mediated through receptors resembling the cholinoceptive sites in autonomic ganglia, and that these receptors are located in the CNS.

Surprisingly, the occurrence of vomiting in response to nicotine seems not to be correlated with its positive reinforcing or its aversive effects in animals. The mechanisms of the aversive effects have been dissociated, in part, from emetic mechanisms (Pratt and Stolerman 1984). There is no evidence to date that the receptors mediating the aversive effects of nicotine are different from those mediating its positive reinforcing effect; the doses of nicotine that produce the two types of effect are quite similar; the antagonist mecamylamine blocks both effects with similar potency, whereas hexamethonium has little effect (Kumar *et al.* 1987; Goldberg and Spealman 1983).

The nicotine cue is blocked by certain ganglion-blocking drugs, again suggesting mediation by CNS sites resembling peripheral cholinoceptors. However, recent evidence that mecamylamine, pempidine, and chlorisondamine (when administered directly into the brain), but not hexamethonium or pentolinium, can block discrimination of nicotine, suggests possible slight differences between the CNS and the ganglionic receptors (Kumar *et al.* 1987). Hexamethonium and pentolinium were inactive even when administered by direct intracerebral injections, so their poor penetration to the brain when administered systemically probably does not account for the results.

Both decreases and increases in rates of conditioned responding or locomotor activity are prevented by ganglion-blockers that penetrate well into the CNS, but not by ganglion-blockers that act only on the peripheral nervous system. Thus mecamylamine is unable to distinguish between the receptors mediating increases or decreases in response rates. Little is known about the receptor mechanisms underlying the effects of nicotine on feeding, or the extent to which these effects change with duration of exposure to nicotine, and there appears to be scope for considerably more work.

Recent observations have shown that nicotine enhances the startle response of mice to sudden loud noises, and that this effect of nicotine is prevented by unusually small doses of mecamylamine. It has been suggested, therefore, that the effect of nicotine on startle responses may be mediated through a receptor that is slightly different from that mediating most of the other behavioural effects considered above (Collins *et al.* 1986).

The failure of ganglion-blockers to inhibit the binding of nicotine has to be explained if these binding sites are to be accepted as receptors. One hypothesis is that the ganglion-blockers are not competitive antagonists of nicotine, but instead act at another closely-related site that may be 'downstream' from the nicotine recognition site in the chain of events that leads to an observed response. Evidence supporting this proposal is that these antagonists act non-competitively in peripheral tissues, by blocking the ion channels of nicotinic receptors (for example, Lingle 1983; Ascher *et al.* 1979). Evidence for the presence of ion-channels in central nicotinic receptors is discussed by Wonnacott (chapter 7, this volume). In the CNS, the ability of ganglion-blockers to prevent the more specific behavioural effects of nicotine (for example, the nicotine cue) cannot be reversed by increasing the dose of nicotine. This too suggests a non-competitive mechanism (Spealman *et al.* 1981; Stolerman *et al.* 1983). An alternative hypothesis is that the receptors in the *in vitro* preparations of rat brain membranes used in binding studies are in an abnormal state, with an induced shift in relative sensitivity to agonists and antagonists. There is evidence that such shifts can occur in peripheral tissues (Hucho 1986), but there is little direct evidence relating to the CNS.

Some other compounds that inhibit nicotine binding are also thought to act as antagonists. These include κbungarotoxin and neosurugatoxin; the latter is a particularly interesting compound since there is evidence that it can block the activity of nicotine in a test for analgesic activity (Yamada *et al.* 1986), as well as the effects of nicotine on an *in vitro* preparation of brain tissue (Wonnacott, chapter 7, this volume). Dihydroβerythroidine potently inhibits the high-affinity binding of [^{3}H]nicotine, but it does not produce or block the nicotine cue when administered intraventricularly in rats (Reavill and Stolerman, unpublished data). Further work with these interesting agents may help to clarify the nature of nicotine receptors in the CNS.

The fact that the effects of nicotine are blocked by ganglion-blocking drugs supports the view that the pertinent receptors resemble the cholinergic receptors in autonomic ganglia rather than those at the neuromuscular junction. The latter type of receptor is blocked by muscle relaxants such as curare and by αbungarotoxin. These agents are difficult to use in behavioural experiments since, when administered systemically, they produce paralysis, whereas central injections are convulsant. However, central injections of subconvulsant doses of such drugs fail to block the convulsant or the discriminative stimulus effect of nicotine, in contrast to central injections of at least some ganglion-blocking agents (Caulfield and Higgins 1983; Kumar *et al.* 1987; Reavill and Stolerman, unpublished data). In view of the limited selectivity of the blocking agents for the ganglionic and neuromuscular sites, and the suggestion of Collins *et al.* (1986) that the convulsant effects of nicotine are mediated through binding sites for αbungarotoxin, these questions should be pursued further.

8.5.3 *Brain concentrations of nicotine*

Although there have been variations in published results, many recent studies agree that the concentration needed to occupy half of the high-affinity binding sites for [^{3}H]nicotine (K_D) is typically about 6 nM (Wonnacott, chapter 7, this volume). Few behavioural effects of nicotine are seen in animal experiments with doses less than 0.1 mg kg^{-1}; with the very sensitive drug discrimination procedure as an example, the ED_{50} for nicotine is about 0.05 mg kg^{-1}. This dose is associated with a peak plasma concentration of nicotine of 17 ng ml^{-1} (100 nM), and this may underestimate the concentration in the brain. Thus it appears that the doses of nicotine that produce behavioural effects in even the most sensitive of procedures are 20 times greater than the doses that would be expected to saturate all the high-affinity binding sites. This discrepancy needs explanation if the high-affinity site is to be retained as a receptor.

The most plausible explanation to date is that prolonged exposure to agonist, and other aspects of *in vitro* binding techniques, puts the sites into an agonist-selective state. There is precedent for this in studies of the neuromuscular junction (Hucho 1986). A different view would be that the low-affinity sites for nicotine have K_D values in the range correlating with behaviourally effective doses, and that these are the important sites. An important low-affinity site would also be compatible with the hypothesis that the effects of nicotine significant for smokers are produced by large local concentrations in arterial blood ('boli') associated with single puffs on cigarettes (Russell and Feyerabend 1978). Difficulties with accepting the low-affinity sites described to date as receptors have been mentioned above. It is difficult to use brain concentrations as a criterion in relation to the αbungarotoxin binding site due to the lack of knowledge of behavioural effects of this material.

8.5.4 *Regional distribution of brain sites*

There have been few attempts to localize brain regions through which the behavioural effects are mediated. Microinjections of nicotine into or around the dorsal hippocampus can produce discriminative effects like those produced by systemic administration of nicotine (Rosecrans and Meltzer 1981; Reavill and Stolerman, unpublished data). Initial studies (Hunt and Schmidt 1978) of the binding of αbungarotoxin suggested a fairly large number of sites on certain cells in the hippocampus (dentate gyrus). Subsequently, it emerged that the hippocampus contained only a moderate concentration of high-affinity nicotine sites. Negative results are also awaited for microinjections into other brain areas, without which the data are difficult to interpret.

The largest numbers of high-affinity binding sites for [^{3}H]nicotine in the rat are in the cortex, thalamus, and striatum, with the interpenduncular nucleus showing a particularly large concentration (Clarke *et al.* 1985; see Clarke, chapter 5, this volume). The regional distribution of nicotine sites overlaps only to a small extent with that for αbungarotoxin sites. There is a consensus emerging that microinjections of nicotine into the ventral tegmental area of the forebrain can increase locomotor activity, whereas similar injections into several other brain regions do not increase activity (Pert and Clarke 1987; Reavill and Stolerman 1988). Lesions of the mesolimbic dopamine system weakened the ability of nicotine to increase locomotor activity, confirming the involvement of this system (Clarke *et al.* 1988). Although the ventral tegmental area is rich in binding sites for nicotine, positive results from microinjection studies involving areas with small numbers of binding sites have also been reported.

Recently, it has been reported that sites in the vestibular nucleus of the cerebellum mediate prostration produced by nicotine, and it was argued that ascending pathways from the cerebellum modulate limbic areas thus generating the reinforcing effect of nicotine (Maiti *et al.* 1986). These observations need to be considered in relation to the small numbers of binding sites for [^{3}H]nicotine in the cerebellum.

The regional distribution of binding sites may ultimately prove to be a useful guide for regions mediating behavioural effects of nicotine. However, with the exception of the experiments involving the ventral tegmental area, a correlation has not established by the studies carried out to date.

8.5.5 *Upregulation of binding related to behaviour*

One of the most surprising aspects of the high-affinity binding of nicotinic agonists is the increase in number of sites with chronic exposure to nicotine (Schwartz and Kellar 1985). Changes in the number of sites cannot, therefore, explain tolerance to nicotine since decreasing numbers of sites would be required for such an effect. Ksir *et al.* (1987) have suggested that there may be a correlation between the development, with chronic exposure to nicotine, of increased locomotor activity, and numbers of nicotinic-cholinergic binding sites. This is a particularly impressive demonstration since the doses of nicotine included some within the 'smoking' range (see Swedberg *et al.*, Fig. 2.10, this volume). Other studies have shown concurrent development of increased binding for both nicotine and αbungarotoxin while tolerance develops to several behavioural effects (Marks *et al.* 1987; see Wonnacott, Fig. 7.5 this volume).

The interpretation of upregulation of binding raises a surprising point: it has been suggested that upregulation occurs because nicotine acts as a blocking agent upon chronic administration (Schwartz and Kellar 1985). This view receives some support from the observation that increasing the synaptic concentration of acetylcholine, by administering an anticholinesterase chronically, produces the expected down-regulation of nicotine sites. Under this hypothesis most, if not all, of the effects of chronic exposure to nicotine are mediated through inhibition of cholinergic systems. Traditionally, it has been thought that smoking doses of nicotine are insufficient to desensitize receptors. In order for the new hypothesis to be accepted, it will be necessary to show that desensitization occurs at smoking doses, and to explain why ganglion blocking drugs antagonize rather than mimic the effects of nicotine.

8.5.6 *Mediation through the release of acetylcholine*

The preceding sections dealt with evidence relating to direct actions of nicotine on cholinergic receptors. It has sometimes been suggested that nicotine acts, at least in part, indirectly by enhancing the release of acetylcholine. Such effects have been shown in certain pharmacological preparations, notably those involving assays on fluid collected from the surface of the cerebral cortex (Armitage *et al.* 1968). Warburton (chapter 3, this volume) suggests that improvements in performance in human subjects are mediated in this way, and that the acetylcholine so released acts upon both muscarinic and nicotinic receptors. Warburton notes that scopolamine has effects in some ways opposite to those of nicotine. This is an interesting notion that could be tested further. However, muscarinic antagonists do not block some of the behavioural effects of nicotine; this suggests that the release of acetylcholine is not a necessary step in the mediation of all the psychological effects of nicotine (see above, section 8.2).

8.5.7 *Involvement of monoamine and opioid systems*

Nicotine can enhance the release of dopamine in *in vitro* pharmacological preparations, as outlined by Wonnacott (chapter 7, this volume). There is clear evidence for nicotinic receptors located presynaptically on dopamine nerve terminals in the striatum (Reavill, chapter 9, this volume), and for involvement of dopamine in effects of nicotine in certain neuroendocrine, electrophysiological, and behavioural systems (Fuxe, chapter 6, this volume; Clarke, chapter 5, this volume). These and other observations are drawn together by Reavill (chapter 9, this volume) and suggest that dopamine mechanisms may play a role in mediating certain key behavioural effects of nicotine, including its ability to serve as a positive reinforcer and as a discriminative stimulus.

Certain physiological effects of nicotine have been linked to opioid systems, and opioids have also been postulated to play a role in behavioural responses to nicotine (Pomerleau and Pomerleau 1984). There have also been attempts to link the effects of nicotine to serotonin systems. Drugs acting as antagonists on certain serotonin and opioid receptors have no effect on discrimination of nicotine (review by Stolerman 1987). Experiments attempting to test the role of serotonin in relation to unconditioned behaviours have yielded very mixed results (Balfour *et al.* 1986; Fitzgerald *et al.* 1985). However, a wide range of drugs thought to act selectively as agonists or antagonists at various subtypes of opioid or serotonin receptor are now available, and relatively few of these quite novel compounds have been used as probes in studies of the mode of action of nicotine.

8.6 Conclusions

The positive reinforcing effect of nicotine is firmly established as the strongest candidate for the behavioural basis of addiction to tobacco. The psychological nature of this effect, and the neuropharmacological mechanisms underlying it, are less well characterized. Nicotine produces a wide-ranging and complex set of behavioural changes: none of these have been clearly identified as correlated with its ability to serve as a positive reinforcer. However, its proposed effects on certain psychological processes may contribute to its reinforcing efficacy. In particular, improvements in aspects of learning and memory, and changes in arousal, may be important factors. However, changes in these processes need further evaluation with a wider range of techniques. The animal studies usefully complement human experiments that are severely limited by difficulties in administering controlled amounts of pure nicotine, especially to non-smokers. Laboratory experiments have lent support to the clinical evidence that nicotine may help to keep weight down, and that withdrawal from nicotine may be associated with increases in weight. Changes in appetite and weight may therefore be seen as possible further factors that may contribute to, and modulate, the positive reinforcing effect of nicotine. In contrast, proposed tranquillizing effects of nicotine have yet to receive much support from studies of its behavioural effects.

Little is known of the neuropharmacological basis for the positive reinforcing effect of nicotine apart from the involvement of a cholinoceptive site, and a possible linkage to a brain reward mechanism is which dopamine is a transmitter. The mesolimbic dopamine system is involved in the locomotor activation produced by nicotine, and its importance in the positive reinforcing effect is suggested by evidence that similar lesions weaken the self-administration of nicotine (Clarke *et al.* 1988; Singer *et al.* 1982). It is noteworthy that some other addictive drugs, including morphine and cocaine, may produce positive reinforcement through this dopamine system. It remains unknown whether direct effects of nicotine upon reward mechanisms can produce addiction independently of detectable beneficial effects on psychomotor performance, or perhaps independently from changes in subjective state that are of perceivable magnitude. It is also customary to examine the effects of nicotine in isolation from other widely practised forms of drug self-administration, but there is a correlation between the use of nicotine on the one hand, and beverages containing ethanol or caffeine, on the other (Istvan and Matarazzo 1984). Some recent evidence raises the possibility that there may be pharmacological interactions between these agents, and such interactions may contribute to the association between their use (Lowe 1986; White 1988).

The animal experiments also reveal that nicotine can serve as a potent aversive stimulus. It is necessary to find out whether a human analogue of this effect sets an upper limit to normal smoke exposure; the development of techniques for potentiating this aversive effect might then provide a basis for lowering the level at which smoking behaviour stabilizes. Information will be needed on whether tolerance develops to either the positive reinforcing or the aversive effects of nicotine. It is widely accepted that tolerance develops to nausea and vomiting that people experience on their initial exposure to nicotine, but the aversive effects in animal experiments seem to assess another dimension of nicotine's action. The ability to produce vomiting seems to be dissociated from the occurrence of positive reinforcing or aversive effects, which must therefore involve some other, as yet unidentified, process.

Discriminative stimulus effects of nicotine have been used to study correlations of psychological with neurochemical processes. While there is evidence to suggest that the discriminative effects may be mediated by a saturable, stereoselective, binding site for nicotine, several aspects of this relationship need further study. The mode of action of existing antagonists in the CNS is unclear, and there is no competitive antagonist available. The lack of behavioural activity of lobeline remains a puzzle, although it accords well with the failure of lobeline as an aid to smoking cessation.

Evidence that the effects of nicotine are mediated through multiple receptors remains patchy, and progress is slowed by the extremely limited number of nicotinic compounds that have been available for study. There is a marked contrast with opioid receptors; the numerous synthetic opioids developed in the search for clinically useful agents have provided powerful investigative tools. In contrast, developing novel nicotinic agonists with different profiles of effect has been a minority interest, and yet any substantial broadening of therapeutic application seems contingent upon finding nicotinic agonists or antagonists with more selective effects.

References

Abood, L.G., Lowy, K., Tometsko, A., and Booth, H. (1978). Electrophysiological, behavioral, and chemical evidence for non cholinergic, stereospecific site for nicotine in rat brain. *Journal of Neuroscience Research* **3**, 327–33.

Abood, L.G., Maiti, A., Grassi, S., Salles, K.S. (1987). The biochemical and functional nature of the nicotine receptor in rat brain. In *Tobacco smoking and nicotine* (W.R. Martin, G.R. Van Loon, E.T. Iwamoto, and L. Davis ed.), pp. 401–11. Plenum, New York.

Armitage, A.K., Hall, G.H., and Morrison, C.F. (1968). Pharmacological basis for the tobacco smoking habit. *Nature* **217**, 331–4.

Ascher, P., Large, W.A., and Rang, H.P. (1979). Studies on the mechanism of action of acetylcholine antagonists on rat parasympathetic ganglion cells. *Journal of Physiology* **295**, 139–70.

Ashton, H., Marsh, V.R., Millman, J.E., Rawlins, M.D., Telford, R., and Thompson, J.W. (1980). Biphasic dose-related responses of the CNV (contingent negative variation) to I.V. nicotine in man. *British Journal of Clinical Pharmacology* **10**, 579–89.

Balfour, D.J.K., Graham, C.A., and Vale, A.L. (1986). Studies on the possible role of brain 5-HT systems and adrenocortical activity in behavioural responses to nicotine and diazepam in an elevated X-maze. *Psychopharmacology* **90**, 528–32.

Battig, K. (1970). The effect of pre- and post-trial application of nicotine on the 12 problems of the Hebb-Williams-test in the rat. *Psychopharmacologia* **18**, 68–76.

Bovet, D., Bovet-Nitti, F., and Oliverio, A. (1966). Effects of nicotine on avoidance conditioning of inbred strains of mice. *Psychopharmacologia* **10**, 1–5.

Caulfield, M.P., and Higgins, G.A. (1983). Mediation of nicotine-induced convulsions by central nicotinic receptors of the 'C_6' type. *Neuropharmacology* **22**, 347–51.

Clarke, P.B.S. (1987). Nicotine and smoking: a perspective from animal studies. *Psychopharmacology* **92**, 135–43.

Clarke, P.B.S., and Fibiger, H.C. (1987). Apparent absence of nicotine-induced conditioned place preference in rats. *Psychopharmacology* **92**, 84–8.

Clarke, P.B.S., Fu, D.S., Jakubovic, A., and Fibiger, H.C. (1988). Evidence that mesolimbic dopamine activation underlies the locomotor stimulant action of nicotine in rats. *Journal of Pharmacology and Experimental Therapeutics* **246**, 701–8.

Clarke, P.B.S., and Kumar, R. (1983*a*). The effects of nicotine on locomotor activity in non-tolerant and tolerant rats. *British Journal of Pharmacology* **78**, 329–37.

Clarke, P.B.S., and Kumar, R. (1983*b*). Nicotine does not improve discrimination of brain stimulation reward by rats. *Psychopharmacology* **79**, 271–7.

Clarke, P.B.S., and Kumar, R. (1984*a*). Some effects of nicotine on food and water intake in undeprived rats. *British Journal of Pharmacology* **82**, 233–9.

Clarke, P.B.S., and Kumar, R. (1984*b*). Effects of nicotine and *d*-amphetamine on intracranial self-stimulation in a shuttle box test in rats. *Psychopharmacology* **84**, 109–14.

Clarke, P.B.S., Schwartz, R.D., Paul, S.M., Pert, C.B., and Pert, A. (1985). Nicotinic binding in rat brain: autoradiographic comparison of [^{3}H]acetylcholine, [^{3}H]nicotine and [^{125}I]-α-bungarotoxin. *Journal of Neuroscience* **5**, 1307–15.

Collins, A.C., Evans, C.B., Miner, L.L., and Marks, M.J. (1986). Mecamylamine blockade of nicotine responses: evidence for two brain nicotinic receptors. *Pharmacology, Biochemistry and Behavior* **24**, 1767–73.

de Wit, H., and Stewart, J. (1983). Drug reinstatement of heroin-reinforced responding in the rat. *Psychopharmacology* **79**, 29–31.

Domino, E.F. (1965). Some behavioral actions of nicotine. In *Tobacco alkaloids and related compounds* (U.S. von Euler ed.), pp. 145–62. Pergamon, Oxford.

Erickson, C.K. (1971). Studies on the mechanism of avoidance facilitation by nicotine. *Psychopharmacologia* **22**, 357–68.

Evangelista, A.M., Gattioni, R.C., and Izquierdo, I. (1970). Effect of amphetamine, nicotine and hexamethonium on performance of a conditioned response during acquisition and retention trials. *Pharmacology* **3**, 91–6.

Fitzgerald, R.E., Oettinger, R., and Battig, K., (1985). Reduction of nicotine-induced hyperactivity by pCPA. *Pharmacology, Biochemistry and Behavior* **23**, 279–84.

Flood, J.F., Landry, D.W., and Jarvik, M.E. (1981). Cholinergic receptor interactions and their effects on long-term memory processing. *Brain Research* **215**, 177–85.

Fudala, P.J., and Iwamoto, E.T. (1986). Further studies on nicotine-induced conditioned place preference in the rat. *Pharmacology, Biochemistry and Behavior* **25**, 1041–9.

Fudala, P.J., and Iwamoto, E.T. (1987). Conditioned aversion after delay place conditioning with nicotine. *Psychopharmacology* **92**, 376–81.

Garg, M. (1969). The effect of nicotine on two different types of learning. *Psychopharmacologia* **15**, 408–14.

Garg, M., and Holland, H.C. (1968). Consolidation and maze-learning. A further study of post-trial injections of a stimulant drug. (nicotine). *International Journal of Neuropharmacology* **7**, 55–9.

Gilbert, D.G. (1979). Paradoxical tranquilizing and emotion-reducing effects of nicotine: *Psychological Bulletin* **86**, 643–61.

Goldberg, S.R., Risner, M.E., Stolerman, I.P., Reavill, C., and Garcha, H.S. (1989). Nicotine and some related compounds: effects on schedule-controlled behavior and discriminative properties in rats. *Psychopharmacology* **97**, 295–302.

Goldberg, S.R., and Spealman, R.D. (1983). Suppression of behavior by intravenous injections of nicotine or by electric shocks in squirrel monkeys: effects of chlordiazepoxide and mecamylamine. *Journal of Pharmacology and Experimental Therapeutics* **224**, 334–40.

Goldberg, S.R., and Stolerman, I.P. (1986). *Behavioral analysis of drug dependence.* Academic Press, Orlando.

Grunberg, N.E. (1982). The effects of nicotine and cigarette smoking on food consumption and taste preferences. *Addictive Behaviors* **7**, 317–31.

Grunberg, N.E., Bowen, D.J., and Morse, D.E., (1984). Effects of nicotine on body weight and food consumption in rats. *Psychopharmacology* **83**, 93–8.

Grunberg, N.E., Bowen, D.J., and Winders, S.E. (1986). Effects of nicotine on body weight and food consumption in female rats. *Psychopharmacology* **90**, 101–5.

Grunberg, N.E., Winders, S.E., and Popp, K.A. (1987). Sex differences in nicotine's effects on consummatory behavior and body weight in rats. *Psychopharmacology* **91**, 221–5.

Henningfield, J.E., and Goldberg, S.R. (1984). Stimulus properties of nicotine in animals and human volunteers: a review. In *Behavioral pharmacology: the current status* (L.S. Seiden, and R.L. Balster, ed.), pp. 433–449. Alan R Liss, New York.

Hucho, F. (1986). The nicotinic acetylcholine receptor and its ion channel. *European Journal of Biochemistry* **158**, 211–26.

Hunt, S., and Schmidt, J. (1978). Some observations on the binding patterns of α-bungarotoxin in the central nervous system of the rat. *Brain Research* **157**, 213–32.

Hunter, B., Zornetzer, S.F., Jarvik, M.E., and McGaugh, J.L. (1977). Modulation of learning and memory: effects of drugs influencing neurotransmitters. In *Handbook of psychopharmacology*, Vol 8, (L.L. Iversen, S.D. Iversen, and S.H. Snyder ed.), pp. 531–577. Plenum, New York.

Istvan, J., and Matarazzo, J.D. (1984). Tobacco, alcohol, and caffeine use: a review of their interrelationships. *Psychological Bulletin* **95**, 301–26.

Iwamoto, E.T., and Williamson, E.C. (1984). Nicotine-induced taste aversion: characterization and preexposure effects in rats. *Pharmacology, Biochemistry and Behavior* **21**, 527–32.

Ksir, C., Hakan, R.L., and Kellar, K.J. (1987). Chronic nicotine and locomotor activity: influences of exposure dose and test dose. *Psychopharmacology* **92**, 25–9.

Kumar, R., Pratt, J.A., and Stolerman, I.P. (1983). Characteristics of conditioned taste aversion produced by nicotine in rats. *British Journal of Pharmacology* **79**, 245–53.

Kumar, R., Reavill, C., and Stolerman, I.P. (1987). Nicotine cue in rats: effects of central administration of ganglion-blocking drugs. *British Journal of Pharmacology* **90**, 239–46.

Lingle, C. (1983). Blockade of cholinergic channels by chlorisondamine on a crustacean muscle. *Journal of Physiology* **339**, 395–417.

Lowe, G. (1986). State-dependent learning effects with a combination of alcohol and nicotine. *Psychopharmacology* **89**, 105–7.

Lupien, J.R., and Bray, G.A. (1988). Nicotine increases thermogenesis in brown adipose tissue in rats. *Pharmacology, Biochemistry and Behavior* **29**, 33–7.

Maiti, A., Salles, K.S., Grassi, S., and Abood, L.G. (1986). Barrel rotation and prostration by vasopressin and nicotine in the vestibular cerebellum. *Pharmacology, Biochemistry and Behavior* **25**, 583–8.

Marks, M.J., Stitzel, J.A., and Collins, A.C. (1987). Influence of kinetics of nicotine administration on tolerance development and receptor levels. *Pharmacology, Biochemistry and Behavior* **27**, 505–12.

Martin, B.R. (1987). Stereoselectivity of nicotine's central effects and its relationship to pain and cardiovascular function. In *Tobacco smoking and nicotine.* W.R. Martin, G.R. Van Loon, E.T. Iwamoto, and L. Davis ed.), pp. 301–16. Plenum, New York.

McNair, E., and Bryson, R. (1983). Effects of nicotine on weight change and food consumption in rats. *Pharmacology, Biochemistry and Behavior* **18**, 341–4.

Morrison, C.F. (1967). Effects of nicotine on operant behavior of rats. *International Journal of Neuropharmacology* **6**, 229–40.

Morrison, C.F., and Stephenson, J.A. (1969). Nicotine injections as the conditioned stimulus in discrimination learning. *Psychopharmacologia* **15**, 351–60.

Morrison, C.F., and Stephenson, J.A. (1970). Drug effects on a measure of unconditioned avoidance in the rat. *Psychopharmacologia* **18**, 133–43.

Morrison, C.F., and Stephenson, J.A. (1972). The occurrence of tolerance to a central depressant effect of nicotine. *British Journal of Pharmacology* **46**, 151–6.

Mundy, W.R., and Iwamoto, E.T. (1988). Actions of nicotine on the acquisition of an autoshaped lever-touch response in rats. *Psychopharmacology* **94**, 267–74.

Murrin, L.C., Ferrer, J.R., Wanyun, Z., and Haley, N.J. (1987). Nicotine administration to rats: methodological considerations. *Life Sciences* **40**, 1699–1708.

Nelsen, J.M., and Goldstein, L. (1972). Improvement of performance on an attention task with chronic nicotine treatment in rats. *Psychopharmacologia* **26** 347–60.

Pert, A., and Clarke, P.B.S (1987). Nicotine modulation of dopaminergic neurotransmission: functional implications. In *Tobacco smoking and nicotine* (W.R. Martin, G.R. Van Loon, E.T. Iwamoto, and L. Davis ed.), pp. 169–89. Plenum, New York.

Peters, R., and McGee, R. (1982). Cigarette smoking and state-dependent memory. *Psychopharmacology* **76**, 232–5.

Pomerleau, O.F., and Pomerleau, C.S. (1984). Neuroregulators and the reinforcement of smoking: towards a biobehavioral explanation. *Neuroscience and Biobehavior Reviews* **8**, 503–13.

Pradhan, S.N., and Bowling, C. (1971). Effects of nicotine on self-stimulation in rats. *Journal of Pharmacology and Experimental Therapeutics* **176**, 229–43.

Pratt, J.A., and Stolerman, I.P. (1984). Pharmacologically specific pretreatment effects on apomorphine-mediated conditioned taste aversions in rats. *Pharmacology, Biochemistry and Behavior* **20**, 507–11.

Pratt, J.A., Stolerman, I.P., Garcha, H.S., Giardini, V., and Feyerabend, C. (1983). Discriminative stimulus properties of nicotine: further evidence for mediation at a cholinergic receptor. *Psychopharmacology* **81**, 54–60.

Reavill, C., Jenner, P., Kumar, R., and Stolerman, I. P. (1988). High-affinity binding of [^{3}H](−)-nicotine to rat brain membranes and its inhibition by analogues of nicotine. *Neuropharmacology* **27**, 235–41.

Reavill, C., Spivak, C. E., Stolerman, I. P., and Waters, J. A. (1987). Isoarecolone can inhibit nicotine binding and produce nicotine-like discriminative stimulus effects in rats. Neuropharmacology **26**, 789–92.

Reavill, C., and Stolerman, I. P. (1987). Interaction of nicotine with dopaminergic mechanisms assessed through drug discrimination and rotational behavior in rats. *Journal of Psychopharmacology* **1**, 264–73.

Reavill, C., and Stolerman, I. P. (1988). Locomotor activity in rats after intracerebral administration of nicotine. *Journal of Psychopharmacology* **2**, Proceedings supplement.

Risner, M. E., Goldberg, S. R., Prada, J. A., and Cone, E. J. (1985). Effects of nicotine, cocaine and some of their metabolites on schedule-controlled responding by beagle dogs and squirrel monkeys. *Journal of Pharmacology and Experimental Therapeutics* **234**, 113–9.

Rosecrans, J. A., Kallman, M. J., and Glennon, R. (1978). The nicotine cue: an overview. In *Stimulus properties of drugs* F. C. Colpaert, and J. A., Rosecrans ed.), pp. 69–81. Elsevier, Amsterdam.

Rosecrans, J. A., and L. T. Meltzer. (1981). Central sites and mechanisms of action of nicotine. *Neuroscience and Biobehavioral Reviews* **5**, 497–501.

Russell, M. A. H., and Feyerabend, C. (1978). Cigarette smoking: a dependence on high-nicotine boli. *Drug Metabolism Reviews* **8**, 29–57.

Sahley, T. L., and Berntson, G. G. (1979). Antinociceptive effects of central and systemic administrations of nicotine in the rat. *Psychopharmacology* **65**, 279–83.

Schaeffer, G. J., and Michael, R. P. (1986). Task-specific effects of nicotine in rats. Intracranial self-stimulation and locomotor activity. *Neuropharmacology* **25**, 125–31.

Schuster, C. R., and Thompson, T. (1969). Self-administration of and behavioral dependence on drugs. *Annual Reviews of Pharmacology* **9**, 483–502.

Schwartz, R. D., and Kellar, K. J. (1985). *In vivo* regulation of [^{3}H]acetylcholine recognition sites in brain by nicotinic cholinergic drugs. *Journal of Neurochemistry* **45**, 427–33.

Singer, G., Wallace, M., and Hall, R. (1982). Effects of dopaminergic nucleus accumbens lesions on the acquisition of schedule induced self injection of nicotine in the rat. *Pharmacology, Biochemistry and Behavior* **17**, 579–81.

Spealman, R. D. (1983). Maintenance of behavior by postponement of scheduled injections of nicotine in squirrel monkeys. *Journal of Pharmacology and Experimental Therapeutics* **227**, 154–9.

Spealman, R. D., and Goldberg, S. R. (1982). Maintenance of schedule-controlled behavior by intravenous injections of nicotine in squirrel monkeys. *Journal of Pharmacology and Experimental Therapeutics* **223**, 402–8.

Spealman, R. D., Goldberg, S. R., and Gardner, M. L. (1981). Behavioral effects of nicotine: schedule-controlled responding by squirrel monkeys. *Journal Pharmacology and Experimental Therapeutics* **216**, 484–491.

Stolerman, I. P. (1987). Psychopharmacology of nicotine: stimulus effects and receptor mechanisms. In *Handbook of psychopharmacology* Vol 19, (L. L. Iversen, S. D. Iversen, and S. H. Snyder ed.), pp. 421–65. Plenum, New York.

Stolerman, I. P., and D'Mello, G. D. (1981). Oral self-administration and the relevance of conditioned taste aversions. In *Advances in behavioral pharmacology*, Vol 3, (T.

Thompson, P.B. Dews, and W.A. McKim, ed.), pp. 169–214. Academic Press, New York.

Stolerman, I.P., Fink, R., and Jarvik, M.E. (1973). Acute and chronic tolerance to nicotine measured by activity in rats. *Psychopharmacologia* **30**, 329–42.

Stolerman, I.P., Goldfarb, T., Fink, R., and Jarvik, M.E (1973). Influencing cigarette smoking with nicotine antagonists. *Psychopharmacologia* **28**, 247–59.

Stolerman, I.P., Pratt, J.A., Garcha, H.S., Giardini, V., and Kumar, R. (1983). Nicotine cue in rats analysed with drugs acting on cholinergic and 5-hydroxytryptamine mechanisms. *Neuropharmacology* **22**, 1029–37.

Tripathi, H.L., Martin, B.R., Aceto, M.D. (1982). Nicotine-induced antinociception in rats and mice: correlation with nicotine brain levels. *Journal of Pharmacology and Experimental Therapeutics* **221**, 91–6.

White, J.M. (1988). Behavioral interactions between nicotine and caffeine. *Pharmacology, Biochemistry and Behavior* **29**, 63–6.

Yamada, S., Kagawa, Y., Takayanagi, N., Hayashi, E., Tsuji, K., and Kosuge, T. (1986). Inhibition by neosurugatoxin of nicotine-induced antinociception. *Brain Research* **375**, 360–2.

9 Action of nicotine on dopamine pathways and implications for Parkinson's disease

C. Reavill

9.1 Introduction

Two questions can be posed with respect to the relationship between smoking and nicotine and Parkinson's disease: does smoking have a shielding action against the pathogenesis of Parkinson's disease, and does nicotine have a palliative effect on the neurological deficits of patients who are already showing symptoms? Over half a century has passed since Moll (1926) observed that the administration of nicotine to patients suffering from post-encephalitic Parkinsonism caused temporary relief from muscular rigidity. Since then, most epidemiological studies have demonstrated an inverse relationship between cigarette smoking and idiopathic Parkinson's disease, and some of these studies have suggested that cigarette smoking may have a protective action against the pathogenesis of the disease (Kahn 1966; Kessler and Diamond 1971). Also, although the smoking habit has been associated with increased prevalence of tardive dyskinesias in psychiatric patients (Binder *et al.* 1987; Yassa *et al.* 1987), nicotine has been reported to abolish dystonia and athetoid movements in one case (Lees 1984). However, these reports are not numerous such that the *Extrapharmacopoeia* (Martindale 1982) describes nicotine as 'having no therapeutic uses but is of considerable pharmacological interest'.

Because nicotine has been reported to protect against Parkinson's disease and ameliorate other movement disorders of central origin, there are obvious advantages for knowing the mechanisms by which nicotine may exert its actions. This review shall concentrate on the relationship between nicotinic and dopaminergic mechanisms, as disturbances of central dopamine action have been implicated in the pathophysiology of movement disorders such as Parkinson's disease. Also, independent evidence has shown close anatomical, physiological, and biochemical relationships between dopaminergic and nicotinic mechanisms in brain areas of rodents. In reviewing the evidence, it is important to bear in mind that the effects of smoking and of nicotine may not always be synonymous because, apart from nicotine, cigarette smoke contains more than 4000 different compounds most of which have not been studied in depth (Jaffe 1985).

9.2 Epidemiology

Most surveys have shown that people who have smoked cigarettes are less likely to develop Parkinson's disease than abstainers (Kahn 1966; Kessler and Diamond 1971; Kessler 1972*a*, 1972*b*; Doll and Peto 1976; Baumann *et al.* 1980; Godwin-Austen *et al.* 1982). For example, in the study by Godwin-Austen *et al.* (1982) it was shown that smokers had a relative risk of 0.40 for having Parkinson's disease compared with a risk of 1.00 for people who had never smoked (Table 9.1). Furthermore, Parkinson's disease patients are more likely to have given up smoking than non-sufferers (Kessler and Diamond 1971; Marttila and Rinne 1980), and the Parkinsonian patients that did smoke tended to smoke fewer cigarettes than controls (Kessler 1972*b*). The reason for this negative correlation is unknown but several hypotheses have been advanced to explain the phenomenon. The importance of selective mortality as a factor for having an apparent protective effect against the development of Parkinson's disease cannot be ignored. It is possible that Parkinsonian patients who do smoke die of smoking-related diseases earlier than Parkinsonian non-smokers, thus resulting in depleted numbers of smokers among the Parkinson's disease group. However, Baumann *et al.* (1980) showed that there was not an excess of smokers among a group of 48 deceased Parkinsonian patients, and there was no deficit of smokers among patients who had the disease for a long time. The results did not support the selective mortality hypothesis.

Other studies have addressed the possibility that patients stopped smoking as a consequence of developing Parkinson's disease, thus resulting in a spuriously low number of smokers among the disease group. There is no doubt that smoking requires some manual dexterity which is affected in Parkinson's disease. Although this may be a contributory factor, studies have shown that there was a reduced rate of smoking among Parkinsonian patients up to 20 years before onset of the disease (Table 9.1) (Kessler and Diamond 1971; Baumann *et al.* 1980; Godwin-Austen *et al.* 1982; Bharucha *et al.* 1986).

In some surveys, control subjects were drawn from hospital patients, yet Nefzger *et al.* (1968) have reported that there is a higher than the average number of smokers among hospital patients. However, in subsequent surveys in which subjects were taken from the population at large, smoking was still found to be associated with a lower incidence of Parkinson's disease (Kessler 1972*b*; Baumann *et al.* 1980; Marttila and Rinne 1980; Ward *et al.* 1983).

Attention also has been drawn to the question of the diagnosis of Parkinson's disease, and whether its presence may be overlooked in smokers due to death being caused by smoking-related diseases (Baron, 1986). Such occurrences would lead to a misleadingly low number of smokers with Parkinson's disease at death.

One approach to testing the hypothesis that smoking protects against Parkinson's disease has been to examine whether the age of onset of the

Table 9.1. Distribution of smoking habits in cases and controls.

Smoking habits	Men (179 pairs)			Women (171 pairs)			Combined
	Cases %	Controls %	Relative risk	Cases %	Controls %	Relative risk	Relative risk
Never smoked	22.9	12.3	1.00	69.0	58.5	1.00	1.00
Given up smoking:							
more than 10 years before onset of Parkinson's disease	19.6	13.4	0.78	7.6	7.0	0.92	0.84
1–10 years before onset of Parkinson's disease	14.0	12.9	0.58	4.7	5.9	0.68	0.62
after onset of Parkinson's disease	15.1	19.6	0.41	7.6	7.0	0.92	0.58
total	48.6	45.8	0.57	19.9	19.9	0.85	0.71
Still smoking at time of interview	28.5	41.9	0.36	11.1	21.6	0.44	0.40

383 patients with Parkinson's disease, and a corresponding number of controls matched for age and sex, were interviewed. The greatest difference in risk of Parkinson's disease was between those who were still smoking at the time of interview and those who had never smoked. Those who had given up smoking had an intermediate risk, with a trend towards risk being lower in those who gave up later than earlier. (From Godwin-Austen *et al.* 1982.)

disease was related to the history and degree of past smoking history (dose–response effect). If the hypothesis was valid, then one would expect a delayed onset of disease in people who had smoked the most cigarettes. In an extensive survey where several parameters including smoking intensity, Parkinson's disease severity, and mode of onset (tremor or rigidity) were studied, there was virtually no evidence of a correlation between Parkinson's disease and smoking (Golbe *et al.* 1986). A further refinement in that survey was to exclude respondents who had ceased smoking after the onset or diagnosis of the first symptoms as those patients would have been more likely to have given up smoking for health reasons. In this selected subgroup there was found to be no correlation between smoking and Parkinson's disease severity. These results confirmed the earlier study of Haack *et al.* (1981). However, there has been argument as to how the epidemiological data should be interpreted with regard to age distribution and differential mortality (Baron and Greenberg 1987; Golbe *et al.* 1987) as these factors may have a bigger influence on the data than has often been reported previously.

Other surveys have evaluated the concordance rate for Parkinson's disease among twins in order to determine whether genetic factors are implicated in the aetiology (Ward *et al.* 1983; Bharucha *et al.* 1986). These studies found that the concordance rate for monozygotic twins was low, and the frequency of Parkinson's disease in monozygotic co-twins was probably the same as expected in unrelated controls. It was concluded that Parkinson's disease was not genetically determined. Examination of the life histories of the twins for such factors as diet, or exposure to drugs, alcohol, infection, and vaccination indicated that there was no significant association between these factors and Parkinson's disease. The only significant finding was confirmation of an inverse relationship between smoking and Parkinsonism. Since genetic factors do not seem to play a role in the pathogenesis of Parkinson's disease, environmental agents have been mooted as possible causative agents (Barbeau 1984). A recent survey has shown localized concentrations of Parkinson's disease in rural areas and it is these regions which use the highest levels of pesticides and herbicides (Barbeau *et al.* 1987). The authors do not claim that this is proof of a causal relationship, but the evidence is consistent with environmental factors playing a role in pathogenesis. Another survey investigating the association of environmental factors and early onset idiopathic Parkinson's disease showed that there was a high prevalence rate of the disease among people who used drinking water from wells predominantly in rural areas (Rajput *et al.* 1986). The involvement and mechanisms of action of putative dopamine neurotoxic agents and environmental factors are covered more extensively in section 9.6.

Some evidence suggests that the low smoking rate among patients with Parkinson's disease is related to the neuropathology; the underlying changes which cause the motor dysfunction also cause the aversion to smoking. At present there is no direct proof of this, but some clinical observations indicate that this hypothesis is worth further investigation. These include a number of reports suggesting that Parkinsonian patients tend to exhibit a particular personality profile including introspection, inflexibility, and anhedonia (Duvoisin *et al.* 1981; Ward *et al.* 1983; Todes and Lees 1985; Rajput *et al.* 1987), and it is conceivable that this personality trait is incompatible with smoking. In one study (Golbe and Duvoisin 1985), it was found that respondents who gave reasons for not smoking, such as not liking the taste, or as describing it as a filthy habit, tended to have more rapidly progressing disease or earlier ages of diagnosis. In relation to these observations it has been claimed that people with ulcerative colitis exhibit obsessional personality traits (Bellini and Tansella 1976). Smoking has been reported to protect against this condition also, with one survey showing that among 25 patients who were suffering from both ulcerative colitis and Parkinson's disease, twenty had never smoked, and another three had not smoked for at least 10 years before onset of Parkinson's disease (Bihari and Lees, 1987). These authors suggested the interesting possibility that people suffering from these two diseases shared the same pre-morbid personality.

Idiopathic Parkinson's disease is a progressive disease, but it is not until 70–80 per cent of the striatal dopamine is depleted that the symptoms of akinesia, rigidity, and tremor become apparent (Bernheimer *et al.* 1973). As the personality traits are apparent before the symptoms of Parkinson's disease appear, it would be valuable to know whether they are associated with pathological changes in dopamine neurons which pre-date the onset of abnormal movements.

9.3 Anatomy

9.3.1 *Nigrostriatal and mesolimbic systems*

Originally, autoradiographic analysis of nicotinic-cholinergic binding sites utilised [^{3}H]ACh as ligand together with a muscarinic antagonist to inhibit binding to muscarinic cholinergic sites (Rainbow *et al.* 1984). However, the availability of radiolabelled nicotine with high specific activity has made possible the direct and detailed examination of nicotine binding sites. Autoradiographic studies which have used either ligand have shown that nicotinic-cholinergic binding sites are concentrated in discrete areas of rat brain (Rainbow *et al.* 1984; London *et al.* 1985; Clarke *et al.* 1985*b*; Harfstrand *et al.* 1988). Moreover, several of these sites such as the substantia nigra zona compacta (SNC), ventral tegmental area (VTA), striatum, and nucleus accumbens have the characteristic of containing a substantial proportion of the cell bodies or nerve terminals of brain dopamine pathways (see Fuxe *et al.*, Fig. 6.9, this volume, for a schematic representation of the mesolimbic pathway).

Since the most pronounced lesion in Parkinson's disease is degeneration of the nigro-striatal pathway, it is therefore not surprising that in animal models of nigro-striatal degeneration the loss of dopamine neurons is associated with a reduced number of nicotine binding sites. In one study (Clarke and Pert 1985) the neurotoxin 6-hydroxydopamine, which destroys catecholamine neurons, was injected unilaterally into the medial forebrain bundle of rats in order to cause degeneration of the ascending dopaminergic nigrostriatal and mesolimbic pathways. Five weeks later, several forebrain areas were analyzed for [^{3}H]nicotine binding sites by autoradiography. The lesion caused a 75–95 per cent reduction of dopamine in the striatum and a 30–40 per cent reduction of [^{3}H]nicotine binding in the striatum and nucleus accumbens in the hemisphere on the same side as the lesion (ipsilateral side). [^{3}H]nicotine binding sites were also greatly reduced in the ipsilateral SNC and VTA (Table 9.2). Thus it appears that a proportion of the binding sites are located on the dopamine nerve terminals in the striatum and nucleus accumbens and on the cell bodies and dendrites of dopamine neurons in the SNC and VTA.

These results agree with a previous *in vitro* binding study (Schwartz *et al.* 1984) in which injection of 6-hydroxydopamine into the cerebral ventricles 7

Table 9.2. [^{3}H]nicotine labelling and striatal dopamine content following unilateral 6-hydroxydopamine lesion of the medial forebrain bundle.

Site	Displaceable OD × 1000		Ratio lesioned/ unoperated (%)
	Unoperated side	Lesioned side	
Anterior striatum	140.3 ± 11.3	95.7 ± 12.6	67.3 ± 4.3
Midstriatum	118.3 ± 7.7	86.8 ± 11.4	70.6 ± 5.4
Posterior striatum	107.2 ± 5.0	81.3 ± 9.2	75.0 ± 5.6
Nucleus accumbens	117.9 ± 12.5	72.3 ± 8.8	61.4 ± 3.9
Frontoparietal cortex	144.4 ± 7.9	144.0 ± 8.3	99.8 ± 2.6
Anterior SN	168.1 ± 23.2	75.3 ± 17.2	45.0 ± 8.3
Posterior Sn	119.8 ± 18.8	82.4 ± 10.6	72.6 ± 7.6
pg D A/μg protein	166.2 ± 13.9	32.8 ± 18.5	19.6 ± 11.5

Five weeks after lesion, micropunches were taken from the striatum and dopamine content was measured by H P L C and electrochemical detection. Brain sections were incubated in [^{3}H](–)nicotine in the absence or presence of excess non-radioactive nicotine bitartrate (10 μM) to define displaceable binding. Regional levels of [^{3}H](–)nicotine binding were assessed by computer-assisted densitometry. Optical density (O D) values were calculated in individual rats (n = 6) by subtracting the mean O D of 4 sections labelled in the presence of excess non-radioactive nicotine from the mean O D of 4 adjacent sections labelled in the absence of non-radioactive nicotine. Lesion-induced changes in [^{3}H](–)nicotine binding were assessed by comparison with the unoperated side in each rat. Experimental values are means ± S E M. (From Clarke and Pert 1985.)

days prior to assay resulted in a 30 per cent reduction of striatal [^{3}H]acetylcholine binding to nicotinic sites. Here the binding decrement was due to loss of sites rather than reduced binding affinity. Another *in vitro* binding study has shown that 1 year after a unilateral 6-hydroxydopamine lesion of the medial forebrain bundle, [^{3}H](−)nicotine binding sites were reduced by 42 per cent in the ipsilateral striatum as compared with the control side (Reavill and Stolerman, unpublished data). These results indicate that there is a persistent loss of presynaptic nicotine binding sites associated with nigrostriatal degeneration. This contrasts with dopamine receptors which appear to proliferate on postsynaptic membranes in the striatum after a similar lesion (Reisine *et al.* 1979).

The loss of striatal [^{3}H]nicotine binding sites is unlikely to be due to toxin-induced postsynaptic degeneration of cholinergic interneurons because the cholinergic marker cholineacetyltransferase is unaltered, and the GABA marker glutamic acid decarboxylase is increased in the striatum after a 6-hydroxydopamine lesion (Reisine *et al.* 1979). These observations are consistent with the hypothesis that nicotine interacts with sites that lie on the cell bodies, dendrites, and terminals of the nigrostriatal and mesolimbic pathways.

9.3.2 *Hippocampal formation and superior colliculus*

Other areas which contain both nicotinic-cholinergic and dopamine markers are the hippocampal formation and the superficial layer of the superior colliculus. In the hippocampal formation nicotine sites are present in CA1 and the molecular layer of the dentate gyrus (Clarke *et al.* 1985*b*; London *et al.* 1985; Yamada *et al.* 1987). Also, anatomical and biochemical techniques have suggested that there is a dopamine-containing projection from the ventral tegmental area and substantia nigra to the dentate gyrus (Simon *et al.* 1979; Scatton *et al.* 1980; Verney *et al.* 1985). Autoradiographic mapping of brain dopamine receptors has demonstrated the presence of receptors of the D2 subtype in the stratum lacunosum moleculare (Bouthenet *et al.* 1987), and receptors of the D1 subtype in the molecular layer of the dentate gyrus and CA1 of the dorsal hippocampus (Savasta *et al.* 1986; Dawson *et al.* 1986). In this respect, it is interesting that in a recent post-mortem study on brains from patients with Parkinson's disease and Alzheimer's disease, it was shown that numbers of [^{3}H]nicotine binding sites were reduced in the hippocampus in both disorders, and furthermore, the phenomenon was observed in both demented and non-demented cases of Parkinson's disease (Perry *et al.* 1987; see Kellar and Wonnacott, Table 10.1, this volume).

9.3.3 *Pedunculopontine nucleus*

Finally, one other brain area worthy of mention is the nucleus tegmenti pedunculopontinus (pedunculopontine nucleus). The pedunculopontine nucleus has been proposed to play an important role in the extrapyramidal circuitry, and has extensive reciprocal connections with the basal ganglia including the substantia nigra and intermediate grey layer of the superior colliculus (Jackson and Crossman 1983; Saper and Loewy 1982; Beninato and Spencer 1986). The ascending projections to several forebrain areas including the substantia nigra are probably cholinergic (Fig. 9.1) (Kasa 1986; Woolf and Butcher 1986; Beninato and Spencer 1987), and it has been suggested that the pedunculopontine–nigral pathway provides an excitatory nicotinic-cholinergic input on to the dopaminergic neurons of the SNC (Clarke *et al.* 1987). However, others believe that the tegmental-nigral pathway originates in a group of non-cholinergic cells in the midbrain extrapyramidal area adjacent to the pedunculopontine nucleus (Rye *et al.* 1987). The projection to the superior colliculus is cholinergic and since the pedunculopontine nucleus has excitatory effects on other basal ganglia structures this pathway also may be excitatory (Beninato and Spencer 1986).

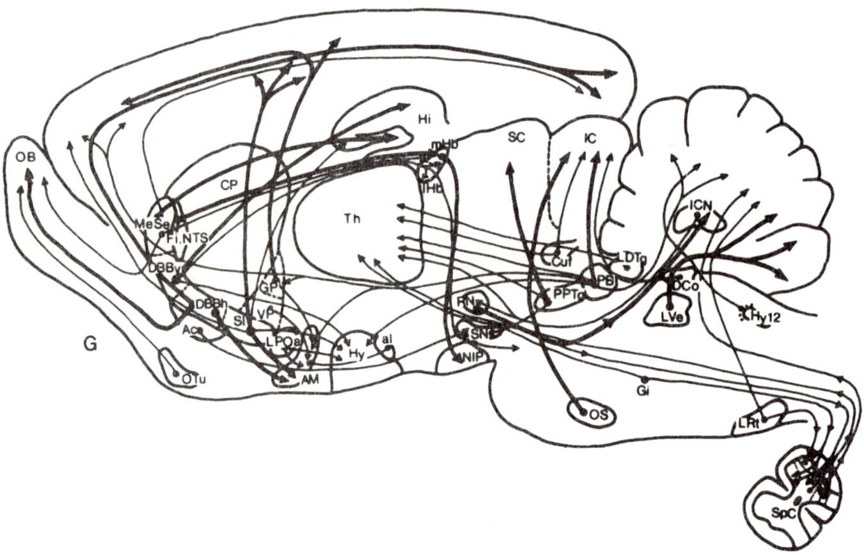

Fig. 9.1. Cholinergic pathways in the rat brain and spinal cord. Well-established routes are indicated by thick arrows; suspected routes by thin arrows. Abbreviations: al, ansa lenticularis; C P, caudate putamen; Cuf, cuneiform nucleus; D B Bh, nucleus of the horizontal limb of the diagonal band (Broca); D B Bv, nucleus of the vertical limb of the diagonal band (Broca); Fi, fimbriatus nucleus; Gi, gigantocellular reticular nucleus; G P, globus pallidus; Hi, hippocampus; Hy, hypothalamus; Hy12, hypoglossal nucleus; I C, inferior colliculus; I C N, intracerebral nuclei; L D Tg, laterodorsal tegmental nucleus; lHb, lateral habenula; L P Oa, lateral preoptic area; L Rt, lateral reticular nucleus; L Ve, lateral vestibular nucleus: MeSe, medial septum; mHb, medial habenula; N I P, interpeduncular nucleus; N T S, triangular septal nucleus; O B, olfactory bulb; O S, superior olive; O Tu, olfactory tubercle; P B, parabrachial nucleus; P P Tg, pedunculopontine tegmental nucleus; R N, red nucleus; S C, superior colliculus; S I, substantia innominata; S N, substantia nigra; SpC, spinal cord; Th, thalamus; V P, ventral pallidum. (From Kasa 1986.)

In a human post-mortem study of Parkinsonian brains it was found that the pedunculopontine nucleus contained significant numbers of Lewy bodies, which are characteristic of areas of cell loss of the brain in Parkinson's disease (Ohama and Ikuta 1976). Also, in a recent comparative morphometric investigation, the pedunculopontine nucleus was examined in the post-mortem brains of patients with Parkinson's disease, Alzheimer's disease, and progressive supranuclear palsy (Jellinger 1988). There was a much more severe loss of cells in the pedunculopontine nucleus in Parkinson's disease and progressive supranuclear palsy than in Alzheimers disease, and Lewy bodies were observed in 6–39 per cent of the remaining cells in Parkinson's disease but only very infrequently in the other two disorders. It was suggested that the deficits in motor performance in Parkinson's disease and progressive supranuclear palsy may be related to the pathology of cholinergic neurons in the pedunculopontine nucleus. In this respect it is interesting that the resting tremor in Parkinson's disease is possibly related to mechanisms in the pedunculopontine nucleus area (Kaelber 1963), and it is conceivable that dysfunction of efferent nicotinic-cholinergic projections may be important in view of the neuropathology.

Also, the oculomotor deficits, particularly those relating to saccadic eye movements which have been observed in Parkinson's disease (White *et al.* 1983), may be associated with degeneration of the projection from the pedunculopontine nucleus to the superior colliculus (an important visual and motor centre), resulting in imbalance between these and other basal ganglia nuclei (Beninato and Spencer 1986). However, a reappraisal of the contribution of pedunculopontine-collicular neuropathology to oculomotor deficits may be necessary with the recent demonstration of a dopamine pathway from the substantia nigra to the superior colliculus which is sensitive to a dopamine neurotoxin (Takada *et al.* 1988).

Whether disturbances in the pedunculopontine nucleus contribute to or result from pathological changes in ascending dopamine pathways remains to be seen. However, in this respect it is interesting that injection of the excitatory amino acid, kainic acid, into the pontine tegmentum including the pedunculopontine nucleus results in degeneration of dopamine neurons in the substantia nigra and loss of striatal tyrosine hydroxylase activity, so it is conceivable that dysfunction in the pedunculopontine-nigral projection may have a bearing on the aetiology of Parkinson's disease (McGeer and McGeer 1984). The involvement of nicotinic mechanisms in the pedunculopontine-nigral pathway is further discussed in section 9.4.2.

9.4 Physiology and biochemistry

9.4.1 *Pharmacokinetics*

Before reviewing the pharmacological effects of nicotine, it is important to take account of the plasma concentrations of nicotine found in smokers, and to compare these with the doses of nicotine required to attain these concentrations in experimental models. Russell *et al.* (1980) have shown that smokers who inhale have plasma nicotine concentrations of 4–72 ng ml^{-1} (25–444 nM). In experiments with rats, nicotine (0.1 mg kg^{-1} subcutaneously) produced a peak plasma nicotine concentration of 35 ng ml^{-1} (216 nM) which was within the range of plasma nicotine concentrations in smokers (Pratt *et al.* 1983). A larger dose (0.4 mg kg^{-1} subcutaneously) of nicotine resulted in a plasma nicotine concentration of 146 ng ml^{-1} (901 nM) which was above the range of plasma nicotine concentrations present in smokers. These data indicate that in experimental studies where nicotine action in rats is being investigated, the optimum dose of nicotine is 0.1 mg kg^{-1}. However, these comparisons may need further consideration to take account of bolus concentrations of nicotine produced by individual puffs in cigarettes (see Russell, chapter 11, this volume).

9.4.2 *Nigrostriatal system*

Compared to the effects seen after administration of dopamine agonists or antagonists, nicotine and cigarette smoke have small but significant effects on the forebrain dopamine systems. At the cellular level, nicotine has been shown to exert its actions on both cell bodies and nerve terminals of the nigrostriatal pathway, resulting in dopamine release from the nerve endings. In an early investigation in which a large dose of nicotine was injected subcutaneously, there was an increase in the cellular fluorescence intensity in the dopamine neurons of the SNC after 30 min (Lichtensteiger *et al.* 1976). In the same study, nicotine caused an initial transient depression of the activity of single neurons followed by a prolonged increase in the firing rate of SNC neurons, and a concommitant increase in the concentration of the dopamine metabolite homovanillic acid in the striatum, probably reflecting enhanced release of dopamine from the nerve terminals of the nigrostriatal pathway. These results were later confirmed, and the firing neurons of the SNC identified as part of the nigrostriatal pathway, by antidromic stimulation. Also, the increased firing could be blocked by the nicotinic antagonist dihydroβerythroidine, but not by the muscarinic antagonist atropine (Lichtensteiger *et al.* 1982; see Clarke, Fig. 5.6, this volume). The resulting small increase in striatal dopamine turnover was similar to that caused by electrical stimulation of nigral neurons at the same frequency as that seen after nicotine administration.

Similar results were seen by Clarke *et al.* (1985*a*) in which subcutaneous or intravenous administration of nicotine in anaesthetized rats were both found to dose-dependently increase the firing rate of SNC neurons. Here the intravenous dose of nicotine was within the range of plasma nicotine levels present in smokers. Mecamylamine, a ganglion blocker which does enter the brain, blocked this response, but chlorisondamine, which poorly penetrates the brain, was without effect. Results have also shown that the zona compacta neurons may be tonically excited by endogenous acetylcholine because physostigmine increased unit activity (Lichtensteiger *et al.* 1976). A later study by Grenhoff *et al.* (1986) showed that there was no significant effect of mecamylamine on the firing rate of SNC cells in the rat, which would suggest the lack of a tonic nicotinic input.

One of the inputs to the nicotinic-cholinergic receptors of the SNC is that from the pedunculopontine nucleus. Injection of kainic acid into this area in the rat elicits a rapid excitation of SNC dopamine neurons, and this can be prevented by intravenous injection of mecamylamine at a dose which blocks the neuronal excitation caused by nicotine (Clarke *et al.* 1987).

In vitro experiments have also shown that nicotine, or acetylcholine in the presence of atropine and physostigmine, can release [^{3}H]dopamine from brain preparations, and evidence suggests that this mechanism is mediated through interaction with presynaptic receptors. Acetylcholine was able to stimulate the release of newly synthesized [^{3}H]dopamine by a calcium dependent process, and this was partially blocked by pempidine or atropine, indicating the contribution of both nicotinic and muscarinic mechanisms (Giorguieff *et al.* 1977). Also, as calcium is required for the excitation-induced secretion of transmitter after depolarization of the nerve terminal membrane, at least some of the physiological requirements for transmitter release have been met.

In preparations of perfused rat striatal slices, a very large concentration of nicotine (1 mM) evoked release of [^{3}H]dopamine in a calcium dependent manner (Westfall 1974). However, in a later investigation it was concluded that the nicotine-stimulated release of [^{3}H]dopamine from rat striatal slices was not dependent on calcium, and that nicotine was displacing the labelled catecholamine from vesicle stores (Arqueros *et al.* 1978). Both of these studies used high nicotine concentrations, and subsequent experiments where lower concentrations have been used have usually shown that nicotine evoked release of [^{3}H]dopamine does depend on the presence of extracellular calcium.

Giorguieff-Chesselet *et al.* (1979) showed that nicotine (1 μM) stimulated release of newly synthesized [^{3}H]dopamine from rat striatal slices in a calcium-dependent manner, and that this process was completely blocked by nicotinic antagonists such as pempidine. The dopamine release mechanism is probably mediated by presynaptic receptors located on the nerve terminals of the nigrostriatal pathway, because acetylcholine in the presence of atropine stimulates efflux of [^{3}H]dopamine from striatal synaptosomes (De Belleroche and Bradford 1978). In another study in which high concentrations of nicotine were used, the two isomers of nicotine were equipotent at releasing synaptosomal [^{3}H]dopamine, and pempidine but not hexamethonium attenuated this release (Connelly and Littleton 1983) which raises the question of whether this was a receptor-mediated process. However, a recent study has demonstrated that [^{3}H]dopamine release from rat striatal synaptosomes is stimulated by (−)nicotine (1 μM) by a calcium-dependent process, and the

(−)-isomer is one hundred times more potent than the (+)-isomer (Rapier *et al.* 1988). Others have shown that although high doses of nicotine or nicotine agonists stimulated [^{3}H]dopamine release from striatal micropunches and synaptosomes, nicotinic antagonists such as hexamethonium did not block [^{3}H]dopamine efflux (Sakurai *et al.* 1982), and that the release process was not calcium dependent (Takano *et al.* 1983).

Generally, experiments which have utilized low concentrations of nicotine have demonstrated that nicotine-evoked release of dopamine from striatal tissue preparations is a calcium-dependent process, and is blocked by nicotinic antagonists. Another contentious issue is whether the release of neurotransmitter is mediated via exocytosis of vesicular contents through the presynaptic membrane, or by release of extravesicular transmitter through presynaptic membrane channels. Detailed discussion of this question is beyond the scope of this chapter, apart from one study which investigated the origin of presynaptic dopamine released by nicotine. Winch and Balfour (1988) looked at the effect of reserpine on nicotine (1 μM) evoked release of [^{3}H]dopamine from rat striatal slices. They found that reserpine (which releases dopamine from granular stores) had no significant effect on the release of [^{3}H]dopamine elicited by nicotine, and concluded that nicotine displaced [^{3}H]dopamine principally from an extra-vesicular compartment rather than by exocytosis.

One would expect that the nicotine-stimulated release of dopamine would precipitate accelerated turnover and increased levels of metabolites. However, in general results show that the changes in dopamine turnover in the striatum are either nil or small, even with large doses of nicotine. Using histochemical fluorescence analysis Andersson *et al.* (1981) looked at the effect of nicotine on dopamine turnover in rat brain after blocking dopamine synthesis with the tyrosine hydroxylase inhibitor alpha-methyl-para-tyrosine (AMPT). Striatal dopamine depletion was only apparent in the medial sector (see Fuxe *et al.*, chapter 6, this volume). Conversely, mecamylamine has been reported to decrease striatal levels of homovanillic acid in rats treated with AMPT, and to decrease haloperidol induced elevation of homovanillic acid, although this effect was only observed if egress of the metabolite from the brain was blocked by probenecid. This suggested that nicotinic mechanisms fulfil only a minor role in striatal dopamine synthesis and release (Ahtee and Kaakkola 1978). Others have reported little or no change in striatal dopamine turnover. Lichtensteiger *et al.* (1976) found only a modest nicotine-induced rise in striatal homovanillic acid, and others showed that nicotine caused no change in accumulation of the dopamine precursor 3,4-dihydroxyphenylalanine (DOPA) in the striatum of rats which had been pre-treated with a DOPA decarboxylase inhibitor (Mitchell *et al.* 1987).

It is possible that *in vivo* dialysis studies will show that nicotine has a more profound effect on striatal dopamine release in awake rats. In this technique, dialysis tubes are implanted in relevant brain areas. After recovery, Ringer solution is pumped through at a constant rate in the freely moving rats and the effluent is analyzed. One recent study has shown that nicotine produces a small but significant (30 per cent) increase in striatal dopamine release, although the dose of nicotine (nicotine bitartrate) used was 3 μmol kg^{-1} which is equivalent to 0.45 mg kg^{-1} expressed as nicotine base (Damsma *et al.* 1988). However, this peripheral dose of nicotine is one of the lowest to date which has been reported to release striatal dopamine.

Generally, experiments which have investigated the effects of nicotine on [^{3}H]dopamine release in striatal tissue preparations have shown that effects are observed at a threshold concentration of approximately 1 μM nicotine and above. This concentration of nicotine is above that found in the plasma of smokers, and this raises the question of whether striatal nicotine–dopamine interaction is important for physiological responses to smoking. However, this does not exclude the possibility that striatal dopamine release is affected by nicotine action in the substantia nigra because, as discussed above, neurons in this structure change their firing characteristics in response to low doses of nicotine.

9.4.3 *Mesolimbic system*

Investigations into the effects of nicotine on dopamine in the mesolimbic system have not been as extensive as those for the nigrostriatal pathway. This is perhaps surprising given the fact that dopamine mechanisms in the VTA and the nucleus accumbens have been implicated in reward and motor activity. Thus although dopamine dysfunction in the nigro-striatal system may contribute to symptoms such as tremor and rigidity, the mesolimbic system may have implications for the reported inverse correlation between smoking and Parkinson's disease. The behavioural implications of this are further discussed in section 9.5.

Electrophysiological experiments by Grenhoff *et al.* (1986) have demonstrated that nicotine has characteristic effects on the firing properties of dopamine cells of the VTA. A higher proportion of cells in this area were of the burst-firing type compared with cells in the SNC. The effect of nicotine administration was to increase the burst firing without significantly affecting the firing rate, whereas in the SNC both burst firing and firing rate were increased. Mecamylamine decreased the firing rate of neurons in the VTA without altering burst firing, which suggested a tonic nicotinic input. The authors stated that the pattern of firing seen before and after nicotine administration indicated that nicotine may exert both modulatory and excitatory effects on the midbrain neurons, and as the dose of nicotine used (0.16 mg kg^{-1}, nicotine base) was compatible with the intake of nicotine after two cigarettes in humans, the increased burst firing observed after nicotine may underly its stimulatory action.

In another study of the firing activity of midbrain dopamine neurons Mereu *et al.* (1987) found that small doses of intravenous nicotine in rats produced a dose-related increase in the firing rate of cells in the VTA and SNC, and these effects could be prevented by mecamylamine. Moreover, the action of nicotine was more than three times more potent in the VTA than in the substantia nigra. The authors stressed the similarity of the effects between nicotine and abused drugs such as morphine, heroin and cocaine which preferentially stimulate mesolimbic and mesocortical dopaminergic transmission.

As in the case of the striatum, administration of mecamylamine to rats reduced the rate of disappearance of limbic dopamine after AMPT treatment (Ahtee and Kaakkola 1978). Also, nicotine (0.3 mg kg^{-1}) produced a selective reduction of dopamine stores in the anterior nucleus accumbens, but not in other limbic or striatal areas (Andersson *et al.* 1981). It was pointed out that nicotine seemed to preferentially produce its dopamine depleting effect in the dopamine-terminal fields which lacked cholecystokinin immunoreactivity. In a separate experiment in which rats were acutely exposed to unfiltered cigarette smoke after tyrosine hydroxylase inhibition, there was increased dopamine depletion in the anterior nucleus accumbens, and lateral posterior tuberculum olfactorium (cholecystokinin negative areas), but not in the striatum. Exposure to filtered smoke did not produce any changes in dopamine depletion (Fuxe *et al.* 1986; see Fuxe *et al.*, Fig. 6.10, this volume). Furthermore, in a study which looked at the effect of nicotine administration on DOPA concentration after inhibition of DOPA decarboxylase, nicotine significantly enhanced DOPA accumulation in the nucleus accumbens, but not in the striatum (Mitchell *et al.* 1987).

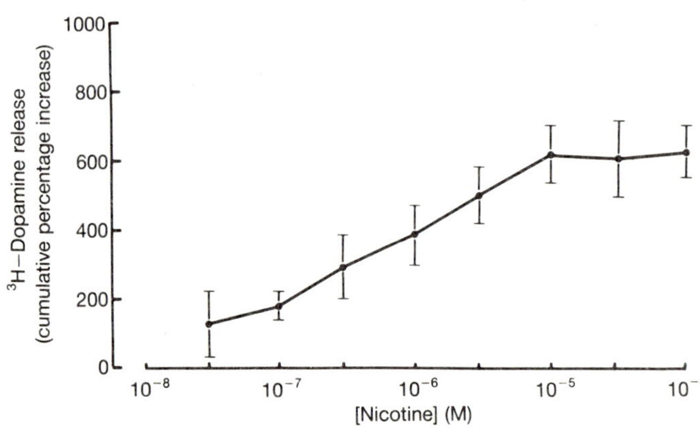

Fig. 9.2. Effect of nicotine on [^{3}H]dopamine release from rat nucleus accumbens *in vitro*. Stimulation of [^{3}H]dopamine release expressed as cumulative percentage increase above baseline during a 10-min incubation with nicotine. Each point represents the mean ± SEM of 4–7 experiments. Nicotine caused stimulation of [^{3}H]dopamine release at concentrations of 30 nM and above. The EC_{50} of nicotine-stimulated [^{3}H]dopamine release was 0.4 μM. (From Rowell *et al.* 1987.)

Few studies have specifically looked at the direct effect of nicotine on dopamine release in the nucleus accumbens. However, in a recent *in vitro* experiment Rowell *et al.* (1987) showed that nicotine in concentrations as low as 30 nM was able to release [^{3}H]dopamine from minced nucleus accumbens tissue, and the process was dose dependent and calcium dependent (Fig. 9.2). Other nicotinic and muscarinic agonists and acetylcholine were less potent than nicotine, and nicotinic antagonists, but not muscarinic antagonists, partially blocked the effect. The ratio of dopamine to its metabolites was the same before and after nicotine. As the dopamine terminal field has been

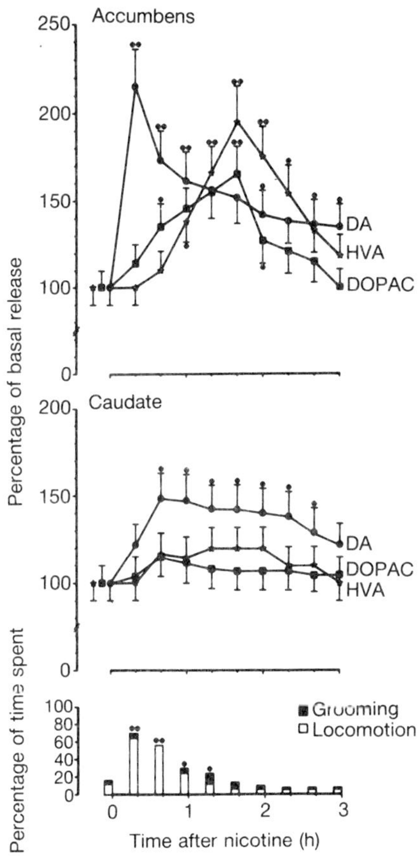

Fig. 9.3. Effect of nicotine (0.6 mg kg^{-1} subcutaneously) on *in vivo* release of dopamine (DA), homovanillic acid (HVA), and DOPAC from nucleus accumbens and from dorsal striatum, and on spontaneous behaviour in the rat. $n = 6$; $*p < 0.05$ $**p < 0.001$ versus results obtained after saline. Nicotine stimulated dopamine release by 100 per cent in the nucleus accumbens and by 50 per cent in the dorsal striatum, whereas HVA and DOPAC output were increased only in the nucleus accumbens. (From Imperato *et al.* 1986.)

separated from its cells of origin in this preparation, presumably nicotine is exerting its effects by acting on the presynaptic nicotine receptors (or on interneurons) in the nucleus accumbens.

The above results agree with the observed effects of nicotine on dopamine release and metabolism as measured by *in vivo* dialysis in freely moving rats (Imperato *et al.* 1986). Peripheral administration of nicotine dose-dependently increased the release of dopamine in the nucleus accumbens and striatum, and increased output of homovanillic acid and dihydroxyphenylacetic acid (DOPAC) from the nucleus accumbens, but not from the *striatum*. Figure 9.3 shows the effects of nicotine (0.6 mg kg^{-1} subcutaneously) on *in vivo* dopamine release and metabolism in the nucleus accumbens and striatum in freely moving rats, and its concurrent effect on spontaneous behaviour. A comparison of the ED_{50}'s revealed that nicotine was five times as potent for releasing dopamine in the nucleus accumbens (0.32 mg kg^{-1} subcutaneously) than in the striatum (1.85 mg kg^{-1} subcutaneously), and that the nicotine evoked dopamine release in the nucleus accumbens was blocked by mecamylamine but not hexamethonium (Imperato *et al.* 1986). Also, the ED_{50} nicotine dose required to elicit dopamine release in the nucleus accumbens was not much higher than that required to produce plasma nicotine levels similar to those found in smokers. This *in vivo* investigation indicates that limbic dopamine release and metabolism is more sensitive to the effects of nicotine than that in the striatum, although it is unknown to what extent nicotine receptors in the nucleus accumbens and VTA contribute to these overall effects.

Overall, these results show that the doses of nicotine which enhance dopamine release in the nucleus accumbens are in close agreement with the doses required to produce behavioural effects, and are broadly equivalent to the levels of plasma nicotine found in smokers.

9.5 Behaviour

The behavioural spectrum displayed by nicotine in several animal models does bear some similarity to that elicited by dopamine releasing agents. In naive rats, subcutaneous nicotine injection initially depresses locomotor activity, and this is followed by a period of locomotor stimulation. With chronic nicotine treatment, tolerance develops so that the initial depressant phase disappears and is replaced by enhanced activity. The motor effects are dose-dependent and can be blocked by central but not peripheral nicotine antagonists (Clarke and Kumar 1983*a*; Clarke and Kumar 1983*b*). Moreover, intraventricular injection of the nicotine agonist cytisine in naive rats produces the characteristic locomotor depression/excitation profile (Pert and Clarke 1987). Dopamine releasing drugs such as amphetamine also increase locomotor activity in rats, and the nucleus accumbens appears to be more important for this response than the striatum (Roberts *et al.* 1975).

Several studies have indicated that the locomotor stimulation produced by nicotine is mediated by central dopaminergic mechanisms. Peripheral nicotine administration increased output of dopamine and its metabolites in the nucleus accumbens concurrently with increased locomotor stimulation (Fig. 9.3, Imperato *et al.* 1986). Therefore the enhanced activity produced by nicotine appears to result from activation of the mesolimbic system.

However, there are some puzzling aspects concerning nicotine–dopamine interaction in the mesolimbic system. A low concentration of nicotine enhances dopamine release from nucleus accumbens tissue *in vitro* (Rowell *et al.* 1987), and direct cytisine injection in to the VTA raises dopamine in the nucleus accumbens and increases locomotor activity (Pert and Clarke 1987; Reavill and Stolerman 1988). This locomotor stimulation is attenuated by mecamylamine or by a 6-hydroxydopamine lesion of the nucleus accumbens (Pert and Clarke 1987). Conversely, direct cytisine injection in to the nucleus accumbens *in vivo* does not increase locomotor activity or affect dopamine levels in this structure (Pert and Clarke 1987; Reavill and Stolerman 1988). Perhaps the apparent conflict between the *in vitro* and *in vivo* data reflects the possibility that the status of neuronal activity within the mesolimbic dopamine pathway has a bearing on nicotine-dopamine interaction within the nucleus accumbens.

The circling rat has also been used to delineate the mechanism of action of nicotine. In this model, unilateral injection of 6-hydroxydopamine into the medial forebrain bundle causes degeneration of the ascending dopamine pathways, and permanent loss of dopamine in the nucleus accumbens and striatum in the ipsilateral hemisphere. While it would be simplistic to regard the circling rat as a model for Parkinson's disease, it has certainly found use as a behavioural test for dopamine agonists. Injection of an indirectly acting dopamine agonist such as amphetamine causes circling towards the lesion (ipsiversive), due to dopamine being released only in the intact striatum and nucleus accumbens, providing postural asymmetry and locomotor drive respectively. Direct dopamine agonists such as apomorphine produce circling away from the side of the lesion (contraversive) due to net action on supersensitive receptors in the lesioned hemisphere (Ungerstedt 1971).

Kaakkola (1981) examined the effect of nicotine on rotation in this behavioural model and showed that nicotine itself did not significantly increase circling, but that it did enhance amphetamine-induced ipsiversive rotation. However, those rats were not tolerant to nicotine at the time of testing so the motor depressant effects of nicotine may have masked the stimulatory action on motor activity, thus reducing the circling response. Recently Lapin *et al.* (1987) suggested that dysfunction of dopamine neurons in Parkinson's disease may pre-date the morbid phase by a long time, and that this interferes with acquisition of the smoking habit. Certainly the dopaminergic mesolimbic pathway is important in reward mechanisms (Wise and Bozarth 1984), and intact dopamine mechanisms in the nucleus accumbens appear to be important for the acquisition of schedule-induced self-injection of nicotine (see Swedberg *et al.*, chapter 2, this volume). A study by Singer *et al.* (1982) showed that a catecholamine depleting 6-hydroxy-dopamine lesion of the nucleus accumbens reduced acquisition of schedule-induced nicotine self-injection in the rat (Fig. 9.4). Although the possibility was not discussed that the lesion had interfered with the ability of the rats to respond rather than reduce the reward value of nicotine, the results were consistent with the hypothesis that viable dopamine mechanisms are important for this behaviour.

Lapin *et al.* (1987) looked at the relationship between changes in dopamine metabolism and the development of tolerance to the behavioural effects of nicotine in rats which had received a unilateral 6-hydroxydopamine lesion of the substantia nigra. In lesioned rats an acute nicotine injection initially produced sedation and hypothermia, but with chronic nicotine administration these effects disappeared and ipsiversive circling and stereotyped behaviour gradually developed. In rats made tolerant to chronic nicotine and then lesioned, the first post-operative dose of nicotine caused ipsiversive rotation. The ratio of dopamine metabolites to dopamine was measured in several areas including the striatum, nucleus accumbens, substantia nigra, and VTA, and

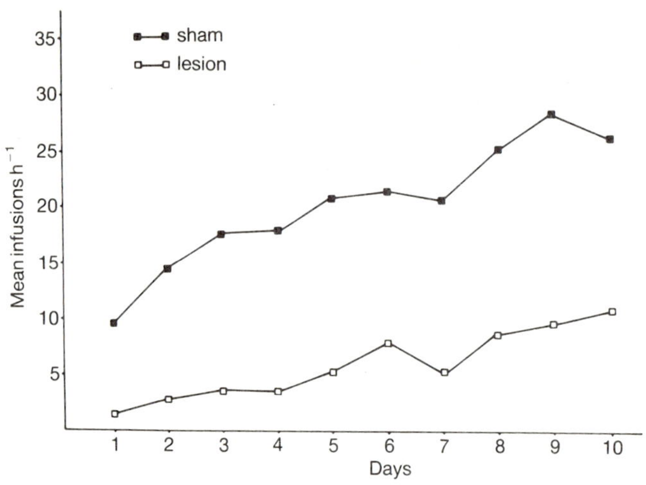

Fig. 9.4. Mean number of nicotine self-infusions for 6 hydroxydopamine lesioned and sham lesioned rats over 10 days. Rats were maintained at 80 per cent of free-feeding weight, catheterized in the right jugular vein, and bilaterally injected with 6 hydroxydopamine (lesion) or ascorbic acid/saline (sham) in to the nucleus accumbens (N A S). After 2 days recovery, rats were given daily 1-h sessions on a fixed-time (60 s) (F T-60) food-delivery schedule in which nicotine(+)tartrate (0.1 mg^{-1} kg^{-1} per infusion) was available. Acquisition of nicotine self-injection behaviour was significantly reduced over the 10-day period in lesioned animals when compared to shams. (From Singer *et al.* 1982.)

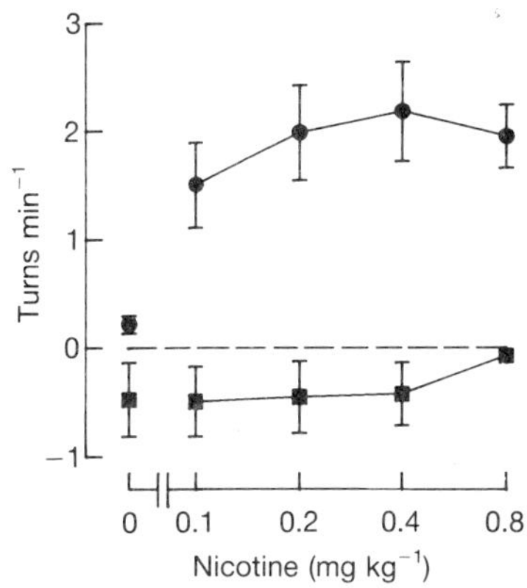

Fig. 9.5. Dose–response relations for rotational behaviour produced by nicotine in rats with unilateral, 6-hydroxydopamine lesions of the nigrostriatal dopamine pathway. After recovery from the lesion, rats were treated with nicotine (0.4 mg kg^{-1} subcutaneously) once daily for at least 2 weeks, and exposed to the test apparatus for 1 h after each injection. The effects of saline (squares) or a range of doses of nicotine (circles) were determined in random order in 1-h tests beginning immediately after injections. (From Reavill and Stolerman 1987.)

no difference was found in controls or rats which had been lesioned prior to or during nicotine injections. It was concluded that the changes which occur that allow disappearance of sedation, and the appearance of rotation and stereotypy, do not seem to depend on changes in dopamine metabolism.

Other experiments have extended these studies of the interaction of nicotinic and dopaminergic mechanisms using the circling model in rats which had been made tolerant to nicotine (Reavill and Stolerman 1987) as distinct from naive rats (Kaakkola 1981). It was found that over a period of 5–60 min after injection, nicotine (0.1–0.8 mg kg^{-1}) did significantly increase ipsiversive rotation (Fig. 9.5), although the circling response was not as strong as that produced by amphetamine. Also, mecamylamine attenuated nicotine-induced circling, and haloperidol blocked both amphetamine- and nicotine-induced circling. In a separate experiment, the effects of nicotine (0.1 and 0.4 mg kg^{-1}) on the dose–response curve for amphetamine-induced rotation were examined. A small dose of amphetamine (0.06 mg kg^{-1}) significantly increased ipsiversive rotation. Circling induced by amphetamine (0.19 mg kg^{-1}) was strongly augmented by nicotine at both doses, whereas maximal rotation caused by amphetamine (0.6 mg kg^{-1}) was not altered by nicotine (Reavill and Stolerman 1987). These results are consistent with the premise that one action of nicotine is to release dopamine in the central nervous system.

The stimulant effect of nicotine on motor behaviour is rather small and the measurement of locomotor activity is rather a crude method to detect the subtle changes produced by some drugs. The drug discrimination assay is a very sensitive technique which allows one to objectively measure the effects of nicotine at smaller doses. In this technique rats are trained to press one of two bars after nicotine injection to gain a food reward, and to press the other bar after saline injection (see Swedberg *et al.*, chapter 2, this volume). Presses on the incorrect bar go unrewarded. Rats are able to learn to discriminate small doses of nicotine to a high degree of accuracy, and studies have shown that a small number of nicotinic agonists produce nicotine-like effects that can be blocked by centrally-acting nicotine antagonists. Moreover, very few non-nicotinic drugs produce nicotine-like effects in this procedure, the exception being partial responses with the dopamine agonists amphetamine, apomorphine, and SKF 38393 (Schechter and Rosecrans 1972; Chance *et al.* 1977; Stolerman *et al.* 1984; Reavill and Stolerman 1987).

Rosecrans and Chance (1977) have shown that rats which have been depleted of cerebral dopamine by neonatal 6-hydroxydopamine treatment are still able to learn to discriminate nicotine, but there was a trend for the mecamylamine dose required to attenuate nicotine-appropriate responding to be reduced. In experiments designed to locate the brain areas responsible for the nicotine-discriminative stimulus, it was found that trained rats pressed the nicotine-appropriate bar when nicotine was injected bilaterally into the dorsal

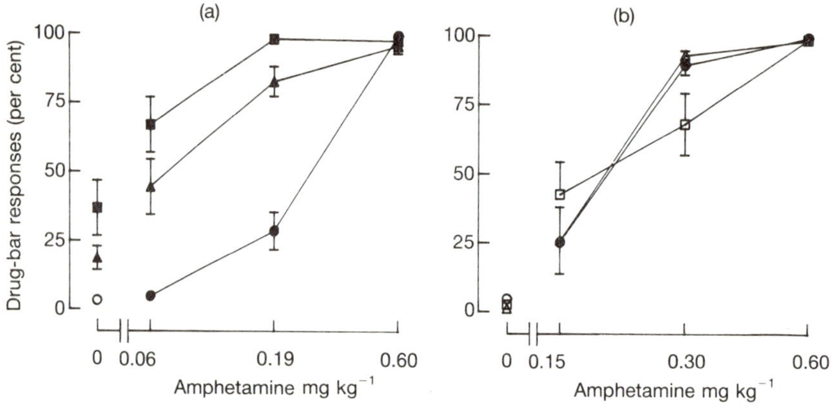

Fig. 9.6. Effect of nicotine and mecamylamine in rats trained to discriminate 0.6 mg kg^{-1} of amphetamine from saline. (a) Dose-response curves for amphetamine in absence (●) and in presence of 0.1 mg kg^{-1} (▲) or 0.2 mg kg^{-1} (■) of nicotine. Injections were given 15 min before tests. Points above zero on horizontal scale show effects of saline (○) and of nicotine in the absence of amphetamine. (b) Dose-response curves for amphetamine in absence (●) and in presence of 0.5 mg kg^{-1} (□) or 1.0 mg kg^{-1} (△) of mecamylamine. Injections were given 30 min (mecamylamine) or 15 min (amphetamine) before tests. Points above zero on horizontal scale show effects of saline (○) and of mecamylamine in the absence of amphetamine. All drugs were administered subcutaneously. Each point represents the mean ± SEM of observations made in 5-min extinction tests, with overlapping SEM omitted for clarity (n = 8). (From Reavill and Stolerman 1987.)

hippocampus, and this effect was blocked by depleting dopamine (Rosecrans *et al.* 1976).

In studies which have looked at the interaction of nicotinic and dopaminergic mechanisms, the effects of nicotine on the amphetamine cue and the effect of neuroleptic treatment on nicotine discrimination have been examined (Reavill and Stolerman 1987). In rats trained to discriminate amphetamine (0.6 mg kg^{-1}) from saline, amphetamine (0.06 mg kg^{-1}) did not

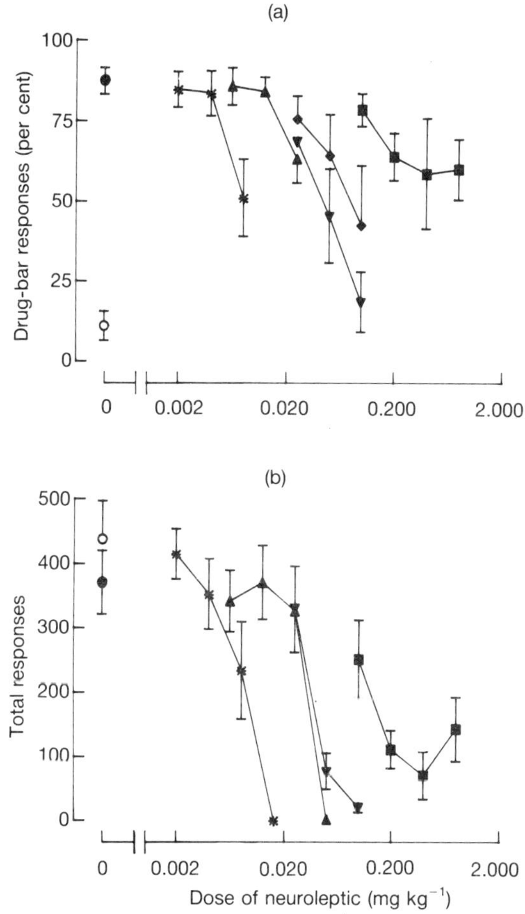

Fig. 9.7. Effects of neuroleptics on discriminative stimulus effect of nicotine (a) and total number of responses (b), in rats trained to discriminate 0.1 mg kg^{-1} of nicotine (●) from saline (○). Injections were given 45 min (S C H 23390), 1 h (droperidol or haloperidol), 3 h (pimozide), or 15 min (nicotine or saline) before tests. Graphs show results for S C H 23390 (*, n = 8), droperidol (▲, n = 6), haloperidol (▼, n = 8), and pimozide (■, n = 6). All drugs were administered subcutaneously. Each point represents the mean ± S E M of observations made in 5-min extinction tests, and overlapping S E M's are omitted for clarity. (From Reavill and Stolerman 1987.)

increase drug-appropriate responding, but nicotine (0.1 and 0.2 mg kg^{-1}) enhanced amphetamine-appropriate responding at that dose. The small amount of drug-appropriate responding with amphetamine (0.19 mg kg^{-1}) was greatly enhanced by nicotine (0.1 and 0.2 mg kg^{-1}) but the discriminative response to amphetamine (0.6 mg kg^{-1}) was not altered by nicotine (Fig. 9.6). As previous work has suggested that a major part of the amphetamine discriminative stimulus is mediated by dopamine mechanisms (Nielsen and Jepsen 1985), these present behavioural results are consistent with nicotine having a dopamine-releasing action.

The study with neuroleptics showed that haloperidol (0.1 mg kg^{-1}) and SCH 23390 (0.08 mg kg^{-1}) reduced the discriminative response to nicotine (0.1 mg kg^{-1}) whereas droperidol and pimozide were without significant effect (Fig. 9.7). It was reasoned that the neuroleptic attenuation of the nicotine cue may be mediated by action on dopamine D1 receptor mechanisms as SCH 23390 is a specific D1 antagonist, and haloperidol is principally a D2 antagonist with some D1 antagonist activity, whereas droperidol and pimozide are almost exclusively D2 antagonists. The overall conclusion from these experiments was that dopamine D1 receptors may mediate a component of the nicotine cue. In this respect it is interesting that nicotine-induced reduction of plasma luteinizing hormone is also blocked by SCH 23390 (Fuxe *et al.* 1987; see Fuxe *et al.*, chapter 6, this volume). Further experiments with dopamine agonists and antagonists are required to delineate the interaction of nicotinic and dopaminergic mechanisms, and the recent availability of selectively acting dopamine D1 and D2 compounds now opens up new areas of investigation.

9.6 Smoking, monoamine oxidase and MPTP

The suggestion that environmental factors play a role in the pathogenesis of Parkinson's disease has assumed a new-found interest with the discovery that a synthetic analogue of meperidine, MPTP (1-methyl-4-phenyl-1,2,3,6-tetrahydropyridine), caused nigro-striatal degeneration, and produced a prolonged clinical syndrome in man and animals very similar to that seen in Parkinson's disease (Langston *et al.* 1983; Burns *et al.* 1983). Endogenous or environmental MPTP has not been detected, but this does not preclude the possibility that other similar molecules exist (Testa *et al.* 1985), because MPTP does have a similar structure to some herbicides such as paraquat (Barbeau *et al.* 1987).

The dopamine neurodegenerative effect of MPTP is not due to this molecule *per se* but to a metabolite (1-methyl-4-phenylpyridine; MPP^+) which is formed via an intermediate (1-methyl-4-phenyl-1,2-dihydropyridine) with the generation of free radicals. The free radicals have been suggested to be responsible for causing cell damage. A major step forward in elucidating the mechanism of action of MPTP was the discovery that monoamine oxidase-B inhibitors, such as deprenyl and pargyline, protected the dopamine neurons against damage by blocking biotransformation of MPTP by monoamine oxidase-B (Cohen *et al.* 1985). Since smokers have been shown to have lower levels of platelet monoamine oxidase-B than non-smokers (Oreland *et al.* 1981), and a monoamine oxidase inhibitor has been demonstrated in human urine (Glover *et al.* 1980), so the reduced ability of smokers to transform any putative endogenous MPTP-like molecules to neurotoxic intermediates would perhaps explain the reduced incidence of Parkinson's disease among smokers. However, in a post-mortem study, there was no difference in monoamine oxidase-B levels in the autopsied brains of control and Parkinsonian patients, although there was less enzyme activity in the platelets of heavy smokers than in non-smokers (Yong and Perry 1986). In this same study, several compounds which are present in the tissues of smokers, namely, nicotine, cotinine, thiocyanate, and hydrazine were tested for their ability to protect against the MPTP-induced depletion of striatal dopamine in mice. Nicotine and cotinine were not effective, but thiocyanate had some protective action, and hydrazine afforded considerable protection. None of these compounds inhibited monoamine oxidase-B in an *in vitro* assay. The inactivity of nicotine was confirmed in another study which showed that amphetamine potentiated MPTP toxicity, while nicotine was without effect (Sershen *et al.* 1986). Additional behavioural and biochemical data have been provided concerning the effects of nicotine in MPTP-treated mice (Sershen *et al.* 1987). In that study, nicotine was found to stimulate locomotor activity in control and MPTP-treated mice, and did not alter the levels of dopamine and its metabolites in the striatum of MPTP-treated mice. This evidence suggested that the locomotor stimulation was not due to increased dopamine release, although concentrations of dopamine and its metabolites were not measured in the nucleus accumbens. However, the results did demonstrate that nicotine was efficacious in this model.

In a previous study, the antioxidants α-tocopherol, β-carotene, ascorbic acid, and N-acetylcysteine were tested for their protective action against MPTP-induced damage (Perry *et al.* 1985). All four compounds diminished striatal dopamine depletion, and α-tocopherol prevented loss of neurons in the substantia nigra. Whether free-radical scavengers present in cigarette smoke protect against damage by reacting with endogenous or exogenous molecules to form less toxic substances remains to be seen, but concentrations of the antioxidant, glutathione, have been reported to be small in the substantia nigras of Parkinsonian brains, and smokers have large concentrations of carbon monoxide and carboxyhaemoglobin which may have a protective action against oxidative damage (Perry *et al.* 1982).

Several endogenous compounds, such as tetrahydroisoquinolines and tetrahydro-β-carbolines, have been suggested as possible causative agents in Parkinson's disease (for review, see Mesnil *et al.* 1984). The former compounds are particularly interesting because their presence has been demonstrated in control and Parkinsonian brains (Niwa *et al.* 1987). One suggestion as to why smoking protects against Parkinson's disease was that nicotine could be metabolized by monoamine oxidase to an iminium ion, and by this process nicotine could act as a competitive inhibitor of the activation of tetrahydroisoquinolines and tetrahydro-β-carbolines by monoamine oxidase (Testa *et al.* 1985). One study has shown that administration of tetrahydroisoquinoline to marmosets renders them Parkinsonian, and post-mortem analysis has revealed that there is reduced tyrosine hydroxylase activity and depleted dopamine concentration in the substantia nigra of these brains (Nagatsu and Yoshida 1988). Also, tetrahydroisoquinoline and MPP^+ inhibit mitochondrial respiratory function in a similar manner (Suzuki *et al.* 1988). Conversely, Perry *et al.* (1988*a*) found that tetrahydroisoquinoline was not toxic to the nigro-striatal system in a strain of mice which is susceptible to damage caused by MPTP. Furthermore, paraquat was found not to damage nigro-striatal neurons (Perry *et al.* 1986), and two analogues of MPTP which are present in tea, namely 2-phenylpyridine and 3-phenylpyridine, did not reduce concentrations of dopamine or its metabolites in mice (Perry *et al.* 1988*b*).

The strongest evidence so far that nicotine protects against damage to dopamine neurons has come from lesion studies in rats (Janson *et al.* 1986; Janson *et al.* 1988*a*; see Fuxe *et al.*, chapter 6, this volume). Here it was shown that in untreated rats, a partial hemitransection of the ascending nigro-striatal dopamine system caused degeneration of dopamine neurons in the substantia nigra. Chronic nicotine (0.125 mg kg^{-1} h^{-1}, 14 days) delivered via inplanted minipumps substantially counteracted the lesion-induced degeneration of tyrosine hydroxylase immunoreactive neurons, dendrites, and nerve terminals. Also, contrary to the results discussed above (Yong and Perry 1986; Sershen *et al.* 1986) it was shown that chronic nicotine treatment did afford some protection against MPTP-induced nigro-striatal degeneration in the mouse (Andersson *et al.* 1988; Janson *et al.* 1988*b*). It is not apparent why there is a discrepancy in these results, although the most obvious methodological differences involved the mode of administration of nicotine. However, it was suggested in this latest work that the protective action of nicotine was due to desensitization of nicotine receptors on the dopamine neurons, and this resulted in reduced energy demands because of the diminished nigral cell firing rate (Janson *et al.* 1988*b*). As the dose of nicotine used in these studies resulted in a serum nicotine concentration of 50 ng ml^{-1} which is within the range of plasma nicotine concentrations found in smokers, it is apparent that tobacco smoke may shield dopamine neurons against damage in man.

9.7 Summary

Although most of the epidemiological surveys have shown that there are less smokers among Parkinsonian patients, there is still some argument as to whether smoking really does protect against Parkinson's discase. When factors such as selective mortality are considered, the inverse relationship between Parkinson's disease and smoking may become less apparent, but this does not necessarily mitigate against the hypothesis.

There is now a large body of experimental evidence that suggests that some of the central actions of nicotine are mediated by dopamine mechanisms. Thus nicotine appears to interact with nicotine receptors on the dopamine nerve terminals in the striatum and nucleus accumbens, and with nicotine receptors on dopamine neurons in the substantia nigra and VTA. However, most studies have shown that the concentrations of nicotine required to modulate striatal dopamine mechanisms by action within the striatum have been above the smoking range, and raises the question of whether smoking directly affects striatal dopamine mechanisms. Conversely, there is substantial evidence that low concentrations of nicotine can modulate dopaminergic activity in the nucleus accumbens, VTA, and substantia nigra. The net outcome of these effects appears to be increased release of dopamine, particularly in the nucleus accumbens. This also coincides with behavioural data which shows that nicotine (at least in tolerant rats) causes motor stimulation but does not produce florid stereotypy, a behavioural syndrome derived from stimulation of striatal dopamine mechanisms.

Until recently, pharmacological investigations have found little or no evidence of a protective action of nicotine on dopamine systems in the MPTP model of neuronal degeneration, although other components of tobacco smoke such as hydrazine do have some activity which may or may not be related to their reducing action. As some evidence points to environmental factors, and not genetic predisposition as the cause of Parkinson's disease, future strategies may lie in arresting the disease process by medical intervention at an early stage after exposure to the putative toxins. Even more important is to identify these supposed toxins, especially at a time when there is such a high use of agricultural chemicals. If these agents do share a common mode of action with MPTP, it is encouraging that drugs exist (for example, deprenyl) which prevent MPTP neurotoxicity in animal models. However, a major hurdle at present is that there is no early diagnostic test to predict future development of Parkinson's disease. Also, the work of Perry and colleagues suggests that the putative environmental toxins which cause Parkinson's disease do not resemble MPTP, and that other chemicals should be examined for their neurotoxic effects (Perry *et al.* 1988*b*).

The disorder is insidiously progressive and it is not until there has been major irreversible damage to the dopamine neurons that the first overt symptoms appear. Thus it is likely that in most cases the pathological processes have begun many years prior to the appearance of motor symptoms. It is interesting therefore that some Parkinsonian patients have been observed to display characteristic personality traits which pre-date the morbid phase, and the possibility arises that these may be related to the pathology. If in the early stages of the degenerative process there is dysfunction of dopamine pathways which are involved in reward mechanisms, thus removing the incentive to smoke, this may help to explain the proposed inverse correlation between smoking and Parkinson's disease. In view of the fairly subtle presynaptic changes that nicotine produces on dopamine release and metabolism, as compared to the effects produced by dopamine agonists, it is unlikely that nicotine would be of any great treatment value in Parkinson's disease, especially at time when most of the dopamine neurons would have already degenerated.

Whether the MPTP model of nigrostriatal degeneration is a valid model of Parkinson's disease is debatable as there is no obvious environmental MPTP-like compound which is neurotoxic to dopamine neurons in man (Perry *et al.* 1988*b*). However, the discovery by Fuxe and co-workers (Janson *et al.* 1986, 1988*a*, 1988*b*) that chronic nicotine treatment diminishes lesion-induced nigral dopamine cell death concurs with the epidemiological data, and provides the strongest scientific evidence yet that smoking indeed may have a protective action. These findings may provide clues to the reasons why dopamine neurons die in the first place when there are no obvious genetic reasons. However, the hazards associated with cigarettes outweigh the possible benefits and preclude the use of tobacco smoking for preventing Parkinson's disease. Research into the mechanisms underlying the nicotine-dopamine link may provide a novel approach to uncovering the mysteries of Parkinsonian degeneration and may produce a rational basis for its prevention.

References

Ahtee, L., and Kaakkola, S. (1978). Effect of mecamylamine on the fate of dopamine in striatal and mesolimbic areas of rat brain; interaction with morphine and haloperidol. *British Journal of Pharmacology* **62**, 213–8.

Andersson, K., Fuxe, K., and Agnati, L.F. (1981). Effects of single injections of nicotine on the ascending dopamine pathways in the rat. Evidence for increases of dopamine turnover in the mesostriatal and mesolimbic dopamine neurons. *Acta Physiologica Scandinavica* **112**, 345–7.

Andersson, K., Fuxe, K, Janson, A.M., Janson, A., Von Euler, G., Bjelke B., Eneroth, P., Harfstrand, A., and Agnati, L.F. (1988). Central actions of nicotine on monoamine neuron systems: focus on neuroendocrine regulation and on trophic mechanisms. *Proceedings of the British Association for Psychopharmacology* p. 2.

Arqueros, L., Naquira, D., and Zunino, E. (1978). Nicotine-induced release of catecholamines from rat hippocampus and striatum. *Biochemical Pharmacology* **27**, 2667–74.

Barbeau, A. (1984). Etiology of Parkinson's disease: a research strategy. *Canadian Journal of Neurological Sciences* **11**, 24–8.

Barbeau, A., Roy, M., Bernier, G., Campanella, G., and Paris, S. (1987). Ecogenetics of Parkinson's disease: prevalence and environmental aspects in rural areas. *Canadian Journal of Neurological Sciences* **14**, 36–41.

Baron, J.A. (1986). Cigarette smoking and Parkinson's disease. *Neurology* **36**, 1490–6.

Baron, J.A., and Greenberg, E.R. (1987). Parkinson's disease and smoking. *Archives of Neurology* **44**, 1110–1.

Baumann, R.J., Jameson, H.D., McKean, H.E., Haack, D.G., and Weisberg, L.M.

(1980). Cigarette smoking and Parkinson's disease: 1. A comparison of cases with matched neighbors. *Neurology* **30**, 839–43.

Bellini, M., and Tansella, M. (1976). Obsessional scores and subjective general psychiatric complaints of patients with duodenal ulcer or ulcerative colitis. *Psychological Medicine* **6**, 461–7.

Beninato, M., and Spencer, R.F. (1986). A cholinergic projection to the rat superior colliculus demonstrated by retrograde transport of horseradish peroxidase and choline acetyltransferase immunohistochemistry. *Journal of Comparative Neurology* **253**, 525–38.

Beninato, M., and Spencer, R.F. (1987). A cholinergic projection to the rat substantia nigra from the pedunculopontine tegmental nucleus. *Brain Research* **412**, 169–74.

Bernheimer, H., Birkmayer, W., Hornykiewicz, O., Jellinger, K., and Seitelberger, F. (1973). Brain dopamine and the syndromes of Parkinson and Huntingdon. Clinical, morphological and neurochemical correlations. *Journal of Neurological Sciences* **20**, 415–55.

Bharucha, N.E., Stokes, L., Schoenberg, B.S., Ward, C., Ince, S., Nutt, J.G., Eldridge R., Calne, D.B., Mantel, N., and Duvoisin, R. (1986). A case-control study of twin pairs discordant for Parkinson's disease: A search for environmental risk factors. *Neurology* **36**, 284–8.

Bihari, K., and Lees, A.J. (1987). Cigarette smoking, Parkinson's disease and ulcerative colitis. *Journal of Neurology, Neurosurgery and Psychiatry* **50**, 635.

Binder, R.L., Kazamatsuri, H., Nishimura, T., and McNiel, D.E. (1987). Smoking and tardive dyskinesia. *Biological Psychiatry* **22**, 1280–2.

Bouthenet, M.-L., Martres, M.-P., Sales, N., and Schwartz, J.-C. (1987). A detailed mapping of dopamine D-2 receptors in rat central nervous system by autoradiography with [^{125}I]iodosulpiride. *Neuroscience* **20**, 117–55.

Burns, R.S., Chiueh, C.C., Markey, S.P., Ebert, M.H., Jacobowitz, D.M., and Kopin, I.J. (1983). A primate model of parkinsonism: selective destruction of dopaminergic neurons in the pars compacta of the substantia nigra by N-methyl-4-phenyl-1,2,3,6-tetrahydropyridine. *Proceedings of the National Academy of Sciences* **80**, 4546–50.

Chance, W.T., Murfin, D,, Krynock, G.M., and Rosecrans, J.A. (1977). A description of the nicotine stimulus and tests of its generalization to amphetamine. *Psychopharmacology* **55**, 19–26.

Clarke, P.B.S., Hommer, D.W., Pert, A., and Skirboll, L.R. (1985*a*). Electrophysiological actions of nicotine on substantia nigra single units. *British Journal of Pharmacology* **85**, 827–35.

Clarke, P.B.S., Hommer, D.W., Pert, A., and Skirboll, L.R. (1987). Innervation of substantia nigra neurons by cholinergic afferents from pedunculopontine nucleus in the rat: neuroanatomical and electrophysiological evidence. *Neuroscience* **23**, 1011–9.

Clarke, P.B.S., and Kumar, R. (1983*a*). The effects of nicotine on locomotor activity in non-tolerant and tolerant rats. *British Journal of Pharmacology* **78**, 329–37.

Clarke, P.B.S., and Kumar, R. (1983*b*). Characterization of the locomotor stimulant action of nicotine in tolerant rats. *British Journal of Pharmacology* **80**, 587–94.

Clarke, P.B.S., and Pert, A. (1985). Autoradiographic evidence for nicotine receptors on nigrostriatal and mesolimbic dopaminergic neurons. *Brain Research* **348**, 355–8.

Clarke, P.B.S., Schwartz, R.D., Paul, S.M., Pert, C.B., and Pert, A. (1985*b*). Nicotinic binding in rat brain: Autoradiographic comparison of [^{3}H]acetylcholine, [^{3}H]nicotine, and [^{125}I]-α-bungarotoxin. *Journal of Neuroscience* **5**, 1307–15.

Cohen, G., Pasik, P., Cohen, B., Leist, A., Mytilineou, C., and Yahr, M.D. (1985). Pargyline and deprenyl prevent the neurotoxicity of 1-methyl-4-phenyl-1,2,3,6-tetrahydropyridine (MPTP) in monkeys. *European Journal of Pharmacology* **106**, 209–10.

Connelly, M.S., and Littleton, J.M. (1983). Lack of stereoselectivity in ability of nicotine to release dopamine from rat synaptosomal preparations. *Journal of Neurochemistry* **41**, 1297–1302.

Damsma, G., Westerink, B.H.C., de Vries, J.B., and Horn, A.S. (1988). The effect of systemically applied cholinergic drugs on the striatal release of dopamine and its metabolites, as determined by automated brain dialysis in conscious rats. *Neuroscience Letters* **89**, 349–54.

Dawson, T.M., Gehlert, D.R., and Wamsley, J.K. (1986). Quantitative autoradiographic localization of central dopamine D-1 and D-2 receptors. *Advances in Experimental Biology and Medicine* **204**, 93–118.

De Belleroche, J., and Bradford, H.F. (1978). Biochemical evidence for the presence of presynaptic receptors on dopaminergic nerve terminals. *Brain Research* **142**, 53–68.

Doll, R., and Peto, R. (1976). Mortality in relation to smoking: 20 years' observations on male British doctors. *British Medical Journal* **2**, 1525–36.

Duvoisin, R.C., Eldridge, R., Williams, A., Nutt, J., and Calne, D. (1981). Twin study of Parkinson disease. *Neurology* **31**, 77–80.

Fuxe, K., Andersson, K., Eneroth, P., Harfstrand, A., Nordberg, A., and Agnati, L.F. (1987). Effects of nicotine and exposure to cigarette smoke on discrete dopamine and noradrenaline nerve terminal systems of the telencephalon and diencephalon of the rat: relationship to reward mechanisms and neuroendocrine functions and distribution of nicotinic binding sites in brain. In *Tobacco smoking and nicotine. Advances in behavioural biology* Vol. 31, (W.R. Martin, G.R. Van Loon, E.T. Iwamoto, and L. Davis, ed.), pp. 225–62. Plenum Press, New York.

Fuxe, K., Andersson, K., Harfstrand, A., and Agnati, L.F. (1986). Increases in dopamine utilization in certain limbic dopamine terminal populations after a short period of intermittent exposure of male rats to cigarette smoke. *Journal of Neural Transmission* **67**, 15–29.

Giorguieff-Chesselet, M.F., Kemel, M.L., Wandscheer, D., and Glowinski, J. (1979). Regulation of dopamine release by presynaptic nicotinic receptors in rat striatal slices: effect of nicotine in a low concentration. *Life Sciences* **25**, 1257–61.

Giorguieff, M.F., Le Floc'h, M.L., Glowinski, J., and Besson, M.J. (1977). Involvement of cholinergic presynaptic receptors of nicotinic and muscarinic types in the control of the spontaneous release of dopamine from striatal dopaminergic terminals in the rat. *Journal of Pharmacology and Experimental Therapeutics* **200**, 535–44.

Glover, V., Reveley, M.A., and Sandler, M. (1980). A monoamine oxidase inhibitor in human urine. *Biochemical Pharmacology* **29**, 467–70.

Godwin-Austen, R.B., Lee, P.N., Marmot, M.G., and Stern, G.M. (1982). Smoking and Parkinson's disease. *Journal of Neurology Neurosurgery and Psychiatry* **45**, 577–81.

Golbe, L.I., and Duvoisin, R.C. (1985). Smoking and Parkinson's disease. *Neurology* **35**, (suppl. 1) 275.

Golbe, L.I., Cody, R.A., and Duvoisin, R.C. (1986). Smoking and Parkinson's disease. Search for a dose-response relationship. *Archives of Neurology* **43**, 774–8.

Golbe, L.I., Cody, R.A., and Duvoisin, R.C. (1987). Smoking and Parkinson's disease. *Archives of Neurology* **44**, 1111.

Grenhoff, J., Aston-Jones, G., and Svensson, T.H. (1986). Nicotinic effects on the firing pattern of midbrain dopamine neurons. *Acta Physiologica Scandinavica* **128**, 351–8.

Haack, D.G., Baumann, R.J., McKean, H.E., Jameson, H.D., and Turbek, J.A. (1981). Nicotine exposure and Parkinson's disease. *American Journal of Epidemiology* **114**, 191–200.

Harfstrand, A., Adem, A., Fuxe, K., Agnati, L., Andersson, K., and Nordberg, A. (1988). Distribution of nicotinic cholinergic receptors in the rat tel- and diencephalon: a quantitative receptor autoradiographical study using [^{3}H]acetylcholine, [α-^{125}I]bungarotoxin and [^{3}H]nicotine. *Acta Physiologica Scandinavica* **132**, 1–14.

Imperato, A., Mulas, A., and Di Chiara, G. (1986). Nicotine preferentially stimulates dopamine release in the limbic system of freely moving rats. *European Journal of Pharmacology* **132**, 337–8.

Jackson, A., and Crossman, A.R. (1983) Nucleus tegmenti pedunculopontinus: efferent connections with special reference to the basal ganglia, studied in the rat by anterograde and retrograde transport of horseradish peroxidase. *Neuroscience* **10**, 725–65.

Jaffe, J.H. (1985). Drug addiction and drug abuse. In *The pharmacological basis of therapeutics*, (A.C. Gilman, L.S. Goodman, T.W. Rall, and F. Murad, ed.), pp. 532–81. Macmillan, New York.

Janson, A.M., Fuxe, K., Agnati, L.F., Kitayama, I., Harfstrand, A., Andersson, K., and Goldstein, M. (1988*a*). Chronic nicotine treatment counteracts the disappearance of tyrosine-hydroxylase-immunoreactive nerve cell bodies, dendrites and terminals in the mesostriatal dopamine system of the male rat after partial hemitransection. *Brain Research* **455**, 332–45.

Janson, A.M., Fuxe, K., Sundstrom, E., Agnati, L.F., and Goldstein, M. (1988*b*). Chronic nicotine treatment partly protects against the 1-methyl-4-phenyl-1,2,3,6-tetrahydropyridine-induced degeneration of nigrostriatal dopamine neurons in the black mouse. *Acta Physiologica Scandinavica* **132**, 589–91.

Janson, A.M., Fuxe, K., Kitayama, I., Harfstrand, A., and Agnati, L.F. (1986). Morphometric studies on the protective action of nicotine on the substantia nigra dopamine nerve cells after partial hemitransection in the male rat. *Neuroscience Letters* **26**, S88.

Jellinger, K. (1988). The pedunculopontine nucleus in Parkinson's disease, progressive supranuclear palsy and Alzheimer's disease. *Journal of Neurology Neurosurgery and Psychiatry* **51**, 540–3.

Kaakkola, S. (1981). Effect of nicotinic and muscarinic drugs on amphetamine- and apomorphine-induced circling behavior in rats. *Acta Pharmacologica et Toxicologica* **48**, 162–7.

Kaelber, W.W. (1963). Tremor at rest from tegmental lesions in the cat. *Journal of Neuropathology and Experimental Neurology* **22**, 695–701.

Kahn, H.A. (1966). The Dorn study of smoking and mortality among US veterans: report on eight and one half years of observation. In *Epidemiological approaches to the study of cancer and other chronic diseases*. (W. Haenzel, ed.) US Department of Health, Education and Welfare, National Cancer Institute Monograph. 19.

Kasa, P. (1986). The cholinergic systems in brain and spinal cord. *Progress in Neurobiology* **26**, 211–72.

Kessler, I.I. (1972*a*). Epidemiologic studies of Parkinson's disease. II. A hospital-based survey. *American Journal of Epidemiology* **95**, 308–18.

Kessler, I.I. (1972*b*). Epidemiologic studies of Parkinson's disease. III. A community-based survey. *American Journal of Epidemiology* **96**, 242–54.

Kessler, I.I. (1978) Parkinson's disease in epidemiologic perspective. *Advances in Neurology* **19**, 355–84. Raven Press, New York.

Kessler, I.I., and Diamond, E.L. (1971). Epidemiologic studies of Parkinson's disease. I. Smoking and Parkinson's disease: a survey and explanatory hypothesis. *American Journal of Epidemiology* **94**, 16–25.

Langston, J.W., Ballard, P., Tetrud, J.W., and Irwin, I. (1983). Chronic Parkinsonism in humans due to a product of meperidine-analog synthesis. *Science* **219**, 979–80.

Lapin, E.P., Maker, H.S., Sershen, H., Hurd, Y., and Lajtha, A. (1987). Dopamine-like action of nicotine: lack of tolerance and reverse tolerance. *Brain Research* **407**, 351–63.

Lees, A.J. (1984). Hemidystonia relieved by nicotine. *Lancet* **ii**, 871.

Lichtensteiger, W., Felix, D., Lienhart, R., and Hefti, F. (1976). A quantitative correlation between single unit activity and fluorescence intensity of dopamine neurones in zona compacta of substantia nigra, as demonstrated under the influence of nicotine and physostigmine. *Brain Research* **117**, 85–103.

Lichtensteiger, W., Hefti, F., Felix, D., Huwyler, T., Melamed, E., and Schlumpf, M. (1982). Stimulation of nigrostriatal dopamine neurones by nicotine. *Neuropharmacology* **21**, 963–8.

London, E.D., Waller, S.B., and Wamsley, J.K. (1985). Autoradiographic localization of [^{3}H]nicotine binding sites in the rat brain. *Neuroscience Letters* **53**, 179–84.

Marttila, R.J., and Rinne, U.K. (1980). Smoking and Parkinson's disease. *Acta Neurologica Scandinavica* **62**, 322–5.

Martindale, W. (1982). *The Extrapharmacopoeia.* 28th edn. (J.E.F. Reynolds, ed.) The Pharmaceutical Press, London.

McGeer, E.G., and McGeer, P.L. (1984). Substantia nigra cell death from kainic acid or folic acid injections into the pontine tegmentum. *Brain Research* **298**, 339–42.

Mereu, G., Yoon, K-W.P., Boi, V., Gessa, G.L., Naes, L., and Westfall, T.C. (1987). Preferential stimulation of ventral tegmental area dopaminergic neurons by nicotine. *European Journal of Pharmacology* **141**, 395–9.

Mesnil, M., Testa, B., and Jenner, P. (1984). Xenobiotic metabolism by brain monooxygenases and other cerebral enzymes. In *Advances in Drug Research* (B. Testa, ed.) **13**, 95–207. Academic Press, London.

Mitchell, S.N., Brazell, M.P., Joseph, M.H., and Gray, J.A. (1987). Regional differences in the effect of nicotine on catecholamine turnover in the rat CNS. *British Journal of Pharmacology* **91**, 430P.

Moll, H. (1926). The treatment of post-encephalitic parkinsonism by nicotine. *British Medical Journal* **I**, 1079–81.

Nagatsu, T., and Yoshida, M. (1988). An endogenous substance of the brain, tetrahydroisoquinoline, produces parkinsonism in primates with decreased dopamine, tyrosine hydroxylase and biopterin in the nigrostriatal regions. *Neuroscience Letters* **87**, 178–82.

Naquira, D., Zunino, E., Arqueros, L., and Viveros, O.H. (1978). Chronic effects of nicotine on catecholamine synthesizing enzymes in rats. *European Journal of Pharmacology* **47**, 227–9.

Nefzger, M.D., Quadfasel, F.A., and Karl V.C. (1968). A retrospective study of smoking in Parkinson's disease. *American Journal of Epidemiology* **88**, 149–58.

Nielsen, E.B., and Jepsen, S.A. (1985). Antagonism of the amphetamine cue by both classical and atypical antipsychotic drugs. *European Journal of Pharmacology* **11**, 167–76.

Niwa, T., Takeda, N., Kaneda, N., Hashizume, Y., and Nagatsu, T. (1987). Presence of tetrahydroisoquinoline and 2-methyl-tetrahydroisoquinoline in parkinsonian and normal human brains. *Biochemical and Biophysical Research Communications* **144**, 1084–9.

Ohama, E., and Ikuta, F. (1976). Parkinson's disease: Distribution of Lewy bodies and monoamine neuron system. *Acta neuropathologica (Berlin)* **34**, 311–9.

Oreland, L., Fowler, C.J., and Schalling, D. (1981). Low platelet monoamine oxidase activity in cigarette smokers. *Life Sciences* **29**, 2511–8.

Perry, E.K., Perry, R.H., Smith, C.J., Dick, D.J., Candy, J.M., Edwardson, J.A., Fairbairn, A., and Blessed, G. (1987). Nicotinic receptor abnormalities in Alzheimer's and Parkinson's diseases. *Journal of Neurology Neurosurgery and Psychiatry* **50**, 806–9.

Perry, T.L., Godin, D.V., and Hansen, S. (1982). Parkinson's disease: a disorder due to nigral glutathione deficiency? *Neuroscience Letters* **33**, 305–10.

Perry, T.L., Jones, K., and Hansen, S. (1988*a*). Tetrahydroisoquinoline lacks dopaminergic nigrostriatal neurotoxicity in mice. *Neuroscience Letters* **85**, 101–4.

Perry, T.L., Jones, K., Hansen, S., and Wall, R.A. (1988*b*). 2-Phenylpyridine and 3-phenylpyridine, constituents of tea, are unlikely to cause idiopathic Parkinson's disease. *Journal of Neurological Sciences* **85**, 309–17.

Perry, T.L,. Yong, V.W., Clavier, R.M., Jones, K., Wright, J.M., Foulks, J.G., and Wall, R.A. (1985). Partial protection from the dopaminergic neurotoxin N-methyl-4-phenyl-1,2,3,6-tetrahydropyridine by four different antioxidants in the mouse. *Neuroscience Letters* **60**, 109–14.

Perry, T.L., Yong, V.W., Wall, R.A., and Jones, K. (1986). Paraquat and two endogenous analogues of the neurotoxic substance N-methyl-4-phenyl-1,2,3,6-tetrahydropyridine do not damage dopaminergic nigrostriatal neurons in the mouse. *Neuroscience Letters* **69**, 285–9.

Pert, A., and Clarke, P.B.S. (1987). Nicotinic modulation of dopaminergic neurotransmission: functional implications. In *Tobacco smoking and nicotine. Advances in behavioral biology* Vol. 31, (W.R. Martin, G.R. Van Loon, E.T. Iwamoto, and L. Davis, ed.), pp. 169–89. Plenum Press, New York.

Pratt, J.A., Stolerman, I.P., Garcha, H.S., Giardini, V., and Feyerabend, C. (1983). Discriminative stimulus properties of nicotine: further evidence for mediation at a cholinergic receptor. *Psychopharmacology* **81**, 54–60.

Rainbow, T.C., Schwartz, R.D., Parsons, B., and Kellar, K.J. (1984). Quantitative autoradiography of nicotinic [^{3}H]acetylcholine binding sites in rat brain. *Neuroscience Letters* **50**, 193–6.

Rajput, A.H., Offord, K.P., Beard, C.M., and Kurland, L.T. (1987). A case-control study of smoking habits, dementia, and other illnesses in idiopathic Parkinson's disease. *Neurology* **37**, 226–32.

Rajput, A.H., Uitti, R.J., Stern, W., and Laverty, W. (1986). Early onset Parkinson's disease in Saskatchewan – environmental considerations for etiology. *Canadian Journal of Neurological Sciences* **13**, 312–6.

Rapier, C., Lunt, G.G., and Wonnacott, S. (1988). Stereoselective nicotine-induced release of dopamine from striatal synaptosomes: concentration dependence and repetitive stimulation. *Journal of Neurochemistry* **50**, 1123–30.

Reavill, C., and Stolerman, I.P. (1987). Interaction of nicotine with dopaminergic mechanisms assessed through drug discrimination and rotational behavior in rats. *Journal of Psychopharmacology* **1**, 264–73.

Reavill, C., and Stolerman, I.P. (1988). Locomotor activity in rats after intracerebral administration of nicotine. *Journal of Psychopharmacology* **2**, Proceedings section.

Reisine, T.D., Nagy, J.I., Fibiger, H.C., and Yamamura, H.I. (1979). Localization of dopamine receptors in rat brain. *Brain Research* **169**, 209–14.

Roberts, D.C.S., Zis, A.P., and Fibiger, H.C.(1975). Ascending catecholamine pathways and amphetamine-induced locomotor activity: importance of DA and apparent non-involvement of norepinephrine. *Brain Research* **93**, 441–54.

Rosecrans, J.A. (1987). Noncholinergic mechanisms involved in the behavioral and stimulus effects of nicotine, and relationships to the process of nicotine dependence. In *Tobacco smoking and nicotine. Advances in behavioral biology*, Vol. 31, (W.R. Martin, G.R. Van Loon, E.T. Iwamoto, and L. Davis, ed.), pp. 125–39. Plenum Press, New York.

Rosecrans, J.A., and Chance, W.T. (1977). Cholinergic and non-cholinergic aspects of the discriminative stimulus properties of nicotine. In *Discriminative stimulus properties of drugs* (H. Lal, ed.), pp. 155–85. Plenum, New York.

Rosecrans, J.A., Chance, W.T., and Schechter, M.D. (1976). The discriminative stimulus properties of nicotine, *d*-amphetamine and morphine in dopamine depleted rats. *Psychopharmacology Communications* **2**, 349–56.

Rowell, P.P., Carr, L.A., and Garner, A.C. (1987). Stimulation of [^{3}H]dopamine release by nicotine in rat nucleus accumbens. *Journal of Neurochemistry* **49**, 1449–54.

Russell, M.A.H., Jarvis, M., Iyer, R., and Feyerabend, C. (1980). Relation of nicotine yield of cigarettes to blood nicotine concentrations in smokers. *British Medical Journal* **280**, 972–6.

Rye, D.B., Saper, C.B., Lee, H.J., and Wainer, B.H. (1987). Pedunculopontine tegmental nucleus of the rat: cytoarchitecture, cytochemistry, and some extrapyramidal connections of the mesopontine tegmentum. *Journal of Comparative Neurology* **259**, 483–528.

Sakurai, Y., Takano, Y., Kohjimoto, Y., Honda, K., and Kamiya, H-O. (1982). Enhancement of [^{3}H]dopamine release and its [^{3}H]metabolites in rat striatum by nicotinic drugs. *Brain Research* **242**, 99–106.

Savasta, M., Dubois, A., and Scatton, B. (1986). Autoradiographic localization of D_1 dopamine receptors in the rat brain with [^{3}H]SCH 23390. *Brain Research* **375**, 291–301.

Saper, C.B., and Loewy, A.D. (1982). Projections of the pedunculopontine tegmental nucleus in the rat: evidence for additional extrapyramidal circuitry. *Brain Research* **252**, 367–72.

Scatton, B., Simon, H., Le Moal, M., and Bischoff, S. (1980). Origin of dopaminergic innervation of the rat hippocampal formation. *Neuroscience Letters* **18**, 125–31.

Schechter, M.D., and Rosecrans, J.A. (1972). Nicotine as a discriminative cue in rats: inability of related drugs to produce a nicotine-like cueing effect. *Psychopharmacologia* **27**, 379–87.

Schwartz, R.D., Lehmann, J., and Kellar, K.J. (1984). Presynaptic nicotinic cholinergic receptors labelled by [^{3}H]acetylcholine on catecholamine and serotonin axons in brain. *Journal of Neurochemistry* **42**, 1495–8.

Sershen, H., Hashim, A., and Lajtha, A. (1987). Behavioral and biochemical effects of nicotine in an MPTP-induced mouse model of Parkinson's disease. *Pharmacology, Biochemistry and Behavior* **28**, 299–303.

Sershen, H., Mason, M.F., Reith, M.E.A., Hashim, A., and Lajtha, A. (1986). Effect of nicotine and amphetamine on the neurotoxicity of N-methyl-4-phenyl-1,2,3,6-tetrahydropyridine (MPTP) in mice. *Neuropharmacology* **25**, 1231–4.

Simon, H., Le Moal, M., and Calas, A. (1979). Efferents and afferents of the ventral tegmental-A10 region studied after local injection of [^{3}H]leucine and horseradish peroxidase. *Brain Research* **178**, 17–40.

Singer, G., Wallace, M., and Hall, R. (1982) Effects of dopaminergic nucleus accumbens lesions on the acquisition of schedule induced self injection of nicotine in the rat. *Pharmacology, Biochemistry and Behavior* **17**, 579–81.

Stolerman, I.P., Garcha, H.S., Pratt, J.A., and Kumar, R. (1984). Role of training dose in discrimination of nicotine and related compounds by rats. *Psychopharmacology* **84**, 413–9.

Suzuki, K., Mizuno, Y., and Yoshida, M. (1988). Inhibition of mitochondrial NADH-ubiquinone oxidoreductase activity and ATP synthesis by tetrahydroisoquinoline. *Neuroscience Letters* **86**, 105–8.

Takada, M., Li, Z.K., and Hattori, T. (1988). Dopaminergic nigrotectal projection in the rat. *Brain Research* **457**, 165–8.

Takano, Y., Sakurai, Y., Kohjimoto, Y, Honda, K., and Kamiya, H.-O. (1983). Presynaptic modulation of the release of dopamine from striatal synaptosomes: differences in the effects of high K^+ stimulation, methamphetamine and nicotinic drugs. *Brain Research* **279**, 330–4.

Testa, B., Naylor, R., Costall, B., Jenner, P., and Marsden, C.D., (1985). Does an endogenous methylpyridinium analogue cause Parkinson's disease? *Journal of Pharmacy and Pharmacology* **37**, 679–80.

Todes, C.J., and Lees, A.J. (1985). The pre-morbid personality of patients with Parkinson's disease. *Journal of Neurology Neurosurgery and Psychiatry* **48**, 97–100.

Ungerstedt, U. (1971). Striatal dopamine release after amphetamine or nerve degeneration revealed by rotational behavior. *Acta Physiologica Scandinavica* Suppl. **367**, 49–68.

Verney, C., Baulac, M., Berger, B., Alvarez, C., Vigny, A., and Helle, K.B. (1985). Morphological evidence for a dopaminergic terminal field in the hippocampal formation of young and adult rat. *Neuroscience* **14**, 1039–52.

Ward, C.D., Duvoisin, R.C., Ince, S.E., Nutt, J.D., Eldridge, R., and Calne, D.B. (1983). Parkinson's disease in 65 pairs of twins and in a set of quadruplets. *Neurology* **33**, 815–24.

Westfall, T.C. (1974). Effect of nicotine and other drugs on the release of ^{3}H-norepinephrine and ^{3}H-dopamine from rat brain slices. *Neuropharmacology* **13**, 693–700.

Westfall, T.C., Grant, H., and Perry, H. (1983). Release of dopamine and 5-hydroxytryptamine from rat striatal slices following activation of nicotinic cholinergic receptors. *General Pharmacology* **14**, 321–5.

Westfall, T.C. and Perry, H. (1986). The nicotinic-induced release of endogenous dopamine from rat striatal slices from animals chronically exposed to dimethylphenylpiperazinium (DMPP). *Neuroscience Letters* **71**, 340–4.

White, O.B., Saint-Cyr, J.A., Tomlinson, R.D., and Sharpe, J.A. (1983). Ocular motor deficits in Parkinson's disease. II. Control of the saccadic and smooth pursuit systems. *Brain* **106**, 571–87.

Winch, G.J.B., and Balfour, D.J.K. (1988). Evidence that nicotine does not evoke the release of vesicular dopamine from striatal slices. *British Journal of Pharmacology* **94**, 350P.

Wise, R.A., and Bozarth, M.A. (1984). Brain reward circuitry: four circuit elements "wired" in apparent series. *Brain Research Bulletin* **12**, 203–8.

Woolf, N.J., and Butcher, L.L. (1986). Cholinergic systems in the rat brain. III.

Projections from the pontomesencephalic tegmentum to the thalamus, tectum, basal ganglia, and basal forebrain. *Brain Research Bulletin* **16**, 603–37.

Yamada, S., Gehlert, D.R., Hawkins, K.N., Nakayama, K., Roeske, W.R., and Yamamura, H.I. (1987). Autoradiographic localization of nicotinic receptor binding in rat brain using [^{3}H]methylcarbamylcholine, a novel radioligand. *Life Sciences* **41**, 2851–61.

Yassa, R., Lal, S., Korpassy, A., and Ally, J. (1987). Nicotine exposure and tardive dyskinesia. *Biological Psychiatry* **22**, 67–72.

Yong, V.W., and Perry, T. (1986). Monoamine oxidase B, smoking, and Parkinson's disease. *Journal of Neurological Sciences* **72**, 265–72.

10 Nicotinic cholinergic receptors in Alzheimer's disease

Kenneth J. Kellar and Susan Wonnacott

10.1 Introduction

Differences in several neurotransmitter systems have been found in brains from Alzheimer's disease victims compared with brains from controls. But it is not known whether these differences are primary changes in Alzheimer's disease or sequelae of the disease process; nor is it known with certainty that these aberrations in neurotransmitter systems are specific to the disease or that they are responsible for any of the important symptoms, most notably the impaired cognition typical of Alzheimer's disease. Despite this uncertainty about the specific roles of neurotransmitter system deficits in Alzheimer's disease, it is a reasonable first assumption that severe deficits in neurotransmission contribute to the disease symptoms, if not the process. Thus these neurotransmission systems provide objectives for further study, and perhaps first targets for rational therapeutic approaches to Alzheimer's disease and to a better understanding of the disease process.

Histological and cytochemical studies indicate that Alzheimer's disease affects neurons in several areas of the brain relatively selectively. Neurofibrillary tangles in cell bodies, neuritic plaques in axon-terminal areas, and loss of neurones, the histological hallmarks of Alzheimer's disease, are most frequently seen in certain brainstem nuclei, the basal forebrain cell groups, hippocampus, amygdala, and neocortex (for reviews see Terry and Katzman 1983; Price *et al.* 1985; Price 1986). More recently lesions of the olfactory areas of Alzheimer brains have been found (Averback 1983; Esiri and Wilcock 1984; Simpson *et al.* 1984), which raises interesting questions about etiological pathways of the disease (Roberts 1986). These neuropathology studies have provided tentative anatomical boundaries that have narrowed the scope of the search for neurotransmitter system abnormalities likely to be relevant to Alzheimer's disease. However, it should be emphasized that these boundaries are still tentative.

The cholinergic systems are probably the most studied of the neurotransmission systems in Alzheimer's brains (Fig. 10.1). One of the reasons for this is that interruption of cholinergic neurotransmission by drugs, or by lesioning cholinergic pathways, can disrupt performance of memory tasks in animals; and in human studies, muscarinic receptor antagonists (for example, scopolamine) can produce similar decrements on memory tasks (for reviews

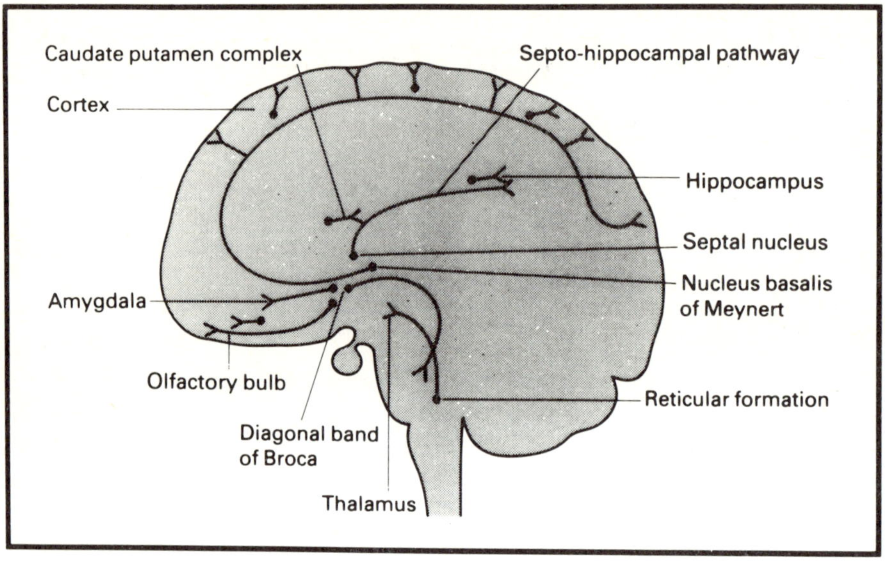

Fig. 10.1. Cholinergic pathways in the brain. The relative contributions of muscarinic and nicotinic transmission in each tract is not known at present. (From Iversen 1988; for a more detailed map of cholinergic projections, see Reavill, Fig 9.1, this volume.)

see Drachman and Leavitt 1974; Bartus *et al.* 1982; Collerton 1986). Over the last decade, a reduction in the activity of choline acetyltransferase, the enzyme that synthesizes acetylcholine, has been well documented in the hippocampus, amygdala, and neocortex of Alzheimer brains (Bowen *et al.* 1976; Davies and Maloney 1976; Perry *et al.* 1977). This deficit appears to be correlated with the severity of dementia and the presence of neuritic plaques (Perry *et al.* 1978; Wilcock *et al.* 1982). The cholinergic innervation to these areas comes primarily from neurons in the medial septum, diagonal band of Broca, and nucleus basalis of Meynert, which together comprise the basal forebrain cholinergic system (see Fig. 10.1). A loss or dysfunction of cholinergic neurons in these basal forebrain nuclear groups is related to the choline acetyltransferase deficit in projection areas of Alzheimer brains (Whitehouse *et al.* 1981, 1982; Tagliavini and Pilleri 1983; McGeer *et al.* 1984).

These discoveries have provided a basis for the notion that disruption of cholinergic neurotransmission is related to at least some of the memory and cognitive impairments in Alzheimer's disease. Therefore it is not surprising that attempts have been made to link some of the wide ranging neurotransmitter deficits in brains from affected patients to a common final pathway involving cholinergic neurotransmission.

In this chapter, some of the evidence indicating changes in cholinergic, adrenergic, and serotonergic neurons and receptors in Alzheimer's brain is reviewed; where data are available, the possible relationship to cholinergic neurotransmission is described. The emphasis of the chapter is on recent studies that have shown that there is a marked loss of brain nicotinic-cholinergic receptors in Alzheimer's disease. The loss of these nicotinic receptors could indicate disruption not only of cholinergic transmission, but of adrenergic, dopaminergic, and serotonergic transmission as well. Changes in neurotransmitter receptors in Alzheimer's disease are of interest not only because they can give an indication of what neurotransmitter systems are affected, but also because the remaining receptors represent potential targets for therapeutic intervention. Cholinergic therapies are evaluated and future prospects for nicotinic treatments are discussed.

10.2 Cholinergic systems in Alzheimer's disease

Although at the present time histological examination is still required for the definitive diagnosis of Alzheimer's disease, a reduction in choline acetyltransferase activity appears to have been unofficially adopted as the gold standard for neurochemical studies; thus it is common for studies measuring various kinds of neurochemical markers in Alzheimer brains to mention comparisons with changes in choline acetyltransferase activity in the same samples. Whether this emphasis on the cholinergic deficit in the disease is profitable or not is arguable, but it is understandable in light of the apparent relationships between decreased enzyme activity, the presence of histological markers for Alzheimer's disease, and the severity of dementia (Perry *et al.* 1978; Wilcock *et al.* 1982), and because of the apparent involvement of cholinergic systems in memory tasks (Drachman and Leavitt 1974; Bartus *et al.* 1982; Collerton, 1986).

10.2.1 *Muscarinic receptors*

In contrast to the invariable loss of choline acetyltransferase activity in cerebral cortex, hippocampus, and amygdala, which is consistent with the presence of the pathological changes in the basal forebrain nuclei that contain the cholinergic neurons that project to these areas, the results of studies of muscarinic cholinergic receptors in brain tissue from Alzheimer patients appear to depend on the brain area examined, the subtype of muscarinic receptor measured, and possibly other factors.

In the cerebral cortex of Alzheimer brains, the total population of muscarinic receptors, which is labelled by antagonist ligands such as [^{3}H]quinuclidinyl benzilate and [^{3}H]*N*-methylscopolamine, has been reported to be increased (Nordberg *et al.* 1987) and decreased (Reinikainen *et al.* 1987), but the majority of studies have found no significant changes (Perry *et al.* 1987; Davies and Verth 1978; Caulfield *et al.* 1982; Lang and Henke 1983; Rinne *et al.* 1984; Shimohama *et al.* 1986*b*; Kellar *et al.* 1987). In contrast to the cerebral cortex, in limbic areas such as hippocampus, amygdala, and accumbens, the total population of muscarinic receptors has consistently been found to be decreased in Alzheimer's disease (Reisine *et al.* 1978; Rinne *et al.*

Table 10.1. Cholinergic receptor binding sites in hippocampal membranes in neurological conditions

	n	Muscarinic receptors Total	M1	M2 (fmol mg^{-1} protein)	Nicotinic receptors
Normal controls	11	230 ± 27	180 ± 20	111 ± 11	10.3 ± 4.0
Alzheimer's disease	8	194 ± 47	153 ± 40[a]	94 ± 18[b]	5.2 ± 3.4[c]
Down's Syndrome					
31 years	1	385	290	188	3.9
> 50 years	4	217 ± 53	165 ± 39	111 ± 29	4.7 ± 4.0[a]
Parkinson's disease					
without dementia	10	273 ± 51[a]	213 ± 37	127 ± 29	6.2 ± 3.2[b]
with dementia	4	231 ± 45	182 ± 35	110 ± 24	6.4 ± 2.6
Alcoholic dementia	7	236 ± 32	186 ± 26	113 ± 13	10.9 ± 2.2
Huntington's disease	4	220 ± 31	178 ± 30	100 ± 11	8.0 ± 1.6
Motor Neurone disease	4	239 ± 23	184 ± 15	107 ± 11	8.8 ± 3.5

Muscarinic receptor binding was measured using [^{3}H]*N*-methylscopolamine (1 nM) in the presence of 0.3 nM carbamylcholine (estimates M1 subtype) or 2 μM pirenzepine (estimates M2 subtype) or in the absence of other drugs (total muscarinic receptors). Non-specific binding was determined in the presence of 1 μM atropine. Nicotinic receptor binding was measured using [^{3}H](−)nicotine (4 nM) in the presence and absence of excess unlabelled nicotine to determine non-specific and total binding respectively. a,b,c: significantly different from normal group: $p < 0.05$, $p < 0.02$, $p < 0.01$ respectively. From Perry *et al.* (1987).

1984; Shimohama *et al.* 1986*b*; Reinikainen *et al.* 1987; Perry *et al.* 1987; see Table 10.1). In the only study to date of the muscarinic receptors in the nucleus basalis of Meynert, the density of receptors was found to be decreased (Shimohama *et al.* 1986*b*).

Taken together, these results indicate that decreased muscarinic receptors in subcortical areas may be a feature of Alzheimer's disease. The loss of receptors in the nucleus basalis of Meynert suggests that some muscarinic receptors are on cholinergic neurons, and possibly on the axons that innervate forebrain areas. In some subcortical forebrain areas innervated by this nucleus the loss of muscarinic receptors could than reflect either the degeneration of cholinergic axons or the loss of cholinoceptive cells, or both. (However, it should be noted that most of the cholinergic axons of the hippocampus originate in the septum and diagonal band, areas in which the muscarinic receptors have not yet been assessed.)

As reviewed above, the density of muscarinic receptors in the cerebral cortex, which receives most of its cholinergic innervation from the nucleus basalis of Meynert, typically appears to be unchanged in Alzheimer's disease. This could indicate that the pathology of cholinergic systems begins in the axon-terminal areas, and that neurons in the cerebral cortex have a greater capacity to compensate for loss of cholinergic input by up-regulating postsynaptic muscarinic receptors than do the cells in the subcortical areas. If this were the case, a decrease in presynaptic muscarinic receptors on degenerating cholinergic axons could be offset by a compensatory increase in postsynaptic receptors on cortical neurons; thus an overall decrease in muscarinic receptors in the cortex would be seen only when the disease progressed enough to involve postsynaptic cortical cells.

Subtypes of muscarinic receptors have been proposed, based on the actions of agonists and antagonists, detailed analyses of agonist binding affinities, the binding of pirenzepine (an atypical antagonist), and differences in the molecular effector systems to which the receptors are coupled (for reviews, see Levine *et al.* 1986). While both M1 and M2 receptor subtypes have high affinity for most muscarinic antagonsits, the M1 subtype displays low affinity for agonists and high affinity for pirenzepine, and the M2 subtype displays high affinity for agonists and lower affinity for pirenzepine. Since muscarinic receptors with high affinity for agonists represent only 10–20 per cent of the total muscarinic receptor population in cerebral cortex, measurements of these receptors by direct binding studies with antagonist ligands such as [^{3}H]quinuclidinyl benzilate are especially insensitive to changes. However, studies of the M2 receptor may be particularly relevant to Alzheimer's disease because in rat cortex and hippocampus these receptors appear to be on cholinergic axons, and to function as autoreceptors which inhibit the release of acetylcholine (Raiteri *et al.* 1984).

Recently, a decrease in muscarinic receptors with high affinity for agonists was found in cerebral cortex from Alzheimer patients, by measuring competition of carbamylcholine for [^{3}H]quinuclidinyl benzilate sites, and by direct binding of the agonist [^{3}H]oxotremorine-M (Mash *et al.* 1985). This suggested that muscarinic receptors with high affinity for agonists (i.e. M2 receptors) are lost from the cerebral cortex in Alzheimer's disease, possibly due to their being located predominantly on cholinergic axons that degenerate (Mash *et al.* 1985). On the other hand, studies with [^{3}H]acetylcholine ([^{3}H]ACh) which also labels muscarinic sites with high affinity for agonists (Kellar *et al.* 1985; Gurwitz *et al.* 1985), indicated that the densities of these sites in frontal and temporal cortex of Alzheimer brains were not significantly different from the densities in control brain (Table 10.2).

The muscarinic site that binds [^{3}H]ACh with high affinity has been well characterized in rat brain. In addition to its high affinity for agonists, it has relatively low affinity for pirenzepine, and it is sensitive to guanyl nucleotides (Kellar *et al.* 1985; Gurwitz *et al.* 1985). These characteristics, as well as the anatomical distribution of the site in rat brain, are those expected of a putative

Table 10.2. Muscarinic receptor sites in the cerbral cortex of control and Alzheimer's disease (A D) brains measured using [^{3}H]ACh

	Density (B_{max}) (fmol mg^{-1} protein)		Affinity (K_d) (nM)	
	Control	AD	Control	AD
Frontal pole	103.4 ± 8.8	118.8 ± 14.7	19.0 ± 2.0	23.0 ± 1.0
Temporal pole	95.8 ± 11.7	83.9 ± 11.9	18.2 ± 2.3	22.3 ± 3.8

The density (B_{max}) and affinity (K_d) of [^{3}H]ACh binding sites were determined from Scatchard plots of binding data. Values are the means ± S E M of independent measurements from 6 brains, and show no statistical difference between control and diseased tissue. Choline acetyltransferase activity was decreased by an average of 46 per cent in both tissues. (From Kellar *et al.* 1987).

M2 receptor as currently defined. The binding characteristics of [^{3}H]ACh in human brain appear to be similar to those in the rat brain (Kellar *et al.* 1987). There are some indications, however, that these high-affinity muscarinic sites are themselves not homogenous (Kellar *et al.* 1985; Gurwitz *et al.* 1985); and there is not yet compelling evidence that high-affinity muscarinic sites are located predominantly on cholinergic axons. Thus the apparent discrepancy between the results of the two studies that examined high-affinity muscarinic receptors in cerebral cortex from Alzheimer brains (Mash *et al.* 1985; Kellar *et al.* 1987) might be due to measurements of different populations of high affinity muscarinic receptors.

10.2.2 *Nicotinic receptors*

Until recently, there were few published studies of nicotinic cholinergic receptors in Alzheimer's disease. Two early studies used the binding of [^{125}I]αbungarotoxin as a measure of nicotinic sites. One study found a decrease in [^{125}I]αbungarotoxin binding in the mid temporal gyrus, but not in the frontal cortex of Alzheimer brains (Davies and Feisullin 1981); the other study found no change in [^{125}I]αbungarotoxin binding in either temporal or cingulate cortex, or in hippocampus (Lang and Henke 1983). While this toxin is a very useful probe of nicotinic receptors in muscle and electric tissues, it does not appear to block nicotinic responses in most studies of neuronal nicotinic receptors, and it is not clear that the binding of [^{125}I]αbungarotoxin is a measure of brain nicotinic cholinergic receptors (Morley *et al.* 1979, 1983; Oswald and Freeman 1981; see Wonnacott, chapter 7, this volume).

[^{3}H]ACh (in the presence of atropine to block muscarinic receptors) and [^{3}H](−)nicotine label apparently identical high-affinity nicotinic cholinergic receptor sites in mammalian brain (Marks *et al.* 1986; Martino-Barrows and Kellar 1987; see Wonnacott, Fig. 7.6, this volume). On the basis of anatomical and pharmacological studies, these agonist binding sites are clearly

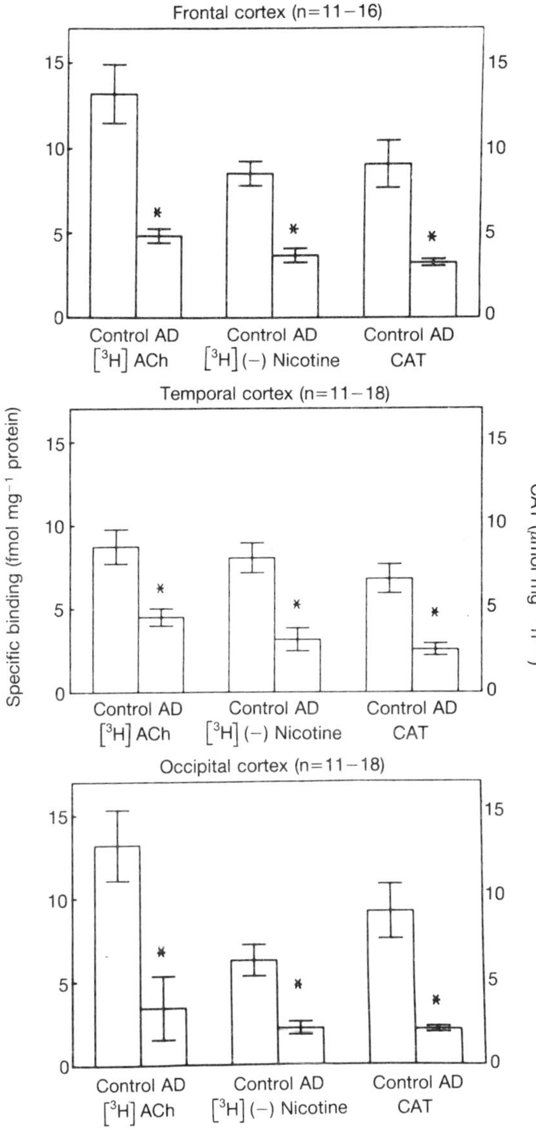

Fig. 10.2. Binding of 6 nM [^{3}H]A Ch and 1 nM [^{3}H](–)nicotine to nicotinic-cholinergic sites, and activity of choline acetyltransferase (C A T) in frontal, temporal, and occipital cortices from control and Alzheimer (A D) brains. Values are the means ± S E M of the number of tissues shown in parentheses. *$P<0.001$ compared to control. (From Whitehouse *et al.* 1986.)

distinct from [^{125}I]αbungarotoxin binding sites in brain (Schwartz *et al.* 1982; Marks and Collins 1982; Clarke *et al.* 1985; Marks *et al.* 1986).

In the last few years, several laboratories have reported that binding of [^{3}H]ACh and [^{3}H](–)nicotine to nicotinic cholinergic sites is significantly decreased in certain areas of Alzheimer brain (Whitehouse *et al.* 1985, 1986; Shimohama *et al.* 1986*b*; Flynn and Mash 1986; Nordberg and Winbald 1986; Nordberg *et al.* 1987; Perry *et al.* 1987; Kellar *et al.* 1987). The most consistent finding has been that binding to nicotinic sites is decreased in the cerebral cortex in Alzheimer's disease. This has been found in the frontal, temporal, and occipital cortex using either [^{3}H]ACh or [^{3}H](–)nicotine as ligands (Fig. 10.2). The decrease in binding is due to a decrease in the density of nicotinic cholinergic binding sites rather than a change in affinity for the ligands used (Fig. 10.3). Autoradiographic studies of nicotinic sites labelled by [^{3}H]ACh indicate that these sites are more concentrated in the deeper laminae (IV–VI) than in the more superficial laminae of the control cerebral cortex; but there appears to be a marked loss of sites in all laminae of cerebral cortex from Alzheimer brains (Whitehouse *et al.* 1988; Fig. 10.4). Significant decreases in nicotinic receptor binding are also seen in the hippocampus (Perry *et al.* 1987; Table 10.1). Furthermore, there appears to be an age-related decrease in nicotinic receptor binding sites in normal human cortex (Flynn and Mash 1986; Kellar *et al.* 1987; Perry *et al.* 1987), but analysis of covariance indicated that the decrease in nicotinic sites in cerebral cortex from Alzheimer patients was due mainly to the presence of the disease (Kellar *et al.* 1987).

Comparison of nicotinic cholinergic binding sites with choline acetyltransferase in three areas of cerebral cortex from control and Alzheimer brains indicated that the nicotinic sites and the enzyme activity were decreased to a similar extent in Alzheimer's disease (Fig. 10.2). Although a close correspondence between decreases in nicotinic cholinergic sites and choline acetyltransferase activity in cerebral cortex from Alzheimer patients has not been an invariable finding (cf., Shimohama *et al.* 1986*b*; Flynn and Mash 1986), a parallel decrease would suggest the possibility that in the cerebral cortex these receptors are located on cholinergic axons that degenerate in Alzheimer's disease. Consistent with this possibility, the nucleus basalis of Meynert, which is a principal source of extrinsic cholinergic innervation of the cortex, has a high density of nicotinic cholinergic sites in human brain (Shimohama *et al.* 1985); and these sites are markedly decreased in Alzheimer's disease (Shimohama *et al.* 1986*b*). In rats, bilateral lesions of the habenula decrease presynaptic cholinergic markers and [^{3}H]nicotine binding in the interpeduncular nucleus, suggesting that nicotinic receptors are on cholinergic axons that innervate the interpenduncular nucleus (Clarke *et al.* 1986). Furthermore, nicotine has been reported to stimulate the release of acetylcholine from rodent cerebral cortex slices and synaptosomes (Balfour 1982; Rowell and Winkler 1984). But in rats, lesions of the nucleus basalis of Meynert, which produce 35–50 per cent decreases in choline acetyltransferase activity, have so far failed to produce measurable changes in nicotinic

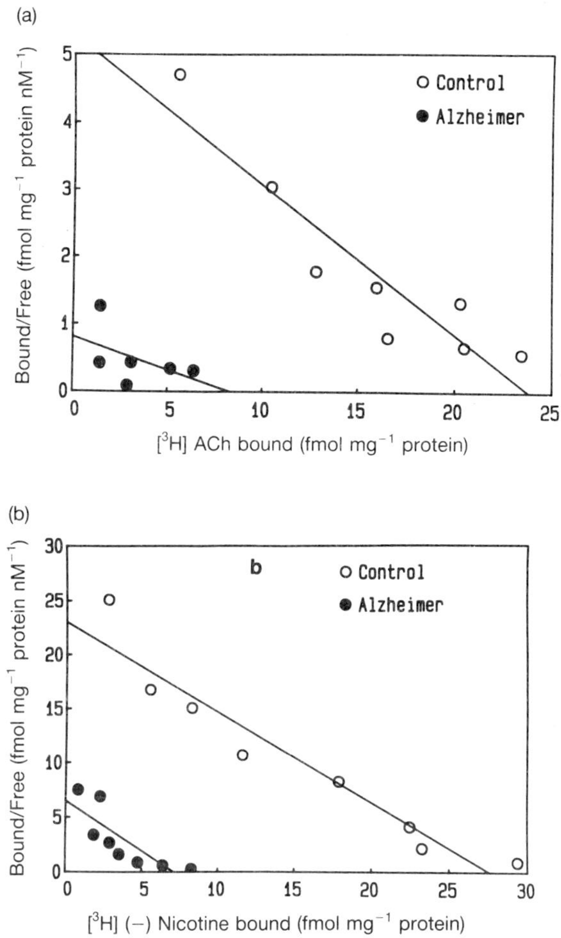

Fig. 10.3. Scatchard plots of (a) [^{3}H]ACh binding and (b) [^{3}H]( )nicotine binding to nicotinic cholinergic sites in frontal cortex from a control and an Alzheimer brain. Saturable binding of nicotinic radioligand was subjected to Scatchard analysis; the slope of the line is equal to $-1/K_d$ (where K_d is the dissociation constant of binding affinity) and the x intercept gives the total number of binding sites under saturating conditions. The slopes of the lines for control and Alzheimer tissue are very similar, whereas the intercepts are markedly changed, indicating that in Alzheimer's disease there is a decrease in the number of binding sites without loss of affinity. (From Whitehouse *et al.* 1986.)

receptors in the cortex (Schwartz *et al.* 1984). The rat, however, is one of the few species in which the cortex contains intrinsic cholinergic neurons, and therefore it is probably not an ideal model in which to study the possibility of nicotinic receptors on extrinsic cholinergic axons of the cortex.

There is also evidence that nicotinic receptors are on catecholamine and

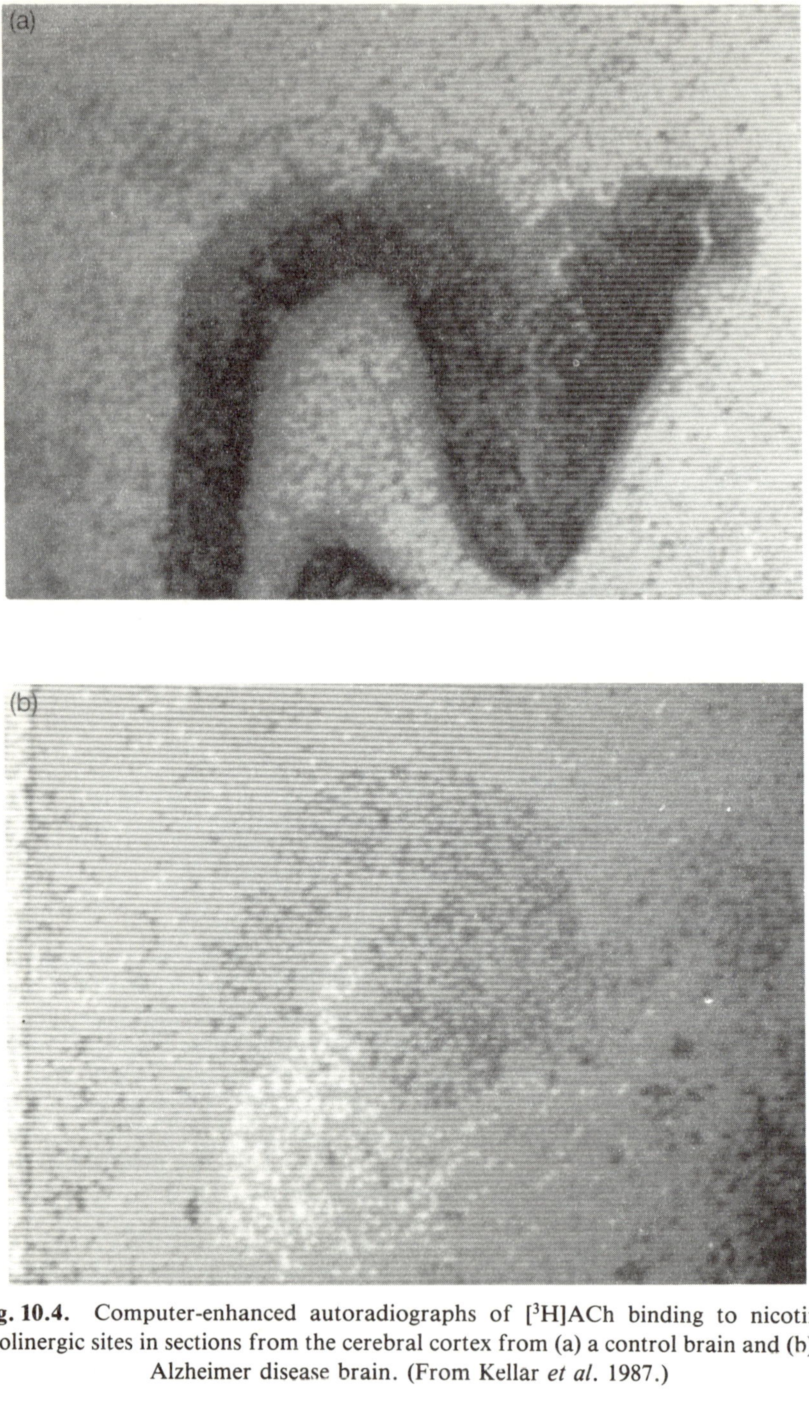

Fig. 10.4. Computer-enhanced autoradiographs of [^{3}H]ACh binding to nicotinic-cholinergic sites in sections from the cerebral cortex from (a) a control brain and (b) an Alzheimer disease brain. (From Kellar *et al.* 1987.)

5-HT neurons in rat brain. Nicotine stimulates or facilitates the release of catecholamines and 5-HT from slices and synaptosomes from rat striatum and hypothalamus (De Belleroche and Bradford 1978; Giorguieff-Chesselet *et al.* 1979; Yoshida *et al.* 1980; Sakurai *et al.* 1982; Rapier *et al.* 1988; reviewed by Balfour 1982). Direct evidence for the existence of nicotinic receptor binding sites on catecholamine and 5-HT neurons in rat brain comes from lesion studies. Lesions of catecholamine or 5-HT axons, by intraventricular injections of 6-hydroxydopamine or 5,7-dihydroxytryptamine respectively, decreased the density of [^{3}H]nicotinic binding sites in the striatum, nucleus accumbens, and the hypothalamus, but not in the cerebral cortex or thalamus (Schwartz *et al.* 1984; Clarke and Pert 1985). This indicates that in the striatum, nucleus accumbens, and hypothalamus, at least some of the nicotinic cholinergic receptors are on catecholamine and 5-HT axons. In the striatum and nucleus accumbens, the catecholamine axons on which the nicotinic sites are located are almost certainly dopamine axons; but in the hypothalamus, the nicotinic sites could be on either dopaminergic or noradrenergic axons, or both (see Fuxe *et al.*, chapter 6, this volume).

The presence of nicotinic cholinergic binding sites on both cholinergic and non-cholinergic axons in specific areas of rat brain, coupled with evidence of nicotine-induced release of certain neurotransmitters (dopamine, noradrenalin, and possibly 5-HT and acetylcholine) suggest that modulation of neurotransmitter release might be an important function of nicotinic-cholinergic receptors (see Wonnacott, chapter 7, this volume). In the basal ganglia and mesolimbic brain areas nicotinic receptors could mediate effects on dopamine neurotransmission (see Reavill, chapter 9, this volume). In the hypothalamus and perhaps other brain areas these receptors could mediate effects on dopamine, noradrenaline, and 5-HT neurotransmission. Furthermore, nicotine has marked effects on pituitary hormone secretion, which may be mediated by nicotinic-cholinergic receptors in the hypothalamus (Fuxe *et al.* 1977; Andersson *et al.* 1981; Sharp and Beyer 1986; see Fuxe, chapter 6, this volume). Thus, endogenous acetylcholine and exogenous nicotine could exert widespread influence on neurotransmission via nicotinic receptors on catecholamine, 5-HT, and cholinergic axons.

10.3 Adrenergic systems in Alzheimer's disease

Within the brainstem nuclei of brains from Alzheimer patients, cell loss and neurofibrillary tangles are seen in the locus coeruleus of the pons (Ishii 1966; Forno 1978; Tomlinson *et al.* 1981; Bondareff *et al.* 1982), which contains the cells of origin of most noradrenergic axons in the brain. The locus coeruleus projects axons to most brain structures; and in the neocortex and hippocampus of Alzheimer brains, neurochemical indices of noradrenergic axons are decreased (Adolfsson *et al.* 1978; Cross *et al.* 1981; Winbald *et al.* 1982; Yates *et al.* 1983).

The density of $\alpha 1$ adrenoceptors in the hippocampus and cerebellar hemisphere, and the density of $\alpha 2$ adrenoceptors in the nucleus basalis of Meynert, have been reported to be decreased (Shimohama *et al.* 1986*a*). The decrease in $\alpha 1$ receptors in the hippocampus and cerebellum in Alzheimer's disease reflects a loss of neurons or neuronal elements within these structures. The decrease in $\alpha 2$ receptors in the nucleus basalis of Meynert is potentially even more interesting. Some $\alpha 2$ receptors are thought to be located on presynaptic noradrenergic axons, where they could function as autoreceptors that regulate the release of noradrenaline; therefore their decreased density in the nucleus basalis of Meynert in Alzheimer brains could be a result of a loss of noradrenergic projections from the locus coeruleus to that nucleus. But it is also possible that the $\alpha 2$ receptors are on cholinergic cells of the nucleus basalis of Meynert, postsynaptic to noradrenergic axons, and that the decreased density of these receptors reflects the well-documented loss of cholinergic cells in this nucleus. Both possibilities are interesting because they imply that a communication link between noradrenergic and cholinergic systems is interrupted in the nucleus basalis of Meynert in Alzheimer's disease.

Two studies have reported that the total population of β adrenergic receptors is not significantly altered in either the neocortex or hippocampus of brains from Alzheimer's patients (Cross *et al.* 1984*a*; Shimohama *et al.* 1987). However, when an attempt was made to resolve the total population of β adrenergic receptors into the $\beta 1$ and $\beta 2$ receptor subtypes, both increases and decreases in both $\beta 1$ and $\beta 2$ receptors were found in various brain areas from Alzheimer patients (Shimohama *et al.* 1987). The interpretation of these changes is unclear, but increases in receptors could reflect up-regulation of receptors by non-diseased target cells following loss of noradrenergic innervation from the locus coeruleus, while decreases in receptors could reflect the independent loss of diseased target cells.

10.4 Serotonergic systems in Alzheimer's disease

In Alzheimer's disease neurofibrillary tangles and degeneration of cells are also seen within the dorsal raphe nucleus (Ishii 1966; Curcio and Kemper 1984; Yamamoto and Hirano 1985), one of the raphe cell groups which contain the cell bodies of 5-HT axons that innervate most forebrain structures. In the neocortex of autopsy or biopsy brain tissue from Alzheimer patients, presynaptic markers of 5-HT axons (such as the concentration of the neurotransmitter and its metabolite, the high-affinity uptake of [^{3}H]5-HT, and the potassium-evoked release of endogenous 5-HT) are reduced (Bowen *et al.*1983; Cross *et al.* 1983; Arai *et al.* 1984; Palmer *et al.* 1987). These deficits are consistent with degeneration of 5-HT neurons in Alzheimer's disease.

Several studies have found that the density of 5-HT receptors is decreased in Alzheimer brains. These changes have been found in the frontal and temporal cortex, cingulate cortex, entorhinal cortex, hippocampus, and amygdala; and they encompass reductions in both 5-HT-1 and 5-HT-2 receptors (Bowen *et al.* 1979, 1983; Cross *et al.* 1984*a*, 1984*b*; Reynolds *et al.* 1984; Perry *et al.* 1984). In one study (Cross *et al.* 1984*b*), there appeared to be substantial reductions in both 5-HT-1 and 5-HT-2 receptors in the septum and in the basal forebrain area that contains the nucleus basalis of Meynert, but in neither area did the decreases in receptors reach statistical significance. Thus any relationships between 5-HT receptors and the origin of the hippocampal and cortical cholinergic systems remain to be determined. On the other hand, there is one report suggesting that, in the rat cortex, some 5-HT-2 receptors are on cholinergic axons (Quirion *et al.* 1985).

5-HT-1 receptors in brain have been subclassified into several subtypes (1-A, 1-B, 1-C) according to differences in pharmacology and anatomical distribution (for review, see Conn and Sanders-Bush 1987). Recently 5-HT-1A receptors were found to be decreased in frontal cortex but not in temporal cortex of Alzheimer brains (Middlemiss *et al.* 1986). This reduction of 5-HT-1A receptors in frontal cortex did not correlate with reductions in the concentrations of 5-HT or its metabolite, indicating that these receptors are not autoreceptors on 5-HT terminals. The integrity of the cholinergic axons was reduced equally in the frontal and temporal cortex of these Alzheimer brains; thus at the present time, any relationships between 5-HT-1A receptors and cholinergic axons are obscure.

10.5 Tobacco smoking and Alzheimer's disease

The loss of nicotinic binding sites in the brains of Alzheimer patients (see section 10.2.2) is the most consistently observed receptor change. Loss of brain nicotinic sites is also associated with Parkinson's disease (Perry *et al.* 1987; Whitehouse *et al.* 1988; see Table 10.1). In view of the evidence for a negative correlation between tobacco smoking and Parkinson's disease (Godwin-Austen *et al.* 1982; see Reavill, chapter 9, this volume), it is reasonable to inquire if a similar relationship exists between smoking and Alzheimer's disease. In an anecdotal report, only 6 out of 30 Alzheimer patients gave a history of having smoked at any time in their lives (Appel 1981), suggesting that an inverse relationship might exist. However, other studies have provided conflicting results. Shalat *et al.* (1987) included cigarette consumption in a case-control study of risk factors for Alzheimer's disease, and observed a significant trend of increased incidence of Alzheimer's disease with increasing numbers of cigarettes smoked per day.

Three other studies have failed to find any correlation between smoking and Alzheimer's disease (Heyman *et al.* 1984; French *et al.* 1985; Jones *et al.* 1987). In the case-control study of Heyman *et al.* (1984) information on numerous environmental, social, genetic, and medical factors was gathered from 40 patients considered to have Alzheimer's disease, and 80 age, sex, and race-matched controls. Approximately 31 per cent of the individuals in both groups had smoked at least 1 pack of cigarettes daily since the age of 40. Thus no correlation (either negative or positive) between the incidence of Alzheimer's disease and a history of smoking was found. The case-control study of French *et al.* (1985) examined 78 patients with a clinical diagnosis of Alzheimer's disease, 76 hospital controls, and 48 neighbourhood controls. These authors reported that 'no significant differences were seen between cases and controls in smoking behaviour'. No further information was given.

In the study by Jones *et al.* (1987), 81 persons with a diagnosis of 'probable dementia of Alzheimer's type' were compared with 112 age-matched controls. 71 per cent of male Alzheimer patients and 77 per cent of male controls gave a history of having smoked. Of the female Alzheimer patients, 32 per cent were (or had been) smokers, compared with 44 per cent of female controls. No significant difference in the lifetime prevalence of smoking between the two groups was found.

More detailed analyses of the average numbers of cigarettes smoked, and the duration of the smoking habit before onset of the disease, are awaited. Other factors such as selective mortality, and physical or social factors that may influence the ability or inclination to smoke, are also important considerations. These are discussed more fully in chapter 9 (see Reavill, this volume). In view of the possible therapeutic value of nicotine in ameliorating some of the cognitive deficits associated with Alzheimer's disease (see section 10.7.4), more detailed smoking histories are warranted to determine if early exposure to nicotine has any influence on the occurrence or progress of the disease.

10.6 Nicotinic receptors in Down's syndrome

Down's syndrome adults who survive into their third or fourth decades display brain changes characteristic of Alzheimer's disease (namely neurofibrillary tangles and neuritic plaques) at autopsy (Masters *et al.* 1985), and commonly develop dementia, i.e. a regression of their cognitive skills. The families of patients with Alzheimer's disease have an increased incidence of Down's syndrome (Heyman *et al.* 1984). A link between Alzheimer's disease and Down's syndrome appeared to be uncovered when the gene encoding the amyloid A4 protein that is deposited in the plaques and tangles was localized to chromosome 21, the aberrant chromosome in Down's syndrome (trisomy 21; St. George-Hyslop *et al.* 1987; Tanzi *et al.* 1987*a*). However, linkage studies have shown that the A4 gene does not segregate with familial Alzheimer's disease (Broeckhoven *et al.* 1987; Tanzi *et al.* 1987*b*); evidently a defect in the amyloid gene cannot explain the brain pathology of all cases of Alzheimer's disease. Nevertheless, the similar neuropathology of brains from Alzheimer patients and aged Down's syndrome individuals has led to a comparison of neurochemical markers (Perry *et al.* 1987). Choline acetyltransferase, acetylcholinesterase, and nicotinic receptor sites were decreased in hippocampal tissue in both conditions; similar changes were found in Parkinson's disease, but not in other neurological diseases (Table 10.1). Alzheimer's and Parkinson's diseases, and Down's syndrome all appear to involve presynaptic cholinergic degeneration, distinguished by the consistently marked decrement in enzyme activities and nicotinic binding. The specificity of these changes encourages specific drug development targeted at the cholinergic system.

10.7 Cholinergic therapies in Alzheimer's Disease

10.7.1 *Strategies*

Because cholinergic deficits constitute the most consistent neurochemical features of Alzheimer's disease, pharmacological treatment of the symptoms of the disease has focussed on the enhancement of cholinergic activity. The synaptic targets, illustrated in Fig. 10.5, comprise:

(1) remaining cholinergic terminals that can be stimulated to increase acetylcholine synthesis and release;

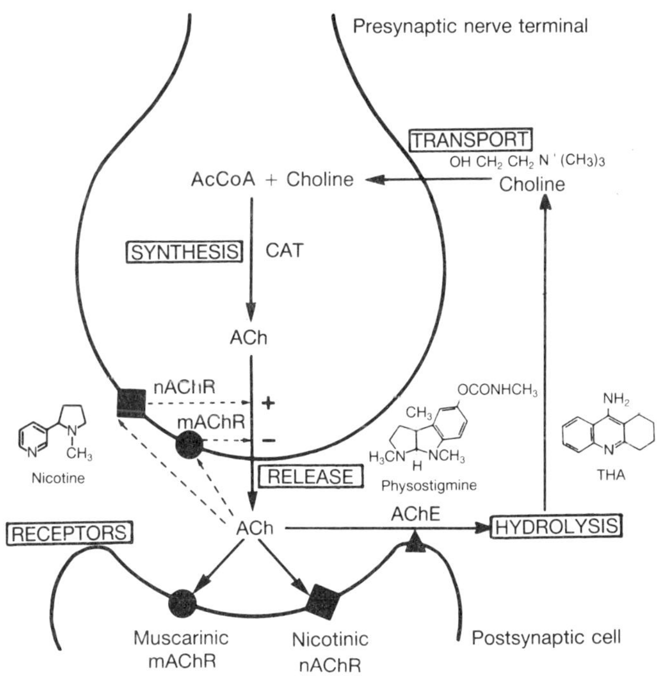

Fig. 10.5. Schematic representation of a cholinergic synapse showing the major targets for pharmacological agents used therapeutically in Alzheimer's disease.

(2) acetylcholinesterase that can be inhibited to limit the breakdown of existing acetylcholine;

(3) postsynaptic cells that can be directly stimulated via receptor activation.

Important considerations in the selection of a drug for clinical trials include a long metabolic half-life (the drug should not be broken down too quickly, so that a single dose will produce a prolonged effect); good penetration of the blood–brain barrier; low side-effects (which may reflect the specificity of the agent; in particular a selectivity for central versus peripheral sites is desirable); and a broad therapeutic window (so that despite individual variation, most patients will respond to the same dose range). An additional complication in the case of Alzheimer's disease is patient selection, because definitive diagnosis still relies on post-mortem histological examination of the brain (see Hollander *et al.* 1986). In the succeeding sections, drug treatments that may include a nicotinic component are briefly reviewed. Muscarinic receptor stimulation is not addressed here; muscarinic agonists have not proved very useful in trials to date and are commonly associated with unacceptable side effects (see Hollander *et al.* 1986).

10.7.2 *Presynaptic agents: choline*

By analogy with the use of the precursor L-dopa to enhance dopamine levels in Parkinson's disease, acetylcholine precursor therapy seemed an obvious means of counteracting the cholinergic deficit in Alzheimer's disease. The supply of substrates appears to be rate-limiting for acetylcholine synthesis (for review, see Jope and Jenden 1980) but clinical trials with choline itself, or lecithin (the principal dietary source of choline), have produced no significant improvement in the symptoms of Alzheimer patients (for review, see Marchbanks 1982). These negative results may be due to the rapid clearance of choline from the plasma, or may indicate that the choline transporter is already saturated. Because transport, synthesis, and release in cholinergic terminals are tightly coupled (Jope and Jenden 1980) choline uptake and acetylcholine synthesis may be enhanced in conditions that promote transmitter release. Thus lecithin has been given in combination with metabolic stimulants such as piracetam (Hollander *et al.* 1986). Lecithin has also been included in other drug trials, for example, with the anti-cholinesterase tetrahydroaminoacridine (Summers *et al.* 1986; see section 10.7.3.2).

In addition to its precursor role in acetylcholine biosynthesis, choline has been reported to have agonist and antagonist activity at nicotinic and muscarinic receptors (Krnjevic and Reinhardt 1979; Ropert and Krnjevic 1982; Sterz *et al.* 1986). Choline competes for [^{3}H]nicotine binding sites in brain membranes, and 50 per cent inhibition is achieved by 200 μM choline (Costa and Murphy 1984). Similar concentrations modified nicotinic receptor responses at the neuromuscular junction (Sterz *et al.* 1986). Choline concentrations in the CSF of Alzheimer patients average 3 μM (Davis *et al.* 1982) and may be raised 5-fold by dietary choline (see Marchbanks 1982). Thus it is unlikely that choline supplements will have any nicotinic effect. On the other hand, choline is reported to be only 5 times weaker than acetylcholine in activating muscarinic receptors in the CNS (Ropert and Krnjevic 1982); activation of inhibitory presynaptic muscarinic autoreceptors could reduce acetylcholine release and counteract the intended benefits of choline therapy.

10.7.3 *Anticholinesterases*

10.7.3.1 *Physostigmine* The cholinesterase inhibitor physostigmine enhances long-term storage of information in young, healthy subjects, without affecting short-term memory or retrieval (Davis *et al.* 1978). Several clinical trials of oral or intravenous physostigmine administered to patients diagnosed as having Alzheimer's disease have reported a modest improvement in long-term memory in these patients (for reviews see Johns *et al.* 1985; Hollander *et al.* 1986).

In addition to its anticholinesterase activity, physostigmine also interacts with nicotinic acetylcholine receptors (Shaw *et al.* 1985; Sadoshima *et al.* 1988). Whereas the enzyme displays a stereospecificy for the (–) enantiomer of physostigmine, the receptor does not discriminate between the two enantiomeric forms (Kawabuchi *et al.* 1988). Physostigmine blocks the ion channel of the receptor at micromolar concentrations (Shaw *et al.* 1985) without competing for the agonist recognition site (Perry *et al.* 1988; see Fig. 10.6.) However, electrophysiological studies revealed some receptor activation by physostigmine at 0.5 μM (Shaw *et al.* 1985). Half-maximal inhibition of acetylcholinesterase in human brain occurs at 50 nM, under optimal conditions (Perry *et al.* 1988). Thus the nicotinic antagonist action of physostigmine is unlikely to be significant in the therapeutic range. Nicotinic inhibition would be undesirable: presynaptically located nicotinic receptors promote transmitter release (see section 10.2.2) and their activation would be desirable in Alzheimer's disease. Postsynaptic nicotinic receptors may have a role in mediating cholinergic transmission.

Physostigmine has the disadvantages of a short half-life and a narrow therapeutic window. Individual variation necesitates careful titration of the drug to determine the optimal dose for each patient. Another consideration in the long-term use of acetylcholinesterase inhibitors is the possibility of provoking a down-regulation of receptors. Administration of irreversible organophosphate cholinesterase inhibitors to rats causes a decrease in the numbers of nicotinic ligand binding sites in the brain (Costa and Murphy 1983; Schwartz and Kellar 1983, 1985; Marks *et al.* 1986; Martino-Barrows and Kellar 1987). Such changes, attributed to the resultant increase in acetylcholine concentrations, could confound the benefits of anticholinesterase treatment, for the reasons indicated above.

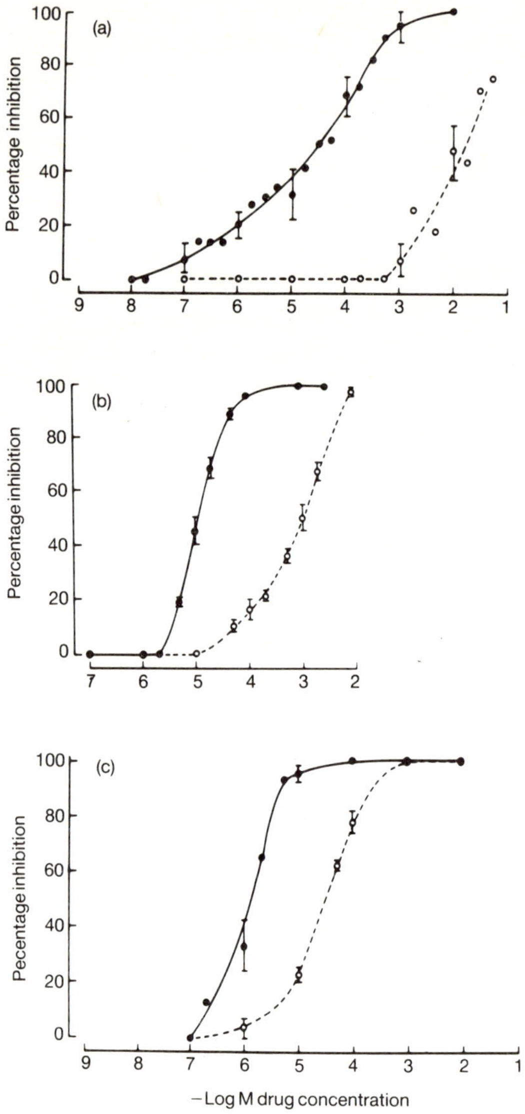

Fig. 10.6. Competition by T H A (solid line) and physostigmine (broken line) for cholinergic ligand binding sites in human cortical membranes. (a) Nicotinic receptor sites labelled by [^{3}H]nicotine (4nM); (b) total muscarinic receptor sites labelled by [^{3}H]*N*-methylscopolamine (1 nM); (c) M1 muscarinic receptor sites labelled by [^{3}H]pirenzepine (2 nM). (From Perry *et al.* 1988.)

10.7.3.2 *Tetrahydroaminoacridine (THA)* Tetrahydroaminoacridine (THA; 9-amino-1,2,3,4-tetrahydroacridine) has recently come to prominence as a potential palliative treatment for Alzheimer's disease. This drug is a centrally-active anticholinesterase (Ho and Freeman 1965) and has a longer duration of action than physostigmine. Publication in the *New England Journal of Medicine* of the results of a long-term trial with oral THA in 12 patients diagnosed as having moderate to severe Alzheimer's disease (Summers *et al.* 1986) has given the drug a high profile and aroused considerable controversy. This study reported an encouraging symptomatic improvement (assessed by memory tests) that was sustained after an average of 12.6 months treatment with THA; the study has been both praised (Editorial 1987) and heavily criticized (Weintraub and Standish 1988).

Although the rationale for using THA stems from its anticholinesterase activity, it appears to have many other actions at the synapse (Albin *et al.* 1988, and references therein), including the competitive inhibition of nicotinic and muscarinic ligand binding to brain tissue (Nilsson *et al.* 1987; Pearce and Potter 1988; Perry *et al.* 1988; Fig. 10.6). THA interacted with nicotinic and muscarinic sites in human cortex at micromolar concentrations (Perry *et al.* 1988; Fig. 10.6) whereas physostigmine was 2 to 3 orders of magnitude less potent. Competition binding assays do not discriminate between agonists and antagonists; a functional assay showed that THA inhibited muscarinic receptor-mediated second-messenger activity (Perry *et al.* 1988), but it is claimed to stimulate presynaptic nicotinic receptors (Nilsson *et al.* 1987).

The relevance of these activities to the therapeutic action of THA depends on several factors, the most crucial is whether receptor interactions occur at clinically effective concentrations. Summers *et al.* (1986) reported the therapeutic range of THA to be between 5 and 70 ng ml^{-1} plasma, equivalent to a concentration range of 0.025–0.35 μM. Half-maximal inhibition of brain acetylcholinesterase occurs at around 1 μM (Pearce and Potter 1988; Perry *et al.* 1988). Phystostigmine was similarly potent (Perry *et al.* 1988), so this parameter does not explain the difference in clinical efficacy between THA and physostigmine, although differences in the half lives of the drugs *in vivo* will contribute to their therapeutic value. Nevertheless, some interaction with nicotinic and M_1 muscarinic sites might be anticipated if brain levels of THA are equivalent to the plasma concentrations (see Fig. 10.6), although muscarinic receptor interactions have been discounted as too weak by Pearce and Potter (1988).

The report that both THA and physostigmine at 100 μM can restore acetylcholine release from Alzheimer brain slices *in vitro*, possibly via a nicotinic mechanism (Nilsson *et al.* 1987), is difficult to interpret in relation to the much lower concentrations of the drugs encountered clinically. Moreover, such functional studies are complicated by the inability to distinguish direct receptor effects of the drugs from their indirect effects via acetylcholinesterase (inhibition of the enzyme will increase synaptic levels of acetylcholine, which will in turn activate receptors) and other targets (Albin *et al.* 1988).

10.7.4 *Nicotine*

10.7.4.1 *Animal and human studies* The loss of nicotinic receptors in Alzheimer's disease (section 10.2.2; Fig. 10.2; Table 10.1) suggests that these receptors may be important in cognitive functioning in normal individuals. This notion is reinforced by the evidence that nicotine improves rapid-information processing performance in healthy individuals (Wesnes and Warburton 1984), and that nicotine is responsible for the improvements in attention and rapid information processing associated with tobacco smoking (see Warburton, chapter 3, this volume). Animal studies of the effects of nicotine on learning and memory have been reviewed by Stolerman (chapter 8, this volume). In general, nicotine has a modest effect on performance in cognitive tasks.

In an animal model of Alzheimer's disease, based on excitotoxin lesions of the nucleus basalis of Meynert, which result in cholinergic losses in the cortex, nicotine (0.1–0.4 mg kg^{-1}, subcutaneously) was found to elicit a greater locomotor response than it does in sham-operated controls (Ksir and Benson 1983). This result suggests that a loss of cholinergic innervation of the cortex may be accompanied by a greater sensitivity to nicotine agonists, perhaps due to an increase in functional nicotinic receptors. However, it is unclear that cortical receptors are involved in this response; recent studies indicate that the mesolimbic dopamine pathway is responsible for the motor stimulation produced by nicotine (Clarke *et al.* 1988).

In a test of short- and long-term memory, lesioned and control rats were compared for their performance in a radial maze task that was learned prior to lesioning. The lesioned animals showed long-lasting deficits in their performance, and choline acetyltransferase activity was reduced by 40 per cent and 25 per cent of control values in the cortex and hippocampus, respectively, of these animals. Low doses of nicotine (0.05 and 0.1 mg kg^{-1} intraperitoneally) produced a dose-dependent improvement in the performance of the lesioned animals, without affecting the control rats (Hodges *et al.* in preparation). This study suggests that nicotine may improve cognitive function in individuals with impaired cholinergic innervation of the cortex.

Nicotine is indeed being considered clinically, and two studies have been published. In the first, Newhouse *et al.* (1988) examined the effects of intravenous infusions of nicotine in 6 elderly patients having a diagnosis of probable Alzheimer's disease and suffering from moderate dementia. Infusions of nicotine base (0.125, 0.25, and 0.5 $\mu g\ kg^{-1}\ min^{-1}$) were given for 60 min, and the patients subjected to cognitive tests and behavioural and physiological assessment. Prominent adverse behavioural effects were encountered, with significant dose-related increases in anxiety and depression; these problems were apparently not seen (at least to such an extent) in most young or middle-aged normal controls (Newhouse *et al.* 1986, 1988). Such mood changes are opposite to the effects of nicotine infusions given to healthy individuals in earlier studies (Murphree 1979; see Warburton, chapter 3, this volume). In the Alzheimer patients, the behavioural effects were sufficiently marked at the highest dose to preclude satisfactory cognitive assessment. However, the intermediate dose (0.25 $\mu g\,kg^{-1}\,min^{-1}$) produced improvements in immediate free recall of lists of words, with fewer errors than observed with placebo or the lowest nicotine dose. The improvement only related to words presented during the infusion, which suggests that nicotine may enhance the acquisition of information, perhaps by increasing attention or vigilance. The authors conclude 'augmenting nicotinic cholinergic function in Alzheimer patients may be useful, but further studies will require careful attention to limiting mood and anxiety effects while maximising possible cognitive benefits.'

In the second published study (Sahakian *et al.* 1989) 7 elderly Alzheimer patients with mild to moderate dementia were compared with similar groups of young, normal adults, and elderly controls, in a number of cognitive tasks. On a rapid information processing task, the elderly controls were slightly impaired, and the Alzheimer patients markedly impaired, relative to the young control group. Low doses of nicotine (0.4, 0.6, and 0.8 mg) administered by subcutaneous injection produced a significant dose-dependent improvement in discriminative sensitivity in both young controls and Alzheimer patients (Fig. 10.7). Nicotine also quickened reaction times in the Alzheimer group, and increased the detection of a flashing or flickering light in a critical-flicker-fusion test, regarded as a measure of cortical arousal. However, the digit span test, considered to be an index of short-term memory capacity, was unaffected by nicotine. The nicotine-induced improvements in attention and information processing (without influencing short-term memory) accord with the effects of nicotine and tobacco smoking in normal subjects (see Warburton, chapter 3, this volume).

10.7.4.2 *Future prospects* Nicotine may exert its beneficial effects by acting at two cellular locations (see Fig. 10.5): firstly, by stimulating presynaptic nicotinic acetylcholine receptors to promote the release of acetylcholine (Rowell and Winkler 1984) and possibly other neurotransmitters (see section 10.2.2), and secondly, by activating postsynaptic receptors. Thus nicotine and other cholinergic drugs may ameliorate the symptoms of Alzheimer's disease, but it must be emphasized that such pharmacological treatments are unlikely to influence the underlying degenerative process. The clear-cut improvements produced by nicotine in the above study (Sahakian *et al.* 1989), together with the positive effects of nicotine infusion on the accuracy of recall (Newhouse *et al.* 1988), suggest that nicotine may be useful in the treatment of Alzheimer's disease, although in the tests employed to date it has not been shown to improve the memory deficit associated with the dementia. However,

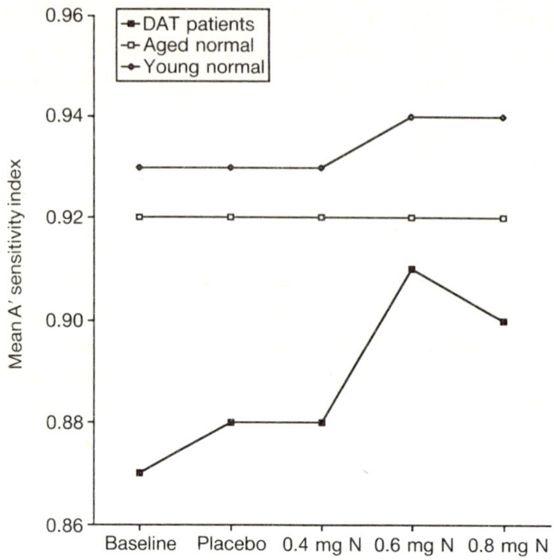

Fig. 10.7. The effect of nicotine (N) on discriminative sensitivity A′ in a computerized task of attention and information processing. 7 patients with a diagnosis of Alzheimer's disease (D A T), 7 young normal controls, and 7 elderly normal controls were tested in 7 successive sessions with the following treatments given by subcutaneous injection: (1) undrugged baseline, (2) placebo (saline), (3) 0.4 mg N, (4) 0.6 mg N, (5) 0.8 mg N, (6) placebo, (7) baseline. Baseline and placebo values were meaned. In the test, digits were presented on a computer screen at the rate of 100 digits min^{-1}, and subjects were required to register the detection of consecutive, ascending odd or even sequences (for example, 2, 4, 6; 5, 7, 9) by pressing a button. Correct detections and false positives were recorded. (From Sahakian *et al.* 1989.)

several factors must be addressed before nicotine can be tried as a therapeutic agent. Some of these issues are summarized below.

1. Patient selection. For the maximum benefit from any pharmacological intervention, drug treatment should be given in the early stages of Alzheimer's disease, before gross structural damage to the brain takes place. For example, in the case of nicotine, degeneration of cholinergic terminals eliminates one locus of action of the drug (namely the presynaptic nicotinic receptors). However, anti-mortem diagnosis of Alzheimer's disease is difficult, especially in the initial stages.

2. Nicotine doses. From studies of human smoking, a great deal is known about the pharmacokinetics of nicotine in man (see Benowitz *et al.*, chapter 4, this volume). These data will assist in selecting appropriate doses and treatment regimes in clinical trials. Nicotine satisfies the majority of the considerations for drug selection outlined in section 10.7.1. Thus it has a reasonably long metabolic half-life (2 h; see Benowitz *et al.*, chapter 4, this volume); it

readily crosses the blood-brain barrier and is most potent at nicotinic receptors in the central and autonomic nervous systems (Paton and Savini 1968), thus reducing the risk of interferance with neuromuscular transmission at centrally effective doses. However, nicotine is likely to have a narrow therapeutic window: data from smokers suggest that nicotine has a low ceiling of tolerated concentrations (see Russell, chapter 11, this volume).

The severe mood changes following the intravenous infusion of nicotine indicate that Alzheimer patients may be markedly more sensitive to nicotine than young, healthy controls (Newhouse *et al.* 1986, 1988). At first, this may seem surprising in view of the loss of nicotinic receptors (for example, 60 per cent decrease in nicotinic binding sites in the cortex, Whitehouse *et al.* 1986). However, the increased sensitivity to nicotine may reflect the lack of acetylcholine and the diminished dual stimulation of both muscarinic and nicotinic receptors by the endogenous transmitter. Nicotine will provide nicotinic stimulation without activating muscarinic receptors; the latter may normally exert an opposing action (for example, at nerve terminals, see Fig. 10.5) and the absence of this counterbalance would result in a greater nicotinic response. Clearly, caution should be exercised in choosing nicotine doses based on data obtained from young, healthy individuals.

3. Route of nicotine administration. The 2 h half-life of plasma nicotine necessitates that the drug is given repeatedly in order to maintain therapeutically useful concentrations. Infusion is impractical for the long-term treatment of mentally disturbed patients. Furthermore, intravenous infusion appears to produce unacceptable behavioural effects (Newhouse *et al.* 1988). Blood nicotine concentrations were not measured in this study, but normal human subjects given a nicotine infusion of $2\,\mu g\,kg^{-1}\,min^{-1}$ for 30 min had peak plasma nicotine concentrations of $30\,ng\,ml^{-1}$; after 3 h plasma levels had declined to $5\,ng\,ml^{-1}$ (Benowitz *et al.* 1982; see Benowitz *et al.*, Fig. 4.14, this volume). Thus plasma concentrations of $1\text{–}10\,ng\,ml^{-1}$ may be anticipated in the clinical study. Such concentrations are at the lower end of the range experienced by smokers, and do not seem unreasonable. However, more work is necessary to characterise the delivery of nicotine by this route in man.

Although subcutaneous injection has produced some beneficial effects in patients (Sahakian *et al.* 1989), repeated injection is also undesirable for long-term treatment. Nicotine has been given orally in dextrose tablets to normal controls (Wesnes and Warburton 1984), and in the form of nicotine gum (polacrylex; see Russell, chapter 11, this volume). The efficiency of buccal absorption of nicotine from gum is discussed more fully in chapter 4. The buccal route has the advantage of ease of use, and the gum lasts for some 30 min, sustaining blood nicotine levels of about $10\,ng\,ml^{-1}$ (from 2 mg nicotine gum) over a longer time course (see Jaffe, Fig. 1.2, this volume). Gum can be given repeatedly throughout the day, but may be less useful for patients with dentures. An alternative means of prolonged nicotine administration is by skin absorption from patches applied to the skin. Some of these considerations would be circumvented by the development of a longer-lasting nicotinic agonist that could be given orally.

4. Consequences of long term nicotine administration. Animals given daily injections of nicotine or chronically infused with the drug develop increased numbers of nicotinic binding sites in the brain (Schwartz and Kellar 1983, 1985; Marks *et al.* 1983, 1985; see Wonnacott, Fig. 7.5, this volume). Such changes are also seen in the brains of smokers (Benwell *et al.* 1988). The functional significance of this increase in binding sites is not understood at present. Chronic administration of low doses of nicotine produces a correlation between increased receptor number and increased psychomotor stimulation by nicotine, but at high nicotine concentrations the even greater increase in binding site density is not accompanied by behavioural stimulation (Ksir *et al.* 1987). Instead, such animals show behavioural tolerance to nicotine (Marks *et al.* 1985; see Swedberg *et al.*, chapter 2, this volume).

This upregulation of nicotinic receptors is considered to reflect the desensitizing properties of nicotine, such that it acts as an antagonist rather than an agonist (Schwartz and Kellar 1985). It is not known if the increased number of binding sites reflects a net increase in functional receptors, or merely maintains the original functional capacity (by compensation for those that are desensitized or functionally inactive). If compensation is incomplete, there could even be a decrease in available functional nicotinic receptors despite the increased number. of binding sites.

These concomitants of chronic nicotine administration have important implications for nicotine therapy. If nicotine is given too frequently or at too high a dose it may inhibit nicotinic receptors rather than providing the stimulation sought. Intermittent nicotine administration (injections or gum) may avoid this problem better than continuous application by infusion or skin absorption. Clearly, there is a need for centrally-active nicotinic agonists that desensitize their receptors less readily than nicotine.

10.8 Concluding remarks

Alzheimer's disease is characterized neurochemically by a loss of cholinergic markers, consistent with the degeneration of cholinergic projections to the neocortex and hippocampus. Nicotinic receptor binding is also markedly reduced whereas muscarinic sites are largely spared; M2 muscarinic receptors which have a predominantly presynaptic localization may also be diminished.

Pharmacological therapies, aimed at relieving the symptoms of Alzheimer's disease, have concentrated on stimulating the cholinergic system. Choline supplements and muscarinic agents have not met with much success to date. Acetylcholinesterase inhibitors, and more recently nicotine, have produced some useful improvements in cognitive performance. A better enhancement of cholinergic transmission may be achieved by administering a cocktail of cholinergic agents, comprising a nicotinic agonist (to promote acetylcholine release by stimulation of presynaptic receptors), lecithin (to provide additional choline to support acetylcholine turnover), and a cholinesterase inhibitor (to maximize the effects of acetylcholine released). Transplantation of embryonic cholinergic neurons into the deafferented areas of the brain is also receiving attention, and has improved passive avoidance memory in an animal model of Alzheimer's disease (Fine *et al.* 1985). Such grafts do not replicate the original innervation patterns but may exert a paracrine function by releasing local concentrations of acetylcholine (Arendt *et al.* 1988).

The current failure to adequately treat the cognitive impairment characteristic of Alzheimer's disease highlights the lack of detailed knowledge about the cholinergic mechanisms underlying learning and memory. The role of nicotinic receptors in the disease process, however, has given these sites a new prominence, and indicates that nicotinic mechanisms play an important role in brain biochemistry other than supporting tobacco smoking. Precisely what that role is remains to be discovered.

Acknowledgements

The studies on Alzheimer's disease brain from Dr Kellar's laboratory were carried out in collaboration with Drs Peter J. Whitehouse and Donald L. Price of The Johns Hopkins University School of Medicine, and with Andrea Martino-Barrows of Georgetown University. These studies were supported by a grant from The Alzheimer's Disease and Related Disorders Association (II-86-100).

References

Adolfsson, R., Gottfries, C.G., Oreland, L., Roos, B.E., and Winblad, B. (1978). Reduced levels of catecholamines in the brain and increased activity of monoamine oxidase in platelets in Alzheimer's disease: Therapeutic implications. In *Alzheimer's disease: senile dementia and related disorders (aging)*, vol. 7, (R. Katzman, R.D. Terry, and K.L. Bick, ed.), pp. 33–34. Raven Press, New York.

Albin, R.L., Young, A.B., and Penney J.B. (1988). Tetrahydro-9-aminoacridine (THA) interacts with the phencyclidine (PCP) receptor site. *Neuroscience Letters* **88**, 303–7.

Andersson, K., Eneroth P., and Agnati, L.F. (1981). Nicotine-induced increases of noradrenaline turnover in discrete noradrenaline nerve terminal systems of the hypothalamus and the median eminence of the rat and their relationship to changes in the secretion of adenohypophyseal hormones. *Acta Physiologica Scandinavica* **113**, 227–31.

Appel, S.H. (1981). Alzheimer's disease. In *Brain neurotransmitters and receptors in aging and age-related disorders* (S.J. Enna, Samorajski, T., and Beer, B. ed.), pp. 203–7. Raven Press, New York.

Arai, H., Kosaka, K., and Iizuka, R. (1984). Changes of biogenic amines and their metabolites in postmortem brains from patients with Alzheimer-type dementia. *Journal of Neurochemistry* **43**, 388–93.

Arendt, T., Allen Y., Sinden J., Schugens M.M., Marchbanks R.M., Lantos, P.L., and Gray J.A. (1988). Cholinergic-rich brain transplants reverse alcohol-induced memory deficits. *Nature* **332**, 448–50.

Averback, P. (1983). Two new lesions in Alzheimer's disease. *Lancet* **2**, 1203.

Balfour, D.J.K. (1982). The effects of nicotine on brain neurotransmitter systems. *Pharmacology and Therapeutics* **16**, 269–82.

Bartus, R.T., Dean, R.L. III, Beer, B., and Lippa, A.S. (1982). The cholinergic hypothesis of geriatric memory dysfunction. *Science* **217**, 408–17.

Benowitz, N.L., Jacob, P., Jones, R.J., and Rosenberg, J. (1982). Interindividual variability in the metabolism and cardiovascular effects of nicotine in man. *Journal of Pharmacology and Experimental Therapeutics* **221**, 368–72.

Benwell, M.E.M., Balfour, D.J.K., and Anderson, J.M. (1988). Evidence that tobacco smoking increases the density of (–)[^{3}H]-nicotine binding sites in human brain. *Journal of Neurochemistry* **50**, 1243–7.

Bondareff, W., Mountjoy, C.Q., and Roth, M. (1982). Loss of neurons of origin of the adrenergic projection to cerebral cortex (nucleus locus coeruleus) in senile dementia. *Neurology* **32**, 164–8.

Bowen, D.M., Allen, S.J., Benton, J.S., Goodhardt, M.J., Haan, E.A., Palmer, A.M., Sims, N.R., Smith, C.C.T., Spillane, J.A., Esiri, M.M., Neary, D., Snowdon, J.S., Wilcock, G.K., and Davison, A.N. (1983). Biochemical assessment of serotonergic and cholinergic dysfunction and cerebral atrophy in Alzheimer's disease. *Journal of Neurochemistry* **41**, 266–72.

Bowen, D.M., Smith, C.B., White, P., and Davison, A.N. (1976). Neurotransmitter-related enzymes and indices of hypoxia in senile dementia and other abiotrophies. *Brain* **99**, 459–96.

Bowen, D.M., White, P., Spillane, J.A., Goodhardt, M.J., Curzon, G., Iwangoff, P., Meirer-Ruge, W., and Davison, A.N. (1979). Accelerated aging or selective neuronal loss as an important cause of dementia? *Lancet* **1**, 11–14.

Broeckhoven, C.V., Genthe, A.M., Vandenberghe, A., Horsthemke, B., Backhovens, H., Raeymaekers, P., VanHul, W., Wehnert, A., Gheurens, J., Cras, P., Bruyland, M., Martin, J., Salbaum, M., Multheup, G., Masters, C., Beyreuther, K., Gurling, H., Mullan, M., Holland, A., Barton, A., Irving, N., Williamson, A., Richards, S., and Hardy, J. (1987) Failure of familial Alzheimer's disease to segregate with the A4-amyloid gene in several European families. *Nature* **329**, 153–5.

Caulfield, M.P., Straughan, D.W., Cross, A.J., Crow, T., and Birdsall, N.J.M. (1982). Cortical muscarinic receptor subtypes and Alzheimer's disease. *Lancet* **ii**, 1277.

Clarke, P.B.S. and Pert, A. (1985). Autoradiographic evidence for nicotine receptors on nigrostriatal and mesolimbic dopaminergic neurons. *Brain Research* **348**, 355–8.

Clarke, P.B.S., Hamill, G.S., Nadi, N.S., Jacobwitz, D.M., and Pert, A. (1986). ^{3}H-Nicotine and ^{125}I-alpha-bungarotoxin labelled nicotinic receptors in the interpeduncular nucleus of rats. II. Effects of habenular deafferentation. *Journal of Comparative Neurology* **251**, 407–13.

Clarke, P.B.S., Schwartz, R.D., Paul, S.M., Pert, C.B., and Pert, A. (1985). Nicotinic binding in rat brain: autoradiographic comparison of ^{3}H-acetylcholine, ^{3}H-nicotine and ^{125}I-alpha-bungarotoxin. *Journal of Neuroscience* **5**, 1307–15.

Clarke, P.B.S., Fu, D.S., Jakubovic, A., and Fibiger, H.C. (1988). Evidence that mesolimbic dopaminergic activation underlies the locomotor stimulant action of nicotine in rats. *Journal of Phamacology and Experimental Therapeutics* **246**, 701–8.

Collerton, D. (1986). Cholinergic function and intellectual declinc in Alzheimer's disease. *Neuroscience* **19**, 1–28.

Conn, P. J. and Sanders-Bush, E. (1987). Central serotonin receptors: effector systems, physiological roles, and regulation. *Psychopharmacology* **92**, 267–77.

Costa, L. G. and Murphy S. D. (1983). ^{3}H-Nicotine binding in rat brain: alteration after chronic acetylcholinesterase inhibition. *Journal of Pharmacology and Experimental Therapeutics* **226**, 392–7.

Costa, L. G. and Murphy, S. D. (1984). Interaction of choline with nicotinic and muscarinic receptors in the rat brain *in vitro*. *Clinical and Experimental Pharmacology and Physiology* **11**, 649–54.

Cross, A. J., Crow, T. J., Ferrier, I. N., Johnson, J. A., Bloom S. R., and Corsellis, J. A. N. (1984*b*). Serotonin receptor changes in dementia of the Alzheimer's type. *Journal of Neurochemistry* **43**, 1574–81.

Cross, A. J., Crow, T. J., Johnson, A., Joseph, M. H., Perry, E. K., Perry, R. H., Blessed, G., and Tomlinson, B. E. (1983). Monoamine metabolism in senile dementia of Alzheimer type. *Journal of Neurological Science* **60**, 383–93.

Cross, A. J., Crow, T. J., Johnson, A., Perry, E. K., Perry, R. H., Blessed, G., Tomlinson, B. E. (1984*a*). Studies on neurotransmitter receptor systems in neocortex and hippocampus in senile dementia of the Alzheimer's type. *Journal of Neurological Science* **64**, 109–17.

Cross, A. J., Crow, T. J., Perry, E. K., Perry, R. H., Blessed, G., and Tomlinson, B. E. (1981). Reduced dopamine-beta-hydroxylase activity in Alzheimer's disease. *British Medical Journal* **282**, 93–4.

Curcio, C. A. and Kemper, T. (1984). Nucleus raphe dorsalis in dementia of the Alzheimer's type: neurofibrillary changes and neuronal packing density. *Journal of Neuropathology and Experimental Neurology* **43**, 359–68.

Davies, P. and Feisullin, S. (1981). Postmortem stability of alpha-bungarotoxin binding sites in mouse and human brain. *Brain Research* **216**, 449–54.

Davies, P. and Maloney, A. J. F. (1976). Selective loss of central cholinergic neurons in Alzheimer's disease. *Lancet* **2**, 1403.

Davies, P. and Verth, A. H. (1978). Regional distribution of muscarinic acetylcholine receptor in normal and Alzheimer's type dementia brain. *Brain Research* **138**, 385–92.

Davis, K. L., Mohs, R. C., Tinklenberg, J. R., Pfefferbaum, A., Hollister, L. E., and Kopell, B. S. (1978). Physostigmine: improvement of long-term memory processes in normal subjects. *Science* **201**, 272–4.

Davis, K. L., Hsieh, J. Y., Levy, M. I., Horvath, T. B., Davis, B. M., and Mohs, R. C. (1982). Cerebrospinal fluid acetylcholine, choline and senile dementia of the Alzheimer type. *Psychopharmacology Bulletin* **18**, 193–5.

De Belleroche, J. and Bradford, H. F. (1978). Biochemical evidence for the presence of presynaptic receptors on dopaminergic nerve terminals. *Brain Research* **142**, 53–58.

Drachman, D. A. and Leavitt, J. L. (1974). Human memory and the cholinergic system. A relationship to aging? *Archives of Neurology* **30**, 113–21.

Editorial (1987). Cholinergic treatment in Alzheimer's disease: encouraging results. *Lancet* **1**, 139–41.

Esiri, M. M. and Wilcock, G. K. (1984). The olfactory bulbs in Alzheimer's disease. *Journal of Neurology Neurosurgery and Psychiatry* **47**, 56–60.

Fine, A., Dunnett, S. B., Bjorklund, A., and Iversen, S. D. (1985). Cholinergic ventral forebrain grafts into the neocortex improve passive avoidance memory in a rat model of Alzheimer's disease. *Proceedings of the National Academy of Sciences USA* **82**, 5227–30.

Flynn, D.D. and Mash, D.C. (1986). Characterization of *L*-[^{3}H]nicotine binding in human cerebral cortex: Comparison between Alzheimer's disease and the normal. *Journal of Neurochemistry* **47**, 1948–54.

Forno, L.S. (1978). The locus coeruleus in Alzheimer's disease. *Journal of Neuropathology and Experimental Neurology* **27**, 614.

French, L.R., Schuman, L.M., Mortimer, J.A., Hutton, J.T., Boatman, R.A., and Christians, B. (1985). A case-control study of dementia of the Alzheimer's type. *American Journal of Epidemiology* **121**, 414–21.

Fuxe, K., Agnati, L., Eneroth, P., Gustafsson, J.-A., Hokfelt, T., Loftstom, A., Skett, B., and Skett, P. (1977). The effect of nicotine on central catecholamine neurons and gonadotropin secretion. I. Studies in the male rat. *Medical Biology* **55**, 148–57.

Giorguieff-Chesselet, M.F., Kemel, M.L., Wandscheer, D., and Glowinksi, J. (1979). Regulation of dopamine release by presynaptic nicotinic receptors in rat striatal slices: effect of nicotine in low concentration. *Life Sciences* **25**, 1257–62.

Godwin-Austen, R.B., Lee, P.N., Marmot, M.G., and Stern, G.M. (1982). Smoking and Parkinson's disease. *Journal of Neurology Neurosurgery and Psychiatry* **45**, 577–81.

Gurwitz, D., Kloog, Y., and Sokolovsky, M. (1985). High affinity binding of [^{3}H]-acetylcholine to muscarinic receptors: regional distribution and modulation by guanine nucleotides. *Molecular Pharmacology* **28**, 297–305.

Heyman, A., Wilkinson, W.E., Stafford, J.A., Helms, M.J., Sigmon, A.H., and Weinberg, T. (1984). Alzheimer's disease: a study of epidemiological aspects. *Annals of Neurology* **15**, 335–41.

Ho, A.K.S. and Freeman, S.E. (1965). Anticholinesterase activity of tetrahydroaminacridine and succinylcholine hydrolysis. *Nature* **205**, 1118–9.

Hollander, E., Mohs, R.C., and Davis, K.L. (1986). Cholinergic approaches to the treatment of Alzheimer's disease. *British Medical Bulletin* **42(1)**, 97–100.

Ishii, T. (1966). Distribution of Alzheimer's neurofibrillary changes in the brain stem and hypothalamus of senile dementia. *Acta Neuropathologica* **6**, 181–7.

Iversen, S.D. (1988). Chemistry of dementia. *Chemistry in Britain*, April, 338–342.

Johns, C.A., Haroutunian, V., Greenwald, B.S., Mohs, R.C., Davis, B.M., Kanof, P., Horvath, T.B., and Davies, K.L. (1985). Development of cholinergic drugs for the treatment of Alzheimer's disease. *Drug Development Research* **5**, 77–96.

Jones, G.M.M., Reith, M., Philpot, M.P., and Sahakian, B.J. (1987). Smoking and dementia of Alzheimer type. *Journal of Neurology Neurosurgery and Psychiatry* **50**, 1383.

Jope, R.S. and Jenden, D.J. (1980). The utilisation of choline and acetyl coenzyme A for the synthesis of acetylcholine. *Journal of Neurochemistry* **35**, 318–25.

Kawabuchi, M., Boyne, A.F., Desphande, S.S., Cintra, W.M., Brossi, A., and Albuquerque, E.X. (1988). Enantiomer (+)physostigmine prevents organophosphate-induced subjunctional damage at the neuromuscular synapse by a mechanism not related to cholinesterase carbamylation. *Synapse* **2**, 139–47.

Kellar, K.J., Martino, A.M., Hall, D.P. Jr., Schwartz, R.D., and Taylor, R.L. (1985). High-affinity binding of [^{3}H]acetylcholine to muscarinic cholinergic receptors. *Journal of Neuroscience* **5**, 1577–82.

Kellar, K.J., Whitehouse, P.J., Martino-Barrows, A.M., Marcus, K., and Price, D.L. (1987). Muscarinic and nicotinic cholinergic binding sites in Alzheimer's disease cerebral cortex. *Brain Research* **436**, 62–8.

Krnjevic, K. and Reinhardt, W. (1979). Choline excites cortical neurons. *Science* **206**, 1321–3.

Ksir, C., and Benson, D.M. (1983). Enhanced behavioural response to nicotine in an animal model of Alzheimer's disease. *Psychopharmacology* **81**, 272–3.

Ksir, C., Hakan, R.L., and Kellar, K.J. (1987). Chronic nicotine and locomotor activity: influences of exposure dose and test dose. *Psychopharmacology* **92**, 25–9.

Lang, W., and Henke, H. (1983). Cholinergic receptor binding and autoradiography in brains of non-neurological and senile dementia of Alzheimer-type patients. *Brain Research* **267**, 271–80.

Levine, R.R., Birdsall, N.J.M., Giachetti, A., Hammer, R., Iverson, L.L., Jenden, D.J., and North, R.A. (ed.) (1986). Subtypes of muscarinic receptors II. *Trends in Pharmacological Science* Suppl. 1–77.

Marchbanks, R.M. (1982). Biochemistry of Alzheimer's disease *Journal of Neurochemistry* **39**, 35–45.

Marks, M.J., Burch, J.B., and Collins, A.C. (1983). Effects of chronic nicotine infusion on tolerance development and nicotinic receptors. *Journal of Pharmacology and Experimental Therapeutics* **226**, 817–25.

Marks, M.J., Stitzel, J.A., and Collins, A.C. (1985). Time course study of the effects of chronic nicotine infusion on drug response and brain receptors. *Journal of Pharmacology and Experimental Therapeutics* **235**, 619–28.

Marks, M.J., and Collins, A.C. (1982). Characterization of nicotine binding in mouse brain and comparison with the binding of alpha-bungarotoxin and quinuclidinyl benzilate. *Molecular Pharmacology* **22**, 554–64.

Marks, M.J., Stitzel, J.A., Romm, E., Wehner, J.M., and Collins, A.C. (1986). Nicotinic binding sites in rat and mouse brain: comparison of acetylcholine, nicotine, and α-bungarotoxin. *Molecular Pharmacology* **30**, 427–36.

Martino-Barrows, A.M., and Kellar, K.J. (1987). [^{3}H]Acetylcholine and [^{3}H](−)nicotine label the same recognition site in rat brain. *Molecular Pharmacology* **31**, 169–74.

Mash, D.C., Flynn, D.D., and Potter, L.T. (1985). Loss of M2 muscarine receptors in the cerebral cortex in Alzheimer's disease and experimental cholinergic denervation. *Science* **228**, 1115–7.

Masters, C.L., Simms, G., Weinman, N.A. Multhaup, G., McDonald, B.L., and Beyreuther, K. (1985). Amyloid plaque core protein in Alzheimer's disease and Down syndrome. *Proceedings of the National Academy of Sciences USA* **82**, 4245–9.

McGeer, P.L., McGeer, E.G., Suzuki, J., Dolman, C.E., and Nagai, T. (1984). Aging, Alzheimer's disease, and the cholinergic system of the basal forebrain. *Neurology* **34**, 741–5.

Middlemiss, D.N., Palmer, A.M., Edel, N., and Bowen, D.M (1986). Binding of the novel serotonin agonist 8-hydroxy-2(di-*n*-propylamino)tetralin in normal and Alzheimer brain. *Journal of Neurochemistry* **46**, 993–6.

Morley, B.J., Farley, G.R., and Javel, E. (1983). Nicotinic receptors in mammalian brain. *Trends in Pharmacological Sciences* **4**, 225–7.

Morley, B.J., Kemp, G.E., and Salvaterra, P. (1979). Alpha-bungarotoxin binding sites in the CNS. *Life Sciences* **24**, 859–72.

Murphree, H.B. (1979). EEG effects in humans of nicotine, tobacco smoking, withdrawal from smoking and possible surrogates. In *Electrophysiological effects of nicotine* (A. Remond and C. Izard, ed.), pp. 227–44. Elsevier, Amsterdam.

Newhouse, P.A., Sunderland, T., Thompson, T., Tariot, P.N., Weingartner, H., Meuller, E.R., Cohen, R.M., and Murphey, D.L. (1986). Intravenous nicotine in a patient with Alzheimer's disease. *American Journal of Psychiatry* **143**, 1494–5.

Newhouse, P.A., Sunderland, T., Tariot, P.N., Blumhardt, C.L., Weingartner, H., Mellow, A., and Murphey, D.L. (1988). Intravenous nicotine in Alzheimer's disease: a pilot study *Psychopharmacology* **95**, 171–5.

Nilsson, L., Adem, A., Hardy, J., Winblad, B., and Nordberg, A. (1987). Do tetrahydroaminoacridine (THA) and physostigmine restore acetylcholine release in Alzheimer brains via nicotinic receptors? *Journal of Neurological Transmission* **70**, 357–68.

Nordberg, A., Adem, A., Nilsson, L., and Winblad, B. (1987).Cholinergic deficits in CNS and peripheral non-neuronal tissue in Alzheimer dementia. In *Cellular and Molecular Basis of Cholinergic Function* (M. Dowdall, ed.), pp. 858–868. Ellis Horwood, Sussex, UK.

Nordberg, A., and Winblad, B. (1986). Reduced number of [^{3}H]nicotine and [^{3}H]acetylcholine binding sites in the frontal cortex of Alzheimer brains. *Neuroscience Letters* **72**, 115–21.

Oswald, R.E., and Freeman, J.A. (1981). Alpha-bungarotoxin binding and central nervous system nicotinic acetylcholine receptors. *Neuroscience* **6**, 1–14.

Palmer, A.M., Francis, P.T., Benton, J.S., Sims, N.R., Mann, D.M.A., Neary, D., Snowden, J.S., and Bowen, D.M. (1987). Presynaptic serotonergic dysfunction in patients with Alzheimer's disease. *Journal of Neurochemistry* **48**, 8–15.

Paton, W.D.M., and Savini, E.C. (1968). The action of nicotine on the motor endplate in the cat. *British Journal of Pharmacology* **32**, 360–80.

Pearce, B.D., and Potter, L.T. (1988). Effects of tetrahydroaminoacridine on M1 and M2 muscarine receptors. *Neuroscience Letters* **88**, 281–5.

Perry, E.K., Gibson, P.H., Blessed, G., Perry, R.H., and Tomlinson, B.E. (1977). Neurotransmitter enzyme abnormalities in senile dementia. *Journal of Neurological Science* **34**, 247–65.

Perry, E.K., Perry, R.H., Candy, J.M., Fairbairn, A.F., Blessed, G., Dick, D.J., and Tomlinson, B.E. (1984). Cortical serotonin-s_2 receptor binding abnormalities in patients with Alzheimer's disease: comparisons with Parkinson's disease. *Neuroscience Letters* **51**, 353–7.

Perry, E.K., Tomlinson, B.E., Blessed, G., Bergmann, K., Gibson, P.H., and Perry, R.H. (1978). Correlation of cholinergic abnormalities with senile plaques and mental test scores in senile dementia. *British Medical Journal* **2**, 1457–9.

Perry, E.K., Perry, R.H., Smith, C.J., Purohit, D., Bonham, J., Dick, D.J., Candy, J.M., Ewardson, J.A., and Fairbairn, A. (1987). Cholinergic receptors in cognitive disorders. *Canadian Journal of Neurological Science* **13**, 521–7.

Perry, E.K., Smith, C.J., Court, J.A., Bonham, J.R., Rodway, M., and Atack, J.R. (1988). Interaction of 9-amino-1,2,3,4-tetrahydroaminoacridine (THA) with human cortical nicotinic receptor binding *in vitro*. *Neuroscience Letters* **91**, 211–6.

Price, D.L. (1986). New perspectives on Alzheimer's disease. *Annual Reviews of Neuroscience* **9**, 489–512.

Price, D.L., Whitehouse, P.J., and Struble, R.G. (1985) Alzheimer's disease. *Annual Reviews of Medicine* **36**, 349–56.

Quirion, R., Richard, J., and Dam T.V. (1985). Evidence for the existence of serotonin type-2 receptors on cholinergic terminals in rat neocortex. *Brain Research* **333**, 345–9.

Raiteri, M., Leardi, R., and Marchi, M. (1984). Heterogeneity of presynaptic muscarinic receptors regulating neurotransmitter release in the rat brain. *Journal of Pharmacology and Experimental Therapeutics* **228**, 209–14.

Rapier, C., Lunt, G.G., and Wonnacott, S. (1988). Stereoselective nicotine-induced release of dopamine from striatal synaptosomes: concentration dependence and repetitive stimulation. *Journal of Neurochemistry* **50**, 1123–30.

Reinikainen, K.J., Reikkinen, P.J., Halonen, T., and Laasko, M. (1987). Decreased muscarinic receptor binding in cerebral cortex and hippocampus in Alzheimer's disease. *Life Sciences* **41**, 453–61.

Reisine, T.D., Yamamura, H.I., Bird, E.D., Spokes, E., and Enna, S.J. (1978). Pre- and postsynaptic neurochemical alterations in Alzheimer's disease. *Brain Research* **159**, 477–81.

Reynolds, G.P., Arnold, L., Rossor, M.N., Iversen, L.L., Mountjoy, C.Q., and Roth, M. (1984). Reduced binding of [^{3}H]-ketanserin to cortical 5-HT_2 receptors in senile dementia of the Alzheimer type. *Neuroscience Letters* **44**, 47–51.

Rinne, J.O., Rinne, J.K., Laasko, K., Paijarvi, L., and Rinne, UK. (1984). Reduction in muscarinic receptor binding in limbic areas of Alzheimer brain. *Journal of Neurology Neurosurgery and Psychiatry* **47**, 651–3.

Roberts, E. (1986). Alzheimer's disease may begin in the nose and may be caused by alumuniosilicates. *Neurobiology of Aging* **7**, 561–7.

Ropert, N., and Krnjevic, K. (1982). Pharmacological characterization of facilitation of hippocampal population spikes by cholinomimectics. *Neuroscience* **7**, 1963–77.

Rowell, P.P., and Winkler, D.L. (1984). Nicotinic stimulation of [^{3}H]acetylcholine release from mouse cerebral cortical synaptosomes. *Journal of Neurochemistry* **43**, 1593–8.

Sadoshima, J., Tokutomi, N., and Akaike, N. (1988). Effects of neostigmine and physostigmine on the acetylcholine receptor-inophore complex in frog isolated sympathetic neurones. *British Journal of Pharmacology* **94**, 620–4.

Sahakian, B., Jones, G., Levy, R., Gray, J., and Warburton, D. (1989). The effects of nicotine on attention, information processing, and short-term memory in patients with dementia of the Alzheimer type *British Journal of Psychiatry*, in press.

Sakurai, Y., Takano, Y., Kohjimoto, Y., Honda, K., and Kamiya, H. (1982). Enhancement of ^{3}H-dopamine release and its ^{3}H-metabolites in rat striatum by nicotinic drugs. *Brain Research* **242**, 99–106.

Schwartz, R.D., and Kellar, K.J. (1983). Nicotinic cholinergic receptor binding sites in the brain: regulation *in vivo*. *Science* **220**, 214 6.

Schwartz, R.D., and Kellar, K.J. (1985). *In vivo* regulation of [^{3}H]acetylcholine recognition sites in brain by nicotinic cholinergic drugs. *Journal of Neurochemistry* **45**, 427–33.

Schwartz, R.D., Lehmann, J., and Kellar, K.J. (1984). Presynaptic nicotinic cholinergic receptors labelled by [^{3}H]acetylcholine on catecholamine and serotonin axons in brain. *Journal of Neurochemistry* **42**, 1495–8.

Schwartz, R.D., McGee, R. Jr., and Kellar, K.J. (1982). Nicotinic cholinergic receptors labelled by [^{3}H]acetylcholine in rat brain. *Molecular Pharmacology* **22**, 56–62.

Shalat, S.L., Seltzer, B., Pidcock, C., and Baker, E.L. (1987). Risk factors for Alzheimer's disease. *Neurology* **37**, 1630–3.

Sharp, B.M., and Beyer, H.S. (1986). Rapid desensitization of the acute stimulatory effects of nicotine on rat plasma adrenocorticotropin and prolactin. *Journal of Pharmacology and Experimental Therapeutics* **238**, 486–91.

Shaw, K.P., Aracava, Y., Akaike, A., Daly, J.W., Rickett, D.L., and Alburquerque, E.X. (1985). The reversible cholinesterase inhibitor physostigmine has channel-blocking and agonist effects on the acetylcholine receptor-ion channel complex. *Molecular Pharmacology* **28**, 527–38.

Shimohama, S., Taniguchi, T., Fujiwara, M., and Kameyama, M. (1985). Biochemical characterization of the nicotinic cholinergic receptors in human brain: Binding of (–)-[^{3}H]nicotine. *Journal of Neurochemistry* **45**, 604–10.

Shimohama, S., Taniguchi, T., Fujiwara, M., and Kameyama, M. (1986*a*). Biochemical characterization of alpha-adrenergic receptors in human brain and changes in Alzheimer-type dementia. *Journal of Neurochemistry* **47**, 1294–301.

Shimohama, S., Taniguchi, T., Fujiwara, M., and Kameyama, M. (1986*b*). Changes in nicotinic and muscarinic cholinergic receptors in Alzheimer's type dementia. *Journal of Neurochemistry* **46**, 228–93.

Shimohama, S., Taniguchi, T., Fujiwara, M., and Kameyama, M. (1987). Changes in beta-adrenergic receptor subtypes in Alzheimer-type dementia. *Journal of Neurochemistry* **48**, 1215–21.

Simpson, J., Yates, C.M., Gordon, A., and St. Clair, Dm. (1984). Olfactory tubercle choline acetyltransferase activity in Alzheimer-type dementia, Down's syndrome and Huntington's disease. *Journal of Neurology Neurosurgery and Psychiatry* **47**, 1138–40.

Sterz, R., Peper, K., Simon, J., Ebert, J.P., Edge, M., Pagala, M., and Bradley, R.J. (1986). Agonist and blocking effects of choline at the neuromuscular junction. *Brain Research* **385**, 99–114.

St George-Hyslop, P.H., Tanzi, R.E., Polinsky, R.J., Haines, J.L., Nee, L., Watkins, P.C., Myers, R.H., Feldman, R.G., Pollen, D., Drachman, D., Growdon, J., Bruni, A., Foncin, J., Salmon, D., Frommelt, P., Amaducci, L., Sorbi, S., Piacentini, S., Stewart, G.D., Hobbs, W.J., Conneally, P.M., and Gusella, J.F. (1987). The genetic defect causing familial Alzheimer's disease maps on chromosome 21. *Science* **235**, 885–890.

Summers, W.K., Majovski, L.V., Marsh, G.M., Tachiki, K., and Kling, A. (1986). Oral tetrahydroaminoacridine in long-term treatment of senile dementia, Alzheimer type. *New England Journal of Medicine* **315**, 1241–5.

Tagliavini, F., and Pilleri, G. (1983). Neuronal counts in basal nucleus of Meynert in Alzheimer's disease and in simple senile dementia. *Lancet* **1**, 469–70.

Tanzi, R.E., Gusella, J.F., Watkins, P.C., Bruns, G.A.P. St.George-Hyslop, P, Van Keuren, M.L., Patterson, D., Pagan, S., Kurnit, D.M., and Neve, R.L. (1987*a*). Amyloid B-protein gene: cDNA, mRNA distribution, and genetic linkage near the Alzheimer locus. *Science* **235**, 880–4.

Tanzi, R.E., St.George-Hyslop, P.H., Haines, J.L., Polinsky, R.J., Nee, L., Foncin, J., Neve, R.L., McClatchey, A.I., Conneally, P.M., and Gusella, J.F. (1987*b*). The genetic defect in familial Alzheimer's disease is not tightly linked to the amyloid B-protein gene. *Nature* **329**, 156–7.

Terry, R., and Katzman, R. (1983). Senile dementia of the Alzheimer type: Defining a disease. In *The neurology of aging*, (R. Katzman and R. Terry, ed.), pp. 51–84. Davis, Philadelphia.

Tomlinson, B.E., Irving, D., and Blessed, G. (1981). Cell loss in the locus coeruleus in senile dementia of Alzheimer type. *Journal of Neurological Science*. **49**, 418–28.

Weintraub, M., and Standish R. (1988). Tetrahydroaminoacridine: a possible treatment for senile dementia of the Alzheimer type. *Hospital Formulary* **23**, 31–5.

Wesnes, K., and Warburton, D.M. (1984). Effects of scopolamine and nicotine on human rapid information processing performance *Psychopharmacology* **82**, 147–50.

Whitehouse, P.J., Martino, A.M., Antuono, P.G., Coyle. J.T., Price, D.L., and Kellar K.J. (1985). Reductions in nicotinic cholinergic receptors measured using [^{3}H]acetylcholine in Alzheimer's disease. *Society for Neuroscience Abstracts* **11**, 134.

Whitehouse, P. J., Martino, A. M., Antuono, P. G., Lowenstein, P. R., Coyle. J. T., Price, D. L., and Kellar K. J. (1986). Nicotinic acetylcholine binding in Alzheimer's disease. *Brain Research* **371**, 146–51.

Whitehouse, P. J., Price, D. L., Clark, A. W., Coyle, J. T., and DeLong M. R. (1981). Alzheimer's disease: Evidence for selective loss of cholinergic neurons in the nucleus basalis. *Annals of Neurology* **10**, 122–6.

Whitehouse, P. J., Price, D. L., Struble, R. G., Clark, A. W., Coyle, J. T., and DeLong, M. R. (1982). Alzheimer's disease and senile dementia: loss of neurons in the basal forebrain. *Science* **215**, 1237–9.

Whitehouse, P. J., Martino, A. M., Wagster, M. V., Price, D. L., Mayeux, R., Atack, J. R., and Kellar, K. J. (1988). Reductions in [^{3}H]nicotinic acetylcholine binding in Alzheimer's disease and Parkinson's disease: an autoradiographic study. *Neurology* **38**, 720–3.

Winblad, B., Adolfsson, R., Carlsson, A., and Gottfries, C.-G (1982). Biogenic amines in brains of patients with Alzheimer's disease. In *Alzheimer's disease: a report of progress in research (aging)*, Vol 19, (S. Corkin, K. L. Davis, J. H. Growdon. E. Usdin, and R. J. Wurtman, ed.), pp. 25–33. Raven, New York.

Wilcock, G. K., Esiri, M. M., Bowen, D. M., and Smith, C. C. T. (1982). Alzheimer's disease. Correlation of cortical choline acetyltransferase activity with the severity of dementia and histological abnormalities. *Journal of Neurological Science* **57**, 407–17.

Yamamoto, T., and Hirano, A. (1985). Nucleus raphe dorsalis in Alzheimer's disease: neurofibrillary tangles and loss of large neurons. *Annals of Neurology* **17**, 573–7.

Yates, C. M., Simpson, J., Gordon, A., Maloney, A. F. J., Allison, Y., Ritchie, I. M., and Urquhart, A. (1983). Catecholamines and cholinergic enzymes in pre-senile and senile Alzheimer-type dementia and Down's syndrome. *Brain Research* **280**, 119–26.

Yoshida, I., Kato, Y., and Imura, H. (1980). Nicotine-induced release of noradrenaline from hypothalamic synaptosomes. *Brain Research* **182**, 361–8.

11 Nicotine intake and its control over smoking

M.A.H. Russell

11.1 Introduction

Since its use began, it has been known that tobacco tends to gain a hold on those who use it. Cigarette smoking, in particular, is highly addictive. Most smokers say they would like to give up, but when they try they usually fail. Yet there has been a resistance to 'official' recognition of compulsive tobacco use as an addiction or dependence disorder (Jaffe, chapter 1, this volume). This may have reflected doubts about the role of nicotine as the underlying cause of the addiction. Until recently such doubts were justifiable. The evidence was largely circumstantial and there were important gaps. Although pharmacologists had long ago identified a host of nicotine effects that are potentially rewarding (Murphree and Weyer 1967), intravenous self-administration studies in animals had failed to provide reliable evidence that nicotine could act as a primary reinforcer. In a study in humans, intravenous bolus doses of nicotine failed to affect ongoing smoking behaviour (Kumar *et al.* 1977), leading the authors to conclude that 'tobacco smoking is generally regarded as a form of nicotine dependence, but the evidence for this is slender.' Examples of addictive behaviours without a pharmacological basis, such as compulsive gambling, nail-biting, and the desire for sweet tastes or high-fat foods, fortified doubts about the pharmacological role of nicotine. In a review of the evidence in 1979 it had to be conceded that the case was not proven (Russell 1979).

In the last few years, progress in three main areas of behavioural research has now firmly established nicotine as the primary and controlling factor in smoking. These include its role as a primary reinforcer, its effects on the tobacco withdrawal syndrome, and its controlling influence in the self-regulation of smoke intake. In a series of studies in several species, Goldberg and his colleagues have shown unequivocally that nicotine can, like other drugs of abuse, act as a primary reinforcer, and that animals will press a lever to self-administer nicotine. This work is reviewed by Swedberg *et al.* (chapter 2, this volume). Why the early studies failed may be explained by use of excessive doses and accumulation of nicotine to aversive levels. Goldberg and colleagues got around this problem by use of schedules which included time-out periods and the requirement of multiple responses to elicit each injection. It is important, however, that this work be replicated in other laboratories and that nicotine doses and blood levels are functionally comparable to those of smokers.

The second important contribution to establishing the pharmacological role of nicotine in the motivation of human smoking has come from the development of nicotine polacrilex gum (Ferno *et al.* 1973). This product and the scientific studies it has generated have made clinicians, psychologists, and the public at large more fully aware of the nature of smoking as a form of drug dependence. Its effects, compared with placebo, on the alleviation of tobacco withdrawal effects and as an aid to smoking cessation have provided new evidence of the role of nicotine in smoking. This work is discussed fully by Jaffe (chapter 1, this volume).

The third line of study stems from the development of reliable methods for measuring nicotine in blood. It has been shown that most smokers absorb sufficient nicotine to obtain pharmacological effects, and that they tend to regulate the way they smoke to maintain their individually characteristic blood nicotine levels, which are kept fairly constant from one day to the next. It has been shown also that blood, and presumably brain, levels of nicotine exercise a controlling influence on smoking behaviour and smoke inhalation, independent of the effects of taste, smell, and the amounts of other smoke components. This work will be discussed in this chapter.

In recent years, research on smoking has tended to focus on whether people smoke mainly for nicotine. The methodology has usually involved study of average effects in relatively small groups of smokers. Now that this has been established, it is appropriate to begin to address the more difficult question of understanding why they smoke for nicotine. This involves consideration of the wide individual differences in nicotine intake and its effects, the role of tolerance and pharmacokinetic factors, of learning and conditioning, as well as the level of dependence, and many other factors. Systematic data are scanty, but attention will be focused on those studies that have administered nicotine or measured blood nicotine levels.

This chapter is divided into three main sections. The first considers the role of nicotine as the controlling factor in the self-regulation of smoke intake. The other two sections are concerned with understanding the mechanisms and processes underlying nicotine's role in smoking behaviour. The second section examines how nicotine tolerance and the pharmacokinetic characteristics of nicotine interact to influence its pharmacological and behavioural effects. The third considers individual differences within the context of the smoking career and level of dependence.

11.2 Self-regulation of nicotine intake

The study of self-regulation of nicotine intake by smokers, or nicotine titration as it is sometimes called, is an important approach for understanding the role of nicotine in smoking. It addresses the issue of the extent to which smoking behaviour is controlled by the consequent blood, and hence brain, nicotine concentrations. In the first instance, the approach was used to demonstrate that smokers alter their smoking pattern to regulate their nicotine intake independently of the intake of tar, carbon monoxide, and other smoke components, and of other factors such as taste and sensory impact. Interest has now shifted to consideration of the precision of intake regulation and its determinants, to the extent to which usual intake levels are exceeded and reductions tolerated, and to the notion of an optimal range above which toxic aversive effects appear, and below which satisfaction is lost and aversive withdrawal effects become manifest.

11.2.1 *Historical background*

In 1968 Armitage and his colleagues showed that with puff duration and frequency kept constant the nicotine content of a puff of cigarette smoke varied according to the volume of the puff, and that the larger the puffs the quicker the cigarette was completed (Armitage *et al.* 1968). Using the blood-pressure response of the anaesthetized cat as a bioassay of nicotine absorption, they showed that the effect of a 25 ml puff of cigarette smoke, containing 170 μg nicotine, introduced into the lungs of the animal was equivalent to that of 120 μg nicotine base injected rapidly into the femoral vein, both responses being sharp and almost instantaneous. The slight dose-effect difference is attributable to retention and metabolism of nicotine introduced to the lungs, 30 per cent compared with 6 per cent of that injected into the pulmonary artery (Turner *et al.* 1975). Cigarette smoke introduced into the mouth had no effect, and the effect of thirty 25 ml puffs of cigar smoke at 30 s intervals over 15 min was slow to develop, and much smaller than that of a single 25 ml puff of cigarette smoke in the lungs (Armitage and Turner 1970). As a result of these studies Armitage drew attention to the puff-by-puff 'literally finger-tip control' that smokers have over their nicotine intake (Armitage *et al.* 1968). His estimates of average nicotine intakes of 50–150 μg per puff, about 0.5–1.5 mg per cigarette, accord well with those using modern pharmacokinetic techniques (Benowitz and Jacob 1984; Feyerabend *et al.* 1985).

Armitage's studies suggested that the rate of nicotine absorption through the lungs is sufficient to imply that in smokers who inhale deeply, systemic delivery of nicotine to their tissues takes the form of an intravenous-like high-nicotine bolus a few seconds after each inhaled puff, and that arterial blood nicotine levels between puffs are lower. Studies on arterial blood nicotine concentrations in dogs after tobacco smoke administration into the lungs showed increases following each 'puff' and, as expected, a drop before the next puff (Isaac and Rand 1969). Finally, one of Armitage's colleagues (Turner 1971) compared the concentrations in carotid arterial blood of cats during and after a series of small intravenous bolus doses of nicotine (4 $\mu g\,kg^{-1}$ every min for 20 min) with those produced by a continuous infusion of the same dose over 20 min. The intermittent bolus dosage produced higher nicotine concentrations (about 102 $ng\ ml^{-1}$ compared with 78 $ng\ ml^{-1}$ for the continuous infusion), but the curves during dosage were similarly smooth, the lack of bolus fluctuations presumably reflecting the fact that only one blood sample was taken between each bolus dose.

These early studies clearly demonstrated the potential that smokers have to regulate and vary their nicotine intake by altering their patterns of puffing and inhalation. They also indicate that it is the smoking patterns of individual smokers rather than the standard machine-smoked deliveries of their cigarettes that will determine nicotine intake. None of these studies, however, inform whether or not smokers do in fact regulate nicotine intake to individually preferred levels. This was demonstrated first by Lucchesi *et al.* (1967) who showed that during prolonged slow intravenous infusions of nicotine smokers lit-up fewer cigarettes, puffed them less frequently, and discarded them earlier than under saline control conditions. Blood nicotine levels were not measured but the results suggest strongly that the administration of nicotine by an alternative route caused the smokers to down-regulate their intake from cigarettes. However, the matter was opened to doubt by subsequent studies in which cigarette consumption was used as the measure of smoke intake.

One example which strongly influenced thought at that time was a study by Waingrow and Horn (1968). They analyzed self-report data from a national sample of 1466 American cigarette smokers interviewed in 1964 and 1966, and found no tendency for smokers who had switched to a lower yield brand to increase their daily cigarette consumption. With hindsight we realize that cigarette consumption is a poor measure of nicotine intake, and that intake and its regulation are more strongly related to the way smokers puff and inhale their cigarettes than to the number they smoke. Thus, while a change in consumption may serve as positive evidence of intake regulation, its absence cannot be construed as negative evidence.

The issue of intake regulation remained unclear until Ashton and Watson (1970) and Frith (1971) independently showed that smokers modified their puffing behaviour according to the tar and nicotine yields of their cigarettes. When smoking higher-yielding cigarettes, they took smaller, less frequent puffs, and when smoking lower-yield cigarettes, puff volume and puff rate were increased. In the case of Ashton and Watson, the finding was serendipitous and was noted during studies on the effect of smoking on perceptual motor-tasks. Frith, however, set out to test the hypothesis, having been inspired by the study of Lucchesi *et al.* (1967).

Numerous brand-switching studies have since confirmed these findings and there are several reviews on the topic (Russell 1978; Gritz 1980; Moss and Prue 1982; McMorrow and Foxx 1983). Most of the studies used indirect measures of smoke intake to the mouth rather than direct measurement of nicotine concentrations in the blood. A further limitation to brand-switching studies stems from the high correlation between tar and nicotine yields, which makes it difficult to infer whether any changes observed are attributable to nicotine rather than tar or other confounding factors. The next section therefore focuses on those more direct studies concerned with the role of nicotine in the regulation of smoke intake.

11.2.2 *Nicotine as a determinant of smoke intake*

To determine the role of nicotine in the regulation of smoke intake from cigarettes requires control of all possible confounding factors (tar yield, draw resistance, flavour, burning characteristics, gas-phase components, and so on). This is difficult to achieve. Various approaches have been tried, including studies of changes in smoking behaviour in response to:

(1) variations in nicotine yield of cigarettes;

(2) nicotine administration by injection, infusion, ingestion, or via nicotine chewing gum;

(3) pharmacological blockade of nicotine actions:

(4) alteration of nicotine excretion by inducing changes in urinary pH.

11.2.2.1 *Nicotine yield and smoking behaviour* It is obviously difficult to manipulate the nicotine yield of cigarettes without at the same time affecting other confounding factors. This has seldom been achieved, but it was done more than twenty years ago by Finnegan *et al.* (1945) in a study for which the experimental cigarettes were prepared by using a tobacco naturally low in nicotine to produce the low-yield cigarette (0.34 mg nicotine), and then adding nicotine to produce the high-yielding cigarette (1.96 mg nicotine). The two cigarettes therefore differed only in their nicotine yields. This study was the first demonstration of the up-regulation of nicotine intake by smokers, although the authors seem to have been unaware of this aspect. Their purpose was to look for the development of withdrawal symptoms after 'blind' substitution of low-nicotine cigarettes. This they found, with 12 of their 24 subjects experiencing moderate to severe withdrawal symptoms on switching from the high- to the low-yield brands. But, in addition, their data show that those who experienced withdrawal symptoms did not change their daily cigarette consumption (mean 25.6 per day on each brand), whereas those who did not suffer withdrawal symptoms increased their consumption by 18 per cent (from 24.3 to 28.7 per day) on switching from the high- to low-yield brand. It is possible that withdrawal effects were prevented by increasing cigarette consumption and thereby maintaining an adequate nicotine intake.

One or two other studies have attempted to examine the effects of cigarettes in which nicotine yields were varied independently of other factors (although with only limited success), and the results suggested that the smoking pattern was affected more by the delivery of nicotine than that of tar (Goldfarb *et al.* 1976; Herning *et al.* 1981).

11.2.2.2 *Nicotine administration.* In 1942, Johnston gave smoking doses of nicotine by hypodermic or intravenous injection to 35 volunteers. He reported that non-smokers found it unpleasant but 'smokers almost invariably thought the sensation pleasant and, given an adequate dose, were disinclined to smoke for some time thereafter.' The nicotine injections also relieved the withdrawal cravings of smokers deprived of tobacco. However, this pioneering study was neither 'blind' nor controlled.

Twenty-five years later, Lucchesi *et al.* (1967) showed, under double-blind conditions, that slow intravenous infusion of nicotine at a rate of 4 mg h^{-1} had an inhibitory effect on ad lib smoking compared with saline control. There was a 27 per cent reduction in the number of cigarettes smoked per session, and subjects also took significantly fewer puffs and discarded their cigarettes earlier. Although the amount of suppression was small in relation to the dose of nicotine given, the slow infusion would not have simulated the intermittent bolus dosage from smoking, and any decrease in inhalation would have been undetected.

The findings of Lucchesi *et al.* (1967) were not confirmed by a subsequent study which failed to show an inhibitory effect on puffing during and after intravenous nicotine (Kumar *et al.* 1977). One explanation is that the subjects in the second study were so hampered by the apparatus for physiological measures that they were unable to hold their cigarettes and smoke naturally. Smoking involved puffing through a holder, which was positioned obtrusively close to the subject's face and contained a continuously lighted cigarette, replaced as necessary by the experimenter. This experimental situation and its demand characteristics may have swamped the factors which normally control smoking. In a recent intravenous nicotine pre-loading study in our laboratory, which is not yet published, the findings support those of Lucchesi *et al.* (1967) in showing a clear inhibitory effect on subsequent smoking behaviour. Furthermore, the down-regulation of nicotine intake was so precise that, despite pre-loading to an average blood nicotine concentration in excess of 25 ng ml^{-1}, there was no tendency for the subjects to overshoot their normal blood nicotine levels when ad lib smoking was permitted.

Studies using other forms of nicotine administration show similar results. One early study showed that ingestion of 10 mg nicotine daily had a small but significant inhibitory effect, and reduced daily cigarette consumption by 8 per cent (Jarvik *et al.* 1970). Bearing in mind that the average nicotine intake from smoking is 20–30 mg per day, and that the systemic bioavailability of oral nicotine is only 30–40 per cent, the small size of the effect is understandable. Nicotine chewing gum also has a small but statistically significant inhibitory effect on ad lib smoking compared with placebo. Fewer cigarettes are smoked after chewing it (Kozlowski *et al.* 1975) and inhalation is reduced as measured by a drop in carboxyhaemoglobin levels (Russell *et al.* 1976*b*; Ebert *et al.* 1984).

Further evidence of the precision of the down-regulation of nicotine intake from smoking is provided by an early study of gum use (Russell *et al.* 1976*b*). A group of 43 smokers were instructed to continue to smoke as they wished, but to also take ten pieces of 2 mg nicotine gum per day at a rate of about 1 h^{-1} (this pattern of gum use generally produces steady-state blood nicotine levels around 10 ng ml^{-1}). Despite the extra nicotine from the gum, blood nicotine levels just after a cigarette were not increased, indicating that the smokers had down-regulated their intake from smoking. The peak blood nicotine just after a cigarette averaged 27.4 ng ml^{-1} when they were taking gum and smoking ad lib, compared with the similar level of 30.1 ng ml^{-1} when smoking normally without the gum.

In a similar study, in 12 subjects, the effects of 2 mg and 4 mg gum on nicotine intake from ad lib smoking were compared with a placebo gum (Ebert *et al.* 1984). There was a systematic reduction in CO intake, but blood nicotine levels averaged 29.5, 30.9, and 40.7 ng ml^{-1} in the placebo, 2 mg, and 4 mg gum conditions respectively ($p < .05$), indicating that down-regulation of the intake from smoking while on 4 mg gum was not very precise. One explanation is that the blood nicotine levels were troughs rather than peaks, and steady-state levels on hourly 4 mg gum could overlap with smoking troughs. It is also possible that down-regulation of trough levels is less precise than for peaks. The subjects were also atypical in being recently treated alcoholics, still in hospital.

In summary, these studies involving the administration of nicotine provide a specific test of the effect of nicotine on the regulation of smoke intake without confounding from other factors. Taken together they provide sound evidence that nicotine is an important factor in down-regulation. However, by their nature they are limited to testing down-regulation and cannot inform about the up-regulation of smoke intake. It has been suggested that the precision of down-regulation may be attributable to avoidance of the aversive toxic effects of excessive doses (Russell 1979).

11.2.2.3 *Nicotinic blockade* A novel approach was the use of a centrally-acting nicotine antagonist, mecamylamine. This increased the number of puffs and the number of cigarettes smoked by 25 per cent and 30 per cent respectively, compared with a placebo (Stolerman *et al.* 1973*b*). In contrast, pentolinium, an antagonist with predominantly peripheral actions, had no effect on smoking behaviour. This study confirms that smoking behaviour is modified to regulate nicotine intake and that this effect is mediated by its actions in the brain. Its actions at peripheral sites appear not to be implicated.

These findings have been confirmed and extended in two recent studies. In one, pre-treatment with mecamylamine caused dose-related increases in number of cigarettes smoked, number of puffs per cigarette and a 76 per cent increase in carbon monoxide intake (Nemeth-Coslett *et al.* 1986). It is of interest that satisfaction from smoking was not impaired even after the highest mecamylamine dose (20 mg). This may have been due to incomplete pharmacological blockade. It also suggests that satisfaction, at least in the short-term, is closely linked with sensory and other conditioned factors. Of interest in the other recent study (Pomerleau *et al.* 1987) is the finding that mecamylamine pre-treatment caused an approximate (numerical data not reported) mean increase of 27 per cent in blood nicotine concentrations from ad lib smoking. Without mecamylamine such increases may have been aversive. Indeed, rapid smoking has been used as an aversive treatment method. This involves smoking and inhaling at a rate of one puff per 6 s until it can no longer be endured. In one study in 15 subjects, tolerance limits were determined by irritation to the mouth and throat in 6 subjects, and by nausea in 9 subjects. Average blood nicotine concentrations were 48.1 $ng\,ml^{-1}$ after rapid smoking compared with 32.4 $ng\,ml^{-1}$ after normal smoking, an increase of 48.5 per cent (Russell *et al.* 1978).

While these studies demonstrate that blockade of nicotinic receptors in the brain (but not peripherally) causes up-regulation of nicotine intake from cigarettes, it is not clear whether up-regulation was mediated by an attempt to overcome a partial blockade to regain positive reinforcement, or whether the blockade enabled the smokers to exceed usual levels that are normally under aversive control.

11.2.2.4 *Urinary nicotine excretion rate* Urinary nicotine excretion can be manipulated by administration of ammonium chloride to acidify the urine, or sodium bicarbonate to make it alkaline. Under acidic conditions, up to 15 per cent or more of nicotine is excreted unchanged in the urine (the remainder being metabolized). However, under alkaline conditions urinary nicotine excretion is negligible (see Benowitz *et al.*, chapter 4, this volume). Schachter showed that when subjects received tablets to acidify their urine, and increase nicotine excretion, their cigarette consumption increased. Conversely, when nicotine excretion was reduced by raising urinary pH, they smoked fewer cigarettes (Schachter *et al.* 1977). This study provides clear evidence that smoking behaviour is modified to regulate nicotine intake independently of other factors. In a recent study, Benowitz and Jacob (1985) manipulated urinary pH and confirmed Schachter's findings. They also showed that blood nicotine concentrations were not increased when urinary nicotine excretion was blocked by sodium bicarbonate (i.e. perfect down-regulation) but were 15 per cent lower during urinary acidification with ammonium chloride (i.e. incomplete up-regulation).

In summary, these four classes of study, employing a variety of approaches, all designed to test specifically the role of nicotine, are remarkly concordant in their findings. They show that nicotine intake is down-regulated with great precision to avoid exceeding the blood nicotine concentrations of usual smoking. On the other hand, up-regulation (or compensation) appears to be less precise, and is partial or incomplete. Blood nicotine concentrations were measured in only four of the studies discussed in this section, so that data on the quantitative aspects of intake regulation are rather limited. Further data on the self-regulation of blood nicotine concentrations are considered in the brand-switching studies in the next section.

11.2.3 *Brand-switching studies*

Experimental studies involving switching to higher- or lower-yield cigarettes should be distinguished from those which make comparisons between self-selected high-yield and low-yield smokers in the population. Unlike the few studies discussed in the preceding section, which involved switching to cigarettes with independently varied tar and nicotine yields, most brand-switching studies lack controls for tar and other confounding factors. Evidence of intake regulation cannot therefore be attributed to nicotine. Nevertheless, they have value in providing additional quantitative data. Out of more than one hundred brand-switching studies attention is focused here on some of the few that included measures of blood nicotine concentrations. Data from four such studies are summarized in Table 11.1. In two cases the switch was short term, 5 h and 24 h. All involved switching to a lower-yield brand, which was either 'low' or 'ultra-low'. Two involved an additional switch to a higher yield brand.

It can be seen in Table 11.1 that there is no tendency for blood nicotine concentrations to increase after short-term switching to high-yield brands. This confirms the picture presented in the preceding section of the precision with which nicotine intake is down-regulated to avoid excessive blood levels. Whether this would continue to hold after long-term switching to a high-yield brand is not known.

The lower sensitivity of up-regulation is also demonstrated (Table 11.1). In all cases blood nicotine concentrations were lower after switching to lower-yield brands, although the drop was less than would have occurred without a compensatory change to more intensive smoking. In no case did the number of cigarettes smoked increase significantly. This indicates that the compensatory changes were confined to altering the way that individual cigarettes were smoked, and suggests that the intake per puff or per cigarette is more important than steady-state blood nicotine level. In the case of the ultra-low brand, complete up-regulation would have been impossible without increasing the number of cigarettes smoked. On the low-yield brands, however, complete up-regulation of nicotine intake would have been relatively easy in mechanical terms. Why this was not accomplished is not known, but it suggests that smokers tolerate a modest drop in nicotine intake more readily than they do an increase above their usual smoking levels. It is noteworthy that the degree of up-regulation appears to become established quickly and to then remain constant, in that blood nicotine concentrations at 10 days and 10 weeks after switching were similar to those at one to 2 days (Table 11.1).

Besides nicotine intake, sensory and mechanical factors may play a part in

Table 11.1. Changes in plasma nicotine concentrations after switching to high, low, or ultra-low yield cigarettes (values shown are group means)

	Brand of cigarette	Tar yield (mg per cig)	Nicotine yield (mg per cig)	Plasma nicotine (ng ml^{-1})
Russell *et al.* (1975)				
(n = 10)	Usual	18.0	1.34	30.1
	Ultra-low	1.0	0.14	8.5
	High	32.0	3.2	29.2
percentage reduction (low vs usual)		94 per cent	90 per cent	72 per cent
Ashton *et al.* (1979)				
(n = 12)	Usual	18.0	1.4	39.6
	Low	6.5	0.6	30.9
	High	26.5	1.84	42.8
percentage reduction (low vs usual)		64 per cent	57 per cent	22 percent
Russell *et al.* (1982)				
(n = 12)	Usual	17.4	1.33	32.4
	Middle	19.0	1.3	35.2
	Usual	17.4	1.33	33.2
	Low (2 days)	10.9	0.7	20.8
	" (8 wks)	"	"	22.7
	" (10 wks)	"	"	22.8
percentage reduction (low vs usual)		39 per cent	47 per cent	33 percent
West *et al.* (1984)				
(n = 11)	Usual	14.2	1.3	22.0
	Ultra-low (1 day)	1.0	0.1	11.2
	" " (3 days)	"	"	7.3
	" " (10 days)	"	"	10.2
percentage reduction		93 per cent	92 per cent	57 per cent

Abstracted from Russell *et al.* (1975), Ashton *et al.* (1979), Russell *et al.* (1982), and West *et al.* (1984d). On switching to lower-yield cigarettes the blood nicotine concentrations are significantly less than on the usual brands. However, the reductions in blood levels are proportionately less than the reductions in nicotine yields indicating that some compensatory increase in puffing and inhalation occurred, although it was not sufficient to compensate completely for the drop in yield (i.e partial up-regulation). Cigarette consumption did not change significantly in any of the studies, except for the switch to the high-yeld cigarette in the study of Russell *et al.* (1975). On switching to a high-yield brand, or a brand with a yield similar to that of the usual brand, the plasma nicotine levels did not change significantly. The drop at 8 and 10 weeks after switching to a low-yield brand was similar to that at 2 days after switching (Russell *et al.* 1982). The plasma nicotine concentrations in the study of West *et al.* (1984d) were 'trough' levels, whereas the other three studies used 'peak' levels 2 min after a cigarette.

determining satisfaction and acceptability of low-yield cigarettes. A change in flavour can affect smokers' reactions to a different brand irrespective of its nicotine delivery. Among the studies shown in Table 11.1, one included a switch to a higher-yield brand (Russell *et al.* 1975), and another to a brand with a yield similar to that of the usual brand (Russell *et al.* 1982). In both cases the smokers reported decreases in satisfaction, although their blood nicotine levels were unchanged. On switching to low-yield cigarettes (0.7 mg nicotine) blood nicotine levels dropped to roughly two thirds of the usual levels, despite partial up-regulation. This was accompanied by some initial loss of satisfaction, but this was no greater than that experienced after switching to another brand with a nicotine yield similar to that of the usual brand (Russell *et al.* 1982). Moreover, the smokers adapted gradually over a few weeks to regain satisfaction despite the lower level of nicotine intake. After switching to ultra-low brands, the more severe drop in nicotine intake to about one third of usual levels (Table 11.1) was accompanied not only by lack of satisfaction but also by mild withdrawal effects: craving, hunger, and drop in heart rate (West *et al.* 1984*d*).

Mechanical factors refer to the difficulties posed by having to smoke more intensively to maintain intake from low-yield cigarettes. With down-regulation no mechanical problems arise, and there is clearly little difficulty in taking smaller puffs and inhaling less deeply after switching to higher-yield cigarettes. Switching to lower-yield cigarettes, on the other hand, may present a number of difficulties. Puff volume and inhalation can be increased to maintain intake and blood nicotine concentrations, but only up to a point. After this, compensatory changes in smoking pattern become progressively more intrusive, then aversive, and eventually, with very low yields, complete compensation becomes impossible. This is affected profoundly by the hyperbolic nature of the mathematical relationship between the nicotine concentration in the smoke and the volume of smoke required to deliver a given dose (Sutton *et al.* 1978).

To illustrate the problem, let the starting point be a medium-nicotine cigarette which delivers 1.2 mg nicotine in 400 ml of mainstream smoke from a machine set to take 10 40-ml puffs. Consider next a series of lower-yield cigarettes having machine-smoked yields, derived from the same machine setting of 10 40-ml puffs, as follows: 0.8, 0.6, 0.4, and 0.2 mg nicotine. To obtain 1.2 mg nicotine from these lower-yield cigarettes, the smoker wouldhave to inhale 600, 800, 1200, and 2400 ml of smoke per cigarette respectively. It is clear that things soon get difficult, and then impossible, as the smoker reaches the steep part of the curve.

The issue is further complicated by various characteristics of the particular brand of lower-yield cigarette, and of the individual smoker. Low-yield cigarettes vary greatly in the extent to which they can be 'oversmoked' to exceed their standard machine-smoked yields. The factors involved include the length of tip overwrap, and the extent to which yield reductions are achieved by filtration, ventilation, or lowering the quantity of tobacco or its nicotine content. On average, heavy smokers take in about 1 mg nicotine per cigarette, but there is wide variation between individual smokers, and individual dosage requirement would determine at what point on the nicotine yield 'league table' a given smoker would hit the steep part of the compensation curve described above.

With UK cigarettes, the majority of smokers appear to first encounter these mechanical problems when nicotine yields fall below about 0.6 mg, and cigarettes with yields below 0.4 mg nicotine are currently almost unsaleable. The longer tip overwraps of low-yield brands in the United States make it easier for smokers to exceed nominal yields which have been artifactually lowered (Grunberg *et al.* 1985). This may account for the lower sales-weighted mean nicotine yield of cigarettes sold in the US compared with the UK (1.0 mg vs 1.3 mg), there being no evidence that the blood nicotine levels of US smokers are any lower than those in the UK. A final point regarding individual differences between smokers is the extent to which they smoke to obtain a given nicotine boost per puff or per cigarette as opposed to a particular steady-state plateau level. In the latter case up-regulation would be easier in that it could be assisted by smoking more cigarettes, a strategy not open to the former.

In summary, the results of these brand-switching studies are consistent with those of the more direct studies in the preceding section in showing the precision with which nicotine intake is down-regulated to avoid the aversive effects of exceeding usual blood nicotine levels. The precision with which this is done indicates that the peak levels of normal smoking are only just below the toxic aversive barrier. The brand-switching studies also confirm that up-regulation to maintain usual nicotine levels is less precise, and that smokers appear to tolerate a drop in blood nicotine to about two thirds of their usual levels. It is not clear how much the initial loss of satisfaction is attributable to changes in flavour and mechanical factors as opposed to the drop in blood nicotine levels. Further reductions in blood nicotine to around one third of usual peak levels result in the emergence of some withdrawal effects. In the next section it will be seen that many withdrawal effects are alleviated by nicotine replacement which maintains levels at one third of those from smoking.

11.2.4 *Switching to nicotine replacement*

Nicotine replacement refers to the administration of nicotine to substitute for that obtained from tobacco use. It has been studied most widely as a means of alleviating the severity of withdrawal during attempts to quit smoking, and in this respect differs from the studies described in the earlier section on the effect of nicotine on ad lib smoking behaviour. The extent to which nicotine alone can provide the satisfaction and reinforcing effects of smoking is one measure of its importance. The extent to which its self-administration is regulated is another. Both no doubt depend crucially on how effectively the nicotine from smoking is replaced.

To administer nicotine as efficiently as occurs with the inhalation of tobacco smoke is difficult. One possibility is a nicotine aerosol, not too irritant, having particles as small as those in smoke to enable them to reach the lung alveoli. Another way would be to inject bolus doses via an arterial cannula with its tip as high as the aortic valve. Even if such studies were done, it could be expected to take several days at least for the conditioned sensorimotor reinforcers of smoking to be replaced by conditioning to sensory and motor cues surrounding the new form of nicotine administration. Indeed, as described in the preceding section, a change in flavour and other factors can cause loss of satisfaction on switching to another brand of cigarette even if nicotine yield and blood nicotine are unchanged.

One of the puzzling features of nicotine (and tobacco use) is to reconcile its high addictiveness with the relatively low intensity, compared with other addictive drugs, of its acute subjective effects and the relatively small changes it produces in mood, performance, and level of arousal. It causes no striking euphoriant effects, and any acute psychic effects it has appear to be short-lived and susceptible to acute tolerance (Jones *et al.* 1978; West and Russell 1987). Two early studies (one uncontrolled, the other giving inadequate details) reported that smokers found nicotine injections pleasant (Johnston 1942; Jones *et al.* 1978), but because the acute drug-like subjective effects which smokers liked were susceptible to acute tolerance, Jones *et al.* (1978) concluded that this 'suggests that the pharmacologic effects of nicotine are not major determinants of smoking behviour throughout the smoking day.' Other effects not susceptible to acute tolerançe appear to have been overlooked. In contrast, the behavioural importance of 'liking' of these acute subjective effects may have been overemphasized by others, including the recent US Surgeon General's Report (1988).

Using the US Addiction Research Center's established questionnaires for testing drugs for their abuse potential, Henningfield *et al.* (1985) found dose-related increases in subjective 'liking' scores and similar changes on the 'euphoria' scale in response to intravenous doses of 0.75, 1.5, and 3.0 mg nicotine administered rapidly over 10 s. It must be said that these are generous doses and that all of the 8 subjects had histories of drug abuse. Indeed, 6 of them identified the nicotine as cocaine, one as amphetamine. This contrasts with the well-known ability of rodents to discriminate nicotine from other drugs with great precision, although cocaine and amphetamine are the two drugs with which some discriminatory overlap with nicotine may occur (Stolerman, chapter 8, this volume).

In a subsequent study, using nicotine gum, Nemeth-Coslett *et al.* (1987) successfully manipulated the systemically available dose by getting subjects to chew 2, 4, or 8 mg nicotine gum in a standard way for 20 min to produce respective mean blood nicotine concentrations of about 27, 31, and

34 ng ml^{-1}, compared with 23 ng ml^{-1} in the placebo condition (data abstracted from figure). In contrast to the intravenous study described above, there were dose-related decreases, rather than increases, in liking and acceptability of nicotine gum. Heart rate and blood pressure were unaffected by the gum, but there was a dose-related decrease in the desire to smoke and an inhibitory effect on subsequent smoking behaviour. The authors attributed the 'attenuation' of the nicotine effects from the gum, compared with intravenous dosage, to the slower rate of administration and absorption. This is no doubt a crucial factor, but another explanation is the confounding effect, in this study, of acute tolerance. Prior to gum dosage, subjects had refrained from smoking for only 50 min, compared with 9 h in the intravenous study described above. Furthermore, the average blood nicotine concentration carried over from prior smoking was about 23 ng ml^{-1} at the end of the placebo-gum condition, which is sufficient to ensure a high degree of acute tolerance, and this might partly explain the lack of subjective 'liking' and lack of heart rate response.

Although there is some evidence that smokers will self-administer intravenous nicotine and regulate their dosage over a 3 h period (Henningfield *et al.* 1983), there are no systematic studies of the response of smokers to switching to nicotine injections for longer periods, let alone days or weeks. There is, however, a mass of data on short- and long-term switching to nicotine polacrilex gum, albeit in a predominantly clinical context. The implications of these data for the role of nicotine in tobacco dependence and its treatment are discussed elsewhere (Jaffe, chapter 1, this volume). The focus here is on the implications for understanding the tendency for smokers to regulate their nicotine intake. A key question posed by the data is why smokers who have switched to the gum do not up-regulate their nicotine intake more completely by taking more of it, even when they are in a state of severe withdrawal and have got over their initial aversion to its taste. A second point of interest concerns how much the gum can do at relatively low blood nicotine levels.

11.2.4.1 *Blood nicotine levels from nicotine gum* Nicotine chewing-gum is a slow and relatively inefficient form of nicotine administration. Besides its design as a slow-release product (it takes 20–30 min of chewing to release 90 per cent of the nicotine content), absorption through the buccal mucosa is also relatively slow, and some of the nicotine is 'wasted' through swallowing or inadequate chewing. The systemic availability of nicotine from the gum is only 30–40 per cent (Benowitz *et al.* 1987). Blood nicotine levels rise slowly to a flat peak at about 30 min after the start of chewing, the average increase in blood nicotine concentration being little over 5 ng ml^{-1} for each piece of 2 mg gum (Jaffe, Fig. 1.2; Benowitz *et al.* Fig. 4.1, this volume). On a fixed-dosage schedule of one piece of 2 mg gum per hour, chewed for 30 min, steady-state blood nicotine concentrations during the afternoon average 10–12 ng ml^{-1} (Russell *et al.* 1976*b*; McNabb *et al.* 1982). This is about one third of the peak levels from smoking. Similarly, during free ad lib clinical use of 2 mg gum, subjects use on average about 8 pieces per day and their blood nicotine levels in several studies have averaged between 8 and 12 ng ml^{-1}, i.e. roughly one third of their former peak smoking levels (Raw *et al.* 1980; McNabb 1984; West *et al.* 1984*a*; Tonneson *et al.* 1988*a*).

On the stronger 4 mg gum, with a fixed dose schedule of 1 h^{-1}, blood nicotine levels average about 30 ng ml^{-1} and 22 ng ml^{-1} for steady-state peaks and troughs respectively, and are similar to those during smoking (Russell *et al.* 1977; McNabb *et al.* 1982). During ad lib clinical use, however, blood nicotine levels are lower and average about two thirds (Tonneson *et al.* 1988*a*) to three quarters (Westling 1976) of those from prior smoking.

11.2.4.2 *Effects of nicotine gum in abstinent smokers* In the previous section on brand-switching it was noted that some withdrawal effects began to emerge when blood nicotine levels dropped to around one third of the usual smoking levels after switching to ultra-low nicotine brands. It is at around these levels that smokers who switch to 2-mg nicotine gum get partial, but far from complete, relief of withdrawal. Despite the lack of significant peaks and troughs, and steady-state blood nicotine levels of only 8–12 ng ml^{-1}, smokers find the gum helpful when they are trying to stop smoking and it increases their chance of success (Jarvis *et al.* 1982; Hjalmarson 1984). It alleviates some of the withdrawal symtoms, particularly the adverse mood changes such as irritability and depression (Hughes *et al.* 1984; Schneider *et al.* 1984; West 1984). It also reverses the impairment of concentration and performance on some cognitive tasks following cigarette withdrawal (Snyder *et al.* 1985), and reverses the drop in heart rate and some of the EEG changes induced by abstinence from smoking (Herning and Pickworth 1985; Pickworth *et al.* 1986). It also appears, in the longer term, to reduce excessive hunger and weight gain after quitting smoking (Stitzer and Gross 1988).

11.2.4.3 *Sources of dissatisfaction on switching to nicotine gum* Despite the wide variety of withdrawal effects alleviated by the gum, it does not appear to provide any of the pleasurable aspects of smoking, and its helpfulness is perceived more readily when taken after stopping smoking. Initially its taste and the local effects of nicotine (such as irritation to the mouth and throat, nausea, dyspepsia, and hiccups) are aversive. As with smoke inhalation, a brief period of sensory adaptation and learning is required before the gum is taken comfortably. Besides adaptation to aversive local side-effects, the switch from smoking and inhalation to the buccal route involves major changes in sensorimotor cues. In the short term, full satisfaction from smoking is closely linked with conditioned sensorimotor reinforcers as well as pharmacological effects (Rose *et al.* 1984). It might be expected that adjustment and conditioning to new cues would be more rapid if the change was less drastic, for example, from tobacco chewing to the gum, or from nasal snuff to a nasal nicotine spray. Finally, as discussed in a later section, the rate of nicotine absorption from the gum is too slow to produce those effects of nicotine that are susceptible to acute tolerance, and would not do so even if high steady-state levels were achieved. As a corollary, it follows that its effects outlined above are mediated by actions of nicotine that are less sensitive to acute tolerance.

11.2.4.4 *Intake regulation on nicotine gum* It must be said that, despite its theoretical interest, the issue of intake regulation by gum users is not understood and data relevant to this question are scanty. What is clear is that clinic subjects well into their third and fourth weeks of using 2 mg gum as an aid to cigarette withdrawal, who have got over the initial aversive side-effects, who find the gum very helpful, and in whom the severity of withdrawal is substantially reduced, continue to use only 7–8 pieces per day on average. Moreover, they take so few pieces despite encouragement from their therapists to take twice as much if necessary, and despite continuing to experience withdrawal effects and difficulty with keeping off cigarettes (West *et al.* 1987). To raise their steady-state blood nicotine from around $10\,\mathrm{ng\,ml^{-1}}$ to $20\,\mathrm{ng\,ml^{-1}}$ by taking more gum would be relatively easy in mechanical and pharmacokinetic terms. Their reluctance to do this may in part be due to fear of becoming addicted to the gum. It is more probable, however, that their failure to up-regulate their nicotine intake more effectively is because it would not increase the pharmacological rewards provided by the gum. This suggests that the gum achieves its maximum reinforcing effect at relatively low blood nicotine levels and that sharper peaks and boli, rather than higher steady-state levels, may be necessary to adequately replace the pharmacological effects of smoking.

Some indirect evidence that abstinent smokers may regulate nicotine intake from the gum, albeit around a lower blood level axis than that of cigarette smoking, comes from a clinical trial by Jarvis *et al.* (1982). The number of 2 mg gums used per day correlated significantly with pre-treatment blood nicotine levels from smoking ($r = 0.48$, $p < .01$) but not with prior cigarette consumption ($r = 0.11$, NS). In contrast, use of placebo gum correlated with cigarette consumption ($r = 0.47$, $p < .05$), but not with pre-treatment blood nicotine levels ($r = 0.17$, NS). Although clinical trials have not shown 4 mg gum to be more effective overall than 2 mg gum (Kornitzer *et al.* 1987; Tonnesen *et al.* 1988*a*), where randomization has been applied to a selected sample of dependent smokers with high blood nicotine levels, 4 mg gum was significantly more effective (Tonnesen *et al.* 1988*b*).

Finally, despite the lack of up-regulation to smoking levels of blood nicotine, the pharmacological effects of 2 mg nicotine gum are sufficient to sustain a high degree of dependence, although the incidence of becoming dependent on the gum is relatively low considering the target population. About 6–7 per cent of those who try it in a clinic setting have difficulty with giving it up and use it continuously for a year or more, while none become dependent on placebo gum (Jarvis *et al.* 1982). This comprises some 25 per cent of those who sustain abstinence from smoking for as long as one year (Wilhelmsen and Hjalmarson 1980; Hajek *et al.* 1988). After placebo substitution or withdrawal of nicotine gum, these long-term users experience withdrawal effects similar to those of cigarette withdrawal (West and Russell 1985*b*; Hughes *et al.* 1986).

11.2.4.5 *Other forms of nicotine replacement* The use of nicotine gum as an aid to smoking cessation has stimulated research into alternative routes for nicotine replacement, such as the skin, the nose, and the lungs. Transdermal absorption from nicotine-containing skin patches is slow and it takes at least 2–3 h to produce blood levels equivalent to those of smoking (Russell 1988). Although this rate of absorption is too slow to produce light-headedness or other subjective effects, the nicotine skin patches appear to reduce craving for cigarettes (Rose *et al.* 1985) and to have some potential as an aid to cessation (Abelin *et al.* 1989).

In contrast with the skin patch and nicotine gum, the rate of absorption through the nasal mucosa is rapid, whether it be from dry tobacco snuff (Russell *et al.* 1981) or a nasal nicotine solution (Jaffe, Fig. 1.2, this volume). A single dose of nasal nicotine, taken as a droplet or spray, produces a sharp rise in blood nicotine which peaks at 5–10 min. The blood nicotine boost averages about 10 ng ml^{-1} for each 2 mg dose, but there is wide variation between subjects and steady-state peak levels during hourly dosage ranged from 8–33 ng ml^{-1}, around a mean of only 16 ng ml^{-1} in 5 subjects (Russell *et al.* 1983; West *et al.* 1984*c*). During ad lib clinical use, averaging 11 doses per day, steady-state blood nicotine trough levels averaged 7.4 ng ml^{-1} in 16 subjects. There were marked differences between subjects, 4 of whom obtained blood nicotine levels equivalent to their trough levels from smoking, around 25 ng ml^{-1} (Jarvis *et al.* 1987). Subjective effects such as calmness, light-headedness, and slight euphoria were reported by two thirds of the subjects, although they did not occur after every dose.

To mimic more closely the rapid absorption of nicotine from cigarette smoke, a smoke-free cigarette was developed which delivered nicotine vapour without having to be lit. However, the rate of nicotine absorption from the device was slow, probably through deposition of nicotine in the mouth, throat, and upper airways before reaching the lung alveoli (Russell *et al.* 1987). In a placebo-controlled study, the device alleviated some of the effects of cigarette withdrawal at average blood nicotine levels of only 6.3 ng ml^{-1} (Hajek *et al.* 1989).

In summary, the effects of nicotine are related to the rate of its absorption as well as to steady-state levels. It seems to require a blood nicotine boost of about 10 ng ml^{-1} in 10 min or less to produce the brief drug-like feeling of light-headedness and slight euphoria which, unlike nicotine injections and nasal spray, skin patches and nicotine gum are unable to provide. However, many important effects are obtained from slow absorption from nicotine gum to steady-state levels equivalent to only one-third of those from smoking. Nicotine gum, for example, can produce EEG changes, improve performance on cognitive and psychomotor tasks, and alleviate many of the adverse mood changes of cigarette withdrawal. There is some suggestive evidence that subjects may down-regulate intravenous and nasal nicotine to avoid excessive levels, but this has yet to be established. The failure of nicotine gum users to up-regulate their intake more effectively may be due partly to the aversiveness of local side-effects and partly to its inability to produce those effects of nicotine that depend on a more rapid rate of absorption.

11.2.5 *Other evidence of intake regulation*

Supportive convergent evidence that smokers and other tobacco users regulate their nicotine intake to individually preferred levels, which are fairly constant from one day to the next, comes from data of various kinds. Historically, tobacco has been used for at least 400 years and its use has fluctuated between chewing, snuffing, and smoking. At no time has one form of use been eliminated without replacement by another. Smoking waned in Britain in the 18th century to be replaced by use of dry nasal snuff for almost 100 years, before the population switched back to smoking. Clearly neither the products of combustion, such as tar, carbon monoxide, and other gases, nor the sensory stimulation to the throat and airways are of primary importance. Nicotine is the one common factor. It is noteworthy also that people have never shown any inclination to compulsively inhale smoke, however aromatic, that does not contain a psychoactive drug. Other kinds of convergent evidence are outlined briefly in this section.

11.2.5.1 *Nicotine intake from self-selected brands* Some support for the controlled brand-switching studies, discussed earlier, comes from several large-scale studies of nicotine intake from self-selected brands. In 1980, we reported a study of the blood nicotine levels of 330 regular smokers who had been smoking their usual brand in their usual way (Russell *et al.* 1980*b*). Some were smokers of low-yield brands, others of middle- or high-yield brands. Blood was sampled, 2 min after a cigarette, during the afternoon or evening, by which time steady-state conditions would have occurred in most cases. The most striking feature was the wide variation between individuals, from 4–72 ng ml^{-1}, the average of 33 ng ml^{-1} being the same in men and women (Fig. 11.1). It was also evident that high blood nicotine levels of 40–60 ng ml^{-1} could be obtained from low-yield cigarettes having nicotine yields of about 0.7 mg, whereas some smokers of high-yield cigarettes with nicotine yields of 1.6 mg or more had blood nicotine levels below 20 ng ml^{-1}. Despite large differences in the machine-smoked yields of the cigarettes, blood nicotine levels were only slightly lower in those on lower-yield brands. The correlation

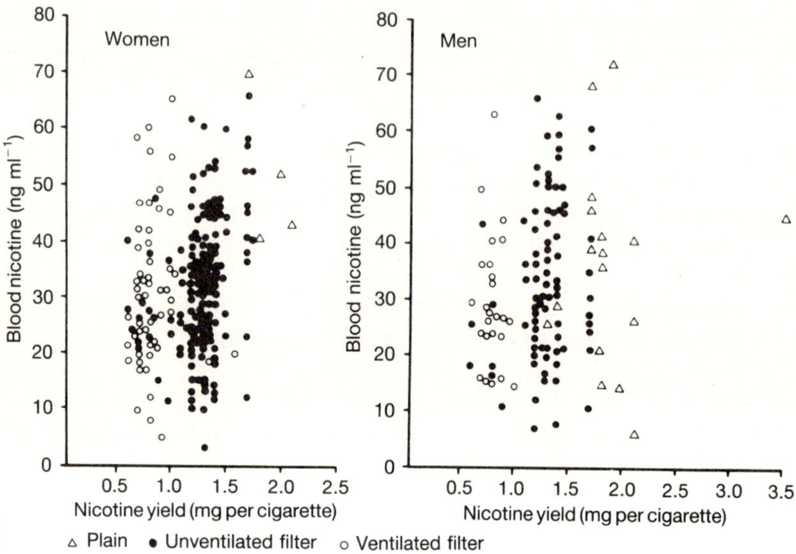

Fig. 11.1. Blood nicotine levels of smokers plotted against the nicotine yields of their cigarettes. All smokers had been smoking their own self-selected brand of cigarette in their usual way. (From Russell *et al.* 1980*b*.)

between blood nicotine and nicotine yield of cigarette, though statistically significant, was low (0.21, $p<.001$), showing that the nicotine yield of the cigarettes accounted for only 4.4 per cent of the variation in blood nicotine levels. The number of cigarettes smoked per day accounted for even less.

These findings have been widely replicated and are also essentially the same in the US where the range of nicotine yields extends well below 0.6 mg, which was the lower limit in our samples of UK smokers (Ebert *et al.* 1983; Benowitz *et al.* 1983; Gori and Lynch 1985; Russell *et al.* 1986). It is clear from all these studies that the nicotine intake of smokers is determined more by the way they puff and inhale than by the nicotine yield of the cigarettes they smoke. The differences in the blood nicotine concentrations of low-yield and higher-yield smokers were proportionally less than the differences between the machine-smoked yields of their brands. This indicates that the lower-yield smokers were smoking and inhaling more intensively, presumably to compensate for the lower yields.

11.2.5.2 *Upregulation after cutting down* Smokers who cut down their daily cigarette consumption compensate by puffing and inhaling more intensively and thereby undermine most of the potential health benefits. In a study of 16 smokers who halved their average consumption form 30 to 15 cigarettes per day, average COHb levels remained unchanged at 7.7 and 7.9 per cent respectively, but peak blood nicotine levels just after a cigarette were increased from 17.0 to 24.8 ng ml^{-1} reflecting the increased intake per cigarette (Ho-yen *et al.* 1982). Their overall daily some intake, therefore, appeared to have been maintained. In a more recent study in hospitalized subjects, up-regulation of smoke intake was confirmed but was incomplete, especially after major restrictions to only 5 cigarettes per day (Benowitz *et al.* 1986). What is clear from these studies is that the nicotine intake of a 10–15 per day smoker who once smoked 20–30 per day is likely to be far higher than that of a 10–15 per day smoker who has never smoked more.

11.2.5.3 *Switching to a pipe or cigars* Unlike the acidic smoke of most cigarettes (pH 5.5), nicotine in the alkaline smoke of pipe and cigar tobacco (pH 8.5) is absorbed through the buccal mucosa. However, the rate of buccal absorption is slow, even under atypical maximal contact conditions (Fig. 11.2). Due to the lack of buffering capacity of smoke, the rate of nicotine absorption is slower than from oral snuff, chewing tobacco or nicotine gum (see Benowitz *et al.*, Fig. 4.1, this volume). Nevertheless, so called 'primary' pipe and cigar smokers who have never smoked cigarettes regularly appear to be satisfied and to derive pleasure from this slow form of nicotine intake. When tested with a large Havana cigar, there was no evidence that they inhaled (Turner *et al.* 1977). Blood nicotine concentrations increased to only 5.1 ng ml^{-1} and COHb levels were unchanged. This contrasted with 'secondary' pipe and cigar

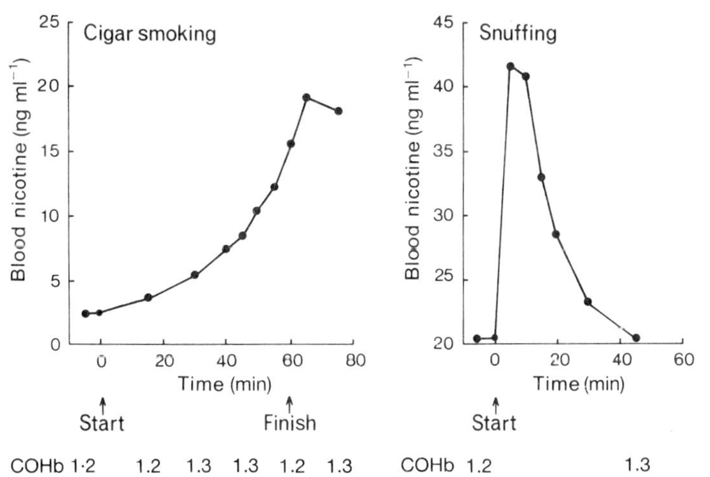

Fig. 11.2. Blood nicotine levels during and after cigar smoking, and after snuffing. The cigar was a large Havana cigar (6.2 g) smoked maximally without inhalation by an occasional cigar smoker who had given up cigarette smoking 12 years ago. The snuff was a single 'pinch' of nasal snuff (S P No. 1) taken by an experienced snuffer. He had been taking snuff as usual before the experiment. The last dose, taken about 1 h previously, accounts for the baseline blood nicotine level of 20.3 ng ml^{-1}. The COHb levels confirm that the cigar smoker did not inhale. (From Russell *et al.* 1980*a*.)

smokers (i.e. those who had previously been regular cigarette smokers) who increased their blood nicotine levels from 12.8 to 36.2 ng ml^{-1} and their COHb from 2.9 to 9.6 per cent (Turner *et al.* 1977). There is now abundant evidence that cigarette smokers who switch to a pipe or cigars continue to inhale, and thereby maintain blood nicotine and COHb levels similar to those obtained from their former cigarette smoking (Castleden and Cole 1973; Jarvis *et al.* 1984; Wald *et al.* 1984). Lower and slower rising levels, it seems, do not satisfy them.

11.2.5.4 *Smokeless tobacco use* Oral 'wet' snuff is widely used in Sweden and the USA. The blood nicotine levels of regular users are similar to those of regular smokers (Gritz *et al.* 1981) and its use induces dependence, the withdrawal syndrome being qualitatively similar to cigarette withdrawal but quantitatively less severe (Hatsukami *et al.* 1987). 'Dry' nasal snuff is still available in the UK although its use is rare. The rate of nicotine absorption from dry nasal snuff is extremely rapid (Fig. 11.2) and appears to be more rapid than from 'wet' oral snuff (see Benowitz *et al.*, Fig. 4.1, this volume). In 11 regular users of dry snuff, the average blood nicotine boost of 12.6 ng ml^{-1} to a peak of 36.1 ng ml^{-1} 10 min after their usual dose was similar to that found in cigarette smokers (Russell *et al.* 1981). Although lacking post-inhalation high-nicotine boli, the similarity of the trough and peak levels of regular snuff users and smokers is striking. It seems unlikely that such similarities resulting from such different rituals is mere coincidence. The most plausible explanation is that the rituals are determined by the nicotine concentrations they produce.

11.3 Nicotine tolerance and pharmacokinetic factors

The first main section addressed the role of nicotine as a major independent factor in the control of smoke intake, and an attempt was made to examine the degree of precision with which blood nicotine levels are regulated. It was apparent that the rate of absorption and trough–peak differences may be as important as the steady-state levels. To attempt to understand these issues the present section will consider how nicotine tolerance and various pharmacokinetic factors may interact to influence the pharmacological and behavioural effects of nicotine.

11.3.1 *Blood nicotine levels from smoking*

Blood nicotine concentrations vary widely between different smokers, as do their smoking habits. Under steady-state conditions during the afternoon of a typical smoking day, peak concentrations of nicotine in venous blood sampled 2 min after a cigarette range from below 5 ng ml^{-1} to just short of 100 ng ml^{-1} around a mean of about 35 ng ml^{-1} (Fig. 11.3), but the levels of individual smokers are fairly constant from one day to the next with a correlation of

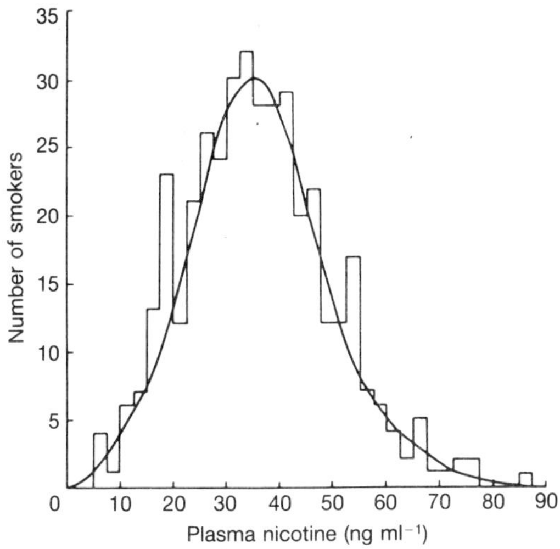

Fig. 11.3. Distribution of peak plasma nicotine concentrations in a sample of 393 heavy smokers (250 women, 143 men) with a mean cigarette consumption of 30.1 per day. Blood was taken 2 min after completing a cigarette during the afternoon of a day of usual smoking. The average plasma nicotine concentration was 35.8 ng ml^{-1} (S D 13.7) and was not significantly different between men and women. (Data from Russell *et al.* 1986.)

about 0.9 (Russell *et al.* 1975, 1982). Levels of 10 ng ml^{-1} are sufficient to produce numerous pharmacological effects and Fig. 11.3 shows that over 95 per cent of smokers obtain these levels. Due to the rapid absorption and distribution of nicotine, the blood nicotine profile of the smoker who inhales is characterized by a series of sharp peaks coinciding with each cigarette. The blood nicotine boost per cigarette, during steady state, averages about 10 ng ml^{-1} (Russell *et al.* 1981) and reflects the average 'smoking dose' of 1.0 mg per cigarette of systemically available nicotine (Benowitz and Jacob 1984; Feyeraband *et al.* 1985). In view of the wide variation between individuals in their long-term nicotine intake one might expect them to have developed different levels of chronic tolerance, but this has not been clearly demonstrated as is discussed below.

11.3.2 *Nicotine tolerance*

The term tolerance refers to the need for larger doses to produce a given effect. Inherited factors are largely responsible for individual differences in the sensitivity of naive subjects. Tolerance applies to those differences in sensitivity that result from prior exposure, and it has several components. *Pharmacokinetic tolerance* arises from the increased rate of drug clearance due to the induction of enzymes, mainly in the liver, which metabolize it. More important is so-called *pharmacodynamic tolerance* which refers to the development of resistance and reduced sensitivity of the nervous system to a given drug concentration, so that higher concentrations are required to produce the given effect. Pharmacodynamic tolerance to nicotine occurs and there are at least two main types: acute tolerance, or tachyphylaxis, which develops within minutes and lasts for a matter of hours, and chronic tolerance which develops within a few days and may last for several days or months, and possibly even longer. Other forms of tolerance that depend on learning and conditioning will not be discussed here. The terms 'acute' and 'chronic', as used here, refer only in part to the rate of acquisition of tolerance, but more importantly to the duration of reduced sensitivity following cessation of drug dosage.

11.3.2.1 *Chronic tolerance* The mechanisms of chronic tolerance are not fully understood. It appears to involve some form of adaptive changes within the nervous sytem, such as changes in the number of drug receptors. It is a general rule that chronic administration of agonist drugs is associated with a decrease in the number of receptors (down-regulation) and that in the case of antagonists they are increased (up-regulation). Once chronic tolerance is acquired, withdrawal or even a sharp drop in drug level may result in rebound effects. Rebound effects are those alterations in activity attributable to drug-induced adaptive changes in the nervous system that remain temporarily intact after the drug is no longer there to counteract them. In so far as they are a consequence of the adaptive changes underlying chronic tolerance they differ fundamentally from, but are not necessarily more important than, simple lack of the primary pharmacological effects of the drug which would be evident after the first dose as its effects wear off.

Chronic tolerance to nicotine has been demonstrated in rodents but not in humans. When doses which initially depress locomotor activity in rats and cause ataxia are repeated daily, tolerance develops to these depressant effects which after a few days are replaced by a stimulant effect (possibly due to unmasking), in that locomotor activity is increased and ataxia is no longer evident (Morrison and Stephenson 1972; Clarke and Kumar 1983; see Stolerman, Fig. 8.5, this volume). Thus the initial effects on naive subjects become reversed on repeated administrations of the same dose. Furthermore, if these 'tolerant' animals are kept without nicotine for 90 days, subsequent administration of the same dose produces stimulant effects, with no evidence of loss of tolerance to the initial depressant effects (Stolerman *et al.* 1973*a*). Such findings illustrate the importance of prior exposure in determining the pharmacological effects of nicotine. Several studies in rodents have shown that nicotine administration for several days gives rise to substantial increases in the number of nicotinic-cholinergic receptors in many parts of the brain (for example, Schwartz and Kellar 1983; Marks and Collins 1985). The up-regulation of receptors suggests that nicotine's actions were predominantly antagonistic at the doses used (see Wonnacott, chapter 7, this volume).

As yet there is no clear evidence that nicotine induces chronic tolerance in humans. This may be due to lack of systematic study and the understandable reluctance of researchers to administer nicotine to non-smokers. The only study to compare abstinent chronic smokers with non-smokers systematically found no difference in their sensitivity to the effects of intravenous nicotine on heart rate, blood pressure, and EEG (Murphree 1979). Similarly, rebound effects after cigarette withdrawal have received little study, but there is some evidence that they do occur (West *et al.* 1984*b*). This issue will be discussed more fully in a later section. Recent evidence that the number of nicotinic receptors in the brains of smokers is greater than in non-smokers (Benwell *et al.* 1988) accords with the findings in rodents, but it is possible that such differences are innate rather than acquired, and they may determine who becomes a smoker. We are forced to rely, therefore, on anecdotal evidence that initial exposure to cigarettes may give rise to palpitations, dizziness, sweating, nausea, or vomiting, and that regular smokers acquire some tolerance to these effects. In pilot studies we have found that some non-smokers develop nausea and slight dysphoria at blood nicotine levels in the range of 10–15 $ng\,ml^{-1}$. This may be accompanied by pallor, faintness, and a drop in blood pressure, but it is not clear how much these effects are due to reflexes associated with nausea or to pharmacological effects emerging at lower doses than is the case with smokers. Finally, the fact that the enhancing effect of nicotine on several performance tasks may be obtained in smokers at relatively low blood nicotine levels suggests that these effects may have little susceptibility to chronic tolerance.

11.3.2.2 *Acute tolerance* Acute tolerance is altogether different from chronic tolerance and occurs in naive subjects (or preparations) and also in non-naive chronically-exposed subjects (for example, regular smokers). It seems to be closely related, in temporal terms, to nicotine's occupancy of its receptors. When they are 'topped up' with nicotine, acute tolerance is evident, and persists to some degree until the nicotine has been eliminated. Its rate of decline could, therefore, be expected to closely reflect the elimination half-life of nicotine in the bloodstream and to be at least as long, i.e. at least 2 h on average (see Benowitz *et al.*, chapter 4, this volume).

Some of nicotine's effects are subject to acute tolerance but others are not. For example, the acute increase in heart rate produced by nicotine is affected by acute tolerance and little effect on heart rate is observed once steady-state has been reached (Russell and Feyerabend 1978; Benowitz *et al.* 1982; West and Russell 1987). Acute tolerance also occurs to the acute drug-like subjective feeling of slight 'dizziness' or 'light-headedness' following a nicotine injection, or occasionally experienced after the first cigarette of the day (Jones *et al.* 1978; West and Russell 1987). On the other hand, the peripheral vasoconstrictor effect of nicotine (Benowitz *et al.* 1982) and its enhancement of finger-tapping performance (West and Jarvis 1986) are unaffected by acute tolerance.

As mentioned above, acute tolerance appears to be closely related to nicotine's occupancy of its receptors, which is related in turn to blood nicotine concentrations. The classical acute action of nicotine is as a nicotinic-cholinergic agonist causing depolarization, and hence stimulation, at the postsynaptic nicotinic receptors of acetylcholine. But unlike the natural transmitter, with nicotine there are no enzymes to remove it rapidly to enable receptor sensitivity to be restored. Its action is therefore biphasic, in that the depolarization is prolonged (tachyphylaxis), and further excitation of receptors is blocked for at least as long as they remain occupied by nicotine. Thus the overall effect of large doses is to produce a depressant or blocking effect after initial stimulation, whereas repeated small doses have a stimulant effect until acute tolerance develops when sufficient receptors are occupied.

It has been suggested that it is the effects of nicotine at postsynaptic receptors that are most susceptible to acute tolerance and that those at presynaptic receptors are less sensitive (Russell 1988). Its actions at presynaptic sites on nerve terminals produce effects through the release of various neurotransmitters, such as dopamine and noradrenaline (Fuxe *et al.*, chapter 6; Wonnacott, chapter 7; Kellar and Wonnacott, chapter 10; all this volume). Although various forms of desensitization and the depletion of neurotransmitters no doubt give rise to some attenuation of the response and short-term tolerance to nicotine's effects at presynaptic sites, these are of a different order and time scale from the acute tolerance caused by prolonged neuronal depolarization and tachyphylaxis that characterize its effects at postsynaptic sites. So little is known of the mechanisms of desensitization processes that to suggest that nicotine tolerance of the tachyphylactic type is confined to its actions at postsynaptic receptors is at present somewhat speculative.

11.3.3 *Pharmacokinetic factors and acute tolerance*

The rapid development of acute tolerance to some of the effects of nicotine affects their dose–response relationships, making them especially susceptible to the rate of change in tissue nicotine concentrations. The efficiency of the modern cigarette as a vehicle for getting nicotine to the brain in the form of a series of high-concentration boli after each inhaled puff has been discussed. Blood nicotine profiles derived from the concentrations in mixed venous blood provide an incomplete picture but much can be inferred from them. It is well known that the nicotine intake from a cigarette depends on the intensity of puffing and the depth of inhalation. Rapid absorption of nicotine through the lungs together with its rapid distribution to the brain and other tissues give inhaling smokers the capacity to obtain a series of sharp blood nicotine peaks coinciding with each cigarette. There are, however, two factors which limit the frequency with which a cigarette can be inhaled deeply. One is pharmacodynamic, the other pharmacokinetic, both subject to marked individual variation. As discussed earlier, even chronic smokers have limited tolerance to nicotine and avoid exceeding their usual peak blood nicotine levels. This suggests that their usual peak levels are set by their individual toxic aversive barriers, and that they regulate their smoking to get as much nicotine as they can without exceeding this barrier.

The decline of nicotine in blood is bi-exponential. The half-life of the initial distributional phase ranges from 7–10 min (mean 9 min), and the terminal or elimination phase averages 2 h but ranges from 1–4 h (Benowitz and Jacob 1984; Feyerabend *et al.* 1985). In view of the 2 h elimination half-life, significant cumulative build-up of nicotine levels occurs even at a modest frequency of smoking (Fig. 11.4). To maintain a profile of such prominent blood nicotine peaks at high frequencies of smoking (around 2–3 cigarettes h^{-1}) would be pharmacokinetically impossible without reaching aversive blood nicotine levels. In consequence, very heavy smokers tend to take in less nicotine from each cigarette and consequently have less prominent blood nicotine peaks (Fig. 11.5). This in turn implies less nicotine intake per puff and less concentrated post-inhalation boli. These constraints imposed by the elimination half-life are compounded by those of the initial distributional phase half-life. For example, at a smoking rate of three cigarettes per h, the interval between cigarettes (allowing for the time taken to smoke them) is less than twice the 9 min initial half-life, resulting in further reduction of the

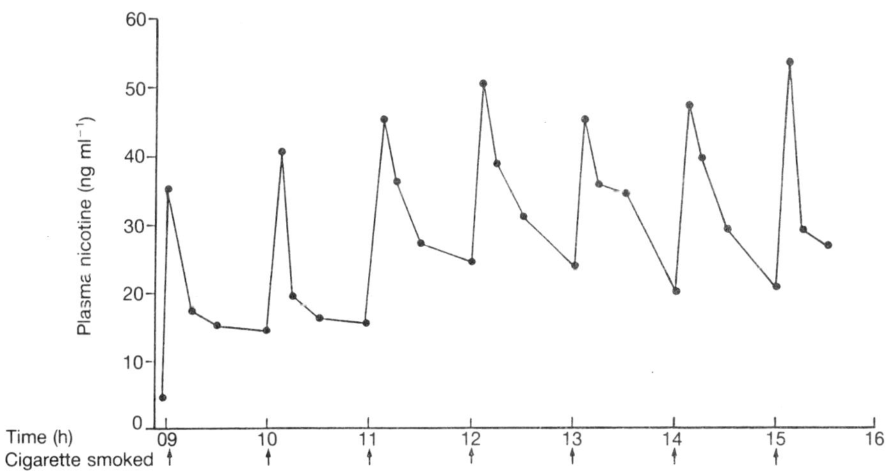

Fig. 11.4. Plasma nicotine levels in an inhaling smoker while smoking at a rate of one cigarette per h for 7 h. Although there is some accumulation of nicotine, the predominant profile is one of prominent blood nicotine peaks following each cigarette. This illustrates the characteristic pattern of the 'peak seeker' type of smoker. From Russell *et al.* 1976*a*.

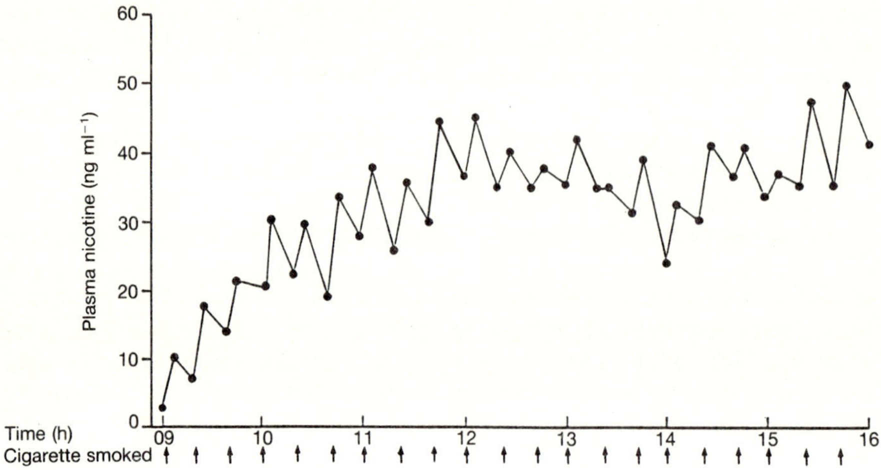

Fig. 11.5. Plasma nicotine levels in an inhaling smoker while smoking at a rate of three cigarettes per h for 7 h. Blood samples were taken just before and 2 min after each cigarette. The blood nicotine builds up to a 'steady state' with peaks which are small relative to the overall level. This illustrates the characteristic pattern of the 'trough-maintainer' type of smoker. (From Russell and Feyerabend 1978.)

prominence of the blood nicotine peaks. Distribution of nicotine from one cigarette will not have been completed by the start of the next one.

The two profiles in Fig. 11.4 and 11.5 are clearly very different and might be expected to have different pharmacological effects. It is unlikely that smokers of the type shown in Fig. 11.5 would get much pharmacological impact from individual cigarettes. This type of profile may characterize smokers who get little positive pleasure and smoke mainly to stave off withdrawal effects (stage 3 of the dependence typology, described later, Section 11.4.3.2). Those at stage 2 who report positive effects from smoking may have blood nicotine profiles more similar to that in Fig. 11.4. The two profiles shown are extreme cases, the average blood nicotine boost per cigarette during steady-state being about 10 ng ml^{-1} as described earlier. Very few data are available on full profiles of different smokers. It is clear from comparing the two profiles contrasted here that a single measure of steady-state peak or trough nicotine (or even cotinine) level would not detect their marked differences. This may in part account for the low correlation ($r = 0.2$) between cigarette consumption and steady-state peak blood nicotine levels found in smokers who smoke 15 or more cigarettes per day (Russell *et al.* 1980*b*).

It is possible that smokers who get prominent blood nicotine peaks may partly overcome some of the effects of acute tolerance despite repeated smoking. Bearing in mind the puff-by-puff intermittent dosage from inhaled smoking, and that the blood nicotine concentrations in post-inhalation boli must be many times higher than those in mixed venous blood, the contrast between bolus levels and the steady-state trough levels may be sufficiently great to enable nicotine to find sufficient numbers of unoccupied receptors to produce an appreciable effect. It is unlikely, however, that smokers with insignificant blood nicotine peaks experience any of those effects of nicotine that are subject to acute tolerance.

Another issue is the potential interaction between acute tolerance and the rate of absorption in determining the pharmacological effects of nicotine. It is suggested that effects that are subject to acute tolerance will not become manifest if the rate of absorption is so slow that acute tolerance develops at the same pace as the rise in blood nicotine concentration. Thus even after prior abstinence for 12 h, the acute drug-like subjective feeling of light-headedness following intravenous or hypodermic injection of nicotine is not experienced if nicotine is infused slowly (Russell 1988). Likewise, it is not experienced following slow absorption from nicotine chewing-gum, but is produced by the more rapid absorption from nasal nicotine solution (see Jaffe, Fig. 1.2, this volume).

A final point is that the pharmacokinetic profile of most smokers is such that for most of the day and much of the night they have high levels of acute tolerance to nicotine. In other words, there is chronic partial blockade of the agonist effects of nicotine at postsynaptic cholinergic receptors. This may explain the apparent paradox between the acute effect of smoking on blood pressure and the lower blood pressure of smokers compared with non-smokers and ex-smokers (Green *et al.* 1986). It may also explain why chronic nicotine exposure gives rise to up-regulation rather than down-regulation of nicotinic receptors (Schwartz and Kellar 1983). Such changes might be expected to be most evident in heavy smokers with high trough blood nicotine levels and relatively small peaks after each cigarette.

11.4 Nicotine and stages of the smoking career

Tobacco smoking is a learnt behaviour determined by a complex interaction between many factors: economic, social, psychological, sensory, motor, and pharmacological. The effects of these factors are in turn mediated by a variety of conscious and unconscious processes including classical and operant conditioning, cognitive mechanisms, and both innate and acquired differences in pharmacokinetic and pharmacodynamic responses to nicotine. This section considers the changing role of nicotine at different stages of the smoking career and at different levels of dependence.

11.4.1 *Natural history of smoking behaviour*

In those who take up smoking, the behaviour has a fairly typical and almost stereotyped natural history. The onset of smoking tends to occur during the teenage years, the proportion who start smoking after the age of 20 being relatively small (McKennell and Thomas 1967). A striking feature of the recruitment process is that it is largely settled after the first few cigarettes. Of those who smoke one cigarette, some 70–75 per cent go on to become regular smokers as adults (McKennell and Thomas 1967; O'Connor and Daly 1985). Similarly, of those who smoke more than one cigarette 80–85 per cent become regular smokers, and of those who smoke at least four cigarettes 90–95 per cent escalate to regular smoking (McKennell and Thomas 1967). Having become a regular smoker, the behaviour is then highly refractory and the majority continue smoking for the next 40 years or more, despite wanting to stop within the first 10 years and, as the years go by and the health risks assume closer personal relevance, making numerous abortive attempts to do so. Of those who ever smoke regularly only about 35 per cent stop permanently before the age of sixty (Jarvis and Jackson 1988). This outline of the typical smoking career testifies to the addictiveness of the behaviour.

11.4.2 *Recruitment to smoking*

The uptake of smoking is mediated by psychosocial factors which are fairly well known. Awareness of the health risks is not a strong deterrent at this stage. They are presumably too remote in time to compete with the more immediate psychosocial motives and pressures to smoke. The sensory and pharmacological effects of nicotine are initially aversive, but sensory adaptation to the local irritancy of nicotine and other smoke components appears to develop rapidly, and skill may be acquired to regulate smoke intake to an acceptable level. Tolerance to aversive effects such as nausea may also develop rapidly, but it is not known whether the absence of nausea after the first few cigarettes is attributable to rapid development of chronic tolerance, to quickly learnt self-titration of nicotine intake, or to selective dropping out. In some novices the predominant subjective nicotine effect is dizziness or light headedness, while in others it is nausea. There is some evidence that the former group are more likely to smoke again than are those who experience nausea or no drug effect (Hirschman *et al.* 1984).

Whatever mechanisms may be involved, the aversiveness of the first one or two cigarettes declines rapidly on repetition. If the psychosocial pressure to smoke is sufficient for the act to be repeated despite its initial aversiveness, it is understandable that smoking will continue once the aversiveness subsides and the effects shift from producing nausea and dizziness to become calming and pleasurable. There are no published data on blood nicotine levels immediately after a cigarette in novice smokers, but estimates based on expired-air carbon monoxide and saliva cotinine levels suggest that the nicotine intake per cigarette of children in the early stages of smoking 2–3 cigarettes per week is similar to that of more regular daily smokers (McNeill *et al.* 1986, 1987).

Table 11.2. Correlation of smoking motivation factors with age and daily cigarette consumption (n = 175)

Motive to smoke	Age	Consumption
Psychosocial	– 0.23*	0.16
Sensorimotor	0.09	0.08
Indulgent	– 0.08	0.27*
Stimulation	0.16	0.50**
Sedative	– 0.11	0.45**
Addictive	0.11	0.63**
Automatic	0.03	0.56**

* $p < .01$, ** $p < .001$ Abstracted from Russell *et al.* (1974)

11.4.3 *Maintenance of smoking*

11.4.3.1 *Expressed motives for smoking* To find out why people smoke it is reasonable to take some account of the reasons they give and the situations in which they are most likely to smoke. Some years ago we searched the literature for self-report data of this kind, and borrowing freely from previously published work and making use of our own clinical experience with smokers, we developed a 34-item questionnaire covering a wide variety of reasons and motives for smoking (Russell *et al.* 1974). The responses of 175 normal smokers were factor analyzed and the seven main smoking motives are shown in Table 11.2, together with their correlations with age and cigarette consumption. The 'sedative' smoking factor was the least robust. It can be seen that the 'stimulation', 'sedative', 'addictive' and 'automatic' factors all correlated most highly with cigarette consumption. They also correlated with each other to form a higher order 'pharmacological-addictive' dimension, and differentiated the main sample of normal smokers from a criterion sample of 103 addicted heavy smokers attending withdrawal clinics (Fig. 11.6). These results are therefore broadly compatible with the known psychopharmacological effects of nicotine.

11.4.3.2 *Three-stage model of dependence* The fact that two thirds of smokers do not stop before the age of sixty, despite awareness of the health risks and despite wanting to stop and trying to stop, underlines the extent to which dependence is a key feature of the maintenance of smoking. The development of dependence can be viewed as proceeding through three main overlapping stages. After a first stage of smoking for psychosocial motives, most smokers progress fairly rapidly to a second stage of smoking for the pleasurable, calming, and mildly stimulant effects of nicotine and its inter-

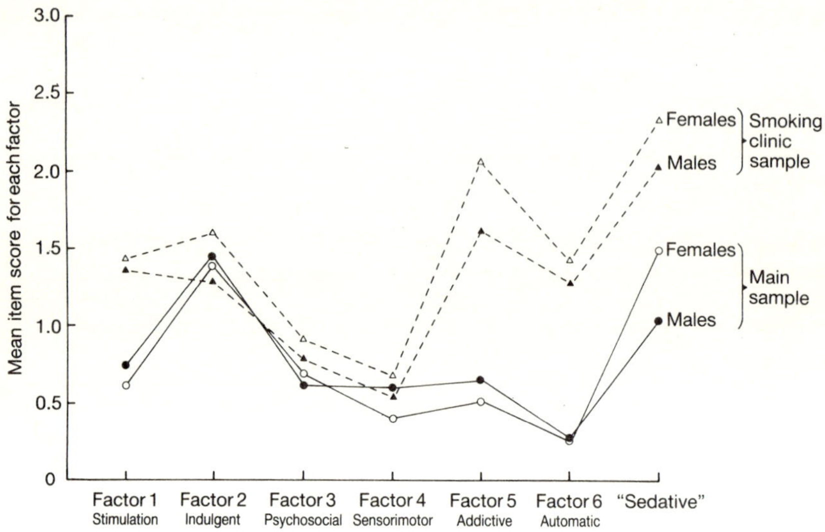

Fig. 11.6. Mean factor scores by sex for main and smoking clinic samples. (From Russell *et al.* 1974.)

action with the sensorimotor components of the habit (positive reinforcement). Some smokers may then progress to a third stage where few of the cigarettes smoked are positively pleasurable, where smoking (when permitted) continues in most situations, and where the behaviour is largely controlled by the need to stave off or avoid withdrawal effects (negative reinforcement). The study described above gives some support to this model.

11.4.3.3 *Primary effects and acquired withdrawal effects.* Although it is unlikely that many remain life-long psychosocial smokers, the proportion of smokers falling predominantly into stages two or three, or between the two, is unknown. Indeed, it is by no means clear to what extent the well-known effects of nicotine on mood and performance represent primary positive effects, or whether they reflect a reversal of the negative effects of withdrawal. The problem is illustrated by comparison of Tables 11.3 and 11.4. The main subjective and behavioural effects of nicotine reported in the literature and summarized by Pomerleau (1986), shown in Table 11.3, can be seen to mirror the main features of the tobacco withdrawal syndrome listed in Table 11.4. Many of the effects of cigarette withdrawal are nicotine-related in that they are reversed or alleviated by nicotine more effectively than by placebos or nicotine-free cigarettes (see earlier section and also Warburton, chapter 3, this volume). Since most of the data on the beneficial effects of nicotine on human mood and performance are derived from studies in smokers after short-term deprivation, it

Table 11.3. Commonly reported behavioural and subjective effects of smoking

Improved concentration/ability to tune out irrelevant stimuli
Improved memory (recall)
Improved psychomotor performance
Increased alertness/arousal
Decreased anxiety/tension
Pleasure/facilitation of pleasure
Decreased body weight (reduced preference for sweet tastes)

(From Pomerleau 1986.)

Table 11.4. Main features of the tobacco withdrawal syndrome

Subjective effects	Physical and objective effects
Craving for tobacco	EEG changes
Increased tension	Drop in heart rate and blood pressure
Restlessness	Increased peripheral circulation
Irritability	Drop in urinary adrenaline, noradrenaline and cortisol
Aggressiveness	Sleep disturbances
Depression	Increased weight
Difficulty with concentration	Decreased performance on:
Increased hunger	Vigilance tasks
	Simulated driving
	Selective attention tasks
	Memory tasks

is not known whether they are primary effects or attributable to the relief of acquired withdrawal effects. Conversely, it is unclear how much the effects of withdrawal are due to simple lack of the primary effects of nicotine, to acquired factors based on learning and conditioning, or to pharmacological 'rebound' phenomena based on neuronal adaptive changes underlying chronic tolerance.

Relatively few studies are available to provide clues to these matters. The limited evidence suggests that some of the effects of nicotine are primary, while others appear to be based on the relief of acquired withdrawal effects. Various approaches have been used. One is to compare the responses or performance of deprived and non-deprived smokers with those of non-smokers. Where this has been done the performance of deprived smokers is inferior to that of non-smokers. When allowed to smoke, or given nicotine, the performance of deprived smokers is improved to reach, but does not surpass, the performance of non-smokers (Heimstra *et al.* 1967). Similar findings have been demonstrated for the ability of non-smokers and smokers to tolerate electric shocks, and their degree of irritation and annoyance about interruptions during viewing a television drama (Schachter 1978). Deprived smokers also have a slower EEG alpha rhythm than do non-smokers, indicating a lower resting level of arousal (Knott and Venables 1977). Such studies show that the major differences between smokers and non-smokers emerge when the smokers are deprived. Whether this is due to acquired withdrawal effects or to innate differences is an interesting question.

A second approach, also seldom done, is to study the effects of nicotine in non-smokers (or possibly long-term ex-smokers). Non-smokers find it difficult to inhale cigarettes or to chew nicotine gum. Crushed nicotine tablets held in the mouth to allow buccal absorption appear to enhance the performance of non-smokers on prolonged symbol recognition tasks (Wesnes and Warburton 1983). The performance of non-smokers on a simple motor task (rate of finger tapping) was increased after taking 2 mg nicotine via a nasal solution (West and Jarvis 1986). This effect was blocked by mecamylamine. With repeated dosage hourly for 6 h there was no evidence of acute tolerance. These two studies of the effects of nicotine in non-smokers indicate that these effects at least are primary and cannot be attributed to the relief of withdrawal.

A third approach used to distinguish between primary effects and those due to reversal of acquired withdrawal effects is to look for evidence of a 'rebound' element. Rebound phenomena following acute withdrawal are believed to depend on various adaptive mechanisms including those that underlie the development of chronic tolerance. In general, therefore, 'rebound' withdrawal phenomena are more likely to occur in relation to those effects of nicotine that are subject to chronic tolerance. To demonstrate a rebound element it is necessary to show that a withdrawal effect, evident in the early stages of abstinence, is followed by a return towards baseline as abstinence is prolonged. Such studies are difficult in free-ranging smokers who find it difficult to abstain for longer than 24 h, and failure to do so confounds the results. It is also selective in that those with more severe withdrawal effects are less likely to comply. Only one study has been reported in which 17 smokers abstained for 10 days (West *et al.* 1984*b*). Urinary adrenaline concentrations dropped significantly in the first three days after stopping smoking. This was followed by a significant rise after 10 days, back to pre-abstinence levels, indicating that this withdrawal effect is mediated by rebound phenomena. It does not follow from this, however, that the effect has motivational significance. Heart rate, finger-skin temperature, and urinary cortisol also showed changes at three days, but there was no significant tendency to return to baseline after 10 days.

Of all the withdrawal effects listed in Table 11.4, it is probable that the urge or craving to smoke is the most important for the smoker. The anatomical sites and neurophysiological mechanisms that mediate pleasure, reward, reinforcement, and so on, are not well understood. Currently prevailing views might place this at the nucleus accumbens and its dopaminergic input from the ventral tegmental area (Wise and Bozarth 1984), both of which are stimulated directly by smoking doses of nicotine, presumably through its action on presynaptic nicotinic-cholinergic receptors (Imperato *et al.* 1986; Reavill, chapter 9, this volume). It is possible that the effects of nicotine at such sites play some part in the pleasure and reinforcement derived from smoking, and that craving is partly mediated by rebound effects, at these sites, following its withdrawal. Although harder for the outside observer to detect and measure than are the florid and dramatically obvious rebound effects in the peripheral autonomic nervous system following opiate withdrawal, similar effects within the mesolimbic dopamine system might be subjectively more important, but manifest only in terms of the severity of craving. To complicate the issue, the pharmacological mechanisms may be closely linked with the effects of learning and conditioning, as is the case with other addictive drugs.

In summary, the subjective and behavioural effects of nicotine are determined by many factors. Despite many similarities there are wide individual differences, some of which are innate, others acquired. A major factor is the smoking history and the pattern of prior exposure and experience which, through interactions between pharmacological and learning mechanisms, determine the stage and level of dependence reached by the subject. Probably through lack of systematic study, chronic tolerance to nicotine has not been demonstrated in humans (acute tolerance is a different matter and will be discussed later). Only one study so far has looked for and found evidence of physiological 'rebound' effects after withdrawal. It is still not clear how much the effects of nicotine on human mood and performance are primary effects or are determined by acquired factors involved in the relief of withdrawal. The answer seems to be both, some are primary, others based on withdrawal relief. As far as the status of smoking as an addiction is concerned, the issue is largely one of semantics. Outmoded notions that compulsive drug use is not an addiction in the absence of chronic tolerance and pharmacological rebound after withdrawal fail to recognize that dependence is essentially a learnt phenomenon. It is the strength of the reinforcement that matters, not its nature. Powerful psychosocial rewards or strongly reinforcing primary pharmacological effects may be more addictive than the relief of pharmacological rebound phenomena that are only mildly aversive.

11.4.4 *Sensitivity to nicotine and severity of withdrawal*

We have recently started to examine the extent to which acute tolerance might affect subjective and other responses to smoking during the course of a normal smoking day (West and Russell 1987). As expected, on smoking a cigarette after a period of abstinence of 24 h there was an increase in heart rate, averaging 14 beats min^{-1}, a drop in finger skin temperature, and just over half the subjects (11 out of 21) experienced clearcut subjective effects from the cigarette. Subjective effects were assessed by asking the subjects whether they had 'felt anything from the cigarette'. All eleven who experienced effects from the post-abstinence cigarette reported feeling 'dizzy' or 'light-headed'. Of these, four also felt nauseous, two felt palpitations, one had a feeling of tremor, one a buzzing in the ears, and one also reported feeling sweaty. In contrast, a cigarette smoked during the course of a normal smoking day caused no change in heart rate and no subjective effects, indicating acute tolerance to these effects of nicotine. Finger skin temperature still dropped significantly, confirming that this effect is less sensitive to acute tolerance. There was a positive correlation between the strength of the subjective effect from the post-abstinence cigarette and the rise in heart rate, which remained when rise in CO was taken into account (partial correlation = 0.54, $p<0.2$). There was also a correlation of 0.28 with drop in skin temperature, but this was not statistically significant. Blood nicotine concentrations were not measured in this study, but none of the responses to the post-abstinence cigarette were related to the CO boost it produced (highest correlation $r = -0.16$ with change in finger skin temperature).

It seems clear from this study that due to the development of acute tolerance these smokers experienced little positive impact from individual cigarettes smoked under steady-state conditions. In the nicotine-depleted state, however, after 24 h abstinence, the striking feature was the marked individual variation in response to a cigarette. Some subjects felt no effect while others felt nauseous. These differences did not appear to be attributable to differences in CO intake (and by inference nicotine intake) from the post-abstinence cigarette, which suggests that the subjects varied considerably in their sensitivity to nicotine. Although further studies will need to be directed specifically at this question, we have some data suggesting that it is the more dependent smokers, who experience more severe withdrawal symptoms, who are also more, rather than less, sensitive to their first cigarette following 24 h abstinence (Table 11.5). Their greater sensitivity to nicotine, indicates a lack of chronic tolerance to these effects once the effects of acute tolerance have been unmasked by 24 hrs abstinence.

Data from two different studies are shown in Table 11.5. One of them (West and Russell 1985*a*) focused on pre-abstinence measures of smoke intake and scores on a smoking motivation questionnaire (SMQ, based on the factors shown in Table 11.2) as predictors of the severity of cigarette withdrawal. The other study (West and Russell 1988) focused on the relation of the severity of withdrawal to the strength of the response to the first post-abstinence cigarette. Severity of withdrawal was assessed by questionnaire ratings of craving and other subjective effects which form the core of the withdrawal syndrome, and which changed significantly during abstinence (for example, depression, irritability, restlessness, hunger, difficulty concentrating).

It can be seen from Table 11.5 that blood nicotine level measured prior to

Table 11.5. Correlations of severity of withdrawal effects during 24-h abstinence from smoking with pre-abstinence variables and responses to the first post-abstinence cigarette

Variable	1985 study (n = 29)	1988 study (n = 21)
Pre-abstinence variables:		
Daily cigarette consumption	0.09	
Plasma nicotine	0.57*	
Expired-air carbon monoxide	0.23	
Smoking motivation scores		
Indulgent	0.33	
Stimulation	0.13	
Sedative	0.15	
Automatic	0.01	
Addictive	0.51*	0.61*
Post-abstinence cigarette:		
Subjective effects		0.56*
Heart rate increase		0.41*

* $p < 0.05$ Correlations with effects of post-abstinence cigarette are partial, the estimated nicotine boost from the cigarette having been partialled out. The strength of the subjective effect from the post-abstinence cigarette was coded as follows: no effect = 0, dizzy or light-headed = 1, any additional effect = 2. No additional effects occurred in the absence of dizziness. (Abstracted from West and Russell 1985*a*, 1988.)

abstaining was the best predictor of severity of withdrawal and that the 'addictive smoking' subscale of the smoking motivation questionnaire also predicted the severity of subsequent withdrawal. In the right-hand column of Table 11.5, the positive correlations between the severity of withdrawal and the strength of response to the first post-abstinence cigarette indicate that the more dependent smokers, who experienced more severe withdrawal effects, appeared to be more sensitive to the effects of nicotine when the confounding effects of acute tolerance had been largely eliminated by 24 h prior abstinence. The increased, rather than decreased, sensitivity to nicotine in the absence of acute tolerance further indicates that these more dependent smokers have not developed a higher degree of chronic tolerance to the acute subjective effects of nicotine, or to its effect on heart rate. These results may at first sight appear inconsistent with those of Hughes and Hatsukami (1986), who found a negative correlation between heart rate boost per unit of nicotine intake from a cigarette smoked prior to abstinence and the severity of subsequent withdrawal. However, their results were confounded by acute tolerance since their subjects had not abstained before smoking the test cigarette.

Our data are consistent with the view expressed earlier that the more addicted type of heavy smokers (stage 3) with high blood nicotine troughs and relatively small peaks would, during most of their smoking day, have a high degree of acute tolerance to the effects of nicotine at postsynaptic sites. This virtually chronic blockade may lead to up-regulation of nicotinic receptors at these sites, which would in turn account for the greater sensitivity of the more dependent smokers to some of the effects of nicotine when in a nicotine-depleted state. On the other hand, individual differences in the responses of naive smokers to their first experiences with nicotine, as alluded to earlier, could be mediated by inherent differences in densities of nicotinic receptors, or to acquired differences resulting from prior exposure *in utero* or to passive smoking. Such differences could in turn determine who becomes dependent on smoking. It is thus unknown at present how much the differences between regular smokers in their degree of dependence, severity of withdrawal, and sensitivity to nicotine 24 h after withdrawal are the consequence or the cause of their smoking behaviour.

The animal studies of Collins, Marks and collaborators are highly relevant to these questions. They have found in studies of mice of different strains that the initial sensitivity to nicotine is related to the number of nicotinic receptors. They have shown further that those strains that are initially more sensitive develop tolerance more rapidly, and that the development of tolerance and its regression are reflected by up-regulation and down-regulation respectively of the density of nicotinic receptors in the brain (for example, Marks *et al.* 1986*a*; 1986*b*).

What about those effects of nicotine that are not subject to acute tolerance of the tachyphylactic type? It was suggested earlier (section 11.3.2.2) that these effects may be mediated by nicotine's action at presynaptic sites on nerve terminals. Since the more addicted, heavier smokers with less prominent blood nicotine peaks are mostly in a state of acute tolerance to nicotine's effects at postsynaptic sites, it is among the effects at presynaptic sites that their main source of reinforcement is likely to be found. It is not yet known whether post-synaptic, as opposed to presynaptic, nicotinic receptors are more responsive to up-regulation, or indeed whether nicotine induces down-regulation at the postsynaptic sites of neurotransmitters released by its action at presynaptic sites.

11.5 Summary and conclusions

This chapter reviews the evidence which establishes nicotine as a major controlling factor in smoking and the self-regulation of smoke intake, and then considers some of the mechanisms and processes that may be involved. In the case of regular smokers, nicotine determines how people smoke, why they smoke, and largely when they smoke. Smoking behaviour is regulated to obtain or maintain the smoker's individually characteristic pattern of nicotine intake, which is in turn determined by the pharmacological effects that ensue.

Associated sensorimotor cues can act as important conditioned reinforcers. Smokers down-regulate their nicotine intake with great precision to avoid the aversive effects of excessive blood levels, but up-regulation to maintain usual nicotine levels is less precise and is influenced by the rate of absorption as well as steady-state level. On average, adequate satisfaction is obtained at two thirds of the blood levels of usual smoking provided the rate of absorption is unchanged. Blood levels as low as one-third of usual smoking levels are sufficient to alleviate some withdrawal effects and to enhance performance on cognitive and psychomotor tasks, even if the rate of absorption is slow.

Despite its addictiveness, the subjective effects of nicotine in 'smoking doses' are of low intensity compared with those of other addictive drugs. Although mildly pleasurable to many regular users, it causes no striking euphoriant effects and its effects on mood, performance, and the level of arousal are relatively small. Better understanding may be gained of the effects of nicotine, and the hold they come to have over those who experience them, by considering some of the multiple sources of individual variation. The subjective and behavioural effects of nicotine vary not only between individuals but also within individuals according to the stage of their smoking career, their level of dependence, and the time since their last few doses. Some of the influences and mechanisms discussed include innate and acquired factors, pharmacokinetic factors, acute and chronic tolerance, learning and conditioning.

It is not clear to what extent the effects of nicotine are primary, or how much they reflect reversal or relief of acquired withdrawal effects. Only one study has found a 'rebound' element in the effects of withdrawal, and although chronic exposure to nicotine induces an increase in the number of nicotinic receptors, chronic tolerance to nicotine has not been demonstrated in humans. Acute tolerance (tachyphylaxis) develops rapidly to many of the effects of nicotine and is completely reversible after nicotine depletion. Other effects of nicotine are less sensitive to acute tolerance. It is suggested that it is the effects of nicotine at postsynaptic receptors that are most susceptible to acute tolerance and that those mediated by its action at presynaptic receptors are less sensitive to it.

Due to accumulation of nicotine and other pharmacokinetic factors, for most of the day and much of the night regular smokers have high levels of acute tolerance to nicotine. In other words, there is a chronic partial blockade of its agonist effects at postsynaptic receptors. This explains why nicotinic receptors are up-regulated rather than down-regulated, and why heavy smokers experience no acute subjective effects from a cigarette smoked during the course of a normal smoking day. When the effects of acute tolerance are unmasked after abstinence for 24 h, it is the more addicted heavy smokers who experience more severe withdrawal effects during abstinence, and who also have stronger subjective and heart rate effects following the first post-abstinence cigarette. Their greater sensitivity to nicotine after abstinence may reflect their higher densities of unoccupied nicotinic receptors. On the other hand, those who have higher innate sensitivity may be more likely to take up smoking and to become more dependent if they do.

Acknowledgements

I thank the Imperial Cancer Research Fund and the Medical Research Council for financial support, Wilhemina Boyle for secretarial help, and Martin Jarvis, Robert West, and my co-editors of this book for valuable comments on earlier drafts of this chapter.

References

Abelin, T., Muller, P., Buehler, A., Versanen, K., and Imhot, P.R. (1989). Controlled trial of transdermal nicotine patch in tobacco withdrawal. *Lancet* **1**, 7–10.

Armitage, A.K., Hall, G.H., and Morrison, C.F. (1968). Pharmacological basis of the tobacco smoking habit. *Nature* **217**, 331–4.

Armitage, A.K., and Turner, D.M. (1970). Absorption of nicotine in cigarette and cigar smoke through the oral mucosa. *Nature* **226**, 1231–2.

Ashton, H., and Watson, D.W. (1970). Puffing frequency and nicotine intake in cigarette smokers. *British Medical Journal* **3**, 679–81.

Ashton, H., Stepney, R., and Thompson, J.W. (1979). Self-titration by cigarette smokers. *British Medical Journal* **2**, 357–60.

Benowitz, N.L., Jacob, P., Jones, R.T., Rosenberg, J. (1982). Inter-individual variability in metabolism and cardiovascular effects of nicotine in man. *Journal of Pharmacology and Experimental Therapeutics* **221**, 368–72.

Benowitz, N.L., Hall, S.M., Herning, R.I., Jacob, P., Jones, R.T., and Osman, A.L. (1983). Smokers of low-yield cigarettes do not consume less nicotine. *New England Journal of Medicine* **309**, 139–42.

Benowitz, N.L., and Jacob, P. (1984). Daily intake of nicotine during cigarette smoking. *Clinical Pharmacology and Therapeutics* **35**, 499–504.

Benowitz, N.L., and Jacob P. (1985). Nicotine renal excretion rate influences nicotine intake during cigarette smoking. *Journal of Pharmacology and Experimental Therapeutics* **234**, 153–5.

Benowitz, N.L., Jacob, P., Kozlowski, L.T., and Yu, L. (1986). Influence of smoking fewer cigarettes on exposure to tar, nicotine, and carbon monoxide. *New England Journal of Medicine* **315**, 1310–3.

Benowitz, N.L., Jacob, P., and Savanapridi, C. (1987). Determinants of nicotine intake while chewing nicotine polacrilex gum. *Clinical Pharmacology and Therapeutics* **41**, 467–73.

Benwell, M.E.M., Balfour, D.J.K., and Anderson, J.M. (1988). Evidence that tobacco smoking increases the density of nicotine binding sites in human brain. *Journal of Neurochemistry* **50**, 1243–7.

Castleden, C.M., and Cole, P.V. (1973). Inhalation of tobacco smoke by pipe and cigar smokers. *Lancet* **2**, 21–2.

Clarke, P.B.S., and Kumar, R. (1983). The effects of nicotine on locomotor activity in non-tolerant and tolerant rats. *British Journal of Pharmacology* **78**, 329–337.

Ebert, R.V., McNabb, M.E., McCusker, K.T., and Snow, S.L. (1983). Amount of nicotine and carbon monoxide inhaled by smokers of low-tar low-nicotine cigarettes. *Journal of the American Medical Association* **250**, 2840–2.

Ebert, R.V., McNabb, M.E., and Snow, S.L. (1984). Effect of nicotine chewing gum on plasma nicotine levels of cigarette smokers. *Clinical Pharmacology and Therapeutics* **35**, 495–598.

Ferno, O., Lichtneckert, S.J.A., and Lundgren, C.E.G. (1973). A substitute for tobacco smoking. *Psychopharmacologia* **31**, 201–4.

Feyerabend, C., Ings, R.M.J., and Russell, M.A.H. (1985). Nicotine pharmacokinetics and its application to intake from smoking. *British Journal of Clinical Pharmacology* **19**, 239–47.

Finnegan, J.K., Larson, P.S., and Haag, H.B. (1945). The role of nicotine in the cigarette habit. *Science* **102**, 94–6.

Frith, C.D. (1971). The effect of varying the nicotine content of cigarettes on human smoking behavior. *Psychopharmacologia* **19**, 188–92.

Goldfarb, T.L., Gritz, E.R., Jarvik, M.E., and Stolerman, I.P. (1976). Reactions to cigarettes as a function of nicotine and tar. *Clinical Pharmacology and Therapeutics* **19**, 762–72.

Gori, G.B., and Lynch, C.J. (1985). Analytical cigarette yields as predictors of smoke bioavailability. *Regulatory Toxicology and Pharmacology* **5**, 314–26.

Green, M.S., Juch, E., and Luz, Y. (1986). Blood pressure in smokers and non-smokers: epidemiologic findings. *American Heart Journal* **111**, 932–40.

Gritz, E.R. (1980). Smoking behavior and tobacco abuse. In *Advances in substances abuse*. (N.K. Mello, ed.), pp. 91–158. JAI Press, Greenwich.

Gritz, E.R., Baier-Weiss, V., Benowitz, N.L., Van Vunakis, H., and Jarvik, M.E. (1981). Plasma nicotine and cotinine concentrations in habitual smokeless tobacco users. *Clinical Pharmacology and Therapeutics* **30**, 201–9.

Grunberg, N.E., Morse, D.E., Maycock, V.A., and Kozlowski, L.T. (1985). Changes in overwrap and butt length of Americal filter cigarettes: an influence on reported tar yields. *New York State Journal of Medicine* **85**, 310–2.

Hajek, P., Jackson, P., and Belcher, M. (1988). Long-term use of nicotine chewing gum: occurrence, determinants and effect on weight gain. *Journal of the American Medical Association* **260**, 1593–6.

Hajek, P., Jarvis, M.J., Belcher, M., Sutherland, G., and Feyerabend, C. (1989). Effect of smoke-free cigarettes on 24 hr cigarette withdrawal: a double-blind placebo-controlled study. *Psychopharmacology* **97**, 99–102.

Hatsukami, D.K., Gust, S.W., and Keenan, R.M. (1987). Physiologic and subjective changes from smokeless tobacco withdrawal. *Clinical Pharmacology and Therapeutics* **41**, 103–7.

Heimstra, N.W., Bancroft, N.R., and Dekock, A.R. (1967). Effects of smoking upon sustained performance in a simulated driving task. *Annals of the New York Academy of Sciences* **142**, 295–307.

Henningfield, J.E., Miyasato, K., and Jasinski, D.R. (1983). Cigarette smokers self-administer intravenous nicotine. *Pharmacology, Biochemistry and Behavior* **19**, 887–90.

Henningfield, J.E., Miyasato, K., and Jasinski, D.R. (1985). Abuse liability and pharmacodynamic characteristics of intravenous nicotine and inhaled nicotine. *Journal of Pharmacology and Experimental Therapeutics* **234**, 1–12.

Herning, R.I., Jones, R.T., Bachman, J., and Mines, A.H. (1981). Puff volume increases when low-nicotine cigarettes are smoked. *British Medical Journal* **283**, 187–9.

Herning, R.I., and Pickworth, W.B. (1985). Nicotine gum improved stimulus processing during tobacco withdrawal. *Psychophysiology* **22**, 594.

Hirschman, R.S., Leventhal, H., and Glynn, K. (1984). The development of smoking behavior: conceptualisation and supportive cross-sectional survey data. *Journal of Applied and Social Psychology* **14**, 184–206.

Hjalmarson, A.I. (1984). Effect of nicotine chewing gum in smoking cessation: a randomized, placebo-controlled, double-blind study. *Journal of the American Medical Association* **52**, 2835–8.

Ho-yen, D.O., Spence, V.A., Moody, J.P., and Walker, W.F. (1982). Why smoke fewer cigarettes? *British Medical Journal* **284**, 1905–7.

Hughes, J.R., Hatsukami, D.K., Pickens, R.W., Krahn, D., Malin, S., and Luknic, A. (1984). Effect of nicotine on the tobacco withdrawal syndrome. *Psychopharmacology* **83**, 82–7.

Hughes, J.R., Hatsukami, D.K., and Skoog, K.P. (1986). Physical dependence on nicotine in gum: a placebo substitution trial. *Journal of the American Medical Association* **255**, 3277–9.

Hughes, J.R., and Hatsukami, D.K., (1986). Signs and symptoms of tobacco withdrawal. *Archives of General Psychiatry* **43**, 289–94.

Imperato, A., Mulas, A., and Di Chiara, G. (1986). Nicotine preferentially stimulates dopamine release in the limbic system of freely moving rats. *European Journal of Pharmacology* **132**, 337–8.

Isaac, P.F., and Rand, M.J. (1969). Blood levels of nicotine and physiological effects after inhalation of tobacco smoke. *European Journal of Pharmacology* **8**, 265–83.

Jarvik, M.E., Glick, S.D., and Nakamura, R.K. (1970). Inhibition of cigarette smoking by orally administered nicotine. *Clinical Pharmacology and Therapeutics* **11**, 574–6.

Jarvis, M.J., Raw, M., Russell, M.A.H., and Feyerabend, C. (1982). Randomized controlled trial of nicotine chewing gum. *British Medical Journal* **285**, 537–40.

Jarvis, M.J., West, R.J., Tunstall-Pedoe, H., and Vesey, C. (1984). An evaluation of the intervention against smoking in the Multiple Risk Factor Intervention Trial. *Preventive Medicine* **13**, 501–9.

Jarvis, M.J., Hajek, P., Russell, M.A.H., West, R.J., and Feyerabend, C. (1987). Nasal nicotine solution as an aid to cigarette withdrawal: a pilot clinical trial. *British Journal of Addiction* **82**, 983–8.

Jarvis, M.J., and Jackson, P.H. (1988). Cigar and pipe smoking in Britain: implications for smoking prevalence and cessation. *British Journal of Addiction* **83**, 323–30.

Johnston, L.M. (1942). Tobacco smoking and nicotine. *Lancet* **2**, 742.

Jones, R.T., Farrell, T.R., and Herning, R.I. (1978). Tobacco smoking and nicotine tolerance. In *Self-administration of abused substances: methods for study* (N.A. Krasnegor, ed.), pp. 202–8. NIDA Research Monograph No. 20, US Government Printing Office, Washington DC.

Knott, V.J., and Venables, P.H. (1977). EEG alpha correlates of non-smokers, smoking and smoking deprivation. *Psychophysiology* **14**, 150–6.

Kornitzer, M., Kittel, F., Dramaix, M., and Bourdoux, P. (1987). A double-blind study of 2 mg versus 4 mg nicotine gum in an industrial setting. *Journal of Psychosomatic Research* **31**, 171–6.

Kozlowski, L.T., Jarvik, M.E., Gritz, E.R. (1975). Nicotine regulation and cigarette smoking. *Clinical Pharmacology and Therapeutics* **17**, 93–7.

Kumar, R., Cooke, E.C., Lader, M.H., and Russell, M.A.H. (1977). Is nicotine important in tobacco smoking? *Clinical Pharmacology and Therapeutics* **21**, 520–9.

Lucchesi, B.R., Schuster, C.R., and Emley, G.S. (1967). The role of nicotine as a determinant of cigarette smoking frequency in man with observations of certain cardiovascular effects associated with the tobacco alkaloid. *Clinical Pharmacology and Therapeutics* **8**, 789–96.

Marks, M.J., and Collins, A.C. (1985). Tolerance, cross-tolerance and receptors after chronic nicotine or oxotremorine. *Pharmacology, Biochemistry and Behavior* **22**, 283–91.

Marks, M.J., Romm, E., Gaffney, D.K., and Collins, A.C. (1986*a*). Nicotine-induced tolerance and receptor changes in four mouse strains. *Journal of Pharmacology and Experimental Therapeutics* **237**, 809–19.

Marks, M.J., Stitzel, J.A., and Collins, A.C. (1986*b*). Dose–response analysis of nicotine tolerance and receptor changes in two inbred mouse strains. *Journal of Pharmacology and Experimental Therapeutics* **239**, 358–64.

McKennell, A.C., and Thomas, R.K. (1967). Adults' and adolescents' smoking habits and attitudes. Government Social Survey, HMSO, London.

McMorrow, M.J., and Foxx, R.M. (1983). Nicotine's role in smoking: an analysis of nicotine regulation. *Psychological Bulletin* **93**, 302–27.

McNabb, M.E., Ebert, R.V., McCusker, K. (1982). Plasma nicotine levels produced by chewing nicotine gum. *Journal of the American Medical Association* **248**, 865–8.

McNabb, M.E. (1984). Chewing gum for 3 months: what happens to plasma nicotine levels? *Canadian Medical Association Journal* **131**, 589–592.

McNeill, A.D., West, R. Jarvis, M.J., and Russell, M.A.H. (1986). Do older children take in more smoke from their cigarettes? Evidence from carbon monoxide levels. *Journal of Behavioral Medicine* **9**, 559–65.

McNeill, A.D., Jarvis, M.J., West, R.J., Russell, M.A.H., and Bryant, A. (1987). Saliva cotinine as an indicator of cigarette smoking in adolescents. *British Journal of Addiction* **82**, 1355–60.

Morrison, C.F., and Stephenson, J.A. (1972). The occurence of tolerance to a central depressant effect of nicotine. *British Journal of Pharmacology* *46*, 141–56.

Moss, R.A., and Prue, D.M. (1982). Research on nicotine regulation. *Behavior Therapy* **13**, 31–46.

Murphree, H.B., and Weyer E.M. (eds) (1967). *The effects of nicotine and smoking on the central nervous system*. Annals of The New York Academy of Sciences, Volume **142**.

Murphree, H.B. (1979). EEG effects in humans of nicotine, tobacco smoking, withdrawal from smoking and possible surrogates. In *Electrophysiological effects of nicotine* (A. Remond and C. Izard, ed.), pp. 227–43. Elsevier Press, Amsterdam.

Nemeth-Coslett, R., Henningfield, J.E., O'Keefe, M.K., and Griffiths, R.R. (1986). Effects of mecamylamine on human cigarette smoking and subjective ratings. *Psychopharmacology* **88**, 420–5.

Nemeth-Coslett, R., Henningfield, J.E., O'Keefe, J., and Griffiths, R.R. (1987). Nicotine gum: dose-related effects on cigarette smoking and subjective ratings. *Psychopharmacology* **92**, 424–30.

O'Connor, J., and Daly, M. (1985). *The smoking habit*. Gill and Macmillan Ltd, Dublin.

Pickworth, W.B., Herning, R.I., and Henningfield, J.E. (1986). Electroencephalographic effects of nicotine chewing gum in humans. *Pharmacology Biochemistry and Behavior* **25**, 879–82.

Pomerleau, O.F. (1986). Nicotine as a psychoactive drug: anxiety and pain reduction. *Psychopharmacology Bulletin* **22**, 865–9.

Pomerleau, C.S., Pomerleau, O.F., and Majchrzak, M.J. (1987). Mecamylamine pretreatment increases subsequent nicotine self-administration as indicated by changes in plasma nicotine level. *Psychopharmacology* **91**, 391–3.

Raw, M., Jarvis, M.J., Feyerabend, C., and Russell, M.A.H. (1980). Comparison of nicotine chewing gum and psychological treatments for dependent smokers. *British Medical Journal* **281**, 481–2.

Rose, J.E., Zinser, M.C., Tashkin, D.P., Newcomb, R., and Ertle, A. (1984). Subjective response to cigarette smoking following airway anesthetization. *Addictive Behaviors* **9**, 211–5.

Rose, J.E., Herskovic, J.E., Trilling, Y., and Jarvik, M.E. (1985). Transdermal nicotine reduces craving and nicotine preference. *Clinical Pharmacology and Therapeutics* **38**, 450–6.

Russell, M.A.H., Peto, J., and Patel, U.A. (1974). The classification of smoking by factorial structure of motives. *Journal of the Royal Statistical Society A*, **137**, 313–33.

Russell, M.A.H., Wilson, C., Patel, U.A., Feyerabend, C., and Cole, P.V. (1975). Plasma nicotine levels after smoking cigarettes with high, medium and low nicotine yields. *British Medical Journal* **2**, 414–6.

Russell, M.A.H., Feyerabend, C., and Cole, P.V. (1976*a*). Plasma nicotine levels after cigarette smoking and chewing nicotine gum. *British Medical Journal* **1**, 1043–6.

Russell, M.A.H., Wilson, C., Feyerabend, C., and Cole, P.V. (1976*b*). Effect of nicotine chewing gum on smoking behaviour and as an aid to cigarette withdrawal. *British Medical Journal* **2**, 391–3.

Russell, M.A.H., Sutton, S.R., Feyerabend, C., Cole, P.V., and Saloojee, Y. (1977). Nicotine chewing gum as a substitute for smoking. *British Medical Journal* **1**, 1060–3.

Russell, M.A.H., and Feyerabend, C. (1978). Cigarette smoking: a dependence on high-nicotine boli. *Drug Metabolism Reviews* **8**, 29–57.

Russell, M.A.H., Raw, M., Taylor, C., Feyerabend, C., and Saloojee, Y. (1978). Blood nicotine and carboxyhaemoglobin levels after rapid-smoking aversion therapy. *Journal of Consulting and Clinical Psychology* **46**, 1423–31.

Russell, M.A.H. (1978). Self-regulation of nicotine intake by smokers. In *Behavioral effects of nicotine* (K. Battig, ed.), pp. 108–22. Karger, Basel.

Russell, M.A.H. (1979). Tobacco dependence: is nicotine rewarding or aversive? In *Cigarette smoking as a dependence process* (N.A. Krasnegor, ed.), pp. 100–22. National Institute on Drug Abuse Research Monograph 23. US Government Printing Office, Washington, DC.

Russell, M.A.H., Jarvis, M.J., and Feyerabend, C. (1980*a*). A new age for snuff? *Lancet* **1**, 474–5.

Russell, M.A.H., Jarvis, M.J., Iyer, R., and Feyerabend, C. (1980*b*). Relation of nicotine yield of cigarettes to blood nicotine concentrations in smokers. *British Medical Journal* **280**, 972–5.

Russell, M.A.H., Jarvis, M.J., Devitt, G., and Feyerabend, C. (1981). Nicotine intake by snuff users. *British Medical Journal* **283**, 814–7.

Russell, M.A.H., Sutton, S.R., Iyer, R., Feyerabend, C., and Vesey, C.J. (1982). Long term switching to low-tar, low-nicotine cigarettes. *British Journal of Addiction* **77**, 145–58.

Russell, M.A.H., Jarvis, M.J., Feyerabend, C., and Ferno, O. (1983). Nasal nicotine solution: a potential aid to giving up smoking. *British Medical Journal* **286**, 683–4.

Russell, M.A.H., Jarvis, M.J., Feyerabend, C., and Saloojee, Y. (1986). Reduction of tar, nicotine and carbon monoxide intake in low-tar smokers. *Journal of Epidemiology and Community Health* **40**, 80–5.

Russell, M.A.H., Jarvis, M.J., Sutherland, G., and Feyerabend, C. (1987). Nicotine replacement in smoking cessation: absorption of nicotine vapor from smoke-free cigarettes. *Journal of the American Medical Association* **257**, 3262–5.

Russell, M.A.H., (1988). Nicotine replacement: the role of blood nicotine levels, their rate of change, and nicotine tolerance. In *Nicotine replacement: a critical evaluation.*

(O.F. Pomerleau and C.S. Pomerleau, ed.), pp. 63–94. Alan. R. Liss Inc, New York.

Schachter, S., Kozlowski, L.T., and Silverstein, B. (1977). Effects of urinary pH on cigarette smoking. *Journal of Experimental Psychology* (Gen). **106**, 13–9.

Schachter, S. (1978). Pharmacological and psychological determinants of smoking. In *Smoking behaviour: physiological and psychological influences.* (R.E. Thornton, ed.), pp 208–28 Churchill Livingstone, London.

Schneider, N.G., Jarvik, M.E., and Forsythe, A.B. (1984). Nicotine versus placebo gum in the alleviation of withdrawal during smoking cessation. *Addictive Behaviors* **9**, 149–56.

Schwartz, R.D., and Keller, K.J. (1983). Nicotinic cholinergic receptor binding sites in the brain: regulation in vivo. *Science* **220**, 214–6.

Stitzer, M.L., and Gross, J. (1988). Smoking relapse: the role of pharmacological and behavioral factors. In *Nicotine replacement: a critical evaluation* (O.F. Pomerleau and C.S. Pomerleau, ed.), pp. 163–84. Alan R. Liss Inc, New York.

Stolerman, I.P., Fink, R., and Jarvik, M.E. (1973*a*). Acute and chronic tolerance to nicotine measured by activity in rats. *Psychopharmacologia* **30**, 329–42.

Stolerman, I.P., Goldfarb, T., Fink, R., and Jarvik, M.E. (1973*b*). Influencing cigarette smoking with nicotine antagonists. *Psychopharmacologia* **28**, 247–59.

Sutton, S.R., Feyerabend, C., Cole, P.V., and Russell, M.A.H. (1978). Adjustment of smokers to dilution of tobacco smoke by ventilated cigarette holders. *Clinical Pharmacology and Therapeutics* **24**, 395–405.

Snyder, F.R., Davis, F.C., and Henningfield, J.E. (1985). Effects of tobacco deprivation on cognitive performance. *Proceedings of the Eastern Psychological Association* **57**, 62.

Tonnesen, P., Fryd, V., Hansen, M., Helsted, J., Gunnersen, A.B., Forchammer, H., and Stockner, M. (1988*a*). Two and four mg nicotine chewing gum and group counselling in smoking cessation: an open, randomized, controlled trial with a 22 month follow-up. *Addictive Behaviors* **13**, 17–27.

Tonnesen, P., Fryd, V., Hansen, M., Helsted, J., Gunnersen, A.B., Forchammer, H., and Stockner, M. (1988*b*). Effect of nicotine chewing gum in combination with group counselling on the cessation of smoking. *New England Journal of Medicine* **318**, 15–18.

Turner, D.M. (1971). Metabolism of small multiple doses of (^{14}C) nicotine in the cat. *British Journal of Pharmacology* **41**, 521–9.

Turner, D.M., Armitage, A.K., Briant, R.H., and Dollery, C.T. (1975). Metabolism of nicotine by the isolated perfused dog lung. *Xenobiotica* **5**, 539–51.

Turner, J.A.M., Sillett, R.W., and McNichol, M.W. (1977). Effect of cigar smoking on carboxyhaemoglobin and plasma nicotine concentrations in primary pipe and cigar smokers and ex-cigarette smokers. *British Medical Journal* **2**, 1387–9.

US Surgeon General's Report on the Health Consequences of Smoking (1988). *Nicotine Addiction.* US Government Printing Office, Washington, DC.

Waingrow, S., and Horn, D. (1968). Relationship of number of cigarettes smoked to tar rating. *National Cancer Institute Monograph* **28**, 29–33.

Wald, N.J., Idle, M., Boreham, J., Bailey, A., and Van Vunakis, H. (1984). Urinary nicotine concentrations in cigarette and pipe smokers. *Thorax* **39**, 365–8.

Wesnes, K., and Warburton, D.M. (1983). Smoking, nicotine and human performance. *Pharmacology and Therapeutics* **21**, 189–208.

West, R.J. (1984). Psychology and pharmacology in cigarette withdrawal. *Journal of Psychosomatic Research* **28**, 379–86.

West, R.J., Russell, M.A.H., Jarvis, M.J., and Feyerabend, C. (1984*d*). Does switching to an ultra-low nicotine cigarette induce nicotine withdrawal effects? *Psychopharmacology* **84**, 120–3.

West, R.J., Jarvis, M.J., Russell, M.A.H., Carruthers, M.E., and Feyerabend, C. (1984*a*). Effect of nicotine replacement on the cigarette withdrawal syndrome. *British Journal of Addiction* **79**, 215–9.

West, R.J., Russell, M.A.H., Jarvis, M.J., Pizzie, T., and Kadam, B. (1984*b*). Urinary adrenaline concentrations during 10 days of smoking abstinence. *Psychopharmacology* **84**, 141–2.

West, R.J., Jarvis, M.J., Russell, M.A.H., and Feyerabend, C. (1984*c*). Plasma nicotine concentrations from repeated doses of nasal nicotine solution. *British Journal of Addiction* **74**, 443–5.

West, R.J., and Russell, M.A.H. (1985*a*). Preabstinence smoke intake and smoking motivation as predictors of severity of cigarette withdrawal symptoms. *Psychopharmacology* **87**, 334–6.

West, R.J., and Russell, M.A.H. (1985*b*). Effects of withdrawal from long-term nicotine gum use. *Psychological Medicine* **15**, 891–3.

West, R.J., and Jarvis, M.J. (1986). Effects of nicotine on finger tapping rate in non-smokers. *Pharmacology, Biochemistry and Behavior* **25**, 727–31.

West, R.J., and Russell, M.A.H. (1987). Cardiovascular and subjective effects of smoking before and after 24 h abstinence from cigarettes. *Psychopharmacology* **92**, 118–21.

West, R.J., Hajek, P., and Belcher, M. (1987). Time course of cigarette withdrawal symptoms during four weeks of treatment with nicotine chewing gum. *Addictive Behaviors* **12**, 199–203.

West, R.J., and Russell, M.A.H. (1988). Loss of acute tolerance and severity of cigarette withdrawal. *Psychopharmacology* **94**, 563–5.

Westling, H. (1976). Experience with nicotine-containing chewing gum in anti-smoking treatment (Swedish). *Lakartidningen* **73**, 2549–52.

Wilhelmsen, L., and Hjalmarson, A. (1980). Smoking cessation experience in Sweden. *Canadian Family Physician* **26**, 737–43.

Wise, R.A., and Bozarth, M.A. (1984). Brain reward circuitry: four circuit elements 'wired' in apparent series. *Brain Research Bulletin* **12**, 203–8.

Index